Smart Innovation, Systems and Technologies

Volume 187

The Smart Innovation, Systems and Technologies book series encompasses the topics of knowledge, intelligence, innovation and sustainability. The aim of the series is to make available a platform for the publication of books on all aspects of single and multi-disciplinary research on these themes in order to make the latest results available in a readily-accessible form. Volumes on interdisciplinary research combining two or more of these areas is particularly sought.

The series covers systems and paradigms that employ knowledge and intelligence in a broad sense. Its scope is systems having embedded knowledge and intelligence, which may be applied to the solution of world problems in industry, the environment and the community. It also focusses on the knowledge-transfer methodologies and innovation strategies employed to make this happen effectively. The combination of intelligent systems tools and a broad range of applications introduces a need for a synergy of disciplines from science, technology, business and the humanities. The series will include conference proceedings, edited collections, monographs, handbooks, reference books, and other relevant types of book in areas of science and technology where smart systems and technologies can offer innovative solutions.

High quality content is an essential feature for all book proposals accepted for the series. It is expected that editors of all accepted volumes will ensure that contributions are subjected to an appropriate level of reviewing process and adhere to KES quality principles.

**** Indexing: The books of this series are submitted to ISI Proceedings, EI-Compendex, SCOPUS, Google Scholar and Springerlink ****

More information about this series at http://www.springer.com/series/8767

Andrey Ronzhin · Vladislav Shishlakov
Editors

Proceedings of 15th International Conference on Electromechanics and Robotics "Zavalishin's Readings"

ER(ZR) 2020, Ufa, Russia, 15–18 April 2020

Zavalishin's
Readings | 2020
15th International Conference
on Electromechanics and Robotics

Springer

Editors
Andrey Ronzhin
St. Petersburg Institute for Informatics
and Automation of the Russian
Academy of Sciences
St. Petersburg, Russia

Vladislav Shishlakov
Saint Petersburg State University
of Aerospace Instrumentation
St. Petersburg, Russia

ISSN 2190-3018 ISSN 2190-3026 (electronic)
Smart Innovation, Systems and Technologies
ISBN 978-981-15-5579-4 ISBN 978-981-15-5580-0 (eBook)
https://doi.org/10.1007/978-981-15-5580-0

This Springer imprint is published by the registered company Springer Nature Singapore Pte Ltd.
The registered company address is: 152 Beach Road, #21-01/04 Gateway East, Singapore 189721, Singapore

Organization

General Chair

Yulia Antokhina

Co-chairs

Ramil Bakhtizin
Sergey Emelyanov
Anatoliy Ovodenko
Vladislav Shishlakov

Committees

Chair of Programme Committee

Andrey Ronzhin

Programme Committee

Karsten Berns, Germany
Nikolay Bolotnik, Russia
Yi-Tung Chen, USA
Sergey Chigvincev, Russia
Alexander Danilov, Russia
Vlado Delic, Serbia
Ivan Ermolov, Russia
Naohisa Hashimoto, Japan
Han-Pang Huang, Taiwan
Shu Huang, Taiwan
Viktor Glazunov, Russia

Mehmet Guzey, Turkey
Oliver Jokisch, Germany
Airat Kalimgulov, Russia
Alexey Kashevnik, Russia
Marat Khakimyanov, Russia
Regina Khazieva, Russia
Pavel Khlyupin, Russia
Sergey Konesev, Russia
Eugeni Magid, Russia
Roman Meshcheryakov, Russia
Zuhra Pavlova, Russia
Vladimir Pavlovskiy, Russia
Francesco Pierri, Italy
Yuriy Poduraev, Russia
Mirko Rakovic, Serbia
Raul Rojas, Germany
Jose Rosado, Portugal
Vitali Shabanov, Russia
Hooman Samani, Taiwan
Yulia Sandamirskaya, Switzerland
Jesus Savage, Mexico
Valery Sapelnikov, Russia
Robert Sattarov, Russia
Vladimir Serebrenny, Russia
Michail Sit, Moldova
Lev Stankevich, Russia
Tilo Strutz, Germany
Georgi Vukov, Bulgaria
Sergey Yatsun, Russia
Arkadiy Yuschenko, Russia
Milos Zelezny, Czech Republic
Lyudmila Zinchenko, Russia

Co-chair of Organizing Committee

Pavel Khlyupin
Sergey Solyonyj
Sergey Yatsun
Andrey Ronzhin

Organizing Committee

Radmir Aflyatunov
Oksana Emelyanova
Natalia Dormidontova
Maksim Ivanov

Nataliya Jarinova
Ilgiza Kaekberdin
Natalia Kashina
Timur Khabibullin
Boris Lushnikov
Alina Matova
Ekaterina Miroshnikova
Anna Motienko
Margarita Avstriyskaya
Irina Podnozova
Elena Reznik
Anton Saveliev
Ekaterina Savelyeva
Sergei Savin
Oksana Solenaya
Dmitry Tyurin
Andrey Yatsun

Foreword

Dmitry Aleksandrovich Zavalishin (1900–1968)—a Russian scientist, corresponding member of the USSR Academy of Sciences, founder of the school of valve energy converters based on electric machines and valve converters energy. The first conference was organized by the Institute of Innovative Technologies in Electromechanics and Robotics of the St. Petersburg State University of Aerospace Instrumentation in 2006.

The purpose of the conference is the exchange of information and progressive results of scientific research work of scientific and pedagogical workers, young scientists, graduate students, applicants and students in the field of: automatic control systems, electromechanics, electric power engineering and electrical engineering, mechatronics, robotics, automation, technical physics and management in the electric power industry.

We express our deepest gratitude to all participants for their valuable contribution to the successful organization of ER(ZR)-2020, hope for and look forward to your attention to the next International Conference on Electromechanics and Robotics "Zavalishin's Readings" in 2021. The conference website is located at: http://suai.edu.ru/conference/zav-read/.

St. Petersburg, Russia
May 2020

Prof. Yulia A. Antokhina
General Chair of 15th International Conference
on Electromechanics and Robotics
"Zavalishin's Readings"—2020
Rector of the St. Petersburg
State University of Aerospace Instrumentation

Preface

This year, the conference The 15th International Conference on Electromechanics and Robotics "Zavalishin's Readings"—2020, ER(ZR)-2020 was organized with XIV International Conference "Vibration-2020. Vibration technologies, mechatronics and controlled machines" and V International Conference "Electric drive, electrical technology and electrical equipment of enterprises" during April 15–18, 2020 in Ufa, Russia. The conferences were organized by St. Petersburg State University of Aerospace Instrumentation (SUAI, St. Petersburg, Russia), St. Petersburg Institute for Informatics and Automation of the Russian Academy of Sciences (SPIIRAS, St. Petersburg, Russia), Southwest State University (SWSU, Kursk, Russia) and Ufa State Petroleum Technical University (USPTU, Ufa, Russia). The conference is held with the financial support of the Russian Foundation for Basic Research, project No. 20–08–20030.

Due to the COVID–19 pandemic in the world, for the first time ER(ZR)-2020 was organized as a virtual conference. The virtual conference in the online format via Zoom service also had a number of advantages including: an increased number of participants, and no costs for travel and accommodation, comfortable home conditions, etc.

During the conference the invited talks were given by Prof. Jesus Savage (National Autonomous University of Mexico, Mexico), Assoc. prof. Lingfei Xiao (Nanjing University of Aeronautics and Astronautics, China), Ilshat Mamaev (Karlsruhe Institute of Technology, Germany), Prof. Oleg Darintsev (Ufa State Aviation Technical University Russia), Prof. Vladimir Fetisov (Ufa State Aviation Technical University, Russia), Assoc. prof. Sergey Konesev (Ufa State Oil Technical University, Russia), Prof Robert Sattarov (Ufa State Aviation University, Russia). More then 173 papers of authors from China, Czech Republic, Mexico, Russia, Taiwan, Turkey, Uzbekistan, Viet Nam and Japan were submitted to the conference and each paper was reviewed by several scientists. Around 30% of the best papers were published in English proceedings by Springer in series Smart Innovation, Systems and Technologies indexed in SCOPUS, Thomson Reuters (Web of Science), Inspec, etc. Due to great efforts of reviewers this book was carefully prepared and consists of 44 contributions.

Special thanks are due to the members of the Local Organizing Committee for their tireless effort and enthusiasm during the conference organization. Hope for and look forward to your attention to the ER(ZR)-2021. The conference website is located at: http://suai.edu.ru/conference/zav-read/.

St. Petersburg, Russia
May 2020

Prof. Andrey L. Ronzhin
Chair of Programm Committee
of 15th International Conference
on Electromechanics and Robotics
"Zavalishin's Readings"—2020
Director of St. Petersburg Institute for Informatics
and Automation of the Russian Academy of Sciences

Prof. Vladislav F. Shishlakov
Co-Chair of 15th International Conference
on Electromechanics and Robotics
"Zavalishin's Readings"—2020
Vice-Rector for Educational Technologies
and Innovative Activities
St. Petersburg State University of Aerospace Instrumentation

Contents

About the Editors

Prof. Andrey Ronzhin is Director of St. Petersburg Institute for Informatics and Automation of the Russian Academy of Sciences (SPIIRAS) and Head of the Department of Electromechanics and Robotics Systems at St. Petersburg University of Airspace Instrumentation. His research focuses on the interaction of autonomous robotic systems and users in a cyber-physical environment. He is a member of Scientific Board of Robotics and Mechatronics of the Russian Academy of Sciences, the Academy of Navigation and Motion Control, Co-Chairman of International Conference Interactive Collaborative Robotics – ICR. He is Deputy Editor-in-Chief of SPIIRAS Proceedings Journal.

Prof. Vladislav Shishlakov is Vice-Rector for Educational Technologies and Innovative Activities, St. Petersburg State University of Aerospace Instrumentation (SUAI) and Head of the Department of Management in Technical Systems. He is Honorary Worker at the Ministry of Education and Science of the Russian Federation since 2009. His research interests are related to the development of methods of synthesis of nonlinear systems of automatic control systems, which are continuous, and with different types of signal modulation, as well as the development and research of electromechanical and electric power systems and complexes based on the effects of high-temperature superconductivity.

Part I
Keynote Lectures

Chapter 1
Aerial Robots and Infrastructure of Their Working Environment

Vladimir Fetisov

Abstract Aerial robots (also known as UAVs—unmanned aerial vehicles) are increasingly being introduced into our life. Today, we can see aerial robots in agriculture, building industry, delivery services, security and monitoring systems and so on. More frequently not single UAVs but their groups are used. And it would be reasonable to control such groups at all functioning stages, including on-ground maintenance, in automatic mode. Development of infrastructure for automatic service and maintenance of aerial robots has appeared on the agenda of many companies specializing in unmanned aerial systems. Some aspects of such infrastructure creation are discussed in this paper with special emphasis on charging stations for UAVs with electrical propulsion system.

1.1 Introduction: What Is AR, UAV, UAS

In robotics the term "aerial robot" (AR) is known from 1998, when Michelson [1] described a new class of highly intelligent, small flying machines. Now the sense covered under the term AR extends much further. In the field of aviation, robotic flying machines are referred to as "unmanned aerial vehicles" (UAVs), or drones, by simply saying.

Unmanned aerial vehicle (UAV) is defined as a pilotless aircraft, which is flown without a pilot-in-command on-board and is either remotely and fully controlled from another place (ground, another aircraft, ship, space) or programmed and fully autonomous [2].

On the other hand, it is known for the following definition of AR: "An aerial robot is a system capable of sustained flight with no direct human control and able to perform a specific task" [3]. According to this definition, any contemporary UAV is AR because UAV's on-board flight controller with embedded navigation equipment

V. Fetisov (✉)
Ufa State Aviation Technical University, Ufa, Russia
e-mail: photonltd@mail.ru

A. Ronzhin and V. Shishlakov (eds.), *Proceedings of 15th International Conference on Electromechanics and Robotics "Zavalishin's Readings"*, Smart Innovation, Systems and Technologies 187, https://doi.org/10.1007/978-981-15-5580-0_1

provides sustainable flight without an operator's participation. Such minimal on-board intelligence of the robot allows to sustain itself in the air with no human intervention.

So notions of UAV and AR are often considered as equivalent. But in recent years, a new trend has appeared to consider AR as an UAV designed to perform special operations in autonomous mode.

In other words, AR is a big class of mobile robots based on UAV for special tasks that can be performed with different degree of autonomy, i.e., AR has a lot of intelligence and self-sufficiency for its special function implementation. However, the UAV operator, as a rule, can control the AR remotely, switching from automatic to manual mode if the situation requires it.

There are many types of UAVs based on different flight principles. This work deals primarily with the rotary-wing type of aerial vehicles (helicopters, multicopters) and other aircraft (convertiplanes and other hybrids) capable of vertical takeoff and landing (VTOL). VTOL UAVs are the closest to common notion of "robots" because of their capability of hovering, which has huge advantages, in comparison with fixed-wing aircrafts, for general versatility. For example, VTOL UAVs can implement various repairs and building-up operations under the object by means of on-board manipulators. VTOL UAVs are capable of hovering and agile at the same time. Their rich sensory and motor abilities allow them to move and work in very different environments: open skies, confined environments, on the ground, on vertical surfaces, in swarms and near humans [4].

ARs are designed for various useful functions: aerial photography, monitoring, construction operations, agricultural works, delivery of small packages and so on. More and more frequently not single UAVs but their groups are used. And it would be reasonable to control such groups at all functioning stages, including on-ground maintenance, in automatic mode. Development of infrastructure for automatic service and maintenance of ARs have appeared on the agenda of many companies specializing in UAS—unmanned aerial systems (or unmanned aircraft systems). UAS is a widely used notion, which is a more complex term than UAV [5, 6]. UAS comprises one or more UAVs, along with the technical equipment necessary to operate them and other components. Full composition of UAS is shown in Fig. 1.1. When UAVs are considered as ARs, UAS provides an infrastructure for working environment of ARs. Let us take a detailed look at all components of UAS.

1.2 Components of Unmanned Aerial System

1.2.1 Main Functional Means

Main functional means of UAS include all components that are closely connected with flights: UAVs, control station (CS), start and landing equipment, means of transportation, navigation and communication equipment and service stations.

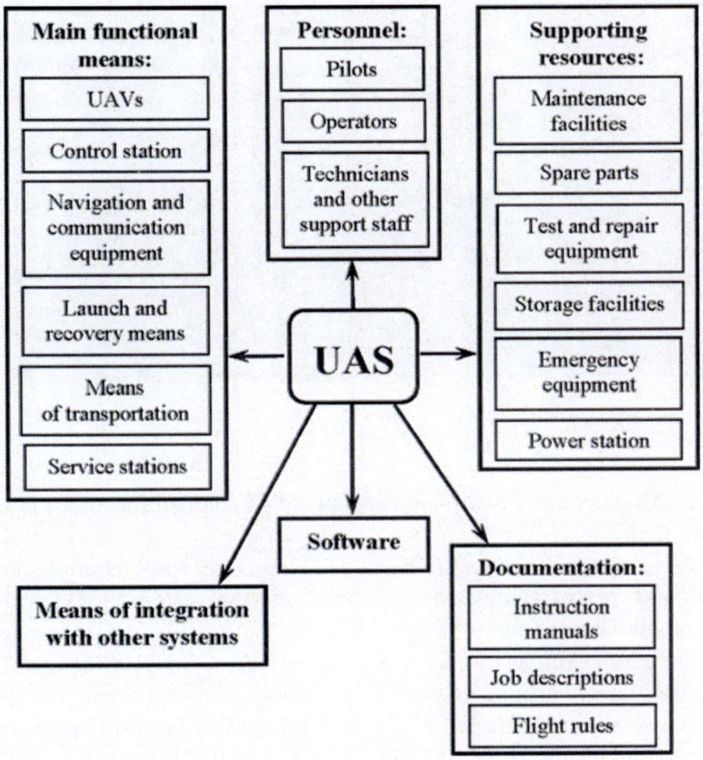

Fig. 1.1 Full composition of UAS

UAVs (*ARs*). One AR is the minimal number of vehicles in the system. The only UAV in UAS is becoming a rarity. In recent times, groups of ARs are used more and more. Various concepts of UAV group control are known, from centralized control of each vehicle to concepts based on artificial intelligence (AI). Among them, swarm intelligence (SI) occupies a special place. The term was introduced by Beni [7], in the context of cellular robotic systems (CRS). A CRS consists of a large number of robots and operates in n-dimensional cellular space under distributed control. Wide centralized mechanism and synchronous clock are not assumed. Limited communication exists only between adjacent robots. On the one hand, these robots operate autonomously; on the other hand, they cooperate to perform predefined global tasks [8]. SI systems consist of simple agents (ARs) interacting locally with one another and with their environment (Fig. 1.2). SI aerial systems are similar to biological systems. The ARs follow very simple rules, and their operations are local and to a certain degree random. There is no centralized control structure defining how individual agents should behave, but interactions between such agents lead to the appearance of smart global swarm behavior, unknown to the individual agents. Examples of SI in natural systems are ant colonies, bird flocking, hawks hunting, animal herding and

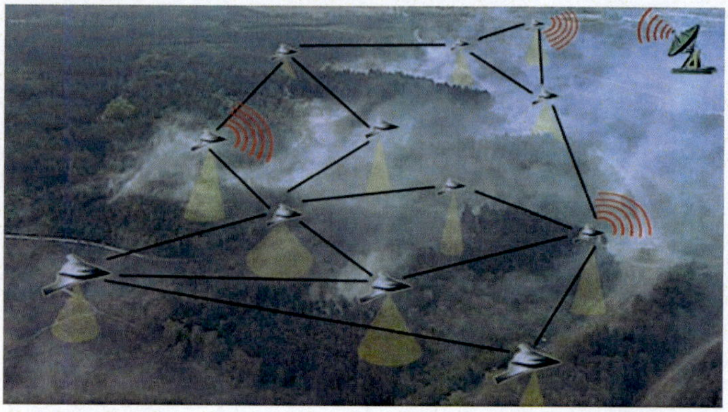

Fig. 1.2 Swarm of UAVs

many others. To date, there are many concepts of SI adapted for different tasks and scenarios [9–12].

ARs of the same type are commonly used in groups, but some examples are known when different types of vehicles are included into the system [13]. Such solution allows to complement functions and possibilities of ARs (Fig. 1.3). Moreover, there are symbiotic aerial ground robotic teams where UAVs are used for aerial-specific tasks, while unmanned ground vehicles (UGVs) aid and assist them [14, 15]. UGVs can provide UAVs with a safe landing area, while UAVs can provide an additional degree of freedom for the UGVs, helping them to avoid obstacles (Fig. 1.4). Such a system may be used, for example, for parcel transportation scenarios.

Fig. 1.3 Heterogeneous UAS with different types of UAVs and the new multi-drone CS for collaborative drone missions (Photo: ECA Group, France, 2017)

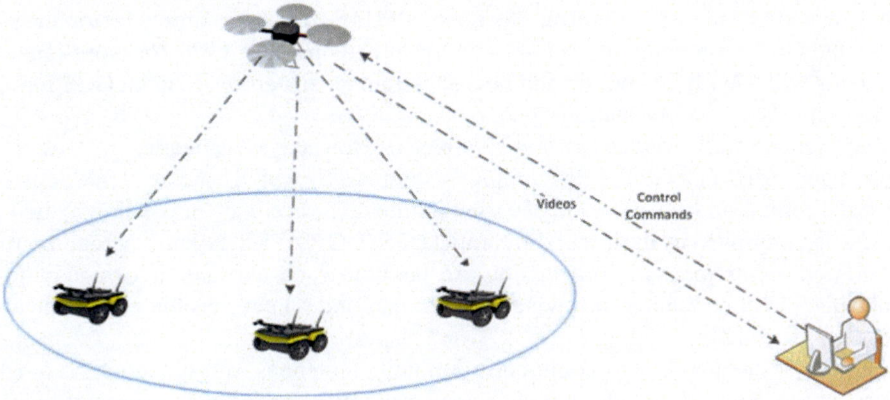

Fig. 1.4 Symbiotic UAV-UGV robotic team (Illustration: [14])

Number of ARs in the system may be huge. In 2018, a Chinese drone company has launched into the sky 1374 drones above the city of Xian to set a Guinness World Record for most UAVs flown at the same time [16]. The quadcopters simultaneously took off and created various colorful 3D formations in the air (Fig. 1.5).

Control Station. The control station (CS) for UAVs may be based on the ground (GCS), aboard ship (SCS) and possibly in a parent aircraft (ACS). It may be simply the control center of a local UAS, but may also be part of a larger system, when it is also interfaced with other components of a network-centric system, sharing information with and receiving information from other elements of the larger system. In this latter case, the mission planning may be carried out in a central command center and retailed to the CS for execution. The CS is the man–machine interface with the UAS. From it the operators may control the flight directly or change a preflight-entered flight program. Depending on the types of UAVs and UAS destination, CS

Fig. 1.5 The biggest formation of UAVs is ready to take off (Photo: RT News [16])

may be realized in different forms. As for small UAS, it may be simple laptop man-portable GCS with communication system and antenna. The GCS for close-range systems will usually be mobile and housed within an all-terrain vehicle. GCS for a large-scale UAS may be stationary.

Navigation and Communication Equipment. Nowadays, navigation in UAS is based generally on a Global Positioning System (GPS) which accesses UAVs positional information from a system of earth satellites. The accuracy of positioning may be further improved by the use of differential GPS (DGPS) [17]. For non-autonomous operation, other means of navigation are possible such as radar tracking, radio tracking, direct reckoning, receiving video frames and relating visible geographical features [5, p. 12].

Many of navigation and communication units (antennas, transponders, homing beacons) are located directly in the CS or in proximity to the CS, but some of them (radiomarkers, weather stations, repeaters) may be distributed throughout the area of flights. Such equipment, where possible, must be autonomous, i.e., must include independent source of energy (mini power station based on solar, wind, hydro or other energy).

Launch and Recovery Means. These are required only for fixed-wing UAVs that are not designed for takeoff and landing using a runway.

Launch Equipment. Usually, this is a rubber or compressed air catapult. This often takes the form of a ramp along which the aircraft is accelerated on a trolley, until the aircraft has reached an airspeed of sustain flight (Fig. 1.6).

Recovery Equipment. It usually takes the form of a parachute, installed within the aircraft, and which is deployed at a suitable altitude over the landing zone. Sometimes, a large net or a carousel apparatus is used, into which the aircraft is flown and caught (Fig. 1.7).

Fig. 1.6 Launching UAV from the compressed air catapult (Photo: Andrew Kendrick, 2012)

Fig. 1.7 Recovery equipment: catching the UAV into the net trap (Photo: Gotham 2015)

Means of Transportation. A UAS system is often required to be mobile. Therefore, transport means must be provided for personnel and all the subsystems are described above. This may vary from one vehicle required to contain and transport a UAS using a small VTOL aircraft to a system using a large and heavier ramp-launched aircraft which needs all the subsystems listed, may have to be dismantled and reassembled between flights and may require, for example, ten crew and six large transport vehicles. Even UAS operating from fixed bases may have specific transport requirements [5, p. 14].

Service Stations. This is specific objects of UAS infrastructure, mainly intended for refueling or recharging UAVs. Service stations are described separately in Sect. 1.3.

1.2.2 Supporting Resources

Besides main functional means described above, many additional resources are necessary for UAS normal functioning. Depending on scale and destination of UAS, such resources may include maintenance and storage facilities, spare parts, test and repair equipment, emergency equipment and power station.

1.2.3 Personnel

In the recent ICAO Regulations [18], among others the following terms have been introduced, concerning UAS personnel:

Remotely piloted aircraft system (RPAS)—a remotely piloted aircraft, its associated remote pilot station(s), the required command and control links and any other components as specified in the type design. RPAS is considered as a subset of UAS.

Operator—a person, organization or enterprise engaged in or offering to engage in an aircraft operation.

Remote pilot—a person charged by the operator with duties essential to the operation of a remotely piloted aircraft and who manipulates the flight controls, as appropriate, during flight time.

Remote pilot-in-command—the remote pilot designated by the operator as being in command and charged with the safe conduct of a flight.

RPA observer—a trained and competent person designated by the operator who, by visual observation of the remotely piloted aircraft, assists the remote pilot in the safe conduct of the flight.

Alongside with the above personnel roles, there is one specific position for military field named "sensor operator" [19]. Responsibilities of sensor operator are as follows: perform intelligence surveillance and reconnaissance; detect, analyze and discriminate between valid and invalid targets; assist in air navigation, fire control planning and determining effective weapons control.

Besides described possible positions, the UAS team may include technicians and other support staff.

1.2.4 Means of Integration with Other Systems

These means may include the following:

- special communication equipment and protocols for interconnecting with state aviation authorities that is charge of monitoring airspace and safe air traffic (i.e., avoidance collision between aircrafts, manned and unmanned);
- special equipment, channels and protocols (outer servers, cloud systems) for provision of data, concerning flight results, for all interested organizations;
- special communication equipment and protocols for tasking from an external command center to the UAV and report back to that or other external source.

Example of UAS with some elements of outer systems is shown in Fig. 1.8.

1.2.5 Software

UAS uses computer software at various levels as follows:

- on-board software of UAVs;
- CS software providing interoperability between the CS and UAVs;

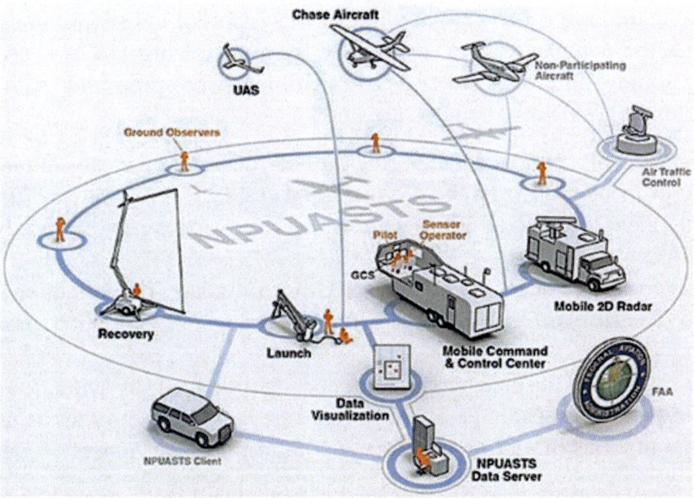

Fig. 1.8 Example of UAS with some elements of outer systems (Illustration: NP UAS TS [20])

- CS software providing interoperability between the CS and other components of the UAS, such as charging stations, remote beacons and so on;
- CS software providing interoperability between the CS and external sources;
- software of other UAS components.

1.2.6 Documentation

UAS documentation includes instruction manuals for pilots and technicians, job descriptions, flight rules and other documentation.

1.3 Service Stations

1.3.1 Motivation

Service stations are intended for executing special operations that are necessary for replenishment of ARs' energy resources and (or) consumables.

The large majority of contemporary VTOL aerial robots is of small, mini or micro-classes [21] with brushless electromotors and rechargeable lithium-polymer (Li-Po) or lithium-ion (Li-Ion) batteries. The limited battery capacity severely impedes their usage for longtime flights. The best Li-Po batteries can provide only 30–40 min of flight. After that the vehicle must replenish its on-board energy supply.

Therefore, the extension of flight duration or support of long continuous missions of UAVs is the actual problem, especially for rotary-wing UAVs. Various ways (including some exotic ones) to solve this problem were proposed. Among these, for example, the following methods are known:

- The usage of solar energy for charging on-board battery by means of photovoltaic cells is distributed on surfaces of wings and fuselage [22, 23]. This approach is rather applicable for fixed-wing vehicles and convertiplanes, but not for rotary-wing UAVs.
- The usage of laser beam energy directed from special ground stations to UAVs [24, 25]. The similar idea concerning directed transmission of microwaves energy is known [26].
- Drastic solution of the problem is to connect directly the UAV with ground energy supply by long-thin cable [27, 28]. In this case, the flight duration is not limited at all, but obvious limitation in the range of motions takes place.

But the most realistic and efficient way to replenish UAV's energy supply is on-board batteries recharging or exchanging on special ground service stations located in the area of interest or along the required flight track. Such energy replenishment stations may be classified by the following way (Fig. 1.9).

Besides refueling the energy source of ARs' propulsion systems, service stations may be used for replenishment of different non-fuel consumables, such as liquid pesticides or fertilizers for spraying plantations, if the ARs, for example, are of agrochemical destination.

Ideally, such stations must be automated and operated without special personnel, i.e., all operations (landing and takeoff, refueling/recharging) must be executed in automatic mode [29].

1.3.2 Classification of Replenishment Service Stations

The main classification criterion is the energy replenishment method which depends on primary energy source for on-board propulsion system. For heat engines of various types, such primary energy source is fuel in liquid, gaseous or solid form. Also fuel is energy source for so-called fuel cells which are in principle chemical sources of electrical energy. After discharging, such elements require refill of fuel (e.g., hydrogen, methanol, etc.) but not charging by electric current as traditional batteries. Some successful projects concerning usage of fuel cells in UAVs are known [30].

Though the operating principles of propulsion systems with heat engines and ones with fuel cells and electromotors are different, the method of refueling is the same. It may be filling of on-board tank with fuel or replacement of empty fuel cartridge with full one [31]. Similar operations may be implemented relative to traditional batteries: exchanging (for voltaic cells or accumulators) and recharging (for accumulators). As a rule, if a replenishment station realizes exchanging of accumulator batteries, it provides also charging of removed batteries. In recent years, technologies of batteries

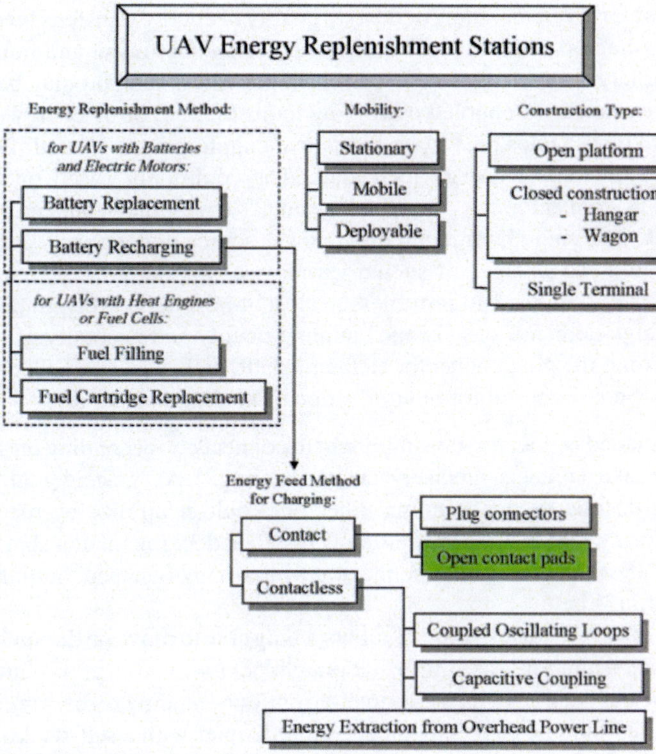

Fig. 1.9 Classification of UAV replenishment stations

exchanging are considerably developed. Many interesting service stations for UAVs batteries swap are proposed [32–34].

Another design conception for replenishment stations, opposite to batteries replacement, is based on charging of on-board batteries without removal of them from aerial vehicles. The advantage of such decision is in simplicity of service terminals, on-board battery units and absence of swapping mechanisms although duration of charging, of course, is longer than swapping. But for many applications, such decision may be better acceptable, especially if the principle of redundancy in configuration of the ground contact system is applied. In such case multiple ground electrodes distributed on large area of a landing platform can provide normal contact with on-board electrodes and charging process even in conditions of inaccurate UAV landing or simultaneous charging for a group of UAVs.

Charging station design may be different depending on contact or contactless energy feed method is used. Both methods have their own advantages and disadvantages.

Contactless energy feed method provides energy transfer from the ground station to the UAV by means of electromagnetic waves without direct galvanic connections. So there is practically no influence of parasitic deposits on terminal contact surfaces (such as thin water layer, ice, snow or dust) to quality of energy transfer. But constraint factor for this method is the low value of transmission coefficient and impossibility of transfer for large values of electric current. Contactless method may be realized, for example, by means of coupled oscillating loops [35, 36] or by energy extraction from overhead power lines [37]. Also capacitive coupling may be used [26].

Realization of contact energy feed method is commonly based on the usage of various plug connectors. A simple two-contact (two-pole) connector provides a common DC charging chain including a power source and an on-board accumulator. The obvious advantages of such a variant are simplicity of charging circuit, minimum of energy losses and proofness against atmospheric precipitation. Disadvantages are high demands of galvanic coupling quality and necessity of very exact UAV landing onto the plug connector element (shaft, jack, nest, slot). The following two solutions were proposed for neutralization of landing inaccuracy:

(1) Funnel-shaped connector. It is important to connect corresponding on-board and ground-based contacts precisely during landing. UAV's accidental deviation from the desired vertical landing trajectory could neutralize by means of the connection system including a vertically installed docking tubular shaft with two contacts above and on-board connector with a funnel-shaped "mouth directed downward [38].
(2) "Crawling" UAV. Various mobile robots, being able to move on the surface, often have a subsystem for searching a stationary docking station, approximation and automatic connection to its wall-mounted or floor-standing recharging terminals [39]. So, it would be easy to combine a multicopter with a self-docking mobile robot [40]. Such a hybrid could be realized, for example, on the basis of four-wheel carriage with a horizontal docking shaft [38]. After landing, the hybrid UAV must begin to search a docking station implementing rotations and "crawls" for approximation to it.

Another way to deliver energy to UAVs being charged is the usage of platforms with open contact pads which are the item of this paper. In this approach, electrical connection between ground and on-board circuits appears right after landing when UAV's undercarriage open electrodes touch and simply lay on ground platform open electrodes (pads, strips). Such solutions provide arrangement of charging process even in conditions of inaccurate UAV landing and charging for a group of UAVs simultaneously. Of course, open contact pads are affected by atmospheric precipitation and dust contamination. So conception of open contact pads usage includes also application of hangars for UAV service stations. Any hangar or other housing (Fig. 1.10) is necessary not only for contact pads protecting but for providing comfort temperature for Li-Po accumulator cells charging. It is well-known that most of Li-Po batteries may be charged only under positive values of temperature. Therefore, covered temperature-stabilized housing is required for arranging of UAV service stations in winter conditions. Besides temperature stabilizing subsystem, such a housing must contain control subsystem for automatic opening/closing a gate in time of a next UAV takeoff/approaching.

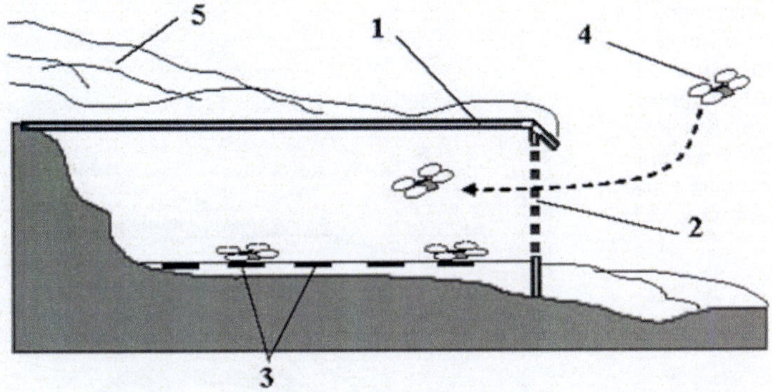

Fig. 1.10 Housing for UAVs: **a** housing cover; **b** automatic gate; **c** charging terminals; **d** UAV; **e** snowdrift

In the present paper, some known technical solutions concerning UAVs charging stations with open contact pads are described. Two original development works are presented. These are the charging station based on arrays of the so-called intelligent contact pads and the charging station based on flat parallel electrodes.

1.3.3 Known Solutions Realizing Open Pads Conception. Classification Criteria

Many various classification criteria may be proposed for systematizing information about charging platforms with open pads. Let us consider some of them.

Number of levels on which contact pads are placed. One- and two-level charging platforms are known. In one-level platforms, all contact pads lie on one plane. It is a trivial solution, further a few examples of such platforms are shown. An example of two-level platform is described in [41]. It is a site with two concentric metal rings, the first ring (little) is positioned on the low (ground) level, and the second one (of bigger diameter) is on the high level (Fig. 1.11). Corresponding UAV electrodes lay on two levels too. In the center of the platform, a LED beacon may be positioned for the short-range navigation. The main advantage of such arrangement of electrodes is that the UAV yaw angle during landing is unrestricted and may be of any value.

Method of multisectional battery charging. The most of UAV on-board batteries are of Li-Ion (Li-Po) type. They require special control of charging process for all battery cells in series coupling. Devices known as balancers are used for this purpose [42]. A balancer for its operation must be connected to all contacts of cells. For example, control of three-cell battery charging requires four points of connection.

The following three variants of charging process arrangement are known as follows:

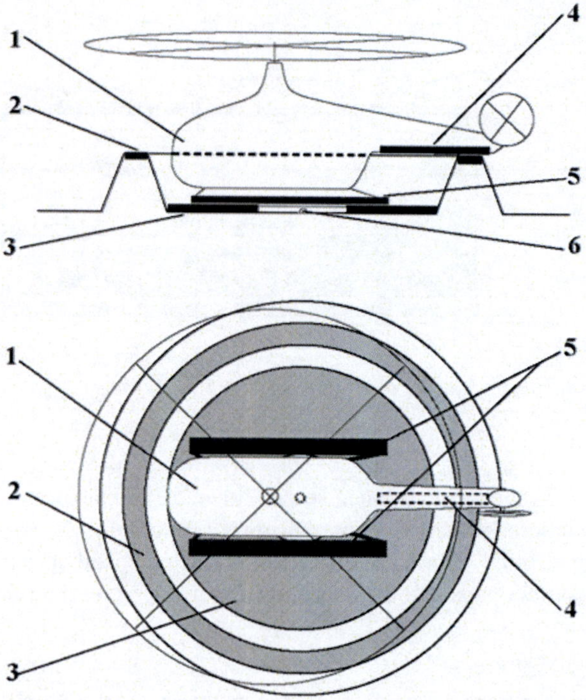

Fig. 1.11 Concentric two-level placement of electrodes on charging station: **a** UAV; **b** upper terminal electrode; **c** lower terminal electrode; **d** upper on-board electrode; **e** lower on-board electrodes; **f** LED beacon

(A) without the control of multisectional battery charging (if the accumulator consists of a single cell or if it is not of Li-Ion type);
(B) multisectional battery charging control is implemented by GCS;
(C) on-board multisectional battery charging control.

In variant B, the balancer is placed on the ground but not on the UAV board. It is a good solution in terms of reducing UAV mass and dimension, but in this case charging station must contain necessary number of contact pads corresponding to battery control points, and the UAV must have the same number of on-board landing electrodes [43]. Landing must be precise enough to implement connection of corresponding on-board and charging station electrodes. Figure 1.12 illustrates one example of such charging station for quadcopter with three-cell battery.

Solutions described below are intended for variant C (the balancer is on-board). Its advantage is that the arrangement of UAV on-board electrodes may be more flexible and less constrained by charging station electrode system. In this case, minimum quantity of electrodes for the UAV and the charging station equals 2 (positive and negative poles). Under the condition of so minimal quantity of electrodes, necessity of precise landing remains as a rule. But precise landing often is awkward, for example, if landing is implemented on a mobile platform or under puffy wind. In addition, the development of such a platform, on which a few UAVs might be serviced simultaneously, would be very useful. It is possible to create such an electrode system,

Fig. 1.12 Charging station with ground-placed balancer for quadcopter with three-cell battery. Good contacting between on-board and station electrodes is provided by means of magnets (Photo: Mulgaonkar [44])

on which many UAVs are charging at the same time without the requirement of their accurate landing and position, if redundancy of contact pads quantity is provided.

Redundancy of contact pads quantity. This factor is the third criterion of classification for charging stations with open contact pads. Under large degree of redundancy, when the quantity of charging station contact pads is much more than the number of on-board landing electrodes, various schemes of feeding contact pads may be realized.

Authors of so-called honeycomb service platform [41] proposed to form landing site with many isolated hexagonal metal cells. The landing UAV has four electrodes corresponding to three-cell on-board battery (each battery terminal has been electrically connected to separate points on the base of the helicopter skid so that they are in contact with the hexagon cells that make up the surface of the platform). The platform control unit scans its constituent hexagon cells to identify which ones are host to a battery terminal (via the skid) and identifies the voltages present on each. With this information, the platform controller connects the terminals of its battery charger to the appropriate hexagon cells (and thus the battery terminals); after that, charging is initiated. The described system contains rather complicated structure and algorithm of interconnections between the UAV and the platform via IR emitters and receivers. The similar system is discussed below in this paper. Its advantages are the comparative simplicity, versatility and bigger square of contacting surfaces of electrodes.

1.3.4 Platforms Based on Intelligent Contact Pads

The idea is that an UAV provided with two undercarriage electrodes (ski-like long and flat metal strips connected with poles of on-board battery) lands onto the field of many small contact pads (so-called intelligent or smart contact pads). Size of each pad and clearances between them are chosen by such a way that one on-board undercarriage electrode could cover and close a few contact pads, but in any event two heteropolar undercarriage electrodes could not close each other.

The control unit for each intelligent contact pad is a functional chain of amplifier, analyzer and commutator that may be simply realized on the basis of standard microcontrollers. One fragment of such a control circuit (DD1, ATMEGA32L-8AU) for switching contact pads is shown in Fig. 1.13. One controller can commutate eight contact pads.

Contact pads X1–X8 are connected to analog inputs PA0–PA7 of microcontroller DD1 via resistor dividers R1R9–R8R16. In standby mode, the microcontroller implements cyclic scanning of these inputs and analyzing voltages on them. After landing UAV on the matrix of contact pads, some of the pads get positive or negative voltage due to connection with one of two on-board electrodes. From this moment, we can consider such pads as activated ones. Numbers of the activated pads and their polarities are saved in the microcontroller's memory. Then inquiries to inputs PA0–PA7 are finished, and the corresponding digital outputs PB0–PB7 and PC0–PC7 eject control signals for switches K1–K8 and K9–K16 which can commutate positive or negative pole of the charging source E through current-limiting resistors to the activated pads. Disconnection commands for switches K1–K8 and K9–K16 are formed by the microcontroller after receiving control signals from other subsystems.

The advantage of such a system comparatively with fully analog control systems is that its measuring and actuating parts are separated and feedback coupling is

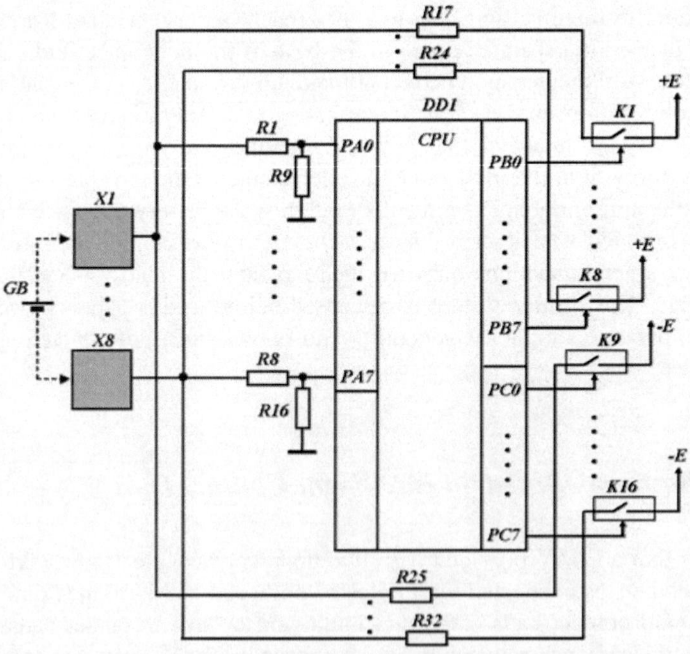

Fig. 1.13 Intelligent contact pads control circuit

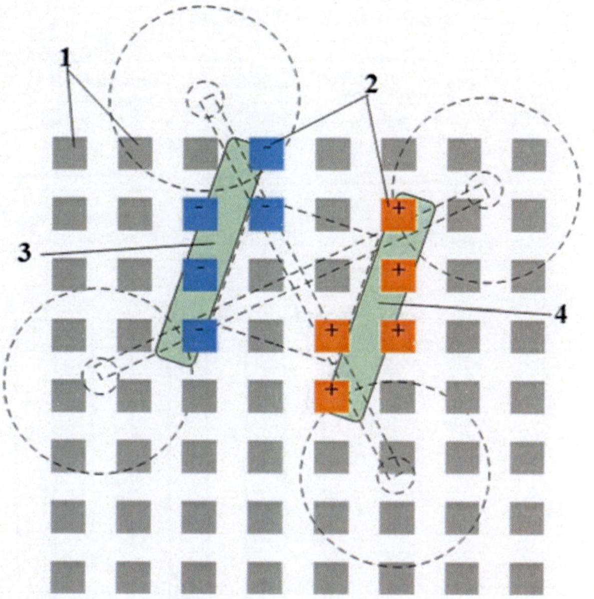

Fig. 1.14 Distribution of activated pads in the smart contact array after UAV's landing: **a** non-activated pads; **b** activated pads; **c** negative undercarriage electrode; **d** positive undercarriage electrode

excluded. Therefore, accidental failures due to feedback factor are excluded. In addition, such a scheme is not limited in quantity of simultaneously commutated contact pads.

Figure 1.14 illustrates the distribution of pads after landing an UAV onto the platform. Note that pads under ski-like on-board electrodes are activated. Each smart contact is able to analyze the voltage (which may be weakened) applied to the pad due to the UAV's landing and touching the pad by undercarriage electrodes. After analyzing phase, the connection to the charging source takes place so that the positive pole of the source is connected to the positive pad and the negative pole is connected if the pad is negative. Thus, we have automatic connection of the ground-charging source to the on-board battery electrodes by the right way even in cases of inaccurate landing.

1.3.5 Charging Stations Based on Flat Parallel Electrodes

The general idea of the second proposed charging station [45] (Fig. 1.15) uses redundancy of platform electrodes as the first one. It consists of a row of contact pads implemented as flat parallel electrodes separated from each other by narrow insulating spacer. One half of platform electrodes are connected with positive pole of the platform power source and another half with negative one, and their polarities are interlaced. The UAV has four on-board landing electrodes positioned at the end of the UAV's legs. Due to special geometrical features of the platform and on-board

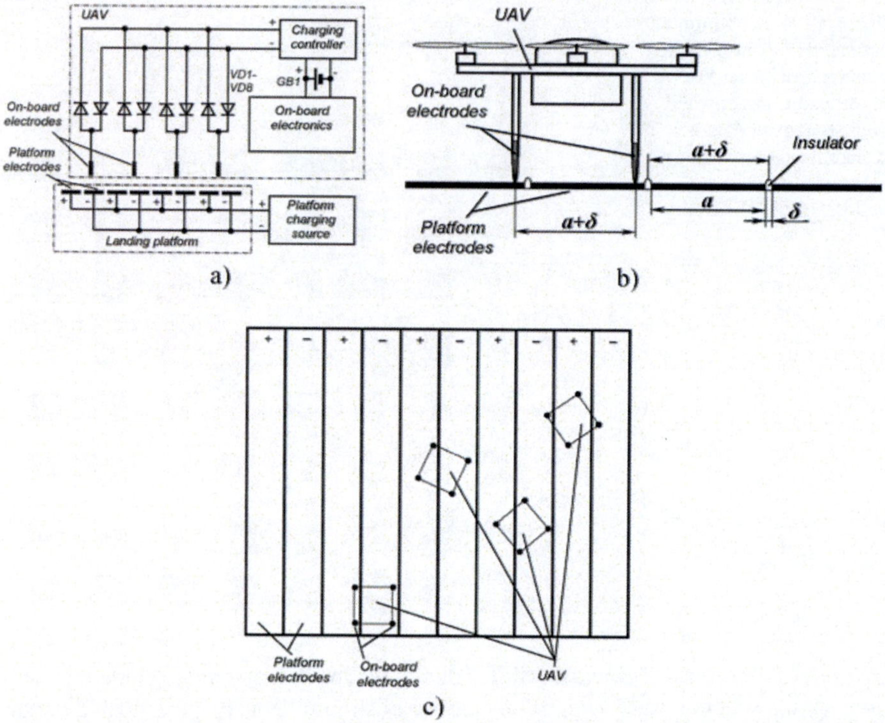

Fig. 1.15 Charging station with flat parallel electrodes: **a** station structure; **b** UAV positioning on the platform flat electrodes; **c** random locations of UAV on-board electrode contact points on the landing platform

electrodes, different polarities of the on-board electrodes would be obtained under any position of the UAV on the station. That is at least one on-board electrode would be of different polarity than others. Special on-board distributing circuit provides right connection of the on-board battery charging controller to the platform power source under any random combination of polarities on on-board landing electrodes.

If the width of the platform electrode is $\boldsymbol{a}$, the width of the insulating spacer is $\boldsymbol{\delta}$, and the length of the side of a square in which vertices correspond to the contact points of on-board electrodes is $\boldsymbol{d}$, then the geometrical condition providing 100% probability of heteropolarity of on-board electrodes is the equality $\boldsymbol{d = a + \delta}$.

Connection of platform electrodes through on-board electrodes to positive and negative terminals of the on-board charging controller is provided by a simple diode distributing circuit (Fig. 1.15a). The charging controller is connected to the accumulator battery GB1. Circuits of balancing battery cells and circuits of disconnection of on-board electronics during charging time are not shown in Fig. 1.15a.

Due to the special form of insulating spacers and ends of on-board landing electrodes, short circuit between neighboring platform electrodes via a landing electrode is practically impossible. The described charging station may be used for service of a few UAVs simultaneously.

1.4 Conclusion

Full composition of UAS with all infrastructure for functioning of ARs is considered.

Automated service stations for replenishment of energy source or other consumables of UAVs are the important component of UAS.

Separately ground-charging stations with open contact pads for electrical UAVs are considered. All known technical solutions concerning charging stations of such a type are systematized. A few classification criteria are proposed. Two original development works are presented. These are the charging station based on arrays of the so-called intelligent contact pads and the charging station based on flat parallel electrodes. Each of these projects has own advantages and disadvantages. Matrices of intelligent contact pads in the aggregate with ski-like flat on-board landing electrodes provide big square of contacting surfaces, and as a result, big values of possible charging current. But such stations are rather expensive and complicate. On the contrary, landing platforms with flat parallel electrodes in the aggregate with on-board landing electrodes legs have small contacting squares, but they are very simple in realization and inexpensive. Both solutions provide arrangement of charging process even in cases of inaccurate UAV landing and also provide simultaneous charging for a group of UAVs.

References

1. Michelson, R.C.: International aerial robotics competition—the world's smallest intelligent flying machines. In: Proceedings of 13th RPVs/UAVs International Conference, pp. 31.1–31.9. UK (1998)
2. Medium Corporation site. Skyeton: What is UAV and what the difference between the terms "Drone" and "UAV". https://medium.com/@skyeton/what-is-uav-and-what-the-difference-between-the-terms-drone-and-uav-9b9c4771cacc. Accessed 06 Jan 2020
3. Feron, E., Johnson, E.N.: Aerial robotics. In: Siciliano, B., Khatib, O. (eds.), Springer Handbook of Robotics, pp. 1009–1029, Springer, Berlin, Heidelberg (2008)
4. EPFL Homepage. https://www.epfl.ch/labs/lis/research/aerial-robotics. Accessed 06 Jan 2020
5. Austin, R.: Unmanned Aircraft Systems—UAVS Design, Development and Deployment. Wiley, UK (2010)
6. Fetisov, V.S., Neugodnikova, L.M., Adamovsky, V.V., Krasnoperov, R.A.: Unmanned Aviation: Terminology, Classification, Current State. Ufa, Photon (2014)
7. Beni, G.: Concept of cellular robotic systems. In: Proceedings of the 3rd IEEE Symposium on Intelligent Control. Arlington, Virginia (1988)
8. Beni, G., Wang, J.: Swarm intelligence in cellular robotic systems. In: Dario, P., Sandini, G., Aebischer, P. (eds.), Robots and Biological Systems: Towards a New Bionics? NATO ASI Series (Series F: Computer and Systems Sciences), vol. 102 (1993)

9. Bonabeau, E., Dorigo, M., Theraulz, G.: Swarm Intelligence. Oxford University Press, Oxford (1999)
10. Hexmoor, H., McLaughlan, B., Baker, M.: Swarm control in unmanned aerial vehicles. In: Proceedings of the 2005 International Conference on Artificial Intelligence (ICAI 2005), vol. 2, pp. 911–917 (2005)
11. Engelbrecht, A.P.: Fundamentals of Computational Swarm Intelligence. Wiley (2006)
12. Osaba, E., Villar, E., Del Ser, J.: Introductory chapter. In: Swarm Intelligence—Recent Advances, New Perspectives, and Applications. IntechOpen (2019). https://doi.org/10.5772/intechopen.90066
13. Kannan, S., Koller, A., Johnson, E.: Simulation and development environment for multiple heterogeneous UAVs. In: Proceedings of AIAA Modeling and Simulation Technologies Conference and Exhibit, p. 5041 (2004). https://doi.org/10.2514/6.2004-5041
14. Arbanas, B., Ivanovic, A., Car, M., Orsag, M., Petrovic, T., Bogdan, S.: Decentralized planning and control for UAV–UGV cooperative teams. Auton Robots **42**, 1601–1618 (2018). https://doi.org/10.1007/s10514-018-9712-y
15. Petrovic, T., Haus, T., Arbanas, B., Orsag, M., Bogdan, S.: Can UAV and UGV be best buddies?—towards heterogeneous aerial-ground cooperative robot system for complex aerial manipulation tasks. In: Proceedings of 12th International Conference on Informatics in Control, Automation and Robotics (ICINCO 2015). Alsace, France (2015). https://doi.org/10.5220/0005548302380245
16. RT News page. 1,374 drones light up the skies to break Guinness World Record in China (2018). https://www.rt.com/news/425666-drones-china-guinness-record. Accessed 08 Jan 2020
17. RaceLogic Support Center site. How Does DGPS Work? https://racelogic.support/01VBOX_Automotive/01General_Information/Knowledge_Base/How_Does_DGPS_(Differential_GPS)_Work%3F. Accessed 06 Jan 2020
18. ICAO Remotely Piloted Aircraft System Concept of Operations for International IFR Operations. https://www.icao.int/safety/UA/Documents/RPAS%20CONOPS.pdf. Accessed 11 Jan 2020
19. US Air Force Recruiting Site. Remotely Piloted Aircraft (RPA) Sensor Operator. https://www.airforce.com/careers/detail/remotely-piloted-aircraft-rpa-sensor-operator. Accessed 11 Jan 2020
20. The Nothern Plains UAS Test Site Homepage. http://www.npuasts.com/research.html. Accessed 11 Jan 2020
21. Nonami, K., Kendoul, F., Suzuki, S., Wang, W., Nakazawa, D.: Autonomous flying robots. In: Unmanned Aerial Vehicles and Micro Aerial Vehicles. Springer (2010)
22. Shiau, J.K., Ma, D.M., Yang, P.Y., Wang, G.F., Gong, J.H.: Design of a solar power management system for an experimental UAV. IEEE Trans. Aerosp. Electron. Syst. **45**(4), 1350–1360 (2009)
23. Betancourth, N.J.P., Parra, J.P., Rios, J.J.V., Cerón-Muñoz, H.D.: Design and manufacture of a solar-powered unmanned aerial vehicle for civilian surveillance missions. J. Aerosp. Technol. Manag. **8**(4), 385–396 (2016)
24. Mason, R.: Feasibility of laser power transmission to a high-altitude unmanned aerial vehicle. RAND Technical Report (1970)
25. Defence Review: Laser motive and lockheed martin skunk works developing "power beaming" in-flight laser-recharging for UAS/UAV/Drone aircraft (2019). http://www.defensereview.com/lasermotive-and-lockheed-martin-skunk-works-developing-power-beaming-in-flight-laser-recharging-for-uasuavdrone-aircraft-are-wireless-poweredlaser-powered-eternal-armedweaponized-unmanned. Accessed 11 Jan 2020
26. Simic, M., Bil, C., Vojisavljevic, V.: Investigation in wireless power transmission for UAV charging. Proc. Comput. Sci. **60**, 1846–1855 (2015)
27. Prior, S.D.: Tethered Drones for Persistent Aerial Surveillance Applications. Southampton University Blogs, http://blog.soton.ac.uk/robotics/files/2015/08/Your-article-pages-78-79.pdf. Accessed 11 Jan 2020
28. Defence Update: Defence Innovation Review: Hovermast—a bird's eye view over the hill (2019). https://defense-update.com/20140616_hovermast.html. Accessed 11 Jan 2020

29. Fetisov, V.S., Artemyev, A.E., Mufazzalov, D.F.: Automoted service stations for electrical unmanned aerial vehicles maintenance. In: Innovative Machinery (2017)
30. González-Espasandín, O., Leo, T.J., Navarro-Arévalo, E.: Fuel cells: a real option for unmanned aerial vehicles propulsion. Sci. World J. Article ID 497642 (2014). https://www.hindawi.com/journals/tswj/2015/419786. Accessed 11 Jan 2020
31. Bennington, S.: Cella Energy Presentation (2014). https://www.h2fc-fair.com/hm14/exhibitors/cellaenergy.html. Accessed 11 Jan 2020
32. Toksoz, T., et al.: Automated battery swap and recharge to enable persistent UAV missions. In: Proceedings of AIAA Infotech@Aerospace Conference, pp. 1405 (2011)
33. Suzuki, K.A.O., Kemper, F.P., Morrison, J.R.: Automatic battery replacement system for UAVs: analysis and design. J. Intell. Robot. Syst. Theory Appl. **65**, 563–586 (2012)
34. Swieringa, K.A., et al.: Autonomous battery swapping system for smallscale helicopters. In: Proceedings of IEEE International Conference on Robotics and Automation, pp. 3335–3340 (2010)
35. Kurs, A., et al.: Wireless power transfer via strongly coupled magnetic resonances. Science **317**, 83–86 (2007)
36. Griffin, B., Detweiler, C.: Resonant wireless power transfer to ground sensors from a UAV. In: Proceedings of 2012 IEEE International Conference on Robotics and Automation, River Centre. Saint Paul, Minnesota, pp. 2660–2665 (2012)
37. Marshall, P.T.: Power line sentry charging. U.S. Patent 7318564, Jan 15 2008
38. Fetisov, V., Dmitriyev, O., Neugodnikova, L., Bersenyov, S., Sakayev, I.: Continuous monitoring of terrestrial objects by means of duty group of multicopters. In: Proceedings of XX IMEKO World Congress "Metrology for Green Grouth", pp. 9–14 (2012)
39. Kartoun, U., Stern, H., Edan, Y., Feied, C., Handler, J.: Vision-based autonomous robot self-docking and recharging. In: Proceedings of World Automation Congress, pp. 1–8 (2006). https://doi.org/10.1109/wac.2006.375987
40. Carlson, S.: Ground Locomotion: Crawling Quadrotors and MultiCopters. http://diydrones.com/profiles/blogs/ground-locomotion-crawling. Accessed 11 Jan 2020
41. Kemper, F.P., Suzuki, K., Morrison, J.: UAV consumable replenishment: design concepts for automated service stations. J. Intell. Rob. Syst. **61**(1–4), 369–397 (2011)
42. Online Instructions for Equinox Li-Po Cell Balancer. http://manuals.hobbico.com/gpm/gpmm3160-manual.pdf. Accessed 11 Jan 2020
43. Dale, D.: Automated ground maintenance and health management for autonomous unmanned aerial vehicles. Thesis (M. Eng.), Department of Electrical Engineering and Computer Science, Massachusetts Institute of Technology (2007)
44. Mulgaonkar, Y.: Automated recharging for persistence missions with multiple micro aerial vehicles. Thesis (M. Eng.), Department of Mechanical Engineering and Applied Mechanics, University of Pennsylvania (2012)
45. Fetisov, V.S., Akhmerov, S.R., Mukhametzyanova, A.I.: Charging system for aerial robot onboard accumulator. Utility model patent 135469 (2013)

Chapter 2
Micrgrippers: Principle of Operation, Construction, and Control Method

Oleg Darintsev

Abstract The micromanipulation operations are a complex problem, so specific approaches are required in the development of microgripper designs and synthesis of its control systems. Different examples of microgrippers are given, and ways to control them are discussed. The problem of performing micromanipulation operations, the main effects acting in the contact zone of parts and a gripper, as well as the features of the implementation of operations to grip objects with dimensions less than 1 mm are considered. The classification of microgripping devices of robots used in the assembly of microsystems or planned for use is given. Particular attention is paid to specific techniques for the design of microgrippers, original technical and technological techniques.

2.1 Introduction

The first articles on a new direction in technology—microrobots, microassembly stations, and microgrippers,—began to appear in large numbers in the 90s of the twentieth century, but the formation and rapid development of a new topic occurred only in the early 2000s, when the cost of MEMS (MEMS) dropped significantly and the volume of their production increased by several orders of magnitude. Around this period, in Russia there has been a sharp increase in the production of microdevices, the scope of their application is expanding, and in 2010, by the decision of interested representatives of the domestic MEMS industry, the "Russian MEMS Association" was formed [1]. MEMS in its classical sense consists of several typical subsystems: control, information, and executive. If the first two subsystems are manufactured using classical technologies for the production of microcircuits, the mechanical components for high-precision and reliable MEMS are manufactured

O. Darintsev (✉)
Ufa State Aviation Technical University, 12, st. K. Marksa, 450008 Ufa, Russia
e-mail: ovd@uimech.org

Mavlyutov Institute of Mechanics, Ufa Centre of the Russian Academy of Sciences, 71, ave. Oktyabrya, 450054 Ufa, Russia

A. Ronzhin and V. Shishlakov (eds.), *Proceedings of 15th International Conference on Electromechanics and Robotics "Zavalishin's Readings"*, Smart Innovation, Systems and Technologies 187, https://doi.org/10.1007/978-981-15-5580-0_2

separately, which require assembly operations at the final stages. With the increasing need for miniature devices, the complexity of the design and the creation of new, more functionally saturated products, automation of their production becomes an urgent problem.

Therefore, at present, a key task in microrobotics used to automate production is the development of specific designs of microgripping devices that take into account the interaction of tools and microscopic objects and guarantee the safety of manipulations with microcomponents.

2.2 Applications and Specifics of Microgripping Devices

Currently, microgrippers are widely used in areas such as:

- production of MEMS and microdevices (assembly, testing, final packaging);
- electronics (microassemblies installation, 3D stack structures formation);
- medical applications and precise manipulation of biological objects;
- materials science (new materials development, manipulations with microvolumes of substances).

The creation of MEMS is a complex task that cannot be reduced simply to a decrease in overall dimensions. When creating microsystems and microdevices for manipulating, moving, or performing any action in the micrometer range, one has to face a number of new problems that need to be addressed at various stages of design and production:

The small size of the components of the gripper and objects requires taking into account the phenomena and forces that were neglected for macroobjects. In the microworld, these phenomena and forces have a significant effect on the static and dynamic characteristics of manipulation systems.

It should be correctly determined what forces (mechanical, optical, electromagnetic, etc.) can be used in a microdevice to achieve (obtain) safe and reliable (efficient) grip, perform elementary operations or movements.

Clarify the requirements for accuracy and the type of technology that can be used in the design and manufacture of the microsystem and its components.

The first problem is related to the adhesion forces arising between the microgripper and the object of manipulation, the actuator, and the surface. Usually, the task of micromanipulation is reduced to the operations of raising (grabbing), moving (carrying), and lowering (releasing) microelements. When the objects that are being manipulated are up to 1 mm in size, gravitational and inertial forces proportional to the volume (mass) of the object have less influence compared to the adhesion forces proportional to the surface area. Cohesive forces (electrostatic, van der Waals, and surface tension) create problems when a microscopic object is released from the gripper.

For example, when capturing a spherical microscopic object (radius of about 50 μm) by capturing with flat jaws, the maximum values for the surface tension force are about 10^{-5} N, van der Waals forces 10^{-7} N, electrostatic force (if ground) 10^{-9} N, and the gravitational force is approximately 2.7×10^{-8} N. Moreover, the surface tension forces can play a more significant role when manipulating with microscopic objects of natural origin (cells) or with wet objects. Another solution to this problem is to reduce the contact surface area between the working part of the gripper and the microscopic object.

Regarding the second problem, various types of physical phenomena and effects can be used to ensure reliable and safe contact. The principles of operation of most developed or manufactured micromanipulation devices and grippers are based on the use of mechanical, hydrodynamic, optical, and electromagnetic forces.

The third problem is related to the task of ensuring high values of accuracy, repeatability, and resolution of micromanipulations, as required by various applications of MEMS.

Each of the applications listed at the beginning of the section is characterized by unique requirements for actuators:

- specific characteristics of the environment in which the manipulations are carried out determine the type of structural materials;
- the type of gripper movements affects the choice of the type of actuators and information support;
- methods of parrying negative effects—a method and means of compensating for disturbances.

The selected technological methods have a significant impact on the design being developed, so almost any microgripper has specific features. The most significant effect on the final appearance is exerted by the choice of power components—microactuators. The need to perform manipulations with objects smaller than 1 mm, as well as movements with an accuracy of fractions of a nanometer, has led to a decrease in the share of application of classical types of actuators. The use of the electric actuator in microexecution is accompanied by the presence in the gripping design of the motion type transducers and/or reduction gears. As a result, in such systems, there are backlashes and gaps that are several times greater than the desired minimum displacements, which mean that the control system is complicated.

Therefore, energy converters are used as the basic representatives of actuators in microgripping mechanisms, which, thanks to the use of specific effects, make it possible to obtain the required type of displacement (shift, rotation, bending, etc.) directly from thermal, electric, and other types of energy. Table 2.1 shows the most common types of microactuators.

Microgrippers using the above types of actuators hold microscopic objects by creating friction in the contact zone. The magnitude of the retention forces directly depends on the size of the contact area of the microscopic object and the working surface of the gripper, therefore, for reliable retention, replaceable jaws of various shapes and sizes are used (Fig. 2.1). The maximum opening size of the microgripper tips can reach several mm with the magnitude of the developed efforts from microNewton to milliNewton.

Table 2.1 Most common types of microgripping actuators

Energy type	Effect	Constructive implementation
Electric (electromagnetic)	Static electricity	Parallel-plate [2]
		Comb actuator [3]
	Piezoelectrics [4]	Bimorph [5]
		Blade
		Stack
		Tube
	Magnetostriction	Crystals [6]
	Electrodynamics	Electric micromotor [7]
Pressure	Hydraulics and pneumatics	Microimplementation of "classic" actuators [8]
		Elastic elements' deformation [9]
Thermal energy	Thermal expansion [10, 11]	Bimorph plates
		Working fluid's expansion (membrane)
		Local expansion
	Shape memory effect (alloys and crystals) [12, 13]	Film coating
		Monolithic (bulk or crystal)
		Wire or foil

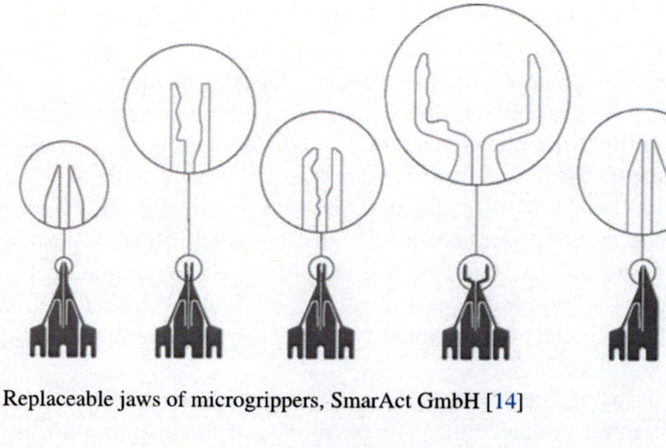

Fig. 2.1 Replaceable jaws of microgrippers, SmarAct GmbH [14]

The use of the frictional method of holding microscopic objects is accompanied by undesirable consequences associated with a significant effect of adhesive effects in the microworld. So, quite often observed: contact electrification (charge accumulation in the oxide film, which covers all objects in contact with the atmosphere); in the presence of polished surfaces—van der Waals force; during manipulations with biological objects—surface tension forces.

Both with full-sized object holding devices and with microgrippers, the problem is working with flat or thin-walled objects, for which even a slight excess of the gripping forces leads to deformation and destruction. Therefore, recently, in addition to frictional grippers, new original designs of microgripping devices have appeared, which use adhesive forces to reliably hold microscopic objects and practically neutralize the negative effects from their manifestation in the contact zone.

2.3 Examples of Microgripper Designs

At USATU, research in the field of microrobototechnics started in 1995 in the form of joint work with the Institute of Process Control Technology, Automation und Robotics Karlsruhe University (currently the Karlsruhe Institute of Technology (KIT)), and subsequently work was carried out under NATO grant No. 972063 "Intelligent Planning and Control for an Automated Microrobot-Based Microassembly Station." Collaboration by German colleagues from 1994 to 1999 was overseen by S. Fatikow. In 2001, he established a new Division for Microrobotics and Control Engineering (AMiR) at the University of Oldenburg, Germany. Since 2001, he is a full professor in the Department of Computing Science and Head of AMiR. His research interests include micro/nanorobotics, industrial robotics and automation at nanoscale, nanohandling inside SEM, AFM-based nanohandling, sensor feedback at nanoscale, and robot control [15–19].

In 1996, a robotics laboratory was organized at USATU, where the MikRob piezoelectric mobile microrobots, a prototype microassembly station with an optical control and feedback channel, were developed [20–22]. The need to perform micromanipulations with an accuracy of 10 nm required the creation of specialized microgrippers and microtools. The first variant of microgripping was synthesized in the form of a finger grip with a sponge opening range from 0.1 to 850–1200 μm based on a Philips multilayer piezoactuator. The maximum force to hold objects in the contact area of the clamping parts was 1.5 N.

Some of the proposed designs are represented by grippers using surface tension forces that occur when a part contacts a liquid film formed on a work surface. Only the methods for producing this film differ: the liquid is initially located in the gripper's cavity from which it enters the surface, or it condenses from the environment on the cooled surface.

So, on the basis of a previously obtained patent [23], a prototype capillary microgripper was made, and a laboratory stand was implemented to create and maintain the required parameters of the working medium (temperature and humidity) (see Figs. 2.2 and 2.3) [24]. The principle of retention of microscopic objects is based on the use of surface tension forces arising from the contact of the surface of a microscopic object and a liquid film. The necessary film of water condenses from

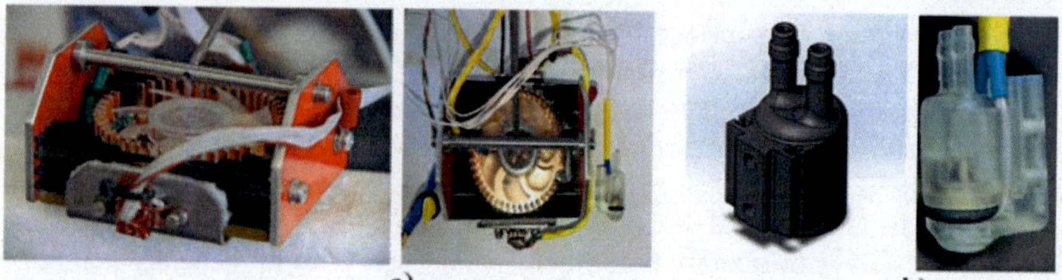

a) b)

Fig. 2.2 Capillary gripper: **a** the first prototype; **b** reduced version

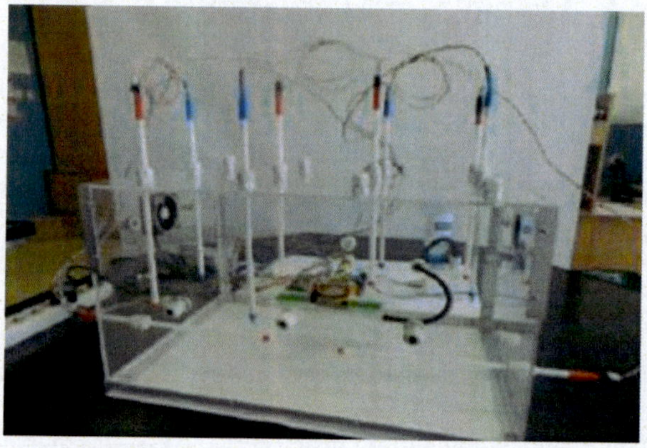

Fig. 2.3 Prototype capillary microgripper and laboratory stand

the working medium on the cooled working surface when temperatures below the dew point temperature are reached.

An approximate estimate of the range of the surface tension force holding the object is according to the following formula:

$$F_L = \frac{\gamma \cdot (\cos\theta_1 + \cos\theta_2) \cdot A}{d}, \tag{1}$$

where γ is the coefficient of surface tension of the liquid (for water $\gamma = 7 \times 10^{-4}$ N/m); A is the contact area; d is the distance between the surfaces; θ_1 and θ_2 are the contact angles between the liquid film and the surface. The inaccuracy of the assessment is related to the problem of determining d.

If we consider the classic finger structures of microgrippers, for their effective operation and the successful implementation of simple assembly operations, such as gripping or releasing assembly objects, it is necessary that the gravitational force prevails over the other forces acting on the microscopic object. According to the results of mathematical modeling and experiments, it is seen that ceteris paribus, capillary forces dominate at the microlevel for microscopic objects smaller than 500 $\div$ 450 nm in size. The effective impact of the forces of van der Waals and Coulomb is much weaker. Gravitational force no longer has the same effect as at the macro-level:

the manifestation of the van der Waals force becomes more significant when the size of a microscopic object does not exceed ~250 ÷ 150 nm, and electrostatic forces—at sizes less than ~70 ÷ 50 nm. The proposed microgripper design is more effective for working at the microlevel, because it using the force of surface tension to hold microscopic objects can reduce the negative effects of other adhesive forces. The effect of electrostatics is insignificant since the grounded structure and the liquid film can neutralize the charge of the object, which is manifested due to contact effects, the presence of an oxide film on the surface, etc. The thickness of the water film far exceeds the minimum value of the effective distance for the manifestation of intermolecular interaction (van der Waals forces). The simulation of the microgripper was carried out using a specialized virtual environment, which is the basic tool for implementing an integrated approach to the design of microdevices and the synthesis of intelligent control systems [25, 26]. The main difference between the applied integrated approaches to the synthesis of the microgripper control system from the system one is that the external environment is presented not as a source of information or external disturbance, but as another subsystem of the control object. Thus, in the synthesis of an intelligent control system, capillary microgripper and the working environment were considered as a single object, in a complex (see Fig. 2.4).

The complexity of the model support of the virtual environment is associated with the need to take into account many factors: the mechanisms of formation of

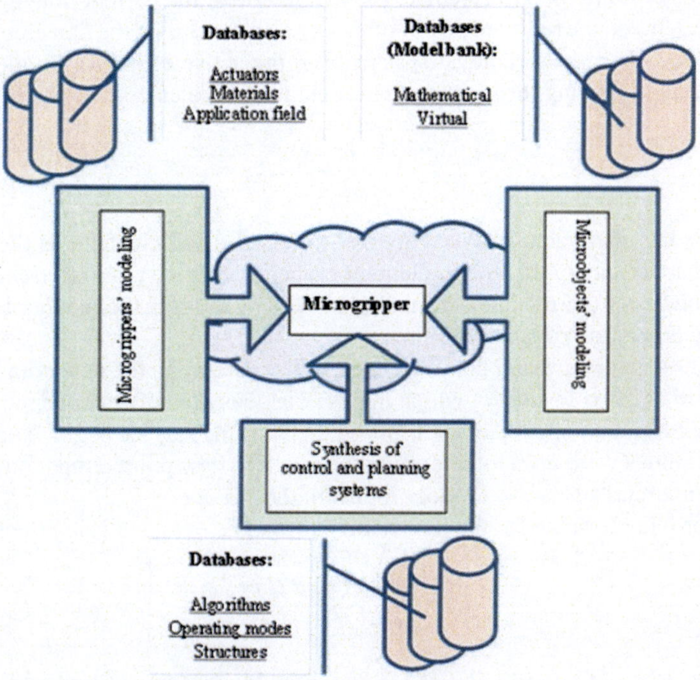

Fig. 2.4 Concept of building intelligent microgripper control systems

an aqueous film on a cooled (cold) surface, the formation of drops, their fusion, the formation of a meniscus at the border with the air, etc. The resulting variety of synthesized models was required to compensate for the "information hunger" that is characteristic of almost any microsystem. The ultra-small dimensions of the structures do not allow the incorporation of sensors with the required resolution into micromanipulation systems. Well-known force sensors and displacements in microperformance (MEMS sensors) are mainly designed to measure values of classical dimension, and the required measurement range for using sensors in microsystems is 2–3 orders of magnitude higher than the instrumental error. Therefore, the information system for capillary microgripping is based on an indirect measurement of the state of the system, i.e., information on the state and progress of the process is synthesized on the basis of data on the working environment, the results of field experiments and expert evaluations [27].

2.4 Information System Architecture

For the correct operation of the gripper's control system, the values of the temperature of the cold and hot gripping surfaces, temperature and humidity of the environment are necessary [4, 5]. The temperature values on the cold and hot sides of the gripper (Peltier element) are used to calculate the magnitude of the current flowing through the active element, the required power of the cooling system (in the first embodiment, the fan speed of the heat removal system from the active element). So the optimal current strength flowing through the Peltier element is calculated by the formula:

$$l_{\text{opt}} = \frac{\alpha \cdot \sigma \cdot S}{l} \cdot T_x, \tag{2}$$

where α is the coefficient of thermopower (Seebeck coefficient); σ is the specific thermal conductivity of the thermoelement branches; S is the cross-sectional area of the semiconductor thermocouple branch; l is the length of the thermocouple branch; T_x is the temperature of the cold side.

Ambient temperature and humidity are used to determine the dew point, i.e., the temperature of the cold side, at which intense condensation of water vapor from the air and the formation of droplets or films on the working surface begin. Nomograms or tabular values were used to determine the required dew point temperature, steam condensation rate, and water evaporation from the surface.

An approximate value for the dew point can also be obtained using the formula:

$$T_d = \frac{b \cdot \gamma(T, RH)}{a - \gamma(T, RH)}, \tag{3}$$

where $a = 17.27$ °C; $b = 237.7$ °C; T is the ambient temperature; $\gamma(T, RH) = \frac{a \cdot T}{b+T} + \ln(RH)$; RH is the ambient humidity in volume fractions ($0 \div 1$).

The calculated dew point temperature for guaranteed condensate production on the working surface of the gripper decreased by $1 \div 2$ degrees and this adjusted value was used to calculate the required current strength.

During the operation of the tong, heat from the hot side is removed into the surrounding working volume, which leads to a change in both temperature and humidity. With increasing ambient temperature, the surface tension coefficient decreases, so two controlled fans were used in the stand. Changing the rotation speed of these fans, and, consequently, the air exchange rate, made it possible to reduce the influence of temperature drift on the performance of the gripper.

During the experiments, when operating the stand fans, a decrease in humidity was noted, which led to an increase in the time of formation of the aqueous film. To correct this drawback, an ultrasonic steam generator was used, the power of which was set by software. From all of the above factors, the variety of components of the stand required the correct synthesis of the information system, taking into account the specifics of the device. Therefore, the number of sensors (measurement points) in the laboratory bench was significantly increased, which made it possible to obtain sufficiently informative training samples, on the basis of which intelligent control algorithms were further synthesized.

The architecture of the information system of the semi-natural stand is implemented in the class of multi-level distributed control systems. When implementing a multi-channel (multi-point) measurement system, we used DS1631S temperature sensors operating according to I2C protocol and SHT75 temperature and humidity sensors with their own non-standard two-wire interface.

Based on the synthesized architecture of the information system, specialized software for a personal computer and controller STM32F407 was developed, which allowed connecting 8 addressable digital sensors and 5 digital humidity sensors with 12-bit resolution. Based on expert data and experimental results, a hybrid intelligent Peltier element operation algorithm was synthesized [28].

2.5 Control System of Intelligent Capillary Microgripper

Experimental verification of the constructed models and synthesized fuzzy control algorithms revealed the need to refine the structure of the control system. Since three basic operating modes of the microgripper were revealed in the course of the experiments—transition to the operating mode; capture of a microscopic object; releasing a microscopic object—calculation (selection) of unique sets of parameters of a fuzzy controller used in a direct control circuit was required. As a result of the work, a new control structure was synthesized on the basis of a hybrid controller (fuzzy rules and a neural network) that identifies the current situation and connects the necessary corrective circuit in real time (see Fig. 2.5).

A neural network consists of two hidden layers and one output layer. The first layer includes 10 neurons, the second—6 neurons, and the output layer is represented by 3 neurons. The structure of the neural network was trained by the method of

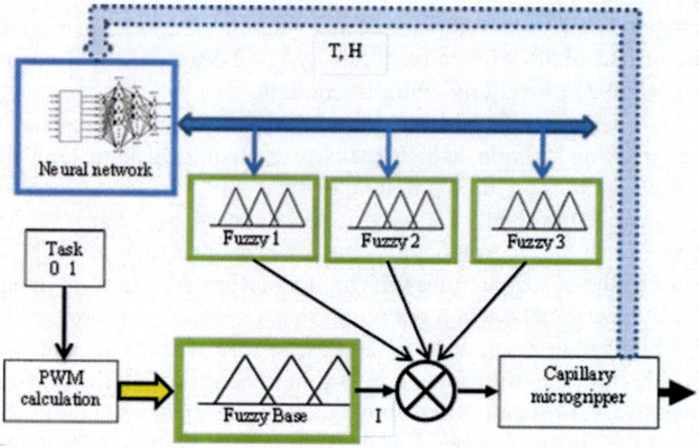

Fig. 2.5 Structure of the hybrid regulator in the microgripper control channel

backpropagation of error; a linear activation function was used in the output layer of the neural network (see Fig. 2.6).

For the correct operation of the developed fuzzy regulators and neural network in the correction loop, it is necessary to obtain data on the parameters of the working medium promptly: H_1, H_2 are measured humidity value directly near the working surface and in the working environment; T_c, T_h are temperature values of the cold and hot sides of the tong; and T_{in1}, T_{in2}, T_{in3} are temperature values at three points of the working environment.

The choice of parameters of fuzzy regulators and training of the neural network were carried out on the basis of data obtained in the course of real and model experiments [26]. To increase the speed of the grip and increase the reliability when holding

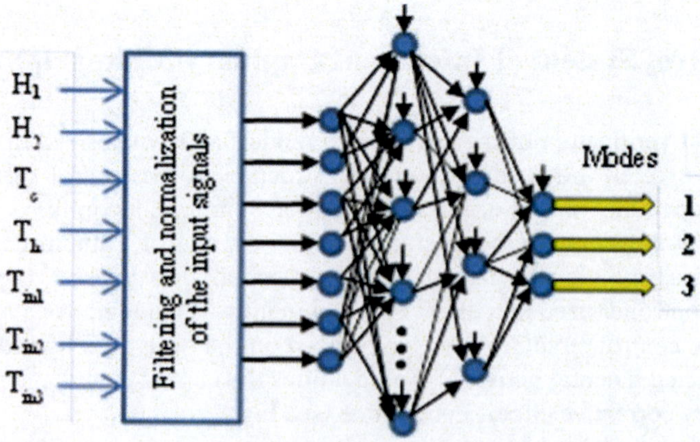

Fig. 2.6 Neural network identification mode architecture

the object, it is necessary to conduct additional experiments to take into account the drift of the characteristics of the working surface of the gripper, changes in the coefficient of surface tension or to identify previously unaccounted parameters of the working process.

2.6 Construction Prospects

When using microgrippers, the original solutions require releasing the part, since the natural evaporation of the liquid occurs slowly: an order of magnitude slower than condensation (the evaporation surface is much smaller than the contact area). Even the suction of liquid from the contact zone is unable to completely remove it and release the microscopic object at the desired point. One of the options for the implementation of capture-release operations is the use of the electro-wetting effect, when under the influence of an electric field (voltage of about 20 V) a drop spreads along the lyophobic material, increasing the wetting angle and creating conditions for holding the object. When the tension is relieved, the lyophobic properties of the surface are restored, a drop of liquid is contracted, the contact area decreases, and the microscopic object is detached from the trap.

A simpler solution to the release problem is to heat the gripper surface or "cold" boiling the liquid while generating ultrasonic vibrations in the liquid. In the first version of capillary gripper, which was created in the laboratory of Mavlyutov Institute of Mechanics, R.A.S., when checking the operability of the idea for releasing objects, an inverse connection of the Peltier element was performed [23, 25]. At present, another miniature version of the gripper has been developed, for which it is planned to use ultrasonic generators based on piezocrystals, which should lead to low-temperature boiling of the liquid and the transformation of the drop into "foam" with loss of conditions for the appearance of surface tension. In this case, unlike a gripper using the effect of electric wetting, the microscopic object will come off without retaining part of the drop on its surface.

2.7 Conclusion

Restrictions on the volume of the article did not allow us to consider other original ways of holding microscopic objects based on the use of the forces of van der Waals. To implement this method of fixing microscopic objects in the microgripper design, either polished plates of fine-grained material or "nanograss" [29] are used, using the principle of object retention, known as the gecko effect.

A "nanocarpet" or "nanograss" is an ordered structure made on a silicon crystal and consisting of cylinders with a diameter of 350 nm and a height of 7 μm, while the distance between them on different samples was from 1 to 4 μm.

The development of the instrumental and technological bases of MEMS production is a powerful incentive for the development of new types of microgripping devices using various principles of reliable retention of microscopic objects. The use of new materials, actuators, and sensors has significantly reduced the dimensions of microgrippers while maintaining their effectiveness and reliability, which can become the basis for solving the problems of full automation of MEMS production.

Based on the results of the work performed and the analysis of publications, the following conclusions can be drawn:

- for the correct operation of microgrippers [30], it is necessary to synthesize specialized architectures of information control systems that most fully take into account the specifics of the control object, the microworld and provide the necessary amount of information;
- virtualization technologies of systems and sensors, which allow reconstructing the state of microsystems using indirect data, can compensate for the negative value of "information hunger," which is typical for most microrobots and microgrippers control systems.

Acknowledgements This research is supported by the Program of the Presidium of the Russian Academy of Sciences and within the framework of state assignment No. 0246-2018-007.

References

1. Russian MEMS Association. http://mems-russia.ru. Accessed on 13 Jan 2020
2. Wenying, M., Changwei, M., Wang, W.: Surface micromachined MEMS deformable mirror based on hexagonal parallel-plate electrostatic actuator. J. Phys. Conf. Ser. **986** (2018)
3. MEMS Video & Image Gallery: (MESA) Sandia National Laboratories. https://www.sandia.gov/mesa/mems_info/movie_gallery.html. Accessed on 13 Jan 2020
4. PI Ceramic. https://www.piceramic.com/en/products/piezoceramic-actuators/. Accessed on 13 Jan 2020
5. Uchino, K.: Multilayer ceramic actuators. In: Encyclopedia of Materials: Science and Technology, pp. 5850–5858 (2001)
6. Magnetic actuators & motors, CEDRAT TECHNOLOGIES. https://www.cedrat-technologies.com/en/technologies/actuators/magnetic-actuators-motors.html. Accessed on 13 Jan 2020
7. Tarazón, R.L.: Robotics in micro-manufacturing and micro-robotics. In: Micro and Nano Technologies, Micromanufacturing Engineering and Technology, 2nd edn. William Andrew Publishing (2015)
8. Su, J., Lucyszyn, S.: Bulk-micromachined hydraulic microactuator (2005)
9. Rehman, T., Mohd Faudzi, A., Nafea, M., Mohamed Ali, M.: PDMS-Based Dual-Channel Pneumatic Microactuator Using Sacrificial Molding Fabrication Technique (2019)
10. Leong, T.G., et al.: Tetherless thermobiochemically actuated microgrippers. Proc. Nat. Acad. Sci. **106**(3), 703–708 (2009)
11. Volland, B., et al.: Duo-action electro thermal micro gripper. Microelectron. Eng. **84** (2007)
12. Shelyakov, A.V., et al: Design of microgrippers based on amorphous-crystalline TiNiCu alloy with two-way shape memory. In: 2019 International Conference on Manipulation, Automation and Robotics at Small Scales (MARSS). Helsinki, Finland, pp. 1–6 (2019)

13. Garcés-Schröder, M., et al.: Shape memory alloy actuators for silicon microgrippers. J. Microelectromech. Syst. 1–13 (2019)
14. Micro-Grippers, SmarAct Products. https://www.smaract.com/micro-grippers. Accessed on 13 Jan 2020
15. Fatikow, S., Rembold, U.: Microsystem Technology and Microrobotics. Springer (1997)
16. Fatikow, S.: Mikroroboter und Mikromontage. Teubner, B.G (2000)
17. Fatikow, S.: Automated Nanohandling by Microrobots. Springer (2008)
18. Fatikow, S., et al.: Nanohandling robot cells. In: Handbook of Nanophysics. Taylor & Francis, CRC Press (2010)
19. Fatikow, S., et al.: Robot-based automation on the nanoscale. In: Encyclopedia of Nanotechnology. Springer (2012)
20. Mardanov, A., Seyfried, J., Fatikow, S.: An automated assembly system for a microassembly station. J. Comput. Ind. **38**, 93–102 (1999)
21. Ilyasov, B.G., et al.: Mobile piezoelectrical microassembly robot: constraction and control. In: Proceedings of the First International Conference on Mechatronics and Robotics, vol. 1, pp. 187–188. Saint-Petersburg (2000)
22. Kusimov, S.T., et al.: Flexible multiagent microrobotic station for microtechnologies and precision assembly (desktop microfactories). In: Proceedings of the International Conference on Life Cycle Approaches to Production Systems Managment, Control and Supervision (ASI' 2000 & IIMB' 2000). Bordeaux, France, pp. 359–365 (2000)
23. Darintsev, O.V., Migranov, A.B.: The capillary microgipper with feedback. Pat. 2261795 of Russia
24. Darintsev, O.V., Bogdanov, D.R., Darintseva, E.O.: Multichannel systems for collecting and processing information for a seminatural stand: architecture and processing algorithms. In: Proceedings of the Mavlyutov Institute mechanics of the Ufa scientific center of the Russian Academy of Sciences, vol. 10, pp. 44–49 (2014)
25. Darintsev, O.V., Migranov, A.B.: The capillary microgipper's control system based on fuzzy logic. In: Proceedings of Artificial Intelligence. Intelligent System (AI-2010), pp. 218–222 (2010)
26. Darintsev, O.V.: The intelligent model basis for capillary microgripper's control system. Artif. Intell. J. **1**, 47–53 (2011)
27. Darintsev, O.V.: Using virtualization technologies in microrobot's and microsystem's control systems. In: Proceedings of the Mavlyutov Institute Mechanics of the Ufa Scientific Center of the Russian Academy of Sciences, vol. 9, pp. 47–52 (2012)
28. Darintseva, E.O., Darintsev, O.V., Bogdanov, D.R.: Information support for an intelligent control system of capillary microgripper. In: Proceedings of ITIDS + RRS '2014, pp. 242–247 (2014)
29. Nano-Optic Endoscope Sees Deep into Tissue at High Resolution. https://www.techbriefs.com/component/content/article/tb/supplements/pit/briefs/33895?m=806. Accessed 13 Jan 2020
30. Zhang, Y., et al.: Autonomous robotic pick-and-place of microobjects. IEEE Trans. Rob. **26**, 200–207 (2010)

Chapter 3
Worm-Like Locomotion Systems for In-Pipe Robots and Its Fuzzy Sliding Mode Controller Design

Robert Sattarov, Xinhao Huang, Cong Lin, and Lingfei Xiao

Abstract In this paper, worm-like locomotion system (WLLS) for in-pipe robots is considered, and a novel fuzzy sliding mode controller is designed for the velocity tracking problem in the WLLS. Because of the strong nonlinearity, an estimator for a friction force is created and it is used in the construction of the sliding mode controller. A sliding mode surface is provided based on the tracking error of the longitudinal displacement and a center of mass velocity. Fuzzy rule is formed to tuning one of the sliding mode designable parameters. Simulation results verify the effectivity of the presented fuzzy sliding mode control method.

3.1 Introduction

Pipes are widely used in normal life. They are used in power plants, chemical plants, gas, oil and sewage system [1]. However, block and leakage often occur in the pipelines. In order to ensure the work conditions of the pipelines and prevent abnormal occurrence, it is necessary to carry out routine maintenance of the pipeline to prolong the lifetime and reduce the expense [2]. In-pipe robots are an attractive research area recently, because of a great number of potential applications of these devices that cannot be performed directly by human operators due to the difficulties and dangers inherent in attaining feasible operating positions inside various pipes [3, 4]. For example, it is difficult for a person to inspect a pressure pipeline buried at high altitude or underground. However, the problem can be solved well by replacing a person with an in-pipe inspection device that can pass through the inside of the pipeline for maintenance and inspection [5].

R. Sattarov (✉)
Electromechanics Department, Ufa State Aviation Technical University, 450008 Ufa, Russia
e-mail: sattar.rb@gmail.com

X. Huang · C. Lin · L. Xiao
College of Energy and Power Engineering, Nanjing University of Aeronautics and Astronautics, Nanjing 210016, China

A. Ronzhin and V. Shishlakov (eds.), *Proceedings of 15th International Conference on Electromechanics and Robotics "Zavalishin's Readings"*, Smart Innovation, Systems and Technologies 187, https://doi.org/10.1007/978-981-15-5580-0_3

Pipeline robot is a mechatronic system that can move inside the pipeline and carry a variety of operators and sensors to perform related pipeline operations. There are many kinds of in-pipe robots. The structure of in-pipe robots may be a wheel, a caterpillar, or a snake. In [6], a compound climbing gait of snake-like robot is shown. A wheeled in-pipe robot that can drive the wheels and roll-joints by only a single actuator is presented in [7]. However, these robots cannot fully explore the space in the pipeline. A fundamental requirement for in-pipe inspection robots should be the ability to fit in pipes of variable size.

New kinds of in-pipe robots that enable locomotion in environments inaccessible for normal in-pipe robots are on agenda. An electromagnetic worm-like robot is proposed in [8]. This robot has a good traction ability, and meets the demands of full contact with pipes, because the maximal possible contact area between the robot and the inner pipe wall is provided. The paper [9] proposes soft actuators for a worm-like robot designed to operate inside constrained tubes. In addition, the slipping problem of worm-like robot is considered in [10].

The worm-like robots are proposed to be used in medicine for motion through narrow channels to an affected organ, and then a diagnostic or surgical operation can be performed [11, 12].

Sliding mode control is essentially a special kind of nonlinear control. Its nonlinearity features for discontinuities during the system control and uncertainties of the system structure. When the mathematical model of the controlled object is known and the sliding mode controller is designed, the system can move in a targeted way in a dynamic process, and the sliding mode state is not affected by changes of object parameters and also independent of disturbances. Hence, variable structure control has the advantages of fast response, insensitivity to parameter changes and disturbances, and simple physical accomplishment, etc. [13]. However, the disadvantage of this method is that after the state trajectory reaches the sliding surface, it is difficult to slide strictly along the sliding surface toward the equilibrium point; instead, it traverses back and forth on both sides of the sliding surface, resulting in chattering.

Fuzzy theory is an intelligent control method based on fuzzy logic. It imitates people's decision-making process in behavior. The method weaves the experience of operators or experts into fuzzy rules. An obvious advantage of fuzzy control is its strong applicability and robustness. As a matter of fact, fuzzy theory can be used under the circumstance of not knowing the mathematical model of the controlled object. However, one disadvantage is that it is not easy to learn and adjust the control parameters due to the adoption of IF-THEN control rules.

In general, a single sliding mode controller cannot meet the control requirements of complex and dynamic systems. Therefore, a fuzzy sliding mode control is proposed and it gives full play to their respective advantages. The accuracy of WLLS is improved and the error is reduced based on the fuzzy sliding mode controller. The simulation results verify that the WLLS has improved dynamic response speed and accuracy after adding a controller.

The remainder of this paper is organized as follows: the composition, the motion principle, and the mathematical description of WLLS are shown in Sect. 3.2. In Sect. 3.3, the design of sliding mode controller is presented. The fuzzy rules for the parameters are in sliding mode controller given in Sect. 3.4. Section 3.5 shows the simulation results of WLLS under the fuzzy sliding mode controller, and the conclusions are drawn in Sect. 3.6.

3.2 The Principle and Mathematical Model of WLLS

3.2.1 Composition and Motion Principle

3.2.1.1 Composition

The general structure of the WLLS is shown in Fig. 3.1 [8]. The WLLS includes two segments: a vibrating ring-like segment and a contact ring-like segment. The longitudinal springs connect two segments and each segment consists of two ferromagnetic semi-rings. In contact segment, the transverse springs connect two semi-rings. There is also a supporting pad in this part to adapt the material of the pipe wall. The exciting coils, which connect with a power supply, are winded on the semi-rings of the vibrating segment [8, 14, 15]. This design can be simplified so that a magnetic system of the WLLS can be very similar to clapping electromagnet [14, 16].

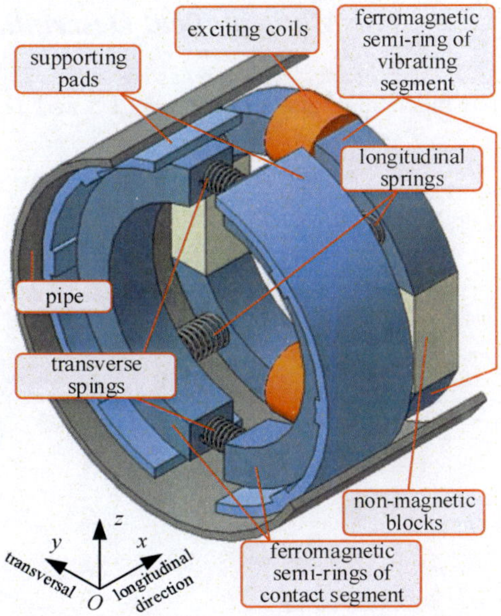

Fig. 3.1 Structure of WLLS [8]

3.2.1.2 Motion Principle

The change of center of mass displacement caused by the mutual change of periodic spring force and electromagnetic force leads to the movement of worm robot. It can be divided into several periods and each period is divided into three stages. Firstly, when the voltage of AC power supply is equal to zero, the electromagnetic force exerted on the worm system is zero. At this moment, the longitudinal spring force is maximum and pushes the support pad against the pipe wall while the longitudinal spring is in a relaxed state. Then, when the voltage of AC power supply gradually increases, the exciting current increases and electromagnetic force rises, attracting the contact and vibration parts and the ferromagnetic half rings of the contact part to each other. The springs are compressed until the peak of electromagnetic force and thus the positive pressure of the contact part support pad on the wall surface of the pipe decreases till a very low value. However, the center of mass does not move. Lastly, when the voltage of AC power supply decreases, the exciting current decreases. Thus, the electromagnetic force decreases and the positive pressure of contact segment increases, which gives rise to the increase of the friction force. The longitudinal spring needs to extend but the increasing friction force limits the motion of the contact segment (via the supporting pad). Hence, the displacement of center of mass moves forward against the contact segment.

3.2.2 *The Mathematical Description*

The mechanical model of the WLLS and a coordinate system are shown in Fig. 3.2.

Fig. 3.2 Mechanical model of the WLLS

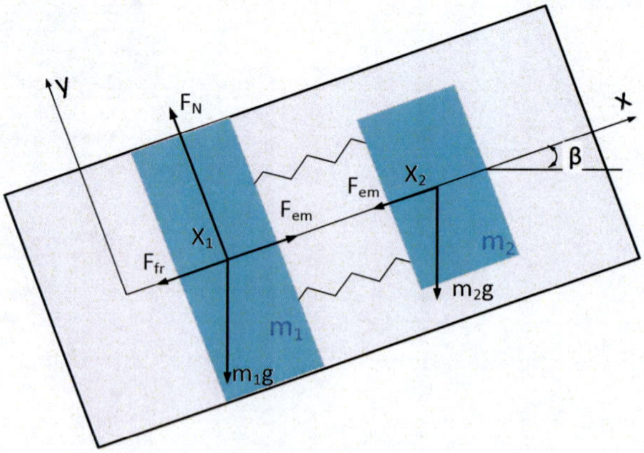

According to Newton's law, the following equations can be derived as follows:

$$\ddot{x}_1 = + \frac{F_{em}}{m_1} - \frac{c}{m_1}(x_1 - x_2) - g \sin\beta + \frac{F_{fr}}{m_1}$$
$$\ddot{x}_2 = - \frac{F_{em}}{m_2} - \frac{c}{m_2}(x_2 - x_1) - g \sin\beta, \tag{3.1}$$

where c is net stiffness of longitudinal springs, F_{em} is the longitudinal electromagnetic force, $m_1 g$ is the gravity force, and F_{fr} is the dry friction force.

$$F_{fr} = \begin{cases} -\mu F_N \operatorname{sign} \dot{x}_1, & \dot{x}_1 \neq 0, \\ -F_a, & \dot{x}_1 = 0 \text{ and } |F_a| \leq F_a, \\ -\mu F_N \operatorname{sign} F_a, & \dot{x}_1 = 0 \text{ and } |F_a| > \mu F_N, \end{cases} \tag{3.2}$$

where F_a is the sum forces without F_{fr}.

The electromagnetic force F_{em} must vary according to a periodic law, the parameters of which can be controlled in a sufficient range. It is easiest to realize electromagnetic force pulses of the required frequency and magnitude when a half-bridge inverter is included in the WLLS and its controller.

Assumption 1 The masses of the elastic contact ring-like segment m_1 and the vibrating solid ring-like segment m_2 are different, namely $m_1 \neq m_2$.

3.2.3 State-Space Model of WLLS

Let $X = \begin{bmatrix} x_1 \; x_2 \; \dot{x}_1 \; \dot{x}_2 \end{bmatrix}^T$ be the state vector, $u = F_{em}$ is input, then (3.1) can be written as:

$$\dot{X} = f(X) + bu, \tag{3.3}$$

where:

$$f(X) = \begin{bmatrix} \dot{x}_1 \\ \dot{x}_2 \\ -\frac{c}{m_1}(x_1 - x_2) - g \sin\beta + \frac{F_{fr}}{m_1} \\ -\frac{c}{m_2}(x_2 - x_1) - g \sin\beta \end{bmatrix}, b = \begin{bmatrix} 0 \\ 0 \\ \frac{1}{m_1} \\ -\frac{1}{m_2} \end{bmatrix}.$$

Remark 1 When ignoring the influence of friction force and linearizing (3.3), it yields:

$$\dot{X} = AX + bu, \tag{3.4}$$

where

$$A = \begin{bmatrix} 0 & 0 & 1\,0 \\ 0 & 0 & 0\,1 \\ -\frac{c}{m_1} & \frac{c}{m_1} & 0\,0 \\ \frac{c}{m_2} & -\frac{c}{m_2} & 0\,0 \end{bmatrix}.$$

Obviously, the (A, b) is uncontrollable. Therefore, the control of WLLS is not trivial.

3.3 The Design of Sliding Mode Controller

The goal of the controller design is to realize the longitudinal velocity tracking for the center of mass. The longitudinal displacement and velocity of the center of mass can be described as:

$$\begin{aligned} x_{coM} &= \frac{m_1 x_1 + m_2 x_2}{m_1 + m_2}; \\ \dot{x}_{coM} &= \frac{m_1 \dot{x}_1 + m_2 \dot{x}_2}{m_1 + m_2}. \end{aligned} \tag{3.5}$$

Let $\bar{X}_1 = \left[\, x_1 \; x_2 \,\right]$,$\bar{X}_2 = \left[\, \dot{x}_1 \; \dot{x}_2 \,\right]^T$, $M_1 = \left[\, \frac{m_1}{m_1+m_2} \; \frac{m_2}{m_1+m_2} \,\right]$, then (3.5) turns to:

$$\begin{aligned} x_{coM} &= M_1 \bar{X}_1, \\ \dot{x}_{coM} &= M_1 \bar{X}_2. \end{aligned} \tag{3.6}$$

Consider the longitudinal velocity problem of electromagnetic WLLS for in-pipe robots. Assume the desired longitudinal velocity of the center of mass is $\dot{x}_{coM_r}$, then the integral of $\dot{x}_{coM_r}$ is x_{coM_r}, which can be viewed as the tracking trajectory for longitudinal displacement.

Assumption 2 The desired longitudinal acceleration $\ddot{x}_{coM_r}$ and velocity $\dot{x}_{coM_r}$ of the center of mass are known. The tracking errors are defined as:

$$\begin{aligned} e_{coM} &= x_{coM_r} - x_{coM}, \\ \dot{e}_{coM} &= \dot{x}_{coM_r} - \dot{x}_{coM}. \end{aligned}$$

According to (3.6), it gives:

$$\begin{aligned} e_{coM} &= x_{coM_r} - M_1 \bar{X}_1, \\ \dot{e}_{coM} &= \dot{x}_{coM_r} - M_1 \bar{X}_2. \end{aligned} \tag{3.7}$$

Design sliding mode function as:

$$s = \sigma_1 e_{coM} + \dot{e}_{coM}, \tag{3.8}$$

where $\sigma_1 > 0$ is a designable parameter which is used to guarantee the stability of sliding mode surface $S = \{e|s(e_{coM}) = 0\}$.

By substituting (3.7) into (3.8), and let $\sigma = \begin{bmatrix}\sigma_1 & 1\end{bmatrix} M$, where $M = \begin{bmatrix} M_1 & 0 \\ 0 & M_1 \end{bmatrix}$, then it gives:

$$s = \left(\sigma_1 x_{coM_r} + \dot{x}_{coM_r}\right) - \sigma X. \tag{3.9}$$

Substituting (3.3) into the derivative of (3.9), it yields:

$$\dot{s} = \left(\sigma_1 \dot{x}_{coM_r} + \ddot{x}_{coM_r}\right) - \sigma(f(X) + bu). \tag{3.10}$$

According to the sliding mode theory, reaching law approach is applied to obtain the control law. The following reaching law is chosen:

$$\dot{s} = -ks - \varepsilon \operatorname{sgn}(s), \tag{3.11}$$

where sgn(s) is sign function, $k > 0$, $\varepsilon > 0$.

By comparing (3.11) with (3.10), it gives:

$$-ks - \varepsilon \operatorname{sgn}(s) = \left(\sigma_1 \dot{x}_{coM_r} + \ddot{x}_{coM_r}\right) - \sigma(f(X) + bu). \tag{3.12}$$

Therefore, the sliding mode control law can be set in the following form:

$$u = (\sigma b)^{-1}\left[\left(\sigma_1 \dot{x}_{coM_r} + \ddot{x}_{coM_r}\right) - \sigma f(X) + ks + \varepsilon \operatorname{sgn}(s)\right]. \tag{3.13}$$

Because the unknown friction force F_{fr} still exists in $f(X)$, (3.13) cannot be carried out in practice. Therefore, an estimator for friction force is constructed. Here, the following estimator $\hat{F}_{fr}$ is employed:

$$\hat{F}_{fr} = -\mu \alpha F_{emm}, \tag{3.14}$$

where α is a ratio of the transverse force to the longitudinal force. μ is the kinetic coefficient of dry friction. The static coefficient force is estimated as the same as the kinetic coefficient force.

The corresponding estimator of $f(X)$ is:

$$\hat{f}(X) = \begin{bmatrix} \dot{x}_1 \\ \dot{x}_2 \\ -c/m_1(x_1 - x_2) - g\sin\beta + \hat{F}_{fr}/m_1 \\ -c/m_2(x_2 - x_1) - g\sin\beta \end{bmatrix}.$$

Hence, the sliding mode control law is:

$$u=(\sigma b)^{-1}\Big[\big(\sigma_1 \dot{x}_{coM_r} + \ddot{x}_{coM_r}\big) - \sigma \hat{f}(X) + ks + \varepsilon \mathrm{sgn}(s)\Big]. \quad (3.15)$$

3.4 The Design of Fuzzy Rules

The simulation shows that the quality m_1 has a great influence on the sliding mode parameters k and ε, so the parameters can be adjusted by fuzzy control rules according to experience. The schematic diagram of sliding mode fuzzy control is shown in Fig. 3.3.

Define input and output fuzzy sets. For input m_1, four fuzzy sets are defined, which are, respectively, represented by "Small" (S), "Small Medium" (SM), "Big Medium" (BM), and "Big" (B). The parameter k is also represented by "Small" (S), "Small Medium" (SM), "Big Medium" (BM) and "Big" (B), respectively. Then, it can be written as $m_1 = \{S, MS, MB, B\}$, $k = \{S, MS, MB, B\}$.. The corresponding fuzzy universes are as follows $m_1 = \{0.24, 0.35, 0.5, 0.7\}$, $k = \{26, 18, 15, 10\}$. The fuzzy rule table is shown in Table 3.1.

For sliding mode parameters ε, after summarizing a large amount of data, it is found that there is an approximate linear relationship between the optimal ε and m_1,

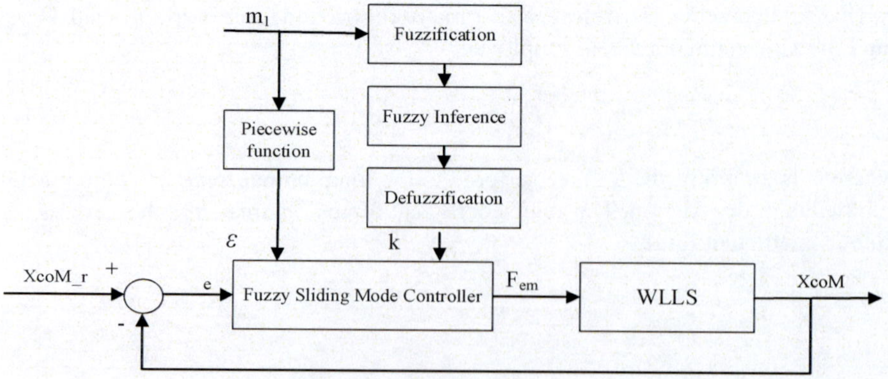

Fig. 3.3 Structure of fuzzy sliding mode controller

Table 3.1 Fuzzy rule table

	m_1			
	S	MS	MB	B
k	B	MB	MS	S

which can be written in the form of piecewise function. Therefore, the nonlinear fuzzy control rule for ε is abandoned, and the optimal value ε obtained by the piecewise function is used.

3.5 Simulation

Table 3.2 gives simulation parameters of WLLS and fuzzy sliding mode controller. Some simulation results are given in the following (Figs. 3.4, 3.5, 3.6, and 3.7).

Figure 3.4 shows the relationship of F_{fr} and F_a with the variation of $\dot{x}$ in 2D graph. The first graph on Fig. 3.4 gives the variation of reference velocity. The second graph on Fig. 3.4 gives the variation of reference force F_a. The third and the last graph on Fig. 3.4 shows the change of friction force with $\dot{x}$ and F_a, respectively. Figure 3.5 demonstrates three-dimensional graph and contour of F_{fr} and F_a with the variation of $\dot{x}$. Figure 3.6 shows the changes of fuzzy sliding mode variable s and control signal u with time. The variable s is very low and the control signal u almost equals to a constant. However, one shortcoming is that there is a peak ($u_{\max} \approx 12$) at the

Table 3.2 Parameters of WLLS and fuzzy sliding mode controller

Parameter	Unit	Value
m_1	Kg	0.3
m_2	Kg	0.1
g	N/kg	9.81
F_{emm}	N	14.7
μ	/	0.3
α	/	1
β	Rad	0
φ	Rad	0
A_a	/	10^5
Δ	/	10^{-3}
k_c	/	20,000
σ_1	/	0.3
k	/	20
ε	/	2

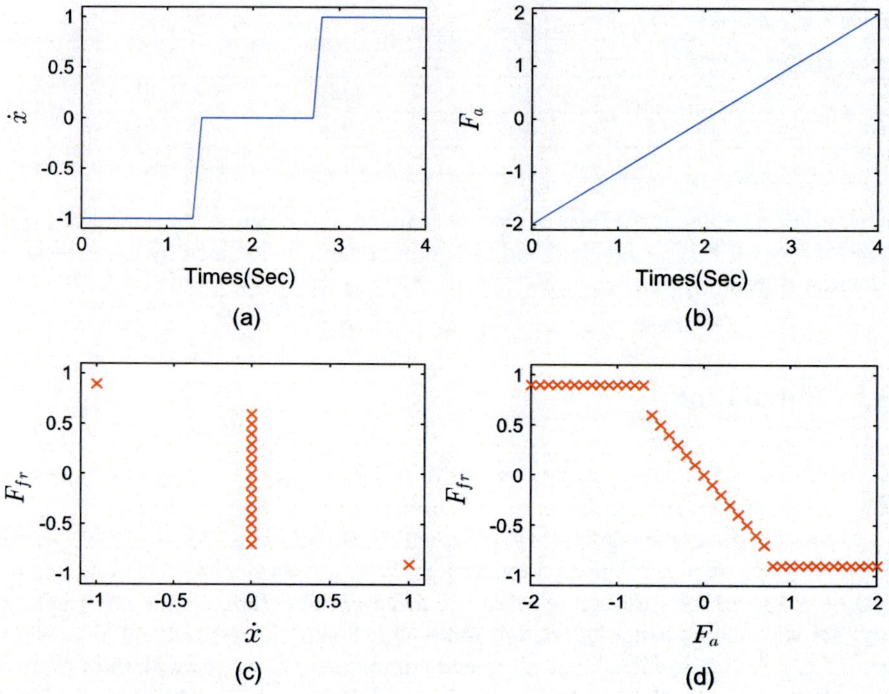

Fig. 3.4 Two-dimensional graph (F_{fr} vs. $\dot{x}$, F_{fr} vs. F_a)

initial time, which may damage WLLS. It can be seen from Fig. 3.7 that the error of $\dot{x}_{coM}$ to $\dot{x}_{coMr}$ is very low, which proves the accuracy of the controller.

3.6 Conclusion

Aimed at the worm-like locomotion system (WLLS) for a class of in-pile robots, a fuzzy sliding mode controller is proposed. Considering that the WLLS is a kind of system with strong nonlinearity, and the friction force varies with respect to both states and control variables, an estimator for friction force is given. In order to determine the sliding mode parameters, a fuzzy rule is presented. The simulation on MATLAB/Simulink demonstrates that the novel fuzzy sliding mode control method reduces velocity error while satisfying the control accuracy of the WLLS.

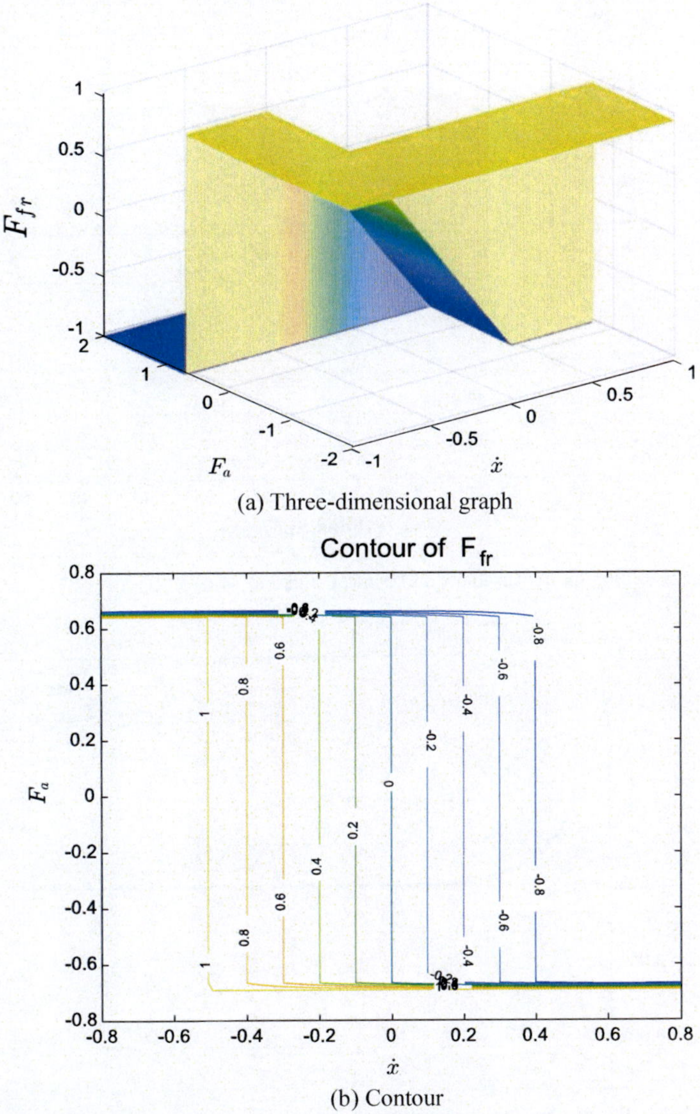

Fig. 3.5 Three-dimensional graph and contour (F_{fr} vs. $\dot{x}$ and F_a)

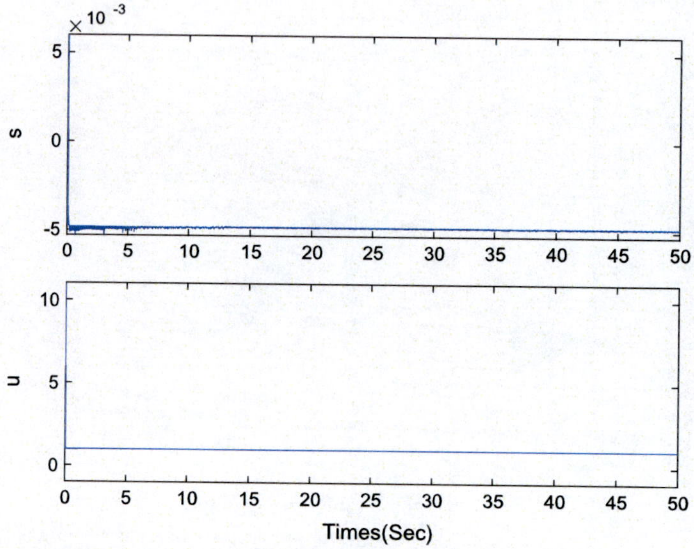

Fig. 3.6 Fuzzy sliding mode variable s and control signal u

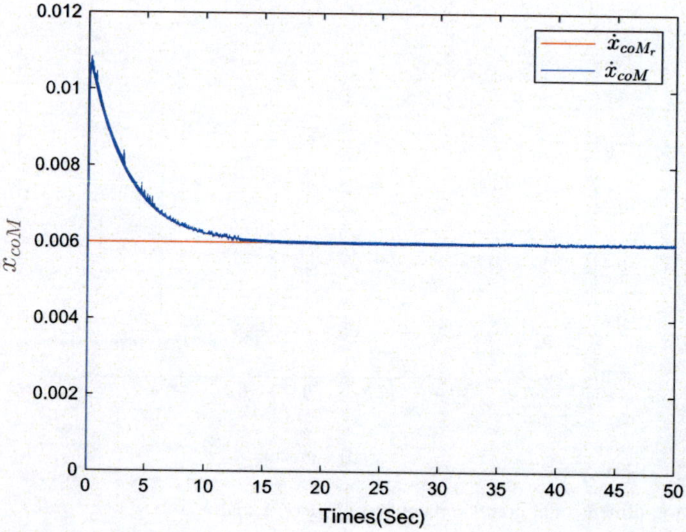

Fig. 3.7 Response of $\dot{x}_{coM}$ to $\dot{x}_{coMr}$

Acknowledgements This paper is partially supported by National Natural Science Foundation of China (No. 51876089).

References

1. Kim, Y.J., Yoon, K.H., Park, Y.W.: Development of the inpipe robot for various sizes. In: IEEE/ASME International Conference on Advanced Intelligent Mechatronic, pp. 1745–1749 (2009)
2. Li, P., Yang, W., Jiang, X., Lyu, C.: Active screw-driven in-pipe robot for inspection. In: 2017 IEEE International Conference on Unmanned Systems (ICUS), pp. 608–613 (2017)
3. Zhang, Y., Yan, G.: In-pipe inspection robot with active pipe-diameter adaptability and automatic tractive force adjusting. Mech. Mach. Theory **42**(12), 1618–1631 (2007)
4. Kim, Y.G., Shin, D.H., Moon, J.I., An J.: Design and implementation of an optimal in-pipe navigation mechanism for a steel pipe cleaning robot. In: 8th International Conference on Ubiquitous Robots and Ambient Intelligence 2011 (URAI), pp. 772–773 (2011)
5. Zhong, H.J., Ling, Z.W., Miao, C.J., Guo, W.C., Tang P.: A new robot-based system for in-pipe ultrasonic inspection of pressure pipelines. In: 2017 Far East NDT New Technology & Application Forum (FENDT), pp. 246–250 (2017)
6. Xiao, S., Bing, Z., Huang, K., Huang, Y.: Snake-like robot climbs inside different pipes. In: 2017 IEEE International Conference on Robotics and Biomimetics (ROBIO), pp. 1232–1239 (2017)
7. Kakogawa, A., Oka, Y., Ma, S.: Multi-link articulated wheeled in-pipe robot with underactuated twisting joints. In: 2018 IEEE International Conference on Mechatronics and Automation (ICMA), pp. 942–947 (2018)
8. Sattarov, R.R., Almaev, M.A.: Electromagnetic worm-like locomotion system for in-pipe robots: design and vibration-driven motion analysis. In: 2017 Dynamics of Systems, Mechanisms and Machines (Dynamics), pp. 1–6. IEEE (2017)
9. Xavier, M.S., Fleming, A.J., Yong, Y.K.: Experimental characterization of hydraulic fiber-reinforced soft actuators for worm-like robots. In: 2019 7th International Conference on Control, Mechatronics and Automation (ICCMA), pp. 204–209 (2019)
10. Huang, Y., Kandhari, A., Chiel, H.J., Quinn, R.D., Daltorio, K.A.: Slip reduction controls of mesh-body worm robot developed from a mathematical model. In: 2017 IEEE International Conference on Robotics and Biomimetics (ROBIO), pp. 1474–1479 (2017)
11. Behn, C., Zeidis, I., Zimmermann, K.: Mechanics of Terrestrial Locomotion. Springer, Berlin Heidelberg (2009)
12. Malchikov, A.V., Jatsun, S.F.: Vibration mobile microrobot. Robot. Tech. Cybern. **3**(4), 64–68 (2014)
13. Young, K.D., Utkin, V.I., Ozguner, U.: A control engineer's guide to sliding mode control. IEEE Trans. Control Syst. Technol. **7**(3), 328–342 (1999)
14. Sattarov, R.R., Almaev, M.A.: Electromagnetic worm-like locomotion system for in-pipe robots: novel design of magnetic subsystem. In: IOP Conference Series: Earth and Environmental Science (2019)
15. Sattarov, R.R., Ismagilov, F.R., Almaev, M.A.: Device to move inside pipeline (versions). Russian patent No. 2419025. (2011)
16. Sattarov, R.R., Almaev, M.A.: Self-moving device for moving inside pipelines. Russian patent No. 2666930 (2018)

Part II
Robotics and Automation

Chapter 4
Tactical Level of Intelligent Geometric Control System for Unmanned Aerial Vehicles

Mikhail Khachumov

Abstract This study considers the tactical level of the intelligent geometric control system designed to solve the cutting-edge scientific problem of controlling unmanned aerial vehicles (UAVs) in unstable conditions. Intelligent geometric theory combines geometric control methods (methods of optimal control, complex motion control and stabilization, formation control, trajectory and target tracking, differential pursuit-evasion games, etc.) with intelligent control methods using tools of artificial intelligence (productions, semantic networks, fuzzy logic, frame-based behavioral micro-programs, frame-based operations, machine learning, genetic algorithms, methods of knowledge acquisition, etc.) and provides reliable and high-performance control techniques for operating in uncertain environments under wind disturbances. Hierarchical architecture of intelligent geometric control system is designed for joint application of precise geometric and adaptive intelligent control methods as parts of a single robotic system. The solution to the problem of controlling a UAV group taking into account mathematical models of an aircraft and wind loads was simulated in MATLAB system.

M. Khachumov (✉)
Federal Research Center "Computer Science and Control" of RAS, 44/2 Vavilova St, Moscow 119333, Russian Federation
e-mail: khmike@inbox.ru; khachumov_mv@rudn.university

Peoples Friendship University of Russia (RUDN University), 6 Miklukho-Maklaya St, Moscow 117198, Russian Federation

A. Ronzhin and V. Shishlakov (eds.), *Proceedings of 15th International Conference on Electromechanics and Robotics "Zavalishin's Readings"*, Smart Innovation, Systems and Technologies 187, https://doi.org/10.1007/978-981-15-5580-0_4

4.1 Introduction

4.1.1 Motivation

One of the cutting-edge study in the field of robotic systems is focused on the development of theory, algorithms, and programs to perform complex group missions under imposed restrictions in an unstable environment. The most challenging problem here is to provide control autonomy both for individual vehicles and their groups. Despite the existence of a massive number of papers dealing with the synthesis of adaptive control algorithms, the problem has not yet been resolved. The complexity of the problem lies in the fundamental impossibility to obtain accurate enough mathematical models of robotic systems as well as an external dynamic environment. The analysis of the worldwide publications showed that proposed methods are scattered and manage to cope with separate tasks.

In this paper, we consider the tactical level of an intelligent geometric control system which is a novel concept designed for joint application of accurate geometric and flexible intelligent control methods complementing each other within an integrated robotic system.

4.1.2 Related Works

The authors of papers [1–3] made a significant contribution to the development of intelligent control theory for robotic systems. Main features of an intelligent system are the high degree of autonomy, ability to sustainably maintain, or achieve necessary system states (goal) under perturbing external factors.

One of the most common approaches related to the representation and processing of knowledge in various intelligent systems is using first-order logic (FOL) [4, 5]. However, an effective implementation of the knowledge model arising from this approach for planning purposeful activities in uncertain conditions faces the following challenges: need to build a detailed model of knowledge representation to derive solutions; inability to use second and higher-order predicates in knowledge models, which significantly reduces the functionality of intelligent problem solvers; complexity of finding solutions to sophisticated problems caused by the fact that knowledge models based on FOL do not involve semantic component to formally describe objects, events, and environment regularities.

A substantial contribution in resolving exponential time problem for deductive inference was made in [6]. In this paper, the authors proposed novel deduction algorithms using semantic networks to provide several types of parallel inference and reduce theorem-proving complexity. Procedures organized in such a way manage to build effective problems solvers, but they do not eliminate the rest of the abovementioned challenges.

One of the promising areas in the field of robotics and group control is a rule-based knowledge representation paradigm. In papers [7, 8] the authors consider intelligent motion control system for mobile robots based on rules, which can rapidly react to changes in the real dynamic environment. Despite the advantages of these methods, in order to successfully implement them onboard, much more attention should be given to accounting for the real external conditions, in particular, wind loads. There is a strong need for developing intelligent and adaptive control algorithms for aerial vehicles with limited computing resources working in disturbed environment.

Geometric control theory is a cutting-edge field that recently has been actively developing [9–11]. It solves one of the major problems of control theory—the controllability problem, which implies finding a control function to drive the state of the system to a prescribed target state at a finite time. In relation to UAV control problems, a geometric approach is applied to optimize trajectory tracking [12], solve stabilization [13], cooperation [14], and other group control problems. We expect that onboard intelligent control algorithms in conjunction with accurate geometric control methods can provide an acceptable accuracy of solutions to control problems for vehicles operating in non-deterministic environments.

This paper deals with the tactical level of a new hierarchical architecture of an intelligent geometric control system designed to operate in cluttered environments with accounting for uncertain external conditions and vehicle's limitations.

4.1.3 Main Contributions

We believe an integration of geometric control methods with artificial intelligence methods can significantly contribute to solving complex dynamic problems, including challenging applied tasks for UAVs, such as flight formation control [15], dynamic planning [16, 17], trajectory optimization [18], monitoring and surveillance [19], pursuit and tracking dynamic goals [20] in a non-deterministic perturbed environment.

The scientific novelty of the study is determined by new models and methods developed in the framework of intelligent geometric control theory to increase the autonomy, reliability, and control efficiency for unmanned vehicles functioning in uncertain conditions. The achievements in the field of geometric control theory are used to formulate and solve optimization problems and to implement complex trajectory motion of vehicles.

4.2 The Principles of Intelligent Geometric Control

4.2.1 The Purpose of Intelligent Geometric Control

Geometric control theory explores the possibilities of applying differential geometric methods to dynamical systems control. The term "geometric" suggests important system concepts, for example, controllability, as geometric properties of the state space or its subspaces. These are the properties that are preserved under coordinate changes, for example, the so-called invariant or controlled invariant subspaces. In the present paper, in addition to the listed set of properties, the concept of "geometric control" also has a slightly different meaning. It is associated with solving several trajectory optimization problems for UAVs. One of them, for example, deals with an accurate geometric calculation of heading to meet with another UAV on a circle or sphere of Apollonius. A similar problem with imposed time constraints takes place, for example, in a pursuit-evasion game, considered by Pontryagin [21]. However, the problem is even more complicated if one considers external disturbances, and, in our opinion, it can be solved by integrating geometric and intelligent control methods. Of significant interest are multi-point path tracking, under disturbances and control restrictions, when the task is to optimize time, path length, and deviation. In such conditions to track "ideal" trajectories obtained by solving optimization problem, it is advisable to apply intelligent control methods. In particular, it is rational to apply production rules that imitate the actions of a human operator, as well as artificial neural or semantic networks. Such a combined principle, covering all hierarchical control levels of an autonomous UAV, is called in the present paper as "intelligent geometric control."

4.2.2 Hierarchical System to Control a Dynamic Object

The principle of hierarchy assumes the presence of three levels of abstraction to solve various-scale control problems (see Fig. 4.1).

Strategic level solves the issues of choosing global goals (goal-setting), control modes, and distribution of tasks and roles. Tactical level solves the tasks given by strategic level to an individual UAV, in particular, tracking a given trajectory under wind disturbances, static and dynamic motion planning in the presence of obstacles under uncertainty, and pursuing a target. At this level, specific control signals (control commands) are generated, which are transmitted to the executive level for processing by a control object.

The integrated database contains necessary facts, current model parameters, and variable values. The knowledge base stores production rules (that define the conditions under which the particular controls are generated in compliance with the current goal), frame-based behavioral microprograms, frame-based operations, semantic networks. Knowledge acquisition is carried out by a special intelligent technology in the process of UAVs functioning.

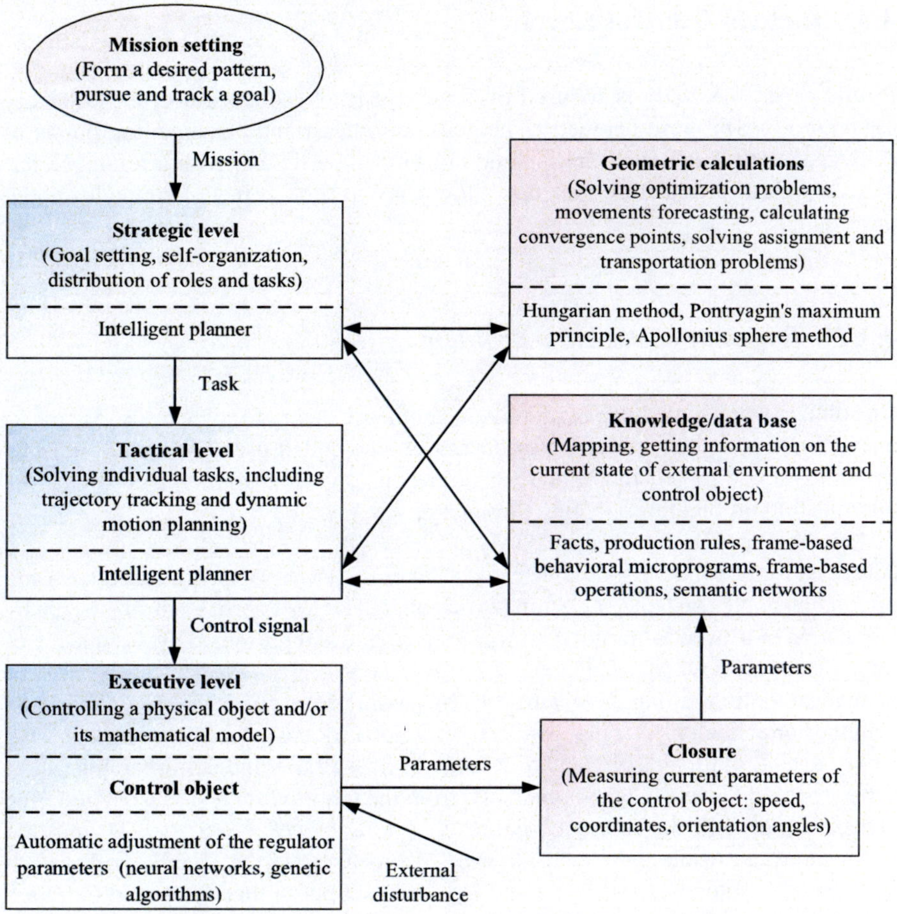

Fig. 4.1 Block diagram of an intelligent geometric control system

The intelligent control unit changes the settings to transit the dynamic system into a new state. System's response time (transients) limits the time to measure and analyze current state parameters and timely update of the database. It is assumed that the intelligent control loop operates in integration with the geometric control loop to cope with conditions of uncertainty and significant external disturbances.

To improve dynamic properties and stabilize the motion of the controlled object (UAV), we introduce a special module for adaptive tuning of PID controller coefficients. The combined control approach is implemented in the framework of a single hierarchical control system.

4.3 Tactical Control Level

In this paper, the study is focused on a tactical level. To track a given trajectory (or target), we propose geometric methods to forecast and calculate the points of convergence between UAV and pursued object (in particular, with a reference target), as well as production rules that allow UAV to form behavior strategies under uncertainties.

4.3.1 Trajectory Tracking Problem

One of the effective methods of intelligent geometric theory to solve trajectory tracking problem is associated with introducing of a so-called pseudo-target or a set of dynamic pseudo-targets that imitate a "reference" trajectory motion and solving the optimization problem of pursuing these targets.

Let the state $Q_{p_i}(t)$ of each UAV p_i, $P = \{p_1, \ldots, p_n\}$ at time instant t is described by the following variables: coordinates $(x_i(t), y_i(t), z_i(t))$, speed $v_i(t)$ and orientation angles $\theta_i(t)$, $\psi_i(t)$. Suppose that an ideal trajectory of each p_i is given by the motion of a"pseudo" target $c_i \in \{c_1, \ldots, c_n\}$ and is represented by a sequence of reference points $(x_{ij}, y_{ij}, z_{ij}), i = 1, \ldots, N, j = 1, \ldots, M$. Pseudo-targets c_i simulate ideal motions along the given paths by passing from one reference point to the neighbor one. Each UAV pursues its target guided by the selected strategy and the ability to control the velocity and direction. As a result of wind perturbations, aerial vehicles could deviate, even appreciably, from the required routes. The desired time of passing through the reference points $t_{ij}, i = 1, \ldots, N, j = 1, \ldots, M$ and that of traveling the whole path T_i are known. We assume for the simplicity that each UAV has two velocities $v_{p_i}^{(1)} > v_{c_i}$ and $v_{p_i}^{(2)} = v_{c_i}$. Suppose that $P_i(t)$, and $C_i(t)$ are the coordinates of UAV p_i and pseudo-target c_i and $d(P_i(t), C_i(t))$ is the distance between them at the instant t. We consider a geometric model of a vehicle as a sphere of radius R (with some safety margin). Safety distance between two vehicles is determined by the value $d(P_i(t), P_j(t)) \geq 2R$.

For each UAV p_i in the group, the problem consists in synthesizing the control $U_{p_i}(t) = (v_{p_i}(t), \theta_{p_i}(t), \psi_{p_i}(t))$ on the time interval $[0, T_i]$, such that

$$\int_{t=0}^{T_i} d(P_i(t), C_i(t))dt \to \min, \tag{4.1}$$

where $\forall i, j, \; i \neq j, \; d(P_i(t), C_i(t)) \geq 2R$.

An approach to solving the problem is based on applying geometric methods (analysis of the location of scene participants; analysis and forecasting of their movements, determination of the point of convergence) in combination with intelligent control methods (strategies imitating behavior of the pilot realized by sets of rules) taking into account imposed control restrictions and disturbances. The mathematical model of UAV motion is defined by the transfer functions that describe a double-circuit control system with an autopilot.

4.3.2 Pontryagin's Maximum Principle

Let us consider the motion of two aerial vehicles (the pursuer and the evader) on the local segments between two points when evader changes its direction in ideal conditions (without disturbances). We determine the following system

$$x_c(t),\ y_c(t),\ z_c(t),\ x_p(t),\ y_p(t),\ z_p(t),\ \theta_p(t),\ \psi_p(t), \tag{4.2}$$

where $x_c(t), y_c(t), z_c(t)$ are the coordinates of the evader, $x_p(t), y_p(t), z_p(t), \theta_p(t)$, $\psi_p(t)$ are the coordinates and orientation angles of the pursuer.

The problem lies in constructing an optimal control $U_p(t) = (\theta_p(t), \psi_p(t))$ for the transition of the system (4.2) from the initial state $(x_c(0), y_c(0), z_c(0), x_p(0), y_p(0), z_p(0), \theta_p(0), \psi_p(0))$ to the final state $(x_c(T), y_c(T), z_c(T), x_p(T), y_p(T), z_p(T), \theta_p(T), \psi_p(T))$ for the minimum time T, such that $x_c(T) = x_p(T)$, $y_c(T) = y_p(T)$, $z_c(T) = z_p(T)$. Velocity v_p of the pursuer, velocity v_c and orientation angles θ_c, ψ_c of the evader are constants. We denote $u_1(t) = \theta_p(t)$, $u_2(t) = \psi_p(t)$.

We give a mathematical description for the motion of the center of mass of flight vehicles in the unperturbed environment.

The simplified model of the motion for the pursuer is given by

$$\begin{aligned}
\dot{x_p} &= v_p \cos u_1 \cos u_2; & x_p &= x_{p0} + \int_0^t v_p \cos u_1 \cos u_2 dt;\\
\dot{y_p} &= v_p \sin u_1; & y_p &= y_{p0} + \int_0^t v_p \sin u_1 dt;\\
\dot{z_p} &= v_p \cos u_1 \sin u_2; & z_p &= z_{p0} + \int_0^t v_p \cos u_1 \sin u_2 dt.
\end{aligned} \tag{4.3}$$

The simplified model of the motion for the evader is given by

$$\begin{aligned}
\dot{x_c} &= v_c \cos\theta_c \cos\psi_c; & x_c &= x_{c0} + v_c \cos\theta_c \cos\psi_c \cdot t;\\
\dot{y_c} &= v_c \sin\theta_c; & y_c &= y_{c0} + v_c \sin\theta_c \cdot t;\\
\dot{z_c} &= v_c \cos\theta_c \sin\psi_c; & z_c &= z_{c0} + v_c \cos\theta_c \sin\psi_c \cdot t,
\end{aligned} \tag{4.4}$$

where $x_c(0) = x_{c0}, y_c(0) = y_{c0}, z_c(0) = z_{c0}, x_p(0) = x_{p0}, y_p(0) = y_{p0}, z_p(0) = z_{p0}$.

We introduce the following system

$$x = x_p - x_c, \quad y = y_p - y_c, \quad z = z_p - z_c. \tag{4.5}$$

Thus, the optimal control problem comes down to the translation of the system (4.5) from the initial state $(x(0), y(0), z(0))$ to the final state $(x(T), y(T), z(T))$ for the minimum time T, such that $x(T) = 0, y(T) = 0, z(T) = 0$.

We obtain the following Mayer optimal control problem

$$\begin{gathered} f_1 = \dot{x}(t) = \dot{x}_p(t) - \dot{x}_c(t) = v_p \cos u_1 \cos u_2 - v_c \cos\theta_c \cos\psi_c; \\ f_2 = \dot{y}(t) = \dot{y}_p(t) - \dot{y}_c(t) = v_p \sin u_1 - v_c \sin\theta_c; \\ f_3 = \dot{z}(t) = \dot{z}_p(t) - \dot{z}_c(t) = v_p \cos u_1 \sin u_2 - v_c \cos\theta_c \sin\psi_c; \\ x(0) = x_{p0} - x_{c0}; \quad y(0) = y_{p0} - y_{c0}; \quad z(0) = z_{p0} - z_{c0}; \\ x(T) = 0; \quad y(T) = 0; \quad z(T) = 0; \\ J = T \to \min. \end{gathered} \tag{4.6}$$

Let us solve the problem (4.6) by applying the L. S. Pontryagin's maximum principle. The Hamiltonian has the form $H = \psi_1 f_1 + \psi_2 f_2 + \psi_3 f_3$, where ψ_i—are functions of time. We find optimal u_1, u_2.

$$\begin{gathered} \frac{dH}{du_2} = -c_1 v_p \cos u_1 \sin u_2 + c_3 v_p \cos u_1 \cos u_2 = 0; \\ \frac{dH}{du_1} = c_2 v_p \cos u_1 - c_1 v_p \sin u_1 \cos u_2 - c_3 v_p \sin u_1 \sin u_2 = 0; \\ \tan u_2 = \frac{c_3}{c_1}; \quad \tan u_1 = \frac{c_2}{c_1 \cos u_2 + c_3 \sin u_2}. \end{gathered} \tag{4.7}$$

Thus, $u_1 = p = \text{const}$, $u_2 = p = \text{const}$, and the model of the pursuer motion (4.3) takes the form

$$\begin{aligned} \dot{x}_p &= v_p \cos u_1 \cos u_2; & x_p &= x_{p0} + v_p \cos u_1 \cos u_2 \cdot t \\ \dot{y}_p &= v_p \sin u_1; & y_p &= y_{p0} + v_p \sin u_1 \cdot t \\ \dot{z}_p &= v_p \cos u_1 \sin u_2; & z_p &= z_{p0} + v_p \cos u_1 \sin u_2 \cdot t. \end{aligned} \tag{4.8}$$

The problem of forecasting the point of pursuer and target convergence can be solved geometrically [22]. When the target changes its direction, we use "parallel approach" [21] strategy based on replanning pursuer's heading. The proposed method of calculating the point of convergence and the "parallel approach" strategy can be applied to solve the problems of dynamic target tracking and path following in a perturbed environment for a group of UAVs. Consider the task of constructing a set of production rules that determine pursuers' strategy.

4.3.3 A Set of Control Rules for Pursuing a Target

Below, we present a set of control rules for pursuer p_i to solve the problem (4.1). Notation (A, B, C) means "the value A is known," "the value B is known," and "the value C is known."

$F_1(d_i(t), Q_{p_i}(t), Q_{c_i}(t))$ is the function that calculates angles of approach $\theta^{(1)}_{p_i}(t)$, $\psi^{(1)}_{p_i}(t)$ by using approximate point of convergence, $v_{p_i} > v_{c_i}$; $F_2(Q_{p_i}(t), Q_{c_i}(t), v_{w_i}(t))$ is the function that calculates tracking angles $\theta^{(2)}_{p_i}(t)$, $\psi^{(2)}_{p_i}(t)$ and takes into account wind loads, $v_{p_i} = v_{c_i}$.

Here d_i is the distance between p_i and the target c_i, $v_{w_i} = (v^{(x)}_{w_i}, v^{(y)}_{w_i}, v^{(z)}_{w_i})$ is the wind velocity. Suppose that Δt is the duration of one clock period (step); ε is the minimum distance at which the task of approaching the target is accomplished.

Closure Rules. These rules are meant for preprocessing of the data required for further calculations.

IF $(Q_{p_i}(t), Q_{c_i}(t))$ THEN $d_i(t) := ((x_{p_i} - x_{c_i})^2 + (y_{p_i} - y_{c_i})^2 + (z_{p_i} - z_{c_i})^2)^{1/2}$;

IF $(d_i(t), Q_{p_i}(t), Q_{c_i}(t))$ THEN $\theta^{(1)}_{p_i}(t) := F^{(\theta)}_1(t)$, $\psi^{(1)}_{p_i}(t) := F^{(\psi)}_1(t)$;

IF $(Q_{p_i}(t), Q_{c_i}(t), v_{w_i}(t))$ THEN $\theta^{(2)}_{p_i}(t) := F^{(\theta)}_2(t)$, $\psi^{(2)}_{p_i}(t) := F^{(\psi)}_2(t)$.

Transition Rules. These rules describe the transition of the system into a new state, which is triggered by a control impulse with the step Δt.

IF $(Q_{p_i}(t), Q_{c_i}(t), v_{w_i}(t))$ THEN

$x_{p_i}(t + \Delta t) := x_{p_i}(t) + v_{p_i}(t)\cos(\theta_{p_i}(t))\cos(\psi_{p_i}(t)) + v^{(x)}_{w_i}(t)$;

$y_{p_i}(t + \Delta t) := y_{p_i}(t) + v_{p_i}(t)\cos(\theta_{p_i}(t))\sin(\psi_{p_i}(t)) + v^{(y)}_{w_i}(t)$;

$z_{p_i}(t + \Delta t) := z_{p_i}(t) + v_{p_i}(t)\sin(\theta_{p_i}(t)) + v^{(z)}_{w_i}(t)$;

Targeting Rules. These rules are meant for selecting the task for the pursuer depending on the current state of the system.

IF $(d_{ij}(t) < 2R)$ THEN task "approach":="active," "tracking":="inactive";

IF $(d_i(t) > \varepsilon)$ THEN task "approach":="active," "tracking":="inactive";

IF $((d_i(t) \le \varepsilon)$ & $(d_{ij}(t) \ge 2R))$ THEN task "approach" := "inactive," "tracking" := "active."

Control Rules. These rules are meant for selecting allowable controls in accordance with the current task.

IF (task "approach"="active") THEN $v_{p_i}(t) := v^{(1)}_p(t)$, $\theta_{p_i}(t) := \theta^{(1)}_{p_i}(t)$, $\psi_{p_i}(t) := \psi^{(1)}_{p_i}(t)$;

IF (task "tracking"="active") THEN $v_{p_i}(t) := v^{(2)}_p(t)$, $\theta_{p_i}(t) := \theta^{(2)}_{p_i}(t)$, $\psi_{p_i}(t) := \psi^{(2)}_{p_i}(t)$.

These rules form the basis of the UAV control system and are experimentally tested in the MATLAB Simulink system.

4.4 Executive Control Level

To conduct experiments we have selected adequate mathematical and simulation models of an aircraft as an object to test geometric and intelligent control methods. We propose the schemes of control and stabilization systems for pitch and yaw angles (Fig. 4.2).

The following designations are accepted on the scheme:

θ_G, θ_C are given and current values of pitch angle,
ψ_G, ψ_C are the given and current values of yaw angle,
$W_{HA}(s)$, $W_{DA}(s)$ are the transfer functions from the height and direction actuators,
$W_{\omega_{gz}}/\Delta\delta_H$ is the transfer function from the elevator to the pitch angular speed,
$W_{\omega_{gy}}/\Delta\delta_D$ is the transfer function from the rudder to the yaw angular speed,
w_x, w_z are the wind components generated by the function $R_t(\tau)$,
w_y is the wind component generated by the function $R_n(\tau)$,
W_{ω_{gz}/w_x}, W_{ω_{gz}/w_y}, W_{ω_{gy}/w_z} are the transfer functions from wind to the angular speeds,
k_θ, k_ψ are the transfer numbers (ratios) of the autopilot from the pitch and yaw angles,
$k_{\omega_{gz}}$, $k_{\omega_{gy}}$ are the transfer numbers of the autopilot from the angular pitch and yaw speed,
s is the Laplace transform parameter.

We denote:

Fig. 4.2 Block diagrams of pitch and yaw stabilization systems

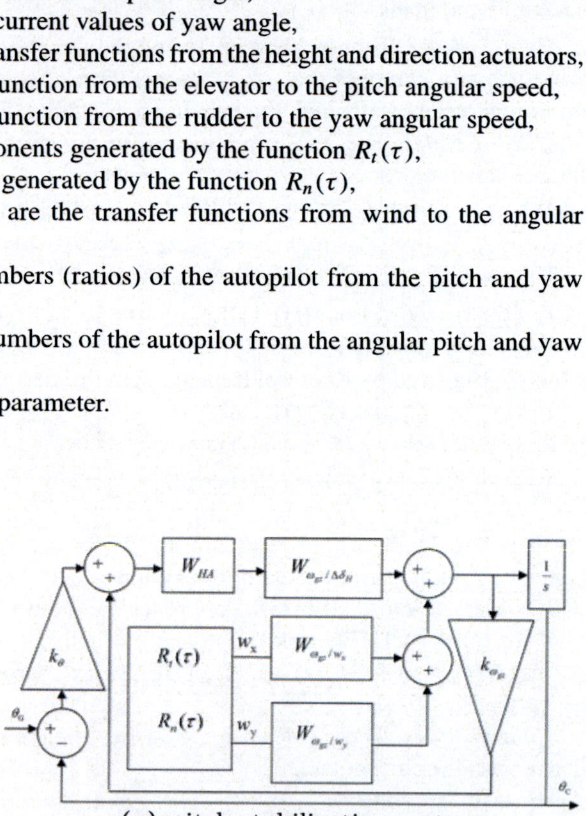

(a) pitch stabilization system

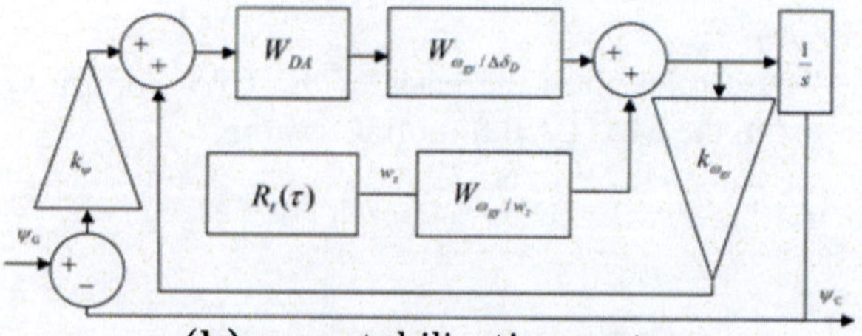

(b) yaw stabilization system

a, b with various subscripts are the parameters of the model of lateral dynamics which depend on the UAV type and flight mode,
c, d, f, T, ζ with various subscripts are the parameters of the model of longitudinal dynamics depending on the UAV type and flight mode.

The transfer functions for the lateral motion of a UAV are as follows:

$$\begin{aligned} W_{\omega_{gy}/\Delta\delta_D}(s) &= \frac{b^{(0)}_{\omega_{gy}} s^3 + b^{(1)}_{\omega_{gy}} s^2 + b^{(2)}_{\omega_{gy}} s + b^{(3)}_{\omega_{gy}}}{\Delta_B}, \\ W_{\omega_{gy}/w_z}(s) &= \frac{b^{(0)}_{\omega_{gy}/w_z} s^3 + b^{(1)}_{\omega_{gy}/w_z} s^2}{\Delta_B}, \\ \Delta_B &= s^4 + a_{B_1} s^3 + a_{B_2} s^2 + a_{B_3} s + a_{B_4}. \end{aligned} \tag{4.9}$$

The transfer functions for the longitudinal motion of UAV are as follows:

$$\begin{aligned} W_{\omega_{gz}/\Delta\delta_D}(s) = \tfrac{f_c(T_1+1)}{T^2s^2+2T\zeta s+1}, \quad W_{\omega_{gy}/w_y}(s) = \tfrac{(c_{\ddot{y}} b_{\dot{y}} - c_{\dot{y}})s}{T^2s^2+2T\zeta s+1}, \\ W_{\omega_{gy}/w_x}(s) = \tfrac{(c_{\dot{x}} - c_{\ddot{y}} d_{\dot{x}})s + c_{\dot{x}} d_{\dot{y}} - c_{\dot{y}} d_{\dot{x}}}{T^2s^2+2T\zeta s+1}. \end{aligned} \tag{4.10}$$

Considered transfer functions are integrated into a single modeling system and serve as the basis for conducting experimental studies in an uncertain environment under wind loads. The developed simulation environment contains special geometric and intelligent control modules that apply strategies and control rules for the rapid response to changes in the external environment.

4.5 Simulation of UAV Movement and Mission Execution

We simulate mission execution for a group of unmanned aerial vehicles operating in a perturbed environment in MATLAB Simulink system. The longitudinal and normal wind actions are modeled by the correlation functions: $R_t(\tau) = \sigma_t^2 \exp^{-|V|\tau/L}$, $R_n(\tau) = \sigma_n^2(1 - 0.5|V|\tau/L) \exp^{-|V|\tau/L}$, where $|V|$ is the mean airspeed; σ_t is the rms value of the longitudinal wind component; σ_n is the rms value of the normal wind component; τ is the period of generation of wind gusts; L is the turbulence scale.

We carry out experimental studies of the pursuit-runaway problem using the selected UAV model. Trajectories of motion of the target (red dashed curve) and the pursuers (gray solid curves) are presented in Fig. 4.3a. Start points are marked with circles. The figure shows that the scene participants take logical steps imitating actions of the pilots for pursuit and evasion.

We also solve the problem of forming the desired pattern in a group of UAVs (Fig. 4.3b). At first vehicles (gray markers) are assigned to uniformly distributed positions (red markers) in the required formation. Next, we obtain ideal UAVs trajectories and simulate formation flight for five vehicles. The figure demonstrates the

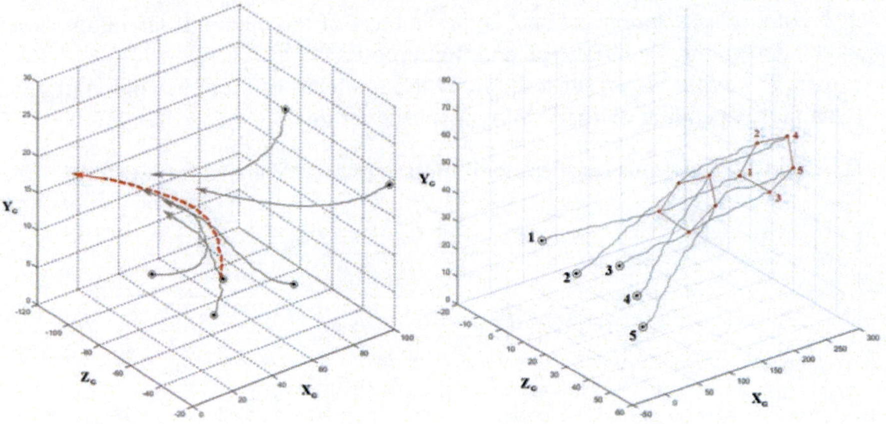

(a) Pursuit-evasion problem (b) Formation flight control

Fig. 4.3 Simulation of mission execution

effect of wind loads on motion trajectories. As one can see, control rules enable UAV to respond promptly to changes in the environment and successfully cope with the tasks.

4.6 Conclusion

This paper describes the tactical level of intelligent geometric control system applied to autonomous UAVs operating in a disturbed environment. To increase the autonomy, reliability, and control efficiency, we combine the methods of geometric and intelligent control theories as parts of a single robotic system. We propose geometric methods and models to forecast the points of convergence between UAV and pursued object (in particular, with a reference target), as well as production rules that allow UAV to form behavior strategies under uncertainties. Novel methods are proposed to solve practical problems of planning and tracking trajectories for unmanned aerial vehicles in a disturbed environment. Conducted experimental studies on solving trajectory tasks confirm feasibility and prospects of the developed theory.

Acknowledgements This research was supported by the Russian Science Foundation (Project No. 16-11-00048).

References

1. Canudas de Wit, C., Siciliano, B., Bastin G.: Theory of Robot Control. Springer Science & Business Media, London (2012)
2. Siciliano, B., Khatib, O.: Handbook of Robotics, 2nd edn. Springer International Publishing, Switzerland (2016)
3. Kelly, A.: Mobile Robotics: Mathematics, Models, and Methods. Cambridge University Press, Cambridge (2013)
4. Kober, J., Peters, J.: Learning Motor Skills: From Algorithms to Robot Experiments. Springer International Publishing, Switzerland (2014)
5. Russell, S., Norvig, P.: Artificial Intelligence: A Modern Approach, 3rd edn. Pearson, London (2009)
6. Vagin, V., Derevyanko, A., Kutepov, V.: Parallel-inference algorithms and research of their efficiency on computer systems. Sci. Tech. Inf. Process. **45**(5), 368–373 (2018)
7. Pandey, A., Parhi, D.: Multiple mobile robots navigation and obstacle avoidance using minimum rule based ANFIS network controller in the cluttered environment. Int. J. Adv. Robot. Autom. **1**, 1–11 (2016)
8. Xavier, J., Selvakumari, S.: Behavior architecture controller for an autonomous robot navigation in an unknown environment to perform a given task. Int. J. Phys. Sci. **10**(5), 182–191 (2015)
9. Agrachev, A., Sachkov, Yu.: Geometric Control Theory. FIZMATLIT, Moscow (2005)
10. Bonnard, A., Chyba, Yu.: Geometric Optimal Control with Applications. Kyushu University, Accelerated Graduate Course, Institute of Mathematics for Industry (2015)
11. Andrew, D.: Fundamental problems in geometric control theory. In: Conference on Decision and Control, IEEE, pp. 1–18 (2012)
12. Lee, T., Leok, M., McClamroch, N.: Geometric tracking control of a quadrotor UAV for extreme maneuverability. IFAC Proc. Vol. **44**(1), 6337–6342 (2011)
13. Goodarzi, F.: Self-tuning geometric control for a quadrotor UAV based on Lyapunov stability analysis. Int. J. Robot. Autom. **5**(3), 1–15 (2016)
14. Lee, T., Sreenath, K., Kumar, V.: Geometric control of cooperating multiple quadrotor UAVs with a suspended payload. In: Conference on Decision and Control, IEEE (2013)
15. Guzey, H.: Adaptive consensus based formation control of unmanned vehicles. Doctoral Dissertation. Missouri University of Science and Technology, Missouri (2016)
16. Radmanesh, M., Kumar, M., Guentert, P.: Overview of path-planning and obstacle avoidance algorithms for UAVs: a comparative study. Unmanned Syst. **6**(2), 95–118 (2018)
17. Khachumov, M.: Controlling the motion of a group of unmanned flight vehicles in a perturbed environment based on the rules. In: Proceedings of the 2017 International Siberian Conference on Control and Communications, IEEE, pp. 1–5 (2017)
18. Santoso, F., Garratt, M., Anavatti, S.: State-of-the-art intelligent flight control systems in unmanned aerial vehicles. IEEE Trans. Autom. Sci. Eng. **15**(2), 613–627 (2018)
19. De Moraes, R., De Freitas, E: Distributed control for groups of unmanned aerial vehicles performing surveillance missions and providing relay communication network services. J. Intell. Robot. Syst. **92**, 645–656 (2018)
20. Daingade, S., Sinha, A: Nonlinear cyclic pursuit based cooperative target monitoring. In: Distributed Autonomous Robotic Systems. The 11th International Symposium, vol. 104, pp. 17–30 (2014)
21. Pontryagin, L., Mishchenko, A: The linear differential game of pursuit (analytic theory). Math. USSR-Sbornik **59**(1), 131–158 (1986)
22. Khachumov, M.: Problems of group pursuit of a target in a perturbed environment. Sci. Tech. Inf. Process. **44**(5), 357–364 (2017)

Chapter 5
Three-Dimensional Consensus-Based Control of Autonomous UAV Swarm Formations

Tagir Muslimov and Rustem Munasypov

Abstract This chapter presents a multi-agent approach to controlling a decentralized swarm three-dimensional (3D) formations of autonomous fixed-wing unmanned aerial vehicles (UAVs). Cooperative control is analyzed within the framework of coordinated rectilinear path following. Focus is made on attaining a pre-specified geometric configuration and maintaining the resulting UAV formation vertically by controlling the altitude difference (distance between aircraft along the vertical). Consensus-based UAV interaction is used, i.e., there is no "leader." Each UAV is assumed to be equipped with a standard autopilot, in which a finite-state machine controls the flight altitude. Thus, an arbitrary preconfigured 3D formation can be attained by combining this strategy with the existing approaches in controlling a formation as projected onto a horizontal plane. The proposed control laws are adjusted to the input constraints arising from the vertical velocity limits of the UAVs. MATLAB/Simulink modeling used complete nonlinear 6 degree-of-freedom (DoF) 12-state models of fixed-wing UAVs equipped with tuned autopilots in two scenarios: a group following a horizontal path, and following a descending path. Modeling showed that the proposed multi-UAV swarm controls were effective, as they could accurately attain and maintain a 3D formation of required shape.

5.1 Introduction

Controlling autonomous multi-UAV (unmanned aerial vehicle) systems is a topic of increasing interest. Formation control is one of the basic problems in cooperative UAV control, as flight at fixed inter-UAV distance has numerous real-world applications. These include co-tracking a target, flight in tight formations to utilize specific aerodynamic effects, coordinated path following, etc. [1–3]. Some tasks require specific three-dimensional (3D) formations. For instance, Coopmans et al.

T. Muslimov (✉) · R. Munasypov
Ufa State Aviation Technical University, K. Marx Str., 12, 450008 Ufa, Russian Federation
e-mail: tagir.muslimov@gmail.com

A. Ronzhin and V. Shishlakov (eds.), *Proceedings of 15th International Conference on Electromechanics and Robotics "Zavalishin's Readings"*, Smart Innovation, Systems and Technologies 187, https://doi.org/10.1007/978-981-15-5580-0_5

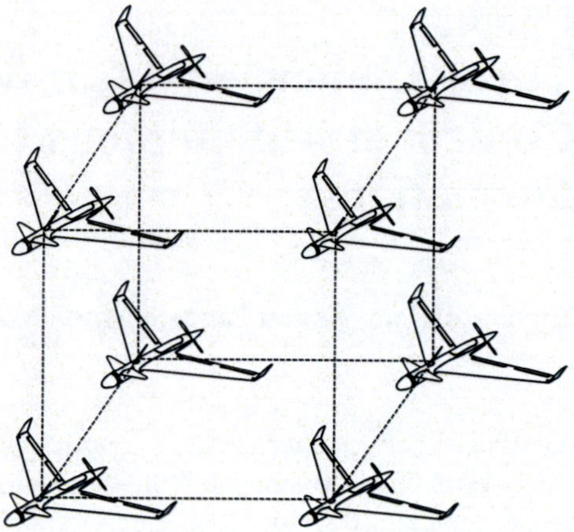

Fig. 5.1 3D swarm formation of fixed-wing UAVs

propose using 3D formations for meteorological studies such as measuring the horizontal and vertical wind profiles [4]. Figure 5.1 shows one such formation. Decentralized approaches based on local interactions are gaining momentum when it comes to controlling large groups of autonomous mobile robots [5, 6]. Such strategies are inspired by the swarms of living organisms as well as by the collective motion in physical and chemical processes [7, 8]. The above applies to cooperative control over autonomous UAV formations, which seeks to create decentralized swarms with consensus-based or leaderless control [9, 10].

Notably, most of the existing papers on decentralized UAV formation controls pay little to no attention to 3D formations, as their authors state and solve 2D problems mainly. At the same time, papers on 3D formations mostly either focus on simplified UAV models or use linear consensus laws [11–14]. Thus, the novelty of this paper arises from obtaining a working consensus-based control strategy for an arbitrarily shaped 3D formation of fixed-wing UAV models with more realistic dynamics. The formation remains stable in any initial UAV position and is therefore not limited to the near-equilibrium initial positions. The devised controls ensure continuous path following without breaking the 3D formation. Besides, unlike earlier studies based on simplified models, the approach proposed herein has been tested on realistic complete-dynamics UAV models using the known finite-state machines to control the flight altitude.

5.2 Preliminary Notes and Used Models

5.2.1 Multi-UAV System Model and UAV Model

Assume that UAVs can use the existing global or local positioning systems (e.g., the optical one) to compute their positions with respect to the adjacent aircraft.

The inertial coordinate system (ICS) in use is an Earth-bound north-east-down (NED) system $F^{\text{NED}}(O, \mathbf{i}^i, \mathbf{j}^i, \mathbf{k}^i)$ [15]. The point of origin O is on the Earth's surface, usually at the beginning of the runway. The unit vector $\mathbf{i}^i$ points in the direction of the north and thus defines the N-axis for the ICS. The unit vector $\mathbf{j}^i$ points in the direction of the east and thus defines the E-axis for the ICS. The unit vector $\mathbf{k}^i$ is a vertical vector perpendicular to the Earth's surface that points in the downward direction; it thus defines the D-axis for the ICS. All the three unit vectors form the right-hand system. The flight altitude h_i and the D-axis coordinate in the ICS of the ith UAV p_i^d are linked via the equation $p_i^d = -h_i$.

Assume that each UAV is equipped with a standard tuned autopilot capable of stabilizing the UAV's pitch angle, bank angle, and yaw angle, while also tracking the commands related to the course angle, the airspeed, and the flight altitude. This paper dwells upon controlling the altitude differences between UAVs to attain and maintain 3D formations in flight. This is why focus is made on UAV dynamics in vertical movement.

An autopilot can use different techniques to control the altitude. Monograph [15] uses a finite-state machine and identifies three flight zones: takeoff, climb, altitude hold, and descent. Assume that using the finite-state machine, a UAV successfully reaches the altitude hold zone where it uses the pitch angle to control the altitude and a throttle to maintain the airspeed. Thus, according to [15], high-level representation of the autopilot loops controlling the altitude can be described as a second-order model:

$$\ddot{h} = \alpha_{\dot{h}}\left(\dot{h}^c - \dot{h}\right) + \alpha_h\left(h^c - h\right), \tag{5.1}$$

where h is the UAV flight altitude, $\dot{h}$ is the vertical velocity, $\ddot{h}$ is the vertical acceleration, h^c is the UAV flight altitude command, $\dot{h}^c$ is the vertical velocity command, and $\alpha_{\dot{h}}$, α_h are positive constants that depend on the autopilot configuration as well as on the UAV specifications.

Due to the physics and safety requirements, a real-world autopilot-UAV system will have vertical movement input constraints. These constraints should be borne in mind when synthesizing control laws; as such, they can be represented as follows:

$$U \triangleq \left\{h^c, \dot{h}^c | h_{\min} \le h^c \le h_{\max}, v_{\min}^{\text{vert}} \le \dot{h}^c \le v_{\max}^{\text{vert}}\right\}, \tag{5.2}$$

where $h_{\min}, h_{\max}$ are the minimum and maximum flight altitude values, $v_{\min}^{\text{vert}}, v_{\max}^{\text{vert}}$ are the minimum and maximum vertical velocity values. Note that $v_{\min}^{\text{vert}}$ is a negative value; besides, $\left|v_{\min}^{\text{vert}}\right| > v_{\max}^{\text{vert}}$ usually holds for aircraft.

Similarly to [10, 16], let a fixed-wing UAV formation be defined as a multi-agent system consisting of N autonomous agents, where $N \ge 2$. Let N_i be the set of all agents. The architecture of their interaction can be represented as a strongly connected graph without loss of generality:

$$G \triangleq (Q, E), \tag{5.3}$$

where the ith UAV agent is the set of vertices in the graph $\eta_i \in Q$, and each arc in the set E which leads from the vertex η_i to the vertex η_j means the agent η_i receives data on the relative position of the agent η_j.

Thus, the set of arcs E demonstrates the configured UAV agent interaction rules, which are herein selected as follows:

$$E = \{(1, 2), (2, 1), (2, 3), \ldots, (N-1, N), (N, N-1)\}.$$

Therefore, the intra-group UAV coordination architecture is an "open chain." This decentralization makes the formation infinitely scalable, as new UAVs can join the group without complicating or compromising the interaction structure. Let $N_j = \left(\eta_j \in Q : \left(\eta_i, \eta_j\right) \in E\right)$ be the set of agents adjacent to the ith one.

Consider controlling a 3D UAV formation tasked to follow a rectilinear path.

Let the rectilinear path P_{line} be represented as the set of vectors $\mathbf{x}$ as in [15]:

$$P_{\text{line}}(\mathbf{r}, \mathbf{q}) = \left\{\mathbf{x} \in \mathbb{R}^3 : \mathbf{x} = \mathbf{r} + \lambda \mathbf{q}\right\},$$

where $\lambda = \pm 1$ defines the direction of travel along the line; $\mathbf{r} = (r_e, r_n, r_d) \in \mathbb{R}^3$ is the start of the path with the coordinates being set in the ICS; $\mathbf{q} = (q_e, q_n, q_d) \in \mathbb{R}^3$ is a unit vector that defines the path direction and thus the final course angle of the UAV formation in the ICS.

5.2.2 *Statement of Problems*

Problem 5.1 To synthesize UAV flight altitude control laws to enable path following at fixed predetermined inter-unit distances while attaining a preconfigured 3D geometry of the UAV formation.

This paper only covers the flight controls designed to attain the required altitude differences, while the subproblem of attaining the required geometry in a horizontal plane is covered in [10, 16, 17]. Combining the controls will fully solve Problem 5.1 to obtain a decentralized 3D UAV formation. Note that the control laws should be adjusted for the kinematic constraints (5.2).

Problem 5.2 To test the effectiveness and performance of the proposed algorithm using computer simulations and nonlinear 6 degree-of-freedom (DoF) 12-state UAV dynamics models outfitted with preconfigured standard autopilot models.

5.2.3 Architecture of Interaction in a Decentralized Multi-UAV System

Assume that the UAVs of a group interact by consensus as shown in graph G (5.3). To describe the architecture of the proposed decentralized control system, let us modify the control-action computation method that was first developed for ground-based robots as linear agents in [18]. This is a consensus-based method that can attain an arbitrary formation geometry using the reverse dynamics model of a group of agents. In papers [10, 16, 17], this architecture was modified to control decentralized UAV formations so as to attain a preconfigured geometry as projected onto a horizontal plane.

Choose a method for calculating the control-action vector defined in terms of some elements of the vector of all possible altitude difference errors $\bar{\mathbf{h}}^e = \left(\hat{h}_{i,j}\right) \in \mathbb{R}^{N(N-1)\times 1}$, where $\mathbb{R}^{N(N-1)\times 1}$ is the space of matrices of size $N(N-1)\times 1$ with components from $\mathbb{R}$, $\hat{h}_{i,j}$ is the value of error for the directly interacting ith and jth agents. The structure of the interaction graph determines which elements are selected from $\bar{\mathbf{h}}^e$ and in which order; in case of (5.3), the control-action vector $\mathbf{h}^e \in \mathbb{R}^{N\times 1}$ is assigned as follows:

$$\mathbf{h}^e = \begin{bmatrix} h_1^e \\ \vdots \\ h_k^e \\ \vdots \\ h_N^e \end{bmatrix} = \begin{bmatrix} \hat{h}_{12} \\ \vdots \\ -\hat{h}_{k-1,\,k} + \hat{h}_{k,\,k+1} \\ \vdots \\ -\hat{h}_{N-1,\,N} \end{bmatrix} = \mathbf{M}\mathbf{h} + \mathbf{D}, \tag{5.4}$$

where $\mathbf{D} = -\mathbf{M}\mathbf{H}^{-1}\left(\mathbf{h}_d^{\mathrm{T}}, \sum_{k=1}^{N} h_k\right)^{\mathrm{T}}$ is the vector to control the system in the space of relative distances ($(N-1)$-dimensional space spanned by the columns of the incidence matrix of the graph G (5.3));

H is a matrix that defines which agents are to interact and to be assigned an altitude difference. This matrix is defined as follows:

$$\mathbf{H} = \begin{bmatrix} \mathbf{q}_1 \\ \mathbf{q}_2 \\ \vdots \\ \mathbf{q}_N \end{bmatrix}, \quad \mathbf{q}_i = \begin{bmatrix} \vdots \\ 1 \\ \vdots \\ -1 \\ \vdots \end{bmatrix}^{\mathrm{T}}, \quad i < N, \quad \mathbf{q}_N = \begin{bmatrix} 1 \\ 1 \\ \vdots \\ 1 \end{bmatrix}^{\mathrm{T}},$$

whereby $\mathbf{H} \in \mathbb{R}^{N\times N}$, $\mathbf{q}_i \in \mathbb{R}^{1\times N}$, the position of “1” and “-1” in $\mathbf{q}_i$ shall be set according to the structure of the graph G (5.3).

$\mathbf{h}_d = \left(h^d_{i,i+1}\right)_{i=\overline{1,N}} \in \mathbb{R}^{(N-1)\times 1}$ is the vector of desired UAV altitude differences;

$\mathbf{h} = (h_i)_{i=\overline{1,N}} \in \mathbb{R}^{N\times 1}$ is the vector of the current flight altitudes in the group;

$\mathbf{M} \in \mathbb{R}^{N\times N}$ is a matrix that describes UAV interaction and in case of a decentralized neighbor-to-neighbor interaction structure defined by the graph G (5.3) is as follows:

$$\mathbf{M} = \begin{bmatrix} -1 & 1 & 0 & \cdots & 0 \\ 1 & -2 & \ddots & \ddots & \vdots \\ 0 & \ddots & \ddots & 1 & 0 \\ \vdots & \ddots & 1 & -2 & 1 \\ 0 & \cdots & 0 & 1 & -1 \end{bmatrix};$$

$\hat{h}_{i,j}, \forall i \in N_i, j \in N_j$ are the elements of the vector of altitude difference errors $\hat{\mathbf{h}} \in \mathbb{R}^{(N-1)\times 1}$.

5.3 Strategy to Control 3D UAV Swarm Formations

Consider two possible UAV group control scenarios: where the path P_{line} is a horizontal line or where it is a descending line.

5.3.1 Formation Control for Horizontal Path Following

This section dwells upon controlling the UAV formation whose path P_{line} is a horizontal line, i.e., has no slope in the ICS. In this case, the vector $\mathbf{h}^s \in \mathbb{R}^{N\times 1}$ of the control law for the altitude commands is proposed as follows:

$$\mathbf{h}^s = \left(h^s_i\right)_{i=\overline{1,N}} = h^{\text{line}}\mathbf{1}_N + \mathbf{F}^{\Delta h}, \tag{5.5}$$

where h^{line} is the altitude of the horizontal path line; $\mathbf{1}_N = \left[1\ 1\ \ldots\ 1\right]^{\mathrm{T}} \in \mathbb{R}^{N\times 1}$;

$\mathbf{F}^{\Delta h}$ can be found from the equation

$$\mathbf{F}^{\Delta h} = \left(k_{hp} h^e_i + k_{hd} \dot{h}^e_i + k_{hi} \int\limits_0^t h^e_i \mathrm{d}t \right)_{i=\overline{1,N}} \in \mathbb{R}^{N\times 1}, \tag{5.6}$$

which defines the altitude difference controller, whereby h^e_i is calculated from (5.4); k_{hp}, k_{hd}, k_{hi} are the tunable coefficients of the controller that determine how smoothly the UAVs will attain the required geometry.

As the control law (5.5) is consensus-based, i.e., there is no "leader" in this UAV group, the final UAV-specific flight altitude is not known in advance. This control strategy is in agreement with the concept of swarm behavior, where each UAV decides on where and how to move on the basis of the adjacent-aircraft data.

Note that to eliminate the noise, the value of the derivative $\dot{h}_i^e$ in (5.6) is calculated either using a smoothing filter or analytically if there is a variometer onboard.

Let us show that the control law (5.5) is subject to the constraints (5.2). The first condition of this constraint is easy to meet if the h^{line} is chosen appropriately. To test the second condition, calculate the derivative:

$$\left|\dot{h}_i^s\right| = \left|k_{hp}\dot{h}_i^e + k_{hd}\ddot{h}_i^e + k_{hi}h_i^e\right|.$$

This formula means that properly chosen coefficients k_{hp}, k_{hd}, k_{hi} shall enable (5.2) to hold true, as apparently, $\left|\dot{h}_i^e\right| \leq 2\left|v_{\text{min}}^{\text{vert}}\right|$ will hold true in case of two-UAV rendezvous due to their physical constraints. A similar inequality can be written for the absolute acceleration $\left|\ddot{h}_i^e\right|$. Note that choosing k_{hi} correctly is particularly important, as h_i^e may have quite large values. However, maximum and minimum possible flight altitudes also limit the values h_i^e.

Let us show similarly to what was shown in [15] for a single UAV that applying the control law (5.5) to the dynamics of altitude maintenance (5.1) enables efficient path following, whereby the errors of altitude difference within the formation converge to zero. The steady-state errors will thereby be zero. Indeed, using a finite-state machine to control the flight altitude of each UAVs, one can rest assured each UAV will reach the altitude hold zone. Then, if the altitude hold zone is sufficiently narrow to apply the theory of linear automatic control, write the outer-loop dynamics (5.1) of every ith UAV as a transfer function:

$$H(s) = h_i / h_i^c = \left(\alpha_{\dot{h}}s + \alpha_h\right) / \left(s^2 + \alpha_{\dot{h}}s + \alpha_h\right).$$

Define the altitude error as

$$e_i^h \triangleq h_i - h_i^s = h_i - h_i^c,$$

to obtain

$$e_i^h / h_i^c = 1 - h_i / h_i^c = s^2 / \left(s^2 + \alpha_{\dot{h}}s + \alpha_h\right).$$

Apply the final value theorem:

$$e_{i,ss}^h = \lim_{s\to 0} s\left(s^2\right) / \left(s^2 + \alpha_{\dot{h}}s + \alpha_h\right)h_i^c = 0, \quad \text{for } h_i^c = H_0 / s,\ H_0 / s^2.$$

Since the UAV autopilot is a rate-feedback one, the steady-state errors will be zero provided that the external wind disturbances are zero or constant in value.

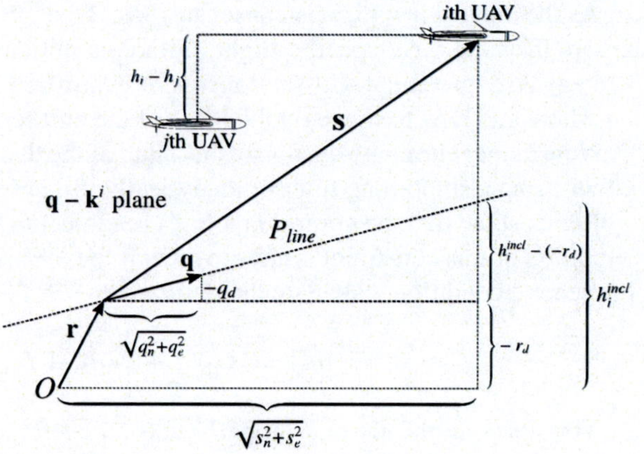

Fig. 5.2 An example of how symbols could be defined for a UAV formation following a descending path. Figure partly adapted from [15]

5.3.2 *Formation Control for Descending Path Following*

This section dwells upon a more difficult case where the path P_{line} is inclined or sloped in the NED ICS. For this scenario, the controls should combine the descending path following strategy for a single UAV presented in [15] with the decentralized UAV formation controls to attain and maintain the desired 3D geometry.

Modify the control law presented in [15] as follows:

$$\mathbf{h}^{\text{incl}} = \left(h_i^{\text{incl}}\right)_{i=\overline{1,N}} = -r_d - \sqrt{s_n^2 + s_e^2}\left(q_d \Big/ \sqrt{q_n^2 + q_e^2}\right) + \mathbf{F}^{\Delta h}, \tag{5.7}$$

where $\mathbf{h}^{\text{incl}} \in \mathbb{R}^{N\times 1}$ is the vector of flight altitude commands for the UAV group; s_n, s_e are the coordinates of the vector $\mathbf{s}^i = \mathbf{e}_p^i - (\mathbf{e}_p^i \cdot \mathbf{n})\mathbf{n}$, which in its turn is calculated from the normal vector $\mathbf{n} = (\mathbf{q} \times \mathbf{k}^i) / \|\mathbf{q} \times \mathbf{k}^i\|$ and the ICS UAV position error vector $\mathbf{e}_p^i \triangleq \left[\, p_n - r_n \;\; p_e - r_e \;\; p_d - r_d \,\right]^{\mathrm{T}}$; vector $\mathbf{F}^{\Delta h}$ is determined as in the Eq. (5.5).

Figure 5.2 shows how symbols could be defined for one possible descending path and UAV group configuration.

Similarly to what was done in Sect. 5.3.1, (5.7) can be shown to comply with the constraints (5.2) while the steady-state errors under the same conditions will be zero. In turn, UAV-specific altitude errors and inter-UAV altitude difference errors converge to zero.

5.4 Simulation Results

Complete nonlinear flying-wing "Zagi" UAV models [15] with 6 degrees of freedom (DoF) and 12 states were used for the numerical testing of the proposed control laws.

Strategies presented in [10, 16, 17] were used to control the formation as projected onto a horizontal plane. Two scenarios were simulated: one as described in Sect. 5.3.1 (Figs. 5.3 and 5.4) and one as in Sect. 5.3.2 (Figs. 5.5 and 5.6). Table 5.1 shows the simulation parameters. For the specifications of the "Zagi" UAVs used for testing, see [15].

Simulation thus showed the stated decentralized UAV control problems solved. Notably, the quality and stability of attaining and maintaining a 3D formation as a unity are largely dependent on whether the formation could be controlled efficiently as projected onto a horizontal plane. The tunable coefficients used in the control laws enable flexible adjustment of the quality of transient flight trajectories.

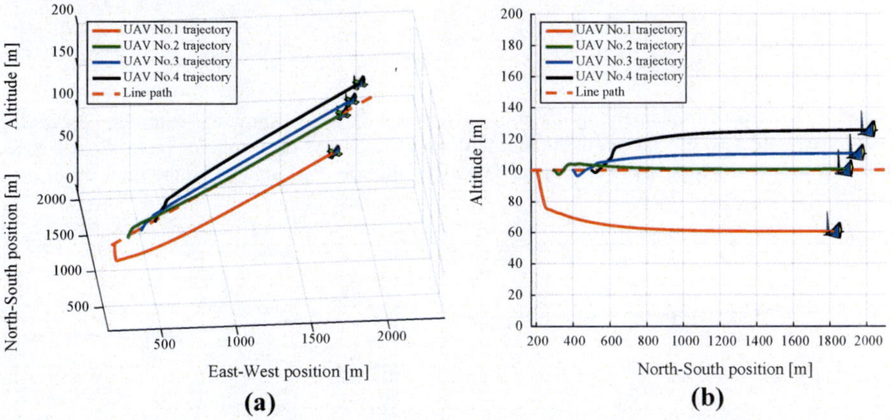

Fig. 5.3 Results of implementing the consensus-based 3D UAV formation controls: horizontal rectilinear path; **a** 3D trajectories of the four UAVs when attaining the formation (at time $t = 150$ s); **b** trajectories of the four UAVs as projected onto the altitude-N-axis plane (at the same time)

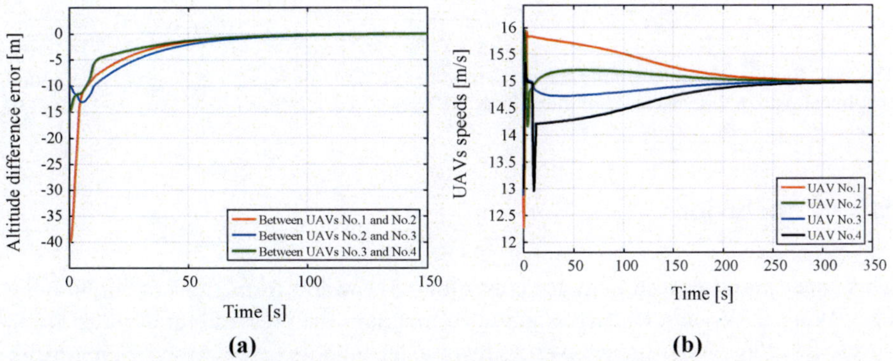

Fig. 5.4 Results of implementing the consensus-based 3D UAV formation controls: horizontal rectilinear path; **a** time response of the altitude difference errors; **b** time response of the UAV flight speeds

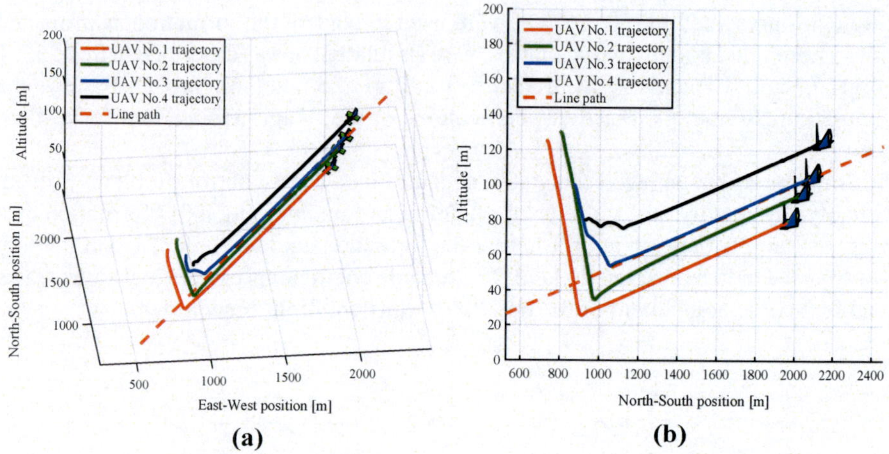

Fig. 5.5 Results of implementing the consensus-based 3D UAV formation controls: descending rectilinear path; **a** 3D trajectories of the four UAVs when attaining the formation (at time $t = 120$ s); **b** trajectories of the four UAVs as projected onto the altitude-N-axis plane (at the same time)

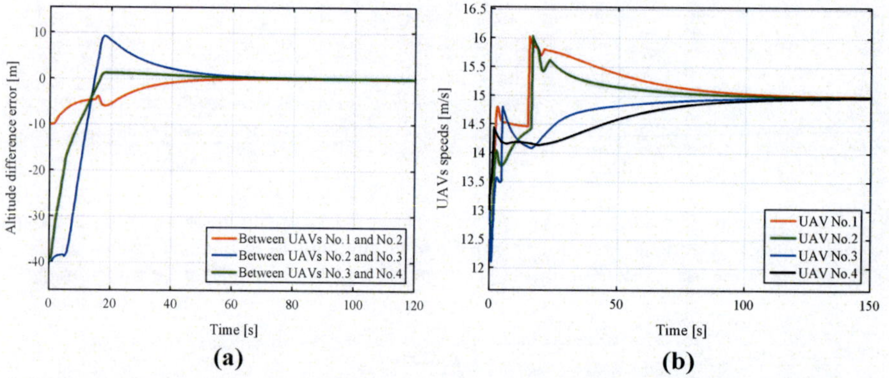

Fig. 5.6 Results of implementing the consensus-based 3D UAV formation controls: descending rectilinear path; **a** time response of the altitude difference errors; **b** time response of the UAV flight speeds

5.5 Conclusions

This paper proposes a strategy for controlling 3D swarm formations of autonomous fixed-wing UAVs in a leaderless group to maintain the preset flight altitude differences. To address coordinated path following, the authors hereof propose consensus-based control laws that, when combined with similar laws for a horizontal projection, can be used to successfully attain an arbitrary 3D geometry of a decentralized UAV

Table 5.1 Simulation parameters

Parameter	Symbol	Values
Initial UAV course vector (rad)	$\boldsymbol{\chi}(0)$	$\begin{bmatrix} 0\ 0\ 0\ 0 \end{bmatrix}^{\mathrm{T}}$
Initial UAV speed vector (m/s)	$\mathbf{v}(0)$	$\begin{bmatrix} 13\ 13\ 13\ 13 \end{bmatrix}^{\mathrm{T}}$
Final cruising speed of the formation (m/s)	v	15
Tunable coefficients	k_{hp}, k_{hd}, k_{hi}	1, 1, 0.1
Vectors defining the rectilinear descending path	$\mathbf{r}$, $\mathbf{q}$	$\begin{bmatrix} 0\ 0\ 0 \end{bmatrix}^{\mathrm{T}}$, $\begin{bmatrix} 1\ 1\ -0.05 \end{bmatrix}^{\mathrm{T}}$
Vector of the UAV starting point in the ICS, scenario 1 (m)	$\mathbf{p}_1(0)$, $\mathbf{p}_2(0)$, $\mathbf{p}_3(0)$, $\mathbf{p}_4(0)$	$\begin{bmatrix} 200\ 200\ -100 \end{bmatrix}^{\mathrm{T}}$, $\begin{bmatrix} 300\ 300\ -100 \end{bmatrix}^{\mathrm{T}}$, $\begin{bmatrix} 400\ 400\ -100 \end{bmatrix}^{\mathrm{T}}$, $\begin{bmatrix} 500\ 500\ -100 \end{bmatrix}^{\mathrm{T}}$
Desired altitude differences, scenario 1 (m)	$\mathbf{h}_d$	$\begin{bmatrix} 40\ 10\ 15 \end{bmatrix}^{\mathrm{T}}$
Vector of the UAV starting point in the ICS, scenario 2 (m)	$\mathbf{p}_1(0)$, $\mathbf{p}_2(0)$, $\mathbf{p}_3(0)$, $\mathbf{p}_4(0)$	$\begin{bmatrix} 730\ 730\ -125 \end{bmatrix}^{\mathrm{T}}$, $\begin{bmatrix} 800\ 800\ -130 \end{bmatrix}^{\mathrm{T}}$, $\begin{bmatrix} 870\ 870\ -100 \end{bmatrix}^{\mathrm{T}}$, $\begin{bmatrix} 930\ 930\ -80 \end{bmatrix}^{\mathrm{T}}$
Desired altitude differences, scenario 2 (m)	$\mathbf{h}_d$	$\begin{bmatrix} 15\ 10\ 20 \end{bmatrix}^{\mathrm{T}}$

formation. Although the approach has been tested for rectilinear paths, it is easily applicable to circular UAV formations.

Further research might address adding the collision avoidance algorithms or support for more complex paths (spirals, etc.), which could not be covered herein.

Acknowledgements The reported study was funded by RFBR according to the research project No. 18-08-01299.

References

1. Ma, L.: Cooperative target tracking by altering UAVs' linear and angular velocities. In: 2019 International Conference on Unmanned Aircraft Systems (ICUAS), pp. 69–78 (2019)
2. Jain, P., Peterson, C.K.: Encirclement of moving targets using relative range and bearing measurements. In: 2019 International Conference on Unmanned Aircraft Systems (ICUAS), pp. 43–50 (2019)
3. Zhang, Q., Liu, H.H.T.: Robust cooperative close formation flight control of multiple unmanned aerial vehicles. In: Lecture Notes in Mechanical Engineering, pp. 61–74 (2019)
4. Coopmans, C., Stark, B., Jensen, A., Chen, Y.Q., McKee, M.: Cyber-physical systems enabled by small unmanned aerial vehicles. In: Handbook of Unmanned Aerial Vehicles, pp. 2835–2860 (2015)
5. Ivanov, D., Kapustyan, S., Kalyaev, I.: Method of spheres for solving 3D formation task in a group of quadrotors. In: Interactive Collaborative Robotics (ICR 2016), LNCS 9812, pp. 124–132 (2016)
6. Panagou, D., Turpin, M., Kumar, V.: Decentralized goal assignment and safe trajectory generation in multi-robot networks via multiple Lyapunov functions. IEEE Trans. Autom. Contr. (2019). https://doi.org/10.1109/tac.2019.2946333
7. Shishika, D., Paley, D.A.: Mosquito-inspired distributed swarming and pursuit for cooperative defense against fast intruders. Auton. Robot. **43**, 1781–1799 (2019)
8. Qiu, H., Duan, H.: Pigeon interaction mode switch-based UAV distributed flocking control under obstacle environments. ISA Trans. **71**, 93–102 (2017)
9. Khachumov, M.: A Rule-based approach for controlling UAVs formation flight. In: Proceedings of 14th International Conference on Electromechanics and Robotics "Zavalishin's Readings", pp. 319–330 (2020)
10. Muslimov, T.Z., Munasypov, R.A.: Consensus-based cooperative circular formation control strategy for multi-UAV system. In: International Russian Automation Conference (RusAutoCon), pp. 1–8 (2019). https://doi.org/10.1109/rusautocon.2019.8867733
11. Zhu, X., Zhang, X.-X., Yan, M.-D., Qu, Y.-H., Lin, H.: Three-dimensional multiple unmanned aerial vehicles formation control strategy based on second-order consensus. In: The Proceedings of the Institution of Mechanical Engineers, Part G: Journal of Aerospace Engineering, vol. 232, pp. 481–491 (2018)
12. Zhang, X., Duan, H.: Altitude consensus based 3D flocking control for fixed-wing unmanned aerial vehicle swarm trajectory tracking. In: The Proceedings of the Institution of Mechanical Engineers, Part G: Journal of Aerospace Engineering, vol. 230, pp. 2628–2638 (2016)
13. Chang, K., Xia, Y., Huang, K.: Coordinated formation control design with obstacle avoidance in three-dimensional space. J. Franklin Inst. **352**, 5779–5795 (2015)
14. Zhu, X., Zhang, X., Yan, M., Qu, Y.: Three-dimensional formation keeping of multi-UAV based on consensus. J. Cent. South Univ. **24**, 1387–1395 (2017)
15. Beard, R.W., McLain, T.W.: Small unmanned aircraft: theory and practice. Princeton University Press (2012)
16. Muslimov, T.Z., Munasypov, R.A.: Decentralized nonlinear group control of fixed-wing UAV formation. Mekhatronika, Avtom. Upr. **21**, 43–50 (in Russian) (2020). https://doi.org/10.17587/mau.21.43-50
17. Muslimov, T.Z., Munasypov, R.A.: UAV formation flight using non-uniform vector field and fuzzy self-tuning PD-control. In: International Russian Automation Conference (RusAutoCon), pp. 1–6 (2018). https://doi.org/10.1109/rusautocon.2018.8501769
18. Yamaguchi, H., Arai, T., Beni, G.: A distributed control scheme for multiple robotic vehicles to make group formations. Robot. Auton. Syst. **36**, 125–147 (2001)

Chapter 6
Approach to UAV Swarm Control and Collision-Free Reconfiguration

Valeriia Izhboldina, Igor Lebedev, and Aleksandra Shabanova

Abstract In this paper, the problem of unmanned aerial vehicle (UAV) swarm control is considered. As the most promising approach to UAV swarm control a composite one was chosen. The proposed conceptual model of the composite method includes operator, base station, and two communication networks: one between agents and another one between base station and agents. UAV trajectory computation is fulfilled on the land-based station using swarm reconfiguration algorithm. Two enhanced 3D reconfiguration algorithms were presented: eased and based on divide and conquer—which differ in used collision avoidance methods. Eased algorithm contains a method of parallel translation of target points, homothety, and target coordinate exchange method. To ensure control of greater swarm size, algorithm based on divide and conquer was developed. Algorithm based on divide and conquer includes parallel translation, delay allocation method, collision graph composition and divide and conquer algorithm. For evaluation algorithm performance tests in virtual environment were performed on swarms of different sizes (from 10 to 100 agents). Greater trajectory calculation time and swarm reconfiguration time are explained by the greater number of used methods. However, the total trajectory and the average number of collisions calculated using algorithm based on divide and conquer have smaller values compared to eased algorithm. Eased algorithm provides UAV movement without collisions with a swarm size of 70 agents, while algorithm based on divide and conquer which has a limit of 90 agents. Therefore, for swarm reconfiguration with a large number of agents, algorithm based on divide and conquer is more preferable.

V. Izhboldina (✉) · I. Lebedev · A. Shabanova
St. Petersburg Institute for Informatics and Automation of the Russian Academy of Sciences, 39, 14th Line, 199178 St. Petersburg, Russia
e-mail: izhboldina.valeriia@gmail.com

I. Lebedev
e-mail: igorlevedev@gmail.com

A. Shabanova
e-mail: iialex.shabanovaii@gmail.com

A. Ronzhin and V. Shishlakov (eds.), *Proceedings of 15th International Conference on Electromechanics and Robotics "Zavalishin's Readings"*, Smart Innovation, Systems and Technologies 187, https://doi.org/10.1007/978-981-15-5580-0_6

6.1 Introduction

Currently, UAVs are actively used as affordable devices for monitoring, reconnaissance, and cargo delivery. The use of a UAV swarm allows to reduce the civilian and military mission time and expand the scope of aircraft application [1–3]. A robotic swarm is a group of UAVs, solving a particular common task. Though as the swarm size increases, the collision-free UAV control becomes more challenging.

The efficiency of robot swarm control largely depends on the chosen control approach. The UAV control approach, in turn, affects the agent control system, the trajectory planning algorithms, and the task distribution among agents.

Approaches to UAV control can be divided into the following categories:

1. swarm control using the base station: swarm agents communicate with the base station via wireless communication network;
2. swarm control with one master agent: All the other robots compute their trajectories, bearing on the master agent;
3. hybrid approach: control with a base station and wireless communication among all agents;
4. manual/non-autonomous approach: UAVs fly along predetermined trajectories with "synchronization points."

An example of a swarm control approach using a base station is shown in [4]. To assess this approach, four UAVs were connected to a base station with communication channels with all UAVs. For the successful mission completion, operator assistance is necessary to coordinate the UAV actions and decide about their take-off, landing, etc.

Swarm control by master agent requires additional devices: computer vision units, lidars. The crucial prerequisite for this approach is wireless communication among UAVs for communication: GPS coordinates, DOP (dilution of precision), and motion vectors. Each agent uses the data to adjust its position. Software based on this approach is shown in [5]. The framework, developed in this research, supports the Internet of things (IoT) paradigm and is intended for autonomous localization of UAVs relative to other agents. Algorithms that can be used for data exchange within the swarm are shown in [6, 7]. These algorithms provide wireless information transfer over long distances without data loss.

Swarm control by master agent can solve the reconfiguration problem in an environment with obstacles [8]. Sensors embedded at master agent allow to reduce weight of wingmen. The proposed approach is designed for a small swarm and does not imply swarm reconfiguration, nor does it guarantee the elimination of collisions in large swarm.

The hybrid approach assumes the presence of a base station and a communication network, embracing all agents. The UAV communication system can be either mounted or embedded, for example, like an AR100-B quadcopter [9]. The hybrid approach requires operator assistance; therefore, to increase the degree of autonomy, an on-board camera can be used, which not only receives information about the environment [11], but also processes the received data [10]. If the UAV is not able to perform the task individually, it searches for the wingman via the communication channel and notifies it of the subtasks.

A non-autonomous approach may serve as a premise for a gesture-based agent control system [12]. In this case, each gesture corresponds to some movement of robot. The control signals are read with a vibrotactile glove equipped with finger vibrating motors to provide vibrotactile feedback. Simultaneously, UAV positions are visually monitored using reflective markers.

Thus, a base station allows to manage a large number of swarm agents due to the ability to use more powerful hardware. High-performance hardware reduces the time needed to compute trajectories, but it makes the control system more expensive. UAV swarm control using a base station can be carried out as by the operator, as without human intervention. The operator only controls the process of UAV motion and facilitates in-mission corrections. Therefore, the main advantage of this approach is its semi-autonomy. The semi-autonomy of the UAV swarm control by using a master agent is achieved through a wireless communication system. However, this approach begins to lag, as the swarm size grows, due to the limitations of on-board computing hardware. The manual control approach is non-autonomous and can only be used with relatively small swarms.

The hybrid approach to swarm control combines the advantages of base station and inter-agent communication system. This approach is semiautonomous and, because of a base station, it allows reducing the time for UAV trajectory calculation and task allocation to agents, compared to the manual approach and master agent control of the swarm. Therefore, a combined approach is most preferable for solving swarm control problem. However, the considered hybrid approaches need to increase the degree of autonomy, which can be achieved with swarm autonomous reconfiguration.

An increase a swarm size can cause in-flight collisions. There are several ways to reduce the impact of UAV collisions on the mission workflow: increasing the structural robustness of an aircraft [13], equip the agents with sensors for dynamic obstacles detection [14], and the development of a swarm reconfiguration algorithm based on the selected control approach [15]. Increasing the structural robustness does not eliminate collisions between UAVs, and, additionally, it makes the devices heavier. Despite the fact that UAVs remain functional after collisions, such accidents cause the agents to deflect from trajectories. Equipping aircraft with sensors to avoid dynamic obstacles also leads to weight increase, which undesirably affects UAV power consumption while the development of the reconfiguration algorithm does not require the installation of additional on-board equipment.

Reconfiguration algorithms may be based on the social behavior principles (ant algorithm, bee algorithm, etc.) [16, 17]. But reconfiguration algorithms can also be based on geometric transformations, graph theory, time delays. The simplest linear algorithms of movement of robots along a straight line and a circle were proposed in [18]. The advantages of the proposed algorithms are its simplicity and convergence for any initial data.

Another example of collision-free reconfiguration algorithm in a dynamic environment is [19]. Collisions were resolved by extending the robot's trajectory into the free region around the collision point. Three approaches to collision avoidance were studied in the work: ignoring collisions, resolving all collisions sequentially for pairs of robots, and resolving collisions by sequentially expanding a group of robots. The best results were shown by the third approach. When using it, the number of robots that reach the goal increased by an average of 25% compared with the first approach on the small and medium volume of filling the map with obstacles on all ranges of map dynamics.

The objective of this paper is to adapt the previously developed reconfiguration algorithm in 2D environment to work in 3D environment in the context of the chosen control approach, and test more advanced algorithms. Two advanced reconfiguration algorithms for 3D environment are presented, which employ different methods of collision detection. The result of the application of these algorithms is straight-line UAV trajectories without collisions.

This paper is organized as follows: The second section describes the approach used to control UAV swarm, the third section describes swarm reconfiguration algorithms, the fourth section describes the experiments and test results.

6.2 UAV Swarm Control Approach

The hybrid approach was chosen to control the UAV swarm. The proposed conceptual model of the hybrid approach is shown in Fig. 6.1. The model includes a base station (BS) with an operator and communication networks, connecting individual agents and BS with agents. In this case, the operator only oversees the mission workflow and does not guide any UAV motions. Trajectory calculation is performed at BS. The advantage of such a system consists in two types of communication: UAV

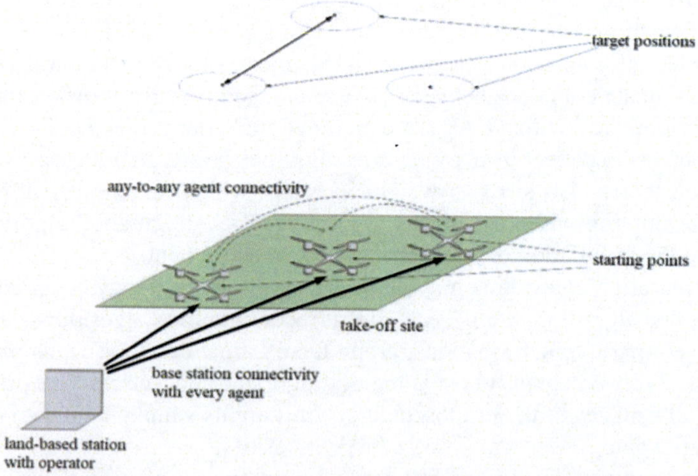

Fig. 6.1 Conceptual model of a combined approach to UAV swarm control

communication with BS and communication among UAVs. Feedback is ensured bypassing agent position coordinates to BS. UAVs should be equipped with GPS beacons, which facilitate UAV detection in case of emergency landing.

Based on the known coordinates of the start and end (target) points, using the reconfiguration algorithms on the BS, the UAV trajectories can be calculated with respect to the minimum allowable distance between them. Thereafter, the obtained trajectories are passed to the agents via BS–UAV communication channel. To simultaneously reach UAVs of all target points, the method of synchronization points was used. This approach is implemented by exchanging information between UAVs about the coordinates of its position on a communication network. The presence of synchronization points implies the expectation of agents who have reached target points, those agents who have not reached target points. UAVs operate in the « hover » mode until all agents reach the target points, after which all the swarm agents will return to their missions.

In the context of the chosen control approach and its conceptual model, the previously developed reconfiguration algorithm was improved. The following are two options for the original 2D algorithm that provides UAV motion without collisions.

6.3 Reconfiguration Algorithms

The previously developed swarm reconfiguration algorithm [15] is intended for constructing agent trajectories in 2D environment. To solve the problem of UAV reconfiguration in real conditions on the basis of a 2D algorithm, two 3D algorithms were developed: simplified (№ 1) and based on divide and conquer (№ 2). The developed algorithms differ in the employed methods, which ensure agent movement without collisions.

Suppose, there are two viable configurations of UAV swarm: the starting one and target one. The initial configuration is the set of UAVs starting points. The target configuration is a geometric shape consisting of many target points. The correspondence between the initial and target configurations is represented by an arbitrary bijection when the starting point *i* of the agent corresponds to the target point *i*. Correspondence is considered short if the total trajectory is minimal. The correspondence between the initial and target configurations is free if the distance between any two trajectories is not less than Dist_{min}. With such arbitrary correspondence, UAVs can simultaneously follow along straight trajectories without collisions.

Reconfiguration algorithm № 1 is a 2D algorithm that includes a parallel translation method, homothety and a target coordinate exchange method. These methods help avoid collisions during flight. Parallel translation (see Fig. 6.2a, b) is based on the search for the central points of the initial and final configurations, presented as points of the initial and final positions, respectively. The central point of such set is a point that satisfies the following condition: every hyperplane containing it divides the set into two equal parts. Parallel translation allows to parallelize translation from the initial configuration to the final one, while aligning their central points.

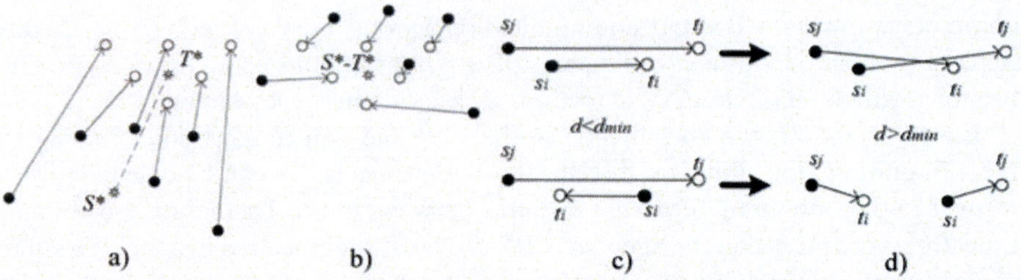

Fig. 6.2 An example of methods: **a** before parallel translation—there are collisions, **b** after parallel translation—no collisions, before target coordinate exchange method—there are collisions, **d** after target coordinate exchange—no collisions

Let us to present the reconfiguration task in the following form: it is necessary to arrange robots $A = \{a_1, a_2, \dots a_i, a_j, \dots a_N\}$ on the three-dimensional figure surface. The initial and final configurations are denoted by S and T, respectively. Then $S*$ and $T*$ are the central points of these configurations. The minimum acceptable distance between the centers of the robots during movement is $Dist_{min}$. If the distance between the UAV becomes less than $Dist_{min}$, then collision occurs. Parallel translation helps avoid collisions during the rectilinear movement of robots.

Collisions can be avoided by choosing the appropriate homothety: It can be applied when transferring the configuration over long distances. However, the use of this method can lead to prolonged duration of path tracing and total reconfiguration time, because small homothety coefficient produces long flight trajectories.

Target coordinate exchange method allows to ensure, that a couple of robots never comes to a collision. Examples of eliminating collisions along a line segment using the method of target points exchange are shown in Fig. 6.2c, d. Flowchart of algorithm № 1 is shown in Fig. 6.3.

At the beginning of the algorithm, the centers of the initial S and target T configurations are determined. Using parallel translation, the center points $S*$ and $T*$ are combined. After that, a new configuration with the center at the point $T*$ is determined and a short correspondence is searched between S' and T using the algorithm [20].

Next, a search of minimum distance between the calculated trajectories is made, assuming that the UAVs begin to move simultaneously and along a straight line with the same speed v. For homothety, the homothety coefficient is calculated using the formula:

$$\lambda = \max\left(1, \frac{\text{Dist}_{\text{min}}}{\delta}\right). \tag{1}$$

Then, using the calculated coefficient λ, homothety is performed. If thereafter collision still remains, the method of target coordinate exchange is applied. The outcome of algorithm № 1 is the trajectory of the UAV, going from the initial to the target points without collisions: $S \to S' \to S'' \to T' \to T$. The trajectories

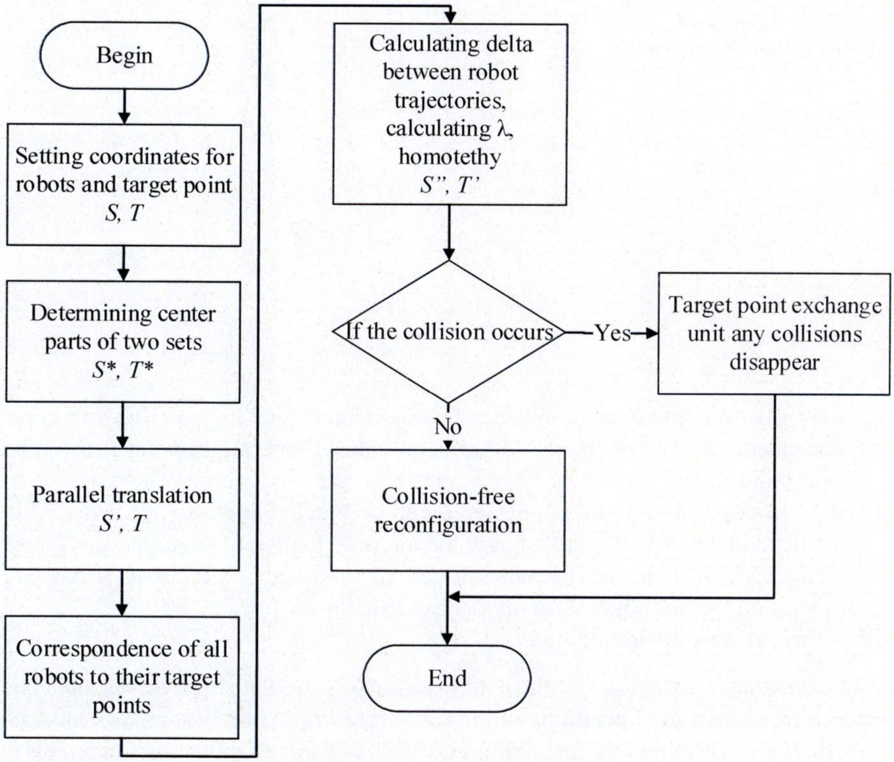

Fig. 6.3 Flowchart of the algorithm № 1

assigned to the robots are linear. The time taken to reconfigure the swarm depends on the initial position of the agents and the target configuration.

With large swarms, the proposed algorithm may be inefficient, the calculated trajectories will cause collisions. Therefore, algorithm № 2 was developed, which includes parallel translation, delay allocation, constructing a collision graph, and the «divide and conquer» algorithm.

A collision graph consists of vertices that correspond to the initial positions of agents with collisions, and edges connecting these vertices. An example of constructing a collision graph is presented in Fig. 6.4.

Delay allocation in initial point produces a collision-free trajectory. Delays are allocated only to those UAVs whose further movement will lead to collision with another agent. The delay is a time slice that postpones the start of UAV movement to the target point. In other words, agent movement will start with a time delay.

The collision graph $\Gamma(A, E)$ is a partially oriented graph with N vertices a_i and a_j. The edges of the graph $e = (a_i, a_j)$ are defined by the following rules:

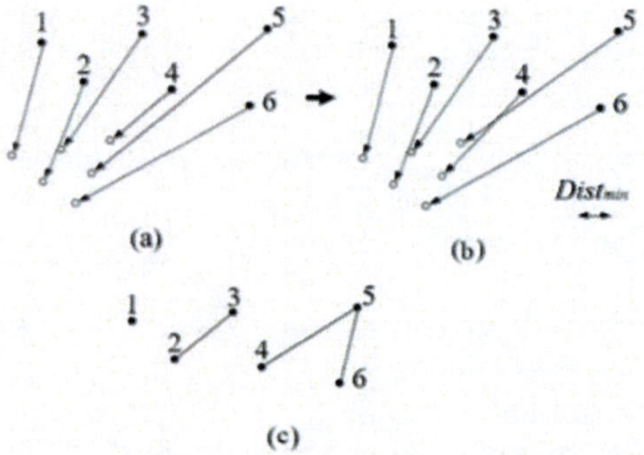

Fig. 6.4 An example of collision graph construction

(1) the vertices a_i and a_j are connected by an oriented edge $\overrightarrow{(a_i, a_j)}$ if a_i and a_j are dependent pair of robots with a collision either at the target point t_i or at the start point s_i;
(2) the vertices a_i and a_j are connected by an undirected edge (a_i, a_j) if a_i and a_j are dependent pair with collision at the internal points of the geometric figure;
(3) in the case of a strongly dependent pair of robots, the vertices a_i and a_j are connected by two edges with opposite orientations;
(4) if there is no collision, there is no edge.

To construct a collision graph, it is necessary to find a short correspondence between initial and final positions of the robot (see Fig. 6.4a). The arrows indicate the path of robot motion to target point. Further, a highly dependent pair of agents is removed by target coordinate exchange method (Fig. 6.4b). Based on the information received, a collision graph is constructed (see Fig. 6.4c).

The idea of the divide and conquer algorithm is to divide the configuration into independent areas of points. After that, reconfiguration is carried out separately in each of these areas. Reconfiguration in each of the areas is synchronized with reconfigurations in the remaining areas. Flowchart of algorithm № 2 is shown in Fig. 6.5.

The first two steps of algorithm № 2 are similar to those of algorithm № 1, and they result in configurations and T with a common center point, as well give a short correspondence between the configurations. After this, the collision graph is constructed and the agents are allocated delays. Next, a check is made for collisions between swarm agents. If collisions persist, the divide and conquer algorithm is applied until all collisions are resolved. The result of the operation of algorithm № 2 is the reconfiguration of the swarm of UAVs without collisions with the assigned delays. To evaluate the performance of the proposed algorithms, as well as identify the most promising of them, a series of experiments was conducted in a virtual environment.

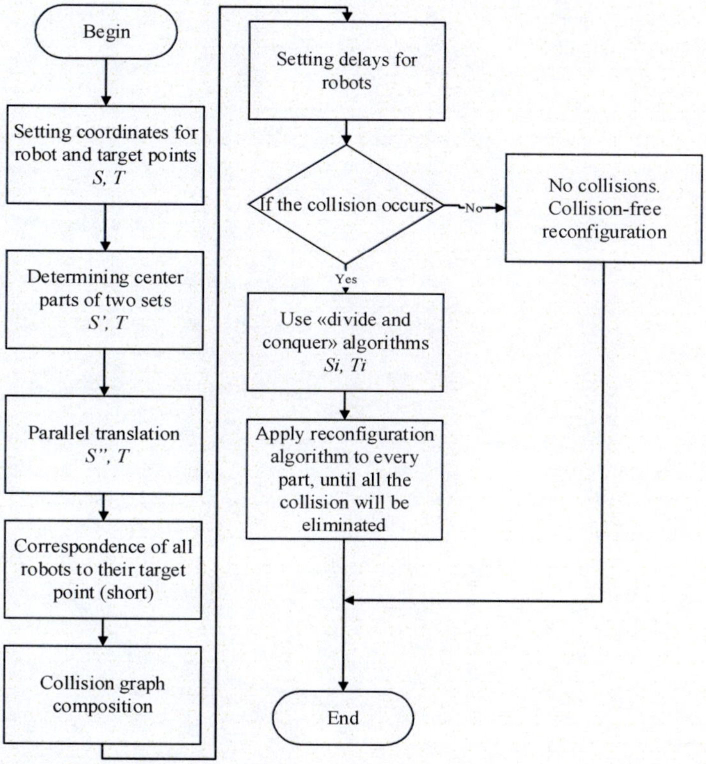

Fig. 6.5 Flowchart of the algorithm № 2

6.4 Results

To evaluate the proposed algorithms, a computer model was developed where the target configuration is a cube surface. Testing was conducted with a swarm of 10–100 agents with increment of 10 at each iteration. For each swarm size, 100 experiments were conducted. The speed of movement of the UAV was set to 1 m/s. Depending on swarm size, the cube edge length and the robot distribution density on surface were calculated.

The parameters used to evaluate the operation of the algorithms were total trajectory length, trajectory calculation time (algorithm running time), average number of collisions and reconfiguration time.

Algorithm № 1 ensured collision-free UAV motion with swarm size from 10 to 70 agents. With a swarm size from 80 to 90 agents, on average, one collision occurs, with 100 agents—2 collisions. Algorithm № 2 provides collision-free motions in a swarm of up to 90 agents. In other words, number of robots, which may collide in the first case, does not exceed 4% of the total number of agents and 2% in the second case. Figures 6.6 and 6.7 show the test results in graphical form.

Fig. 6.6 Dependency graph of total trajectory length on swarm size

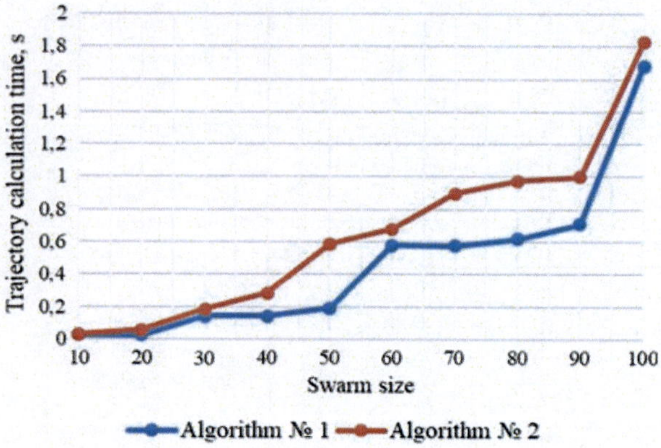

Fig. 6.7 Dependency graph of total trajectory length on swarm size

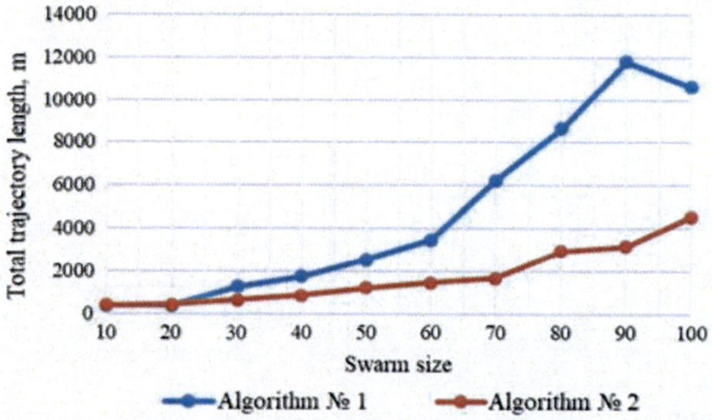

Algorithm running time, total trajectory length and swarm reconfiguration time are directly proportional to a swarm size. The time required to calculate the trajectories of 100 UAVs does not exceed 2 s; therefore, both reconfiguration algorithms can be used in real-time mode. However, the running of algorithm № 2 is more time-consuming than by algorithm № 1, since it contains a larger number of methods.

Large values of the total trajectory length of algorithm № 1 are explained by homothety, which leads to additional movements. Homothety coefficient depends on the minimum acceptable distance between UAV; therefore, when collisions appear, homothety coefficient decreases and number of UAV movements increases. From the point of view of the total trajectory length, algorithm № 2 is more efficient than the former. When using algorithm № 2 for reconfiguring a swarm of 100 agents, this parameter is approximately two times smaller than in the case of algorithm № 1.

The total reconfiguration time was calculated under the condition, that the UAV speed is 1 m/s. But in reality, UAVs can have a higher speed, and reconfiguration will require less time than during testing.

The main advantage of algorithm № 1 is the availability of simple geometric methods that simplify the implementation of the algorithm and increase the calculation speed. Algorithm № 1 allows to produce a trajectory without collisions with a swarm size of 70 agents, while algorithm № 2, because of the use of additional methods, has a size limit of 90 agents. However, because of delay allocation and divide and conquer algorithm, the swarm reconfiguration time increases and the requirements for the hardware computing power increase. Despite the fact that Algorithm № 2 requires more time for the overall execution and swarm reconfiguration, the total trajectory length and the average number of collisions are lower compared to Algorithm № 1. Therefore, Algorithm № 2 is more preferable for swarm reconfiguration, if swarm size is sufficiently large.

6.5 Conclusion

The objective of this work consists in solving of UAV swarm control problem, specifically, eliminate collisions between the agents. A hybrid approach was chosen for swarm control, which assumes the presence of a base station, wireless communication networks, connecting the agents with each other and with the base station. This approach allows to reduce the algorithm running time because to the ability to use more powerful hardware. Based on the hybrid approach of UAV control, 2D reconfiguration algorithm was improved to allow agents to move in 3D environment without collisions. Two advanced algorithms (№ 1 and № 2), which differ in collision avoidance methods, are presented and tested in this paper.

To evaluate the performance of the presented algorithms, they were tested in a virtual environment. The experiments were performed for swarms from 10 to 100 agents. The total trajectory length, trajectory calculation time (algorithm running time), the average number of collisions and the reconfiguration time were used to evaluate the efficiency of the algorithms. Testing results showed that less than 2 s is required to compose trajectories of 100 UAVs using both algorithms. Therefore, algorithms № 1 and № 2 can be used in real-time mode. Thereby, algorithm № 2 has a greater running time than algorithm № 1, because it contains more methods for avoiding collisions. Greater total trajectory length, obtained using algorithm № 1, compared to algorithm № 2, requires homothety introduction. Homothety leads to an increase of number of agent movements and reconfiguration time. Algorithm № 1 provides UAV movement without collisions with a swarm of 70 agents; by comparison, algorithm № 2 has a threshold value of 90 agents. Because of algorithm № 2 has a shorter total trajectory length and an average number of collisions compared to algorithm № 1, it is more preferable for UAV swarm reconfiguration.

Further research will be aimed at evaluating the presented algorithms in real conditions, as well as developing methods for reducing the power consumption of UAVs during reconfiguration.

References

1. Shakhatreh, H., et al.: Unmanned aerial vehicles (UAVs): a survey on civil applications and key research challenges. IEEE Access **7**, 48572–48634 (2019)
2. Stöcker, C., et al.: Review of the current state of UAV regulations. Remote Sens. **9**(5), 459 (2017)
3. Hayat, S., Yanmaz, E., Muzaffar, R.: Survey on unmanned aerial vehicle networks for civil applications: a communications viewpoint. IEEE Commun. Surv. Tutor. **18**(4), 2624–2661 (2016)
4. Ardupilot Swarming/Formation-Flying Interface. http://ardupilot.org/planner/docs/swarming.html. Last accessed 2020/01/10
5. Tahir, A., et al.: Swarms of unmanned aerial vehicles: a survey. J. Ind. Inf. Integr. **16**, 100106 (2019)
6. Asadpour, M., Hummel, K.A., Giustiniano, D., Draskovic, S.: Route or carry: motion-driven packet forwarding in micro aerial vehicle networks. IEEE Trans. Mob. Comput. **16**(3), 843–856 (2016)
7. Pshikhopov, V.K., Medvedev, M.Y.: Group control of autonomous robots motion in uncertain environment via unstable modes. SPIIRAS Proc. **5**, 39–63 (2018)
8. Kremlev, A., Kolyubin, S., Vrazhevsky, S.: Autonomous multi-agent "robot-guide" system to solve area monitoring problems. J. Instrum. Eng. **56**(4), 61–65 (2013)
9. Bürkle, A., Segor, F., Kollmann, M.: Towards autonomous micro UAV swarms. J. Intell. Rob. Syst. **61**(1–4), 339–353 (2011)
10. Korotin, A.A., Kozyrev, G.I., Nazarov, A.V., Blagodyrenko, E.V.: Investigation of reliability of combinatorial-metric algorithm for recognition of n-dimensional group point object in hierarchy features space. SPIIRAS Proc. **18**, 976–1009 (2019)
11. Leuchter, S. et al.: Karlsruhe generic agile ground station. In: Future Security 2nd Security Research Conference, pp. 159–162 (2007)
12. Tsykunov, E., et al.: Guiding a swarm of micro-quadrotors with impedance control using a wearable tactile interface. IEEE Trans. Haptics **12**(3), 363–374 (2019)
13. Mulgaonkar, Y., Cross, G., Kumar, V.: Design of small, safe and robust quadrotor swarms. In: 2015 IEEE International Conference on Robotics and Automation (ICRA), pp. 2208–2215. IEEE (2015)
14. Vedder, B., Eriksson, H., Skarin, D., Vinter, J., Jonsson, M.: Towards collision avoidance for commodity hardware quadcopters with ultrasound localization. In: 2015 International Conference on Unmanned Aircraft Systems (ICUAS), pp. 193–203. IEEE (2015)
15. Vatamaniuk, I.V., Panina, GYu., Ronzhin, A.L.: Reconfiguration of robot swarm formation. Large-Scale Syst. Control **58**, 285–305 (2015)
16. Ahmed, H., Janice, G.: Swarm intelligence: concepts, models and applications. Queens University Technical Report, School of Computing (2012)
17. Malyshev, A.A., Burgov, E.V.: Revisiting parameters of bioinspired behavior models in group foraging modeling. SPIIRAS Proc. **19**, 79–103 (2020)
18. Petrikevich, Y.: Linear algorithms for managing the geometric location of objects in a multi-agent system. Large-scale Syst Control **30**(1), 665–680 (2010)
19. Motorin, D., Popov, S., Kurochkin, L.: An algorithm for collision avoidance in path planning for a group of robots in a spatio-situational indeterminacy. St. Petersburg State Polytech. Univ. J. Comput. Sci. Telecomm. Control Syst. **10**(2), 32–44 (2017)
20. Vaidya, P.M.: Approximate minimum weight matching on points ink-dimensional space. Algorithmica **4**(1–4), 569–583 (1989)

Chapter 7
Approach to Robotic Mobile Platform Path Planning Upon Analysis of Aerial Imaging Data

Egor Aksamentov, Konstantin Zakharov, Denis Tolopilo, and Elizaveta Usina

Abstract The research is focused on solving one of the main problems of robotic system navigation—building an energy-efficient trajectory. A method of trajectory planning is proposed, according to which images obtained using an unmanned vehicle camera are stitched into an orthomosaic image, on which the Mask R-CNN neural network detects all static obstacles. In addition to creating a map of the area, the resulting images are also used to create a 3D model of the area. The developed method enables the robotic to find all static obstacles, as well as to identify all terrain features. Based on the obtained data on the state of the terrain and the presence of static obstacles, the system finds the most energy-efficient path.

7.1 Introduction

One of the key problems of robotics is the autonomous navigation of robotic vehicle (RV) in open space. For successful movement, the RV [1] navigation system should automatically mark all passable sections of the path (where motion is possible), as well areas that are dangerous for RV. For example, places with elevation difference where RV may overturn, and path sections where dimensions of the RV exceed the dimensions of the road. Another task of the autonomous RV motion is to find the optimal route to maximize the performance of the RV after a one-time battery charge.

In this paper, authors propose a methodology for solving the problem of outdoor RV navigation using terrain maps. The proposed method solves the problem of static obstacles detection for RVs, and the problem of image-based path tracing, where the images are captured from UAVs, thereby respecting the terrain features.

E. Aksamentov (✉) · K. Zakharov · E. Usina
St. Petersburg Institute for Informatics and Automation of the Russian Academy of Sciences, 39, 14th Line, 199178 St. Petersburg, Russia
e-mail: Egor.aksamentov.96@mail.ru

D. Tolopilo
State University of Aerospace Instrumentation, 67, St. Bolshaya Morskaya, 190000 St. Petersburg, Russia

A. Ronzhin and V. Shishlakov (eds.), *Proceedings of 15th International Conference on Electromechanics and Robotics "Zavalishin's Readings"*, Smart Innovation, Systems and Technologies 187, https://doi.org/10.1007/978-981-15-5580-0_7

7.2 Related Work

One of the ways for navigating RV is to use a pre-prepared terrain orthophotomap. Orthophotomap is a kind precise representation of the explored terrain on a single image. For orthophotomap preparation, satellite images or aerial images taken from UAVs may be used [2]. The advantage of UAV-aided imaging over satellite-based imagery in terrain map preparation consists of higher quality of the UAV images [3]. The accuracy of terrain mapping directly depends on the source data. For example, due to weather conditions, some images obtained from UAVs may be of poor quality, as well as sufficient illumination is required for accurate terrain mapping. For example, due to weather conditions, some images taken from UAVs may be blurry. Another necessary prerequisite for the accurate construction of the terrain is sufficient lighting.

The movement of RV based on navigation by GPS tags. Two points are enough to build a simple path—a straight line that does not respect dangerous spans of the route. For building of a safe route, it is necessary to accurately identify all sites where RV motion is possible. For example, in [4], a new approach to planning the path of movement using a 3D map is implemented. The constructed 2D path is projected onto a 3D map using the digital terrain model function, where each pair of values (x, y) corresponds to a unique elevation value (z). When projecting a 2D map onto a 3D copy of it, the actual distance on the terrain between adjacent paths increases or decreases, which can cause them to diverge or intersect on slopes. The presented numerical approach for analyzing zones with diverging/intersecting paths ensures full coverage of the map regardless of its geometric features and the selected angle of motion. The results of the experiment on the test zones showed that the developed approaches allow to find 2–14% more passable areas of the terrain map compared to alternative approaches.

In [5], authors present a path planning system for 3D environment. The proposed algorithm is based on the RRT* Goal Limit (RRT* GL) algorithm, which is a combination of two algorithms: RRT* Goal (guides the exploration of the terrain map toward the target point) and RRT* Limits (when found a possible path, the search area for the next iteration will be shifted toward finding exactly this path). The results of performance testing of the algorithms of the RRT family (RRT, RRT*, RRT* Goal, RRT* Limits, RRT* GL) showed that the RRT* GL algorithm generates the optimal path much faster, but requires more computational resources of the system, than other algorithms.

In [6], authors present the heuristic algorithm Z^* for energy-efficient road planning over rugged terrain. Here, energy efficiency is considered in context of robot battery use, which depends on the inclination angle, the surface friction coefficient, and the weight of the robot. If the inclination angle of the surface is too large for the robot to drive directly to a given point, the algorithm generates a zigzag path to it. The authors conducted real-world experiments with robots to find the path using the algorithms D-Dopt [7], D-Eopt [8], basic Z^* [9] and Z^*. Upon the experiments, the shortest path was found by the D-Dopt algorithm, but this algorithm does not consider the energy consumption of robots to go this route. The developed Z^* algorithm showed approximately the same results as the other algorithms, but it allowed to complete the task 1, 5 times faster.

When estimating the mobility and autonomy of the RV, it is importantly to consider reliability and motion in rugged terrain. Movement along difficult paths is accompanied by solving the task of trajectory control. This task is difficult due to problems associated with excessive sliding of the wheelbase in non-deterministic conditions, when there is no data on the nature of the surface that the wheels come into contact with when maneuvering. In [10], the adaptive super twisting path tracking controller (AST) was developed based on the adaptive super twisting (ST) algorithm for a four-wheel mobile robot. The controller adjusts its amplifications dynamically, in order to maintain the necessary control action. Therefore, its practical implementation does not require a continuous approximation of the signal—i.e., functions to reduce vibration, and the properties of stability and finite convergence in time do not fade out. The effectiveness of the developed controller was experimentally proved by using the Pioneer 3AT robotic platform and comparing images taken while moving along paths that correspond to the algorithms being used. Experimental results show that the proposed controller is able to ensure the characteristics of the second-order sliding mode with less rattling compared to the results obtained in [11, 12].

Thus, outdoor navigation of robotic vehicle directly depends on the quality of the data obtained using the UAV. High-resolution images are required to create an accurate orthophotomap and 3D space. The accuracy of zone detection with obstacles in the path of movement of robotic vehicle, as well as the result of finding the optimal path, will depend on the result of 3D surface construction.

7.3 Orthomosaic Image Stitching Using Georeferencing to GPS

To build the orthophotomap of the terrain over which robotic vehicle will move, the UAV has to perform several flights over the territory, photographing it each time. From the obtained image sequence, for two consecutive images, a set of key points with a pronounced neighborhood is created. The resulting sets of points from the images are compared with each other, and point whose neighborhood has a minimum Euclidean distance form pair. A pair of such points represent the same object but photographed from different angles. Based on the found pairs, the images are stitched together so that all the points forming the pairs correspond to each other. The resulting stitched image from two images is compared with another image from the series, and a new iteration of finding the specific points occurs. Image stitching occurs until all photos received from UAV are processed. An example of an orthomosaic image is shown in Fig. 7.1.

To find areas on the orthophotomap impassable for robotic vehicle, the Mask R-CNN neural network (NN) was used, the peculiarity of which is to generate a

Fig. 7.1 Result of creating an orthomosaic image of 7 frames

detected object at the mask output and determine its boundaries [13]. An example of the NN operation is shown in Fig. 7.2.

For accurate positioning of detected objects, it is necessary to reference GPS coordinates to the orthophotomap. When imaging a terrain using a UAV, GPS coordinates corresponding to the central pixel in the image where the picture was taken are recorded for each frame. The resulting stitched orhophotomap contains many such pixels with corresponding GPS coordinates. Based on the known data, the corresponding GPS coordinates for each point of the orthophotomap are proportionally calculated. Further calculations of the robotic vehicle movement paths are performed with respect to GPS coordinates of the boundaries of obstacles.

7.4 Building 3D Map of Area

During robotic vehicle motion, this process is additionally complicated by obstacles, various slopes, potholes, and other areas of the terrain where the robotic vehicle can be turned over is also difficult. To identify all the hills and irregularities of the landscape, it is necessary to build a 3D representation of the terrain. To build a 3D map of the area, one needs to capture several pictures of the same area from different angles [14]. The obtained images, using photogrammetry, form a sparse cloud of points, which exactly corresponds to the terrain. The result of building of a 3D map of the area is shown in Fig. 7.3.

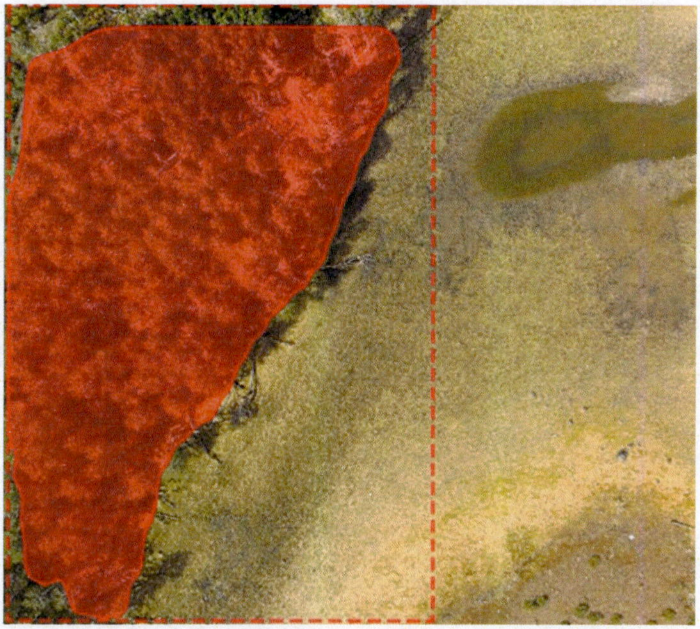

Fig. 7.2 Result of mask R-CNN operation

Fig. 7.3 Example of building a three-dimensional map of the area

7.5 Path Planning Algorithm for a Robotic Vehicle

Having identified all the areas, dangerous for the motion of the robotic vehicle, for the rational use of battery resources, it is necessary to find the optimal path. To move the robotic vehicle along an uneven surface, an algorithm for path planning and robot navigation was developed (Fig. 7.4) based on the A^* algorithm [15].

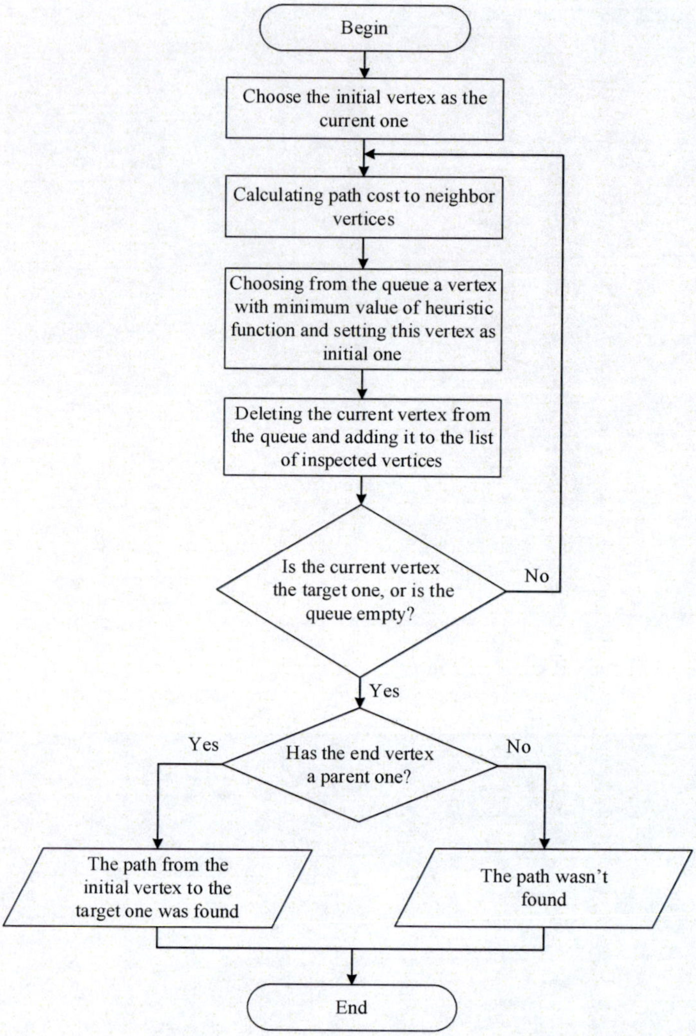

Fig. 7.4 Path planning algorithm

The environment surrounding the robot is represented in the form of a graph $G_v = (V_v, E_v)$, in which the vertices V_v, having unique coordinates (x, y, z), are connected by the edges E_v. At the initial stage of the path planning algorithm, all vertices of the graph are checked for passability, and areas with a very uneven surface and/or a very large difference in elevation are marked as obstacles. Then, weights of all edges of the graph are calculated based on the distance of the vertices to the nearest obstacles, the distance between the vertices, local irregularities and difference in elevation. Planning a path from one vertex of the graph G_v to another with known edge weights is performed using the A^* algorithm [14]. The algorithm starts search from the initial vertex and visits adjacent nodes. For these vertices, the cost of the path is calculated, comprised of the sums of the weights of all the edges along which it is necessary to go to reach this vertex. In addition to the sum of the weights of the edges, a predecessor is assigned for each vertex—the node from which this vertex was reached. The algorithm considers different paths to the same peak, so if the new cost of the path is less than the one that corresponds to the top at the moment, then a new path cost value and new predecessor is assigned to it. When all neighboring nodes have been visited, the visited vertex is recorded into the *closed* list and is no longer visited by the algorithm, and the neighbors of the current vertex are recorded into the *open* list. After that, the algorithm scans that vertex from the *open* list, the Euclidean distance from which to the target vertex will be the smallest and which is not in the *closed* list and visits its neighbors.

The algorithm finishes the search for the path if the current vertex is finite. If during the path planning process, the *open* list becomes empty, then the path from the initial to the target vertex cannot be found. The path is built by sequential extraction of the predecessors of the vertices and adding them to the top of the *path* list. The process of extraction of predecessors begins with the target and ends at the initial vertex.

7.6 Navigation Grid Building

Local terrain irregularity implies a curvature of the surface caused by elevation difference. The roughness of the region $p(v)$ around the vertex v is expressed as the average angle of inclination of the surface for the nearest vertices $N_l(v)$ located at a distance less than l from the vertex v. The angle between the normal of the vertices is calculated by the formula $c_{r^{(u,v)}} = \operatorname{acos}(n(u) * n(v))$, and the average angle $a_{\mathrm{av}(v)}$ is calculated by the formula (1), where $N_d(v)$ are all direct co-neighbors of the vertex v.

$$a_{\mathrm{av}}(v) = \frac{\sum_{u \in N_d(v)} c_r^{(u,v)}}{|N_d(v)|},$$

$$p(v) = \frac{\left(a_{\mathrm{av}}(v) + \sum_{u \in N_l(v)} a_{\mathrm{av}}(u)\right)}{|N_l(v)| + 1}. \tag{1}$$

For finding landscape irregularities, ravines, and soil breaks, it is necessary to use an estimated value of the local elevation difference. The local difference in elevation $b(v)$ reflects the maximum difference in elevation between vertex v and each vertex from $N_l(v)$ and is calculated using the formula (2), where $v(z)$ and $u(z)$ are the z-coordinate values of vertices v and u, respectively.

$$b(v) = \max_{u \in N_l(v)} \left| \frac{p(v) - p(u) * z)}{|z|} \right|. \tag{2}$$

The algorithm analyzes the local unevenness and the maximum difference in elevation differences for each vertex in the graph. If the values exceed a certain upper threshold, then these vertexes are marked as obstacles.

The boundaries of the areas where obstacles are located are extended to a distance comparable to the size of the robot, so that the path planning algorithm builds a route, all points of which are at a safe distance from dangerous areas. The gradual inflation algorithm smoothly reduces the weight of vertices located at a distance from r_i to r_o from the region in which the obstacle is located. The riskiness $r(v)$ reflects the degree of remoteness of the vertex v from the nearest obstacle and is calculated by the formula (1), where $c_{\max}$ is the maximum total cost and d_{lethal} is the smallest distance to the obstacle:

$$r(v) = c_{\max} * \left(\cos\left(\frac{d_{\text{lethal}} - r_i}{r_o - r_i} * \pi\right) + 1\right). \tag{3}$$

To calculate the minimum distance to the obstacle—d_{lethal}, the Li algorithm is used [16].

For global path planning, it is necessary to calculate the final weights of the edges, based on the distance between the vertices, the riskiness (a parameter whose value is directly proportional to the distance from the given vertex to the nearest obstacle), local unevenness, and local difference in elevation. The final edge weight $c(u, v)$ for each edge is calculated by the formula (4), where k_p and k_b are used to determine the influence of the roughness p and the local difference in elevation b on the change in the weight of the edge, and $d(u, v)$ is the Euclidean distance between vertices u and v. Depending on the values of the coefficients k_p and k_b, the path planner will find different paths with different characteristics:

$$\begin{aligned} C^{(u,v)} &= ((k_p * \max(p(u), p(v))) + 1) \\ &* ((k_b * \max(b(u), b(v))) + 1) * d^{(u,v)} + |r(u) - r(v)|. \end{aligned} \tag{4}$$

7.7 Experiments and Results

Experiments were conducted in Gazebo simulation environment. Figure 7.5 shows the results of testing of the path planning algorithm for two different pairs of initial and final vertices of the graph.

White dots in Fig. 7.5 indicate the tops of the path found using the planning algorithm. The route found is close to a straight line, and neighboring peaks run through sections of the map with a relatively small difference in elevation. Based on this information, it can be concluded that the resulting route is energy efficient, since the energy consumption of the RV for mountain climbing will be minimal.

Figure 7.6 shows a graph, representing the dependence of path planning time on

Fig. 7.5 Visualization of the constructed path in Gazebo

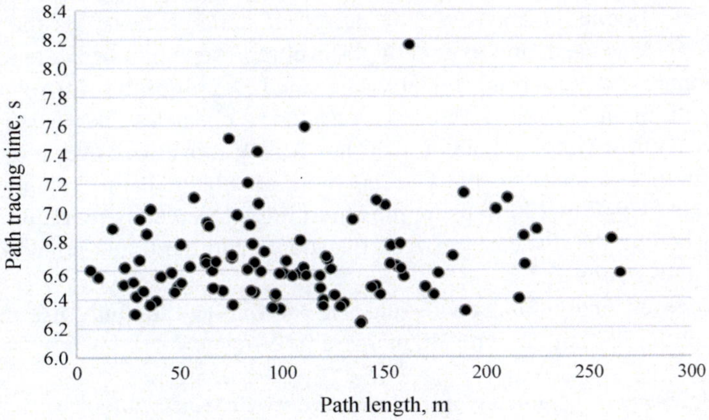

Fig. 7.6 Graph of the dependence of path planning time on path length

the overall path length, which is calculated as the sum of lengths of path segments. 100 experiments were performed б for each of them a random pair of vertices was generated, where every pair consists of a source vertex and a target vertex. As follows from the graph, path planning operation requires from 6 to 8.2 s, whereas path length in different experiments ranges from 7 to 270 m. Consequently, path length is virtually irrelevant in terms of path planning duration.

In addition to the features of the terrain and static obstacles [17], another problem for the movement of the RV over rugged terrain is represented by variable surface properties and, consequently, the possibility of getting bogged down the RV. For example, this problem becomes evident when the RV moves a sand embankment or through marshland. To minimize these factors, it is necessary to foresee the design of the RV, as well as to develop a special control system to solve the problem of wheel slippage. Based on the above factors, it is proposed to use a 4-wheel platform with differential drive (4wdd) as the basis for the RV. The absence of a steering system makes the 4wd platform strength and capable of easily and maneuverable moving over rugged terrain. Thus, the proposed base has the basic properties necessary for the RV, namely maneuverability, permeability, strength, domain of mobile robotics, and navigation problem is of primary importance.

7.8 Conclusion

Domain of mobile robotics, navigation problem is very relevant today. Almost all robotic systems suffer from a common problem—limited battery capacity. Therefore, when constructing the route of the RV movement, it is necessary to consider not only the distance to be passed, but also energy consumption. In this paper, we propose a methodology for solving the problem of planning the optimal RV route in rugged terrain, based on the construction of an orthomosaic map and its 3D model from images obtained using UAVs. To detect all static obstacles on the map, NN Mask R-CNN was used, and to get a high-level representation of the terrain, using photogrammetry, a 3D terrain model was created. An algorithm for robot navigation and path planning was developed, considering obstacles and elevation differences, which allows finding the optimal route for RV movement over rough terrain. The algorithm was tested in the Gazebo robot simulator. Path planning duration for segments from 7 to 270 m using the algorithm, presented here, requires from 6 to 8.2 s, whereby the distance between the source and the target point is virtually irrelevant in this context.

The proposed methodology will make it possible to rationally use the energy resources of the RV, and, as a result, increase its efficiency.

Acknowledgements This work is supported by state research № 0073-2019-0001.

References

1. Ovcharenko, K.L., Eremeev, I.Yu., Sazonov, K.V., Semenjuk, S.S., Abakumov, A.N.: Differential ranging method of locating satellite communication system earth stations using repeater on an unmanned aerial vehicle. SPIIRAS Proc. **18**(1), 176–201 (2019)
2. Samad, A.M., Kamarulzaman, N., Hamdani, M.A., Mastor, T.A., Hashim, K.A.: The potential of unmanned aerial vehicle (UAV) for civilian and mapping application. In: 2013 IEEE 3rd International Conference on System Engineering and Technology, pp. 313–318 (2013)
3. Muchiri, N., Kimathi, S.: A review of applications and potential applications of UAV. In: Proceedings of sustainable research and innovation conference, pp. 280–283 (2016)
4. Hameed, I.A., la Cour-Harbo, A., Osen, O.L.: Side-to-side 3D coverage path planning approach for agricultural robots to minimize skip/overlap areas between swaths. Robot. Autonom. Syst. **76**, 36–45 (2016)
5. Aguilar, W.G., Morales, S.G.: environment mapping using the Kinect V2 and path planning based on RRT algorithms. Electronics **5**, 70 (2016)
6. Ganganath, N., Cheng, C.T., Chi, K.T.: A constraint-aware heuristic path planner for finding energy-efficient paths on uneven terrains. IEEE Trans. Industr. Inf. **11**, 601–611 (2015)

7. Dijkstra, E.W.: A note on two problems in connexion with graphs. Numerische Mathematic **1**, 269–271 (1959)
8. Kim, H., Kim, B.K.: Online minimum-energy trajectory planning and control on a straight-line path for three-wheeled omnidirectional mobile robots. IEEE Trans. Industr. Electron. **61**(9), 4771–4779 (2013)
9. Ganganath, N., Cheng, C.T., Chi, K.T.: Finding energy-efficient paths on uneven terrains. In: 2014 10th France-Japan/8th Europe-Asia Congress on Mecatronics, pp. 383–388 (2014)
10. Matraji, I., Al-Durra, A., Haryono, A., Al-Wahedi, K., Abou-Khousa, M.: Trajectory tracking control of skid-steered mobile robot based on adaptive second order sliding mode control. Control Eng. Pract. **72**, 167–176 (2018)
11. Yu, W., Chuy Jr., O.Y., Collins Jr., E.G., Hollis, P.: Analysis and experimental verification for dynamic modeling of a skid-steered wheeled vehicle. IEEE Trans. Rob. **26**(2), 340–353 (2010)
12. Salgado, I., Cruz-Ortiz, D., Camacho, O., Chairez, I.: Output feedback control of a skid-steered mobile robot based on the super-twisting algorithm. Control Eng. Pract. **58**, 193–203 (2017)
13. Johnson, J.W.: Adapting mask-RCNN for automatic nucleus segmentation. arXiv preprint arXiv:1805.00500 (2018)
14. Remondino, F., Barazzetti, L., Nex, F., Scaioni, M., Sarazzi, D.: UAV photogrammetry for mapping and 3d modeling–current status and future perspectives. Int. Arch. Photogramm. Remote Sens. Spatial Inf. Sci. **38**(1), C22 (2011)
15. Hart, P.E., Nilsson, N.J., Raphael, B.: A formal basis for the heuristic determination of minimum cost paths. IEEE Trans. Syst. Sci. Cybern. **4**(2), 100–107 (1968)
16. Rubin, F.: The Lee path connection algorithm. IEEE Trans. Comput. **100**(9), 907–914 (1974)
17. Miroshnichenko, S.Y., Titov, V.S., Dremov, E.N., Mosin, S.A.: Hough transform application to digitize rectangular spatial objects on aerospace imagery. SPIIRAS Proc. **6**(61), 172–196 (2018)

Chapter 8
Improving Methods of Objects Detection Using Infrared Sensors Onboard the UAV

Peter Trefilov, Mark Mamchenko, Maria Romanova, and Igor Ischuk

Abstract The rapid development of unmanned aerial vehicles (UAVs) has contributed to a proliferation of multispectral aerial survey technologies and services. Aerial reconnaissance allows to obtain the detailed digital map of the area, including the geographical distribution of radiant temperatures. These temperature maps allow to find out the composition of the objects (figure out type of the material), their real dimensions and size. There are studies concerning processing of multispectral aerial survey images (both obtained in visible and infrared ranges) and selecting the UAV optimal flight altitude for detection, recognition, and identification of monitoring objects. However, the issue of developing an integrated algorithm of UAV multispectral aerial survey for classifying monitoring objects (taking into account the choice of optimal flight altitude and camera's resolution parameters) still remains open. This article considers the choice of the flight altitude of the UAVs based on Johnson's criteria for detection, recognition, and identification of monitoring objects. The integrated approach for conducting aerial survey based on mathematical relationship between optimal flight altitude of the UAVs and resolving power of its onboard camera is proposed.

8.1 Introduction

Unmanned aerial vehicles (UAVs) have become widespread. Drones' availability in technology-intensive product markets resulted in their broad application in the

P. Trefilov (✉) · M. Mamchenko · M. Romanova
V. A. Trapeznikov Institute of Control Sciences of Russian Academy of Sciences, 65, St. Profsoyuznaya, 117997 Moscow, Russia
e-mail: petertrfi@gmail.com

I. Ischuk
Zhukovsky–Gagarin Air Force Academy, 54A, St. Bolshevikov, 394064 Voronezh, Russian Federation
e-mail: igorb_ff@mail.ru

A. Ronzhin and V. Shishlakov (eds.), *Proceedings of 15th International Conference on Electromechanics and Robotics "Zavalishin's Readings"*, Smart Innovation, Systems and Technologies 187, https://doi.org/10.1007/978-981-15-5580-0_8

spheres of aerial photography and mapping. The reduction in costs of multispectral aerial survey hardware and high demand for periodic inspections of objects from the air also resulted in wider availability of respective services. Previously, such technologies were available only for large corporations with immense financial resources, but now it is possible to form UAVs from ready-made components with unique requirements and equipment in order to carry out specific tasks [1].

Within smart cities environment, UAVs may be engaged for enabling safe operations of manned and unmanned vehicles, including monitoring urban infrastructure facilities, roads, park areas, etc.

The common requirement for each of the above-described ways of applying UAVs is the detailed and real-time information provided by the UAV. The timeliness of the received information and its level of detail are greatly defined by the flying height H of the UAV. While searching stealthy objects, special attention should be devoted to algorithmic support tasks, designed for detection and recognition of stationary and quasi-stationary objects using multispectral optoelectronic systems (as payloads of the UAVs of different classes). Upon receipt of multispectral images, material characteristics affect the temperature differentials in the infrared (IR) range, on which the detection of objects is based.

Three degrees of detail of video data are used for classification of surface objects: detection, recognition, and identification. This video-based three-tiered process is a probabilistic one.

Once objects are detected, recognized, and identified, they can be classified (clustered) according to different criteria.

8.2 Problem Statement

The difference in the physical and chemical properties of materials affects their IR radiation reflective power and absorbing capacity. This allows to form the map of radiant temperature distribution, relating specific color and each pixel of the image to a certain temperature value. In paper [2], authors proposed an algorithm of co-processing of aerial survey multispectral images (in visible and IR ranges) using numerical solutions for solving direct and inverse problems of radiation thermal physics. This algorithm allows increasing the effectiveness of detection of «stealthy» technogenic ground objects, as well as practically eliminating stochastic factors of aerial survey. However, a significant limitation for the radiation temperature mapping is the need for selection of the UAV optimal flight altitude based on the resolution power of the camera, and this issue was not considered in article [2].

Other researchers proposed various approaches to finding the optimal flight altitude of the UAV. In paper [3], an approach to calculating UAV flight altitude has been proposed. This approach might be used for unmanned aerial surveys of the areas using set detection, recognition, and identification probability values and camera resolution parameters. However, this article does not consider calculating UAV flight altitude for multispectral survey missions. Besides, the problem of classification (clustering) of identified objects (including further stages of processing multispectral images) remains unsolved. Paper [4] addresses the image recognition approach using multi-scale gray difference weighted image entropy, but it allows to recognize the target, not to cluster it. Article [5] considers object detection via heterogeneous sensor, but such an approach implies the use of multiple cameras, significantly increasing the drone's mass-dimensional characteristics. Proposed in papers [6, 7] approaches imply recognition of individual objects rather than the whole image. Finally, articles [8–11] are aimed at human recognition. Thus, we can conclude that the subject of recognition of different materials and ground object using aerial survey images (including multispectral ones) is insufficiently investigated.

Therefore, the scientific novelty of the paper is the combined application of known methods of calculating UAV optimal flight altitude based on camera resolution parameters for conducting efficient multispectral aerial surveys (in visible and IR ranges), detecting, recognition, and identifying monitored objects. We will also try to classify identified objects using radiant temperature distribution maps. In the sphere of photo-interpretation, this will allow to increase the effectiveness of ground objects detection because of the redundant informativeness of the received data. Furthermore, stochastic factors of air photography can be virtually excluded by using the indicator of effective thermal conductivity, which is the determined parameter of the aerial surveys.

8.3 Approach to Solving the Problem

Video-recording systems use Johnson's criteria to determine the probability of the objects detection, recognition, and identification. Curves of the dependence of the objects identification on the number of N pixels (resolution elements) on the video data appear as follows:

$$P_{\text{det}} = 1 - e^{-0,2(N-1)^2}, \tag{1}$$

$$P_{\text{rec}} = 1 - e^{-0,015(N-1)^2}, \tag{2}$$

$$P_{\text{id}} = 1 - e^{-0,005(N-1)^2}, \tag{3}$$

where P_{det}, P_{rec}, P_{id} are the probabilities of detection, recognition, and identification of the objects (respectively).

We will now establish a link between the number of pixels (resolution elements) N and the height H. Let the object with Nm size creates a video image Nu. An image dot with the size of a corresponds to the element A of the object. Camera objective lens with an f focal length is located at the altitude (height) of H and the distance of d (relative to the object), and forms the image of the object. Generally, $d \neq f$ inequality is valid. The following values—H, d, and f—are linked through the lens formula [12].

In case of our survey $H \gg d$, therefore we may take $d \approx f$. Equation (4) is based on this assumption:

$$\frac{1}{H} = \frac{1}{d} + \frac{N_m}{f}. \tag{4}$$

The height H is calculated as the mean geometric sum of the resolving power of the camera lens Ro and the sensing element Re.

Now let us substitute the final number of resolution elements N into Nu. Having solved the equation for H, we shall obtain:

$$H_{\text{det}} = \frac{RN_m f}{1 + \sqrt{-\frac{\ln(1-P_{\text{det}})}{0{,}2}}}, \tag{5}$$

$$H_{\text{rec}} = \frac{RN_m f}{1 + \sqrt{-\frac{\ln(1-P_{\text{rec}})}{0{,}015}}}, \tag{6}$$

$$H_{\text{id}} = \frac{RN_m f}{1 + \sqrt{-\frac{\ln(1-P_{\text{id}})}{0{,}005}}}, \tag{7}$$

where H_{det}, H_{rec}, H_{id} are corresponding values of the UAV flight altitude, which satisfy the probabilities of the objects detection, recognition, and identification based on received aerial survey video data.

Thus, the approach to calculating the UAV flight altitude for carrying out surveillance missions (taking into account the set probabilities of the objects detection, recognition, and identification) has been proposed.

8.4 Earth Remote Sensing Data Processing

Processing of images means, generally, the processing of digital information, as it is currently the basic format of the data to be received from the IR cameras. Digital data is presented as two-dimensional images in IR range [13]. These images may be conceived as the arrays of radiant temperature $T(i, j)$, each of their values represents the intensity of the radiation, received by the sensor from the single element of the surface, to which one image pixel conforms (see Table 8.1).

Table 8.1 Two-dimensional array structure of the digital image

$T(1, 1)$	$T(1, 2)$	…	$T(m, 1)$
$T(1, 2)$	$T(2, 2)$		$T(m, 2)$
…			…
$T(1, n)$	$T(2, n)$	…	$T(m, n)$

The image consists of $n * m$ pixels, each pixel has (i, j) coordinates—the row number and the column number. UAV remote surveillance using IR detector allows to receive determined values of radiant temperature on the surface of the area. Accordingly, formalized problem statement related to detection, recognition, and identification of the objects using UAV multispectral aerial surveillance data is as follows:

$$\hat{\mho} = U\left(\hat{\omega}, \hat{\omega}'\right), \tag{8}$$

where $\hat{\mho}$ is the array of estimations of the multispectral image pixel's relation to the same class of materials, of which the technogenic objects or the background are composed; U is the decision function, based on test of differences for the set of classes; $\hat{\omega}, \hat{\omega}'$ are classes of materials in IR and visible ranges.

Radiant temperature of each pixel is represented as a fixed value $T_r[m, n : k]$, where m, n are spatial intervals, k are time intervals, so that $m \in 1, M;\ n \in 1, N;\ k \in 1, K$. Since the two-dimensional space T_r is time-bound (has the time reference), we may present $\tilde{T} = W(T_{r1}, T_{r2}, \ldots, T_{rk})^T$ as the cuboid of IR images obtained via an operator transformation W of the radiant temperatures into thermodynamic ones for each of the IR image pixels. This transformation is based on the following system of equations:

$$\begin{cases} T_r^4 = \varepsilon \tilde{T}_{\text{td}}^4 + (1 - \varepsilon) T_A^4 \\ \tilde{T} = \tilde{T}_{\text{td}} - \tilde{T}_{\text{td min}} \end{cases}, \tag{9}$$

where $\tilde{T}_{td}$ is the value of thermodynamic temperature of the inspected area surface, calculated using IR detector measurement data T_r with an error δ_r; $\tilde{T}_{\text{td min}}$ is the minimal value of the thermodynamic temperature of the inspected area's surface; $\tilde{T}$ is surface temperature of the inspected area; ε is the radiation capacity factor for the spectral range of the IR detector; T_A is the value of the near-surface temperature of the air.

It is admitted that $\tilde{T}$ is presented only on some subset of Ω, its characteristic function is defined as μ_0.

As a result, the IR images cuboid reduction problem will be solved via minimizing the functional:

$$J(T_r, Q|\lambda) = \frac{\zeta}{2} \int_{\tau 0}^{\tau 1} \int \Omega ||Q(\Psi_1|\lambda) - \hat{Q}^{(0)}|| d\Omega d\tau + \\ + \frac{\zeta}{2} \int_{\tau 0}^{\tau 1} \int \Omega \mu_0 ||T\left(\Psi_{2,Q}|\lambda|_{z=0}\right) - W(T_r)|| d\Omega d\tau, \tag{10}$$

where $Q^{(0)} = q_E^{(0)} + q_v^{(0)}$ is the total quantity of radiation and convection heat; $W(T_r) \equiv \tilde{T}$, $Q^{(0)} = \hat{Q}^{(0)}(x, y, \tau)$ is the set function; ψ is the vector of boundary problem parameters; $\zeta = \text{const} > 0$ is the regularization parameter.

In case of $\zeta > 0$ inequality, the IR image reduction problem has a single solution. The existence of the optimal solution is evident from the classical results of the extreme value theory and is continuous in the Q stream (flow). In case of $\zeta > 1$ inequality, the problem may not always have a solution, but there is a sequence of regularized solutions, that are minimizing the functional (9).

The resulting estimates of the two-dimensional spatial distribution of effective heat conductivity $\hat{\lambda}$ on the surface of the area (*x*, *y*) or the conjugate layer *hl*(*x*, *y*) are additional signatures of stealthy objects. Co-processing of these signatures together with aerial survey images taken in other optical ranges enhances effectiveness of the objects searching mission. A set of two-dimensional distributions of the conjugate layers forms the image of two-dimensional spatial distribution of effective heat conductivity for the ground (soil) of the survey area.

Now we are about to analyze the problem of classification of stealthy objects using additional revealing signature—the effective heat conductivity values for different materials.

Considering that each class ω_g has known conditional probability density function of the heat conductivity values $\lambda p(\lambda|\omega_g)$ and the prior probability $p(\omega_g)$ of class ω_g occurrence, the decision function is being set according with the maximum likelihood $p(\omega_g) \geq p(\lambda|\omega_f * p(\omega_f)$ for $g, g \in \overline{1, \zeta}$.

We use likelihood function $L\left(\hat{\lambda}|\omega\right) : \Xi \to R$ (where $\hat{\lambda} \in R$) for ω parameter highest likelihood estimation and point estimation of the pixel affiliation to the ω_g class on two-dimensional heat conductivity distribution images.

Based on the above-mentioned provisions, the algorithm of processing the multispectral images using UAV aerial surveillance data may be formed. It includes the decision rule concerning the affiliation of the pixel (on visible range image and two-dimensional heat conductivity distribution image) to the $\mho$ class. Assuming that each image pixel may belong to the ω class (for IR wavelength range) and the ω' class (for visible wavelength range), the decision rule U has the following structure:

1. Initial conditions rule in processing the whole area of two-dimensional heat conductivity distribution image. If the location of the class of the material (object) in visible range image and the two-dimensional heat conductivity distribution image has the value that is below the threshold level ($\hat{\omega} < k_{\text{thres}}$), and the simplified medium mathematical model (one-dimensional anisotropic medium $\lambda(\theta) = \lambda = \text{const}$: $C_p \frac{dt}{d\tau} = \lambda \frac{d^2T}{dz^2}$ with $\lambda \frac{dT}{dZ} - B^0 = q_E$ boundary condition) is used, then the $\mho = \hat{\omega}$ equation is true.
2. Decision making. If $\hat{\omega} \geq k_{\text{thres}}$ inequality is valid, then the simplified multilayer (orthotropic) medium mathematical model is used. A hypothesis for the existence of hidden subsurface object under the soil's upper layer shall be accepted, and there may be two outcomes:

   - The subsurface object belongs to the "thermal insulator" class of materials;
   - the subsurface object belongs to the "heat conductor" class of materials.

8.5 Experimental Results

Remote aerial surveys have been conducted as the objects of the full-scale experiment using a multicopter UAV at the altitude of 100 m within 24 h and in 2-h intervals. The electro-optical system comprised the *Sony EXMOR 1/2.3* optical sensor and the *Flir Tau2* IR detector. Twelve IR images (corrected via photogrammetric methods based on points of control for inspected areas) have been formed in order to solve the reduction problem [14, 15].

The luminance distribution (range) for the territory shows that the sequential solution of the problem (8) allows to divide the objects into material classes according to a new indicator—the value of the effective heat conductivity.

Figures 8.1 and 8.2 represent two-dimensional distributions for the estimations of the affiliation of the multispectral image pixel to the same class (in accordance with the set classifier): styrofoam, timber, precast monolithic slab, mass concrete, water, soft ground, hard ground, sunbaked ground with dry grass, sunbaked ground with green grass, wet ground with green grass, sand, and iron.

The proposed algorithm may serve as an alternative or an addition to formerly known methods of objects detection, based on thermal contrast analysis or color clustering. Despite the existence of similar approaches to processing multispectral image data (based on synthesis of IR and other ranges images), the presented algorithm does not have a severe disadvantage, related to ignoring stealthy objects due to UAV's invalid flight altitude. A considerable benefit of the described multispectral image-processing algorithm is its possible software implementation using any known object-oriented programming framework and the applicability for major available workstations.

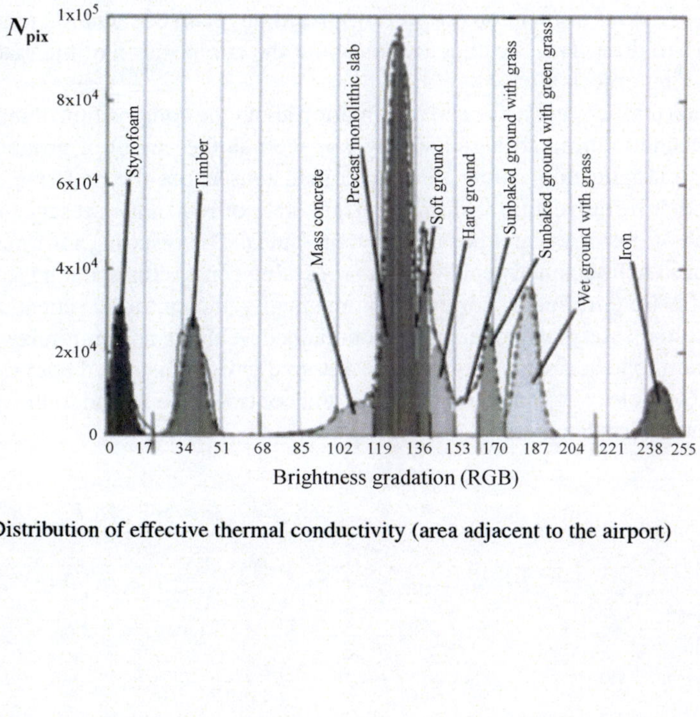

Fig. 8.1 Distribution of effective thermal conductivity (area adjacent to the airport)

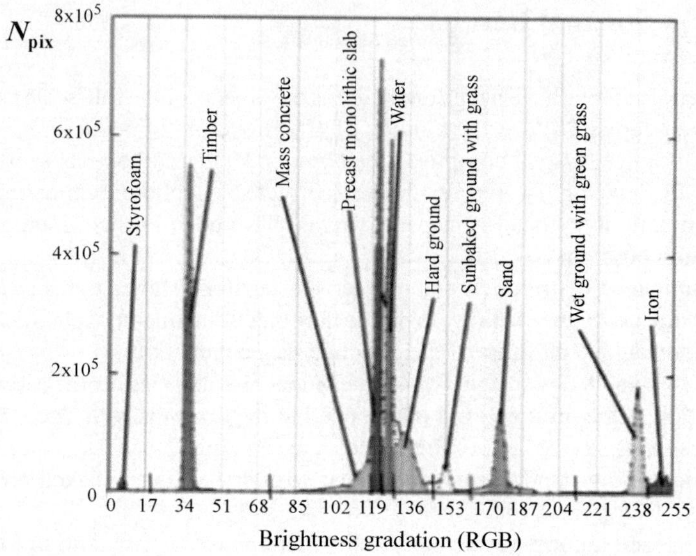

Fig. 8.2 Distribution of effective thermal conductivity (area adjacent to the bridge)

8.6 Conclusion

The article proposes a complex approach that can be used during preparation and conduct of multispectral aerial survey missions of the terrain using UAVs. The optimal flight altitude and camera resolution parameters are taken into account. This makes detection, recognition, and identification of monitoring objects effective. The algorithm of data classification (clustering) based on received radiation temperature map allows to draw accurate conclusions about the composition of the materials of the monitoring objects.

Radiation temperature maps and information about the composition of monitoring objects obtained during IR aerial surveys, as well as the output of ground objects clustering (classification) allow to draw conclusions about the real size of buildings and other infrastructure facilities, current state of roads, the presence or recent appearance of obstacles, and possible detour routes. The resulting information can be transferred to unmanned vehicles data processing centers for updating terrain and ground infrastructure maps, correcting unmanned vehicles traffic routes, etc. This will ensure the safety of movement of unmanned vehicles and minimize possible damage from road accidents. Data can be automatically transmitted both directly to unmanned vehicles automated information and control systems and to the operators of traffic monitoring centers.

As we pursue our work, we plan to apply neural networks and machine learning technologies for analyzing radiant temperature maps and the composition of monitoring facilities in order to obtain a verified list of objects, their real size and dimensions. It is also planned to build a thermal imaging navigation system for UAVs like [16].

Acknowledgements This research is supported by the RFBR Project No. 19-29-06044.

References

1. Meshcheryakov, R.V., et al.: An application of swarm of quadcopters for searching operations. IFAC-PapersOnLine **52**(25), 14–18 (2019)
2. Ischuk, I.N., Filimonov, A.M., Dolgov, A.A., Stepanov, E.A., Tyapkin, V.N.: Algorithm for the joint processing of multispectral image data by aerial shooting with drones. Ind. Autom. Control Syst. Controll. 27–34 (2018)
3. Tishcenko, A.I., Ischuk, I.N., Gromov, Y.Y.: Determination of the flight altitude unmanned aerial apparatus for performing video surveil lance is required the degree of detail of information. Inf. Sensor Syst, Thermophys. Res. **2**, 190–193 (2018)
4. Deng, H., Sun, X., Liu, M., Ye, C., Zhou, X.: Infrared small-target detection using multiscale gray difference weighted image entropy. IEEE Trans. Aerosp. Electron. Syst. **52**(1), 60–72 (2016)
5. Silva, H., et al.: UAV trials for multi-spectral imaging target detection and recognition in maritime environment. In: OCEANS MTS/IEEE Monterey, pp. 1–6 (2016)
6. Wang, X., Zhang, K., Yan, J., Yin, J.: Analysis and fusion algorithm of weak target based on infrared dual-band. In: IEEE 4th International Conference on Computer and Communications (ICCC), pp. 1730–1735 (2018)
7. Wang, P., Wang, W., Wang, H.: Infrared unmanned aerial vehicle targets detection based on multi-scale filtering and feature fusion. In: 3rd IEEE International Conference on Computer and Communications (ICCC), pp. 1746–1750 (2017)
8. Lahouli, I., Chtourou, Z., Haelterman, R., De Cubber, G., Attia, R.: A fast and robust approach for human detection in thermal imagery for surveillance using UAVs. In: 15th International Multi-Conference on Systems, Signals & Devices (SSD), pp. 184–189 (2018)
9. Shetty, A., Disha, B.: Detection and tracking of a human using the infrared thermopile array sensor—"Grid-EYE". In: 2017 International Conference on Intelligent Computing, Instrumentation and Control Technologies (ICICICT), pp. 1490–1495 (2017)
10. Setjo, C., Achmad, B.: Thermal image human detection using Haar-cascade classifier. In: 7th International Annual Engineering Seminar (InAES), pp. 1–66 (2017)
11. Tan T., Teoh, S., Fow, E., Yen, K.: Embedded human detection system based on thermal and infrared sensors for anti-poaching application. In: IEEE Conference on Systems, Process and Control (ICSPC), pp. 37–42. Bandar Hilir (2016)
12. Maxwell, J.C.: A treatise on electricity and magnetism, 2nd edn, pp. 68–73. Oxford, Clarendon (1892)
13. Romanova, M.A., et al.: Simulation of thermal fields in an anisotropic alternating saturated porous medium for environmental monitoring tasks using UAV. In: Proceedings of the 12th International Conference "Management of Large-Scale System Development" (MLSD), pp. 1–3. IEEE (2019)
14. Ishchuk, I.N., Tyapkin, V.N., Dolgov, A.A., Bebenin, A.A.: Method of classification of technogenic objects on the basis of construction of multilayer thermal tomograms. Inf. Sensor Syst. Thermophys. Res. **1**, 251–256 (2018)

15. Gromov, Y.Y., Ishchuk, I.N., Alekseev, V.V., Didrikh, V.E., Tyutyunnik, V.M.: Information support of finding a solution the problem of hidden objects. J. Theoret. Appl. Inf. Technol. **95**(3), 615–620 (2017)
16. Xaud, M.F.S.; Leite, A.C.; From, P.J.: Thermal image based navigation system for skid-steering mobile robots in sugarcane crops. In: International Conference on Robotics and Automation (ICRA), pp. 1808–1814 (2019)

Chapter 9
Integrated Sensor System for Controlling Altitude–Velocity Parameters of Unmanned Aircraft Plane Based on the Vortex Method

Elena Efremova and Vladimir Soldatkin

Abstract The paper considers the original scheme, algorithms for the formation and processing of time–frequency primary informative signals, and determination of the altitude–velocity parameters of unmanned aircraft planes in channels of integrated sensor system with one receiver of primary information. The competitive advantages of the offering sensor system for controlling the altitude–velocity parameters are provided, which determine the prospects for its use on unmanned aircraft planes of various classes.

9.1 Introduction

Flights of large kind of unmanned aircraft planes (AP) occur in the limits of the atmosphere. For controlling of AP and support the fulfillment of flight tasks, reliable information about barometric altitude, indicated speed, components of true airspeed vector, other altitude–velocity parameters which determine the displacement of the AP relative to the surrounding air environment is required [1, 2].

9.2 Formation of Primary Information on the Basis of the Vortex Method

Traditional sensor systems for controlling altitude—velocity parameters of unmanned AP are based on the aerometric, aerodynamic and vane methods. They contain pitot-static tubes (PST), receiver of stagnation temperature (ST), and vane sensors of aerodynamic angles (SAA) installed on the right and left craft and carried out in the incoming air flow [3]. The use of these receivers and sensors of primary

E. Efremova · V. Soldatkin (✉)
Kazan National Research Technical University Named After A.N. Tupolev-KAI, 10 Karl Marks Street, 420111 Kazan, Russian Federation
e-mail: w-soldatkin@mail.ru

A. Ronzhin and V. Shishlakov (eds.), *Proceedings of 15th International Conference on Electromechanics and Robotics "Zavalishin's Readings"*, Smart Innovation, Systems and Technologies 187, https://doi.org/10.1007/978-981-15-5580-0_9

information for controlling of altitude–velocity parameters of unmanned AP is the cause of a significant complication, increase in weight and cost of the sensor system. In addition, the input informative parameters of sensory systems, implementing aerometric, aerodynamic, and vane methods are amplitude signals in the form of magnitude of the pressure, pressure differential, resistance of the thermistor and the angle of rotation of the weather vane, the conversion of which into electrical signals accompanied by additive and multiplicative errors. Reduction of this error also complicates the sensor system of unmanned AP.

The possibility to reduce the errors when the allocation and conversion of time–frequency primary signals [4] and obtain all output signals with use a single receiver in digital form determines the prospects of sensor system for controlling altitude–velocity parameters of unmanned AP with use of the vortex method [5].

The vortex method is based on the empirical relationship between the frequency f of formation, disruption of vortices, and formation of Karman's vortex paths behind poorly streamlined bodies installed in the incoming gas or liquid flow [5]:

$$f = \frac{\mathrm{Sh}}{l} V, \tag{1}$$

where Sh—the Struhal's number. Karman's vortex paths occur in limit of velocities V of incoming flow, determined by interval $10^3 < \mathrm{Re} < 3 \times 10^5$ Reynold's numbers $\mathrm{Re} = \frac{V}{\nu l}$, where ν—kinetic viscosity of gas or liquid. In this case, Karman's vortex paths are formatted only at subsonic velocities of the incoming air flow.

Frequency f of vortex formation behind plate with width l and thickness h set at angle φ to the direction of incoming air flow will be determined by expression in form:

$$f = \mathrm{Sh} \frac{V}{l \sin \phi + h \cos \phi}. \tag{2}$$

For poorly streamlined body in form of wedge-shaped pyramid, with size of base l is set counter to incoming air flow at angle φ, the frequency of vortices will be equal to:

$$f = \mathrm{Sh} \frac{V}{l \sin \phi}. \tag{3}$$

Behind wedge-shaped pyramid, the amplitude of pulsations of velocity $U \approx 0,2\ V$ and the amplitude of pulsations of pressure near the body will be equal to by expression [6]:

$$P_m \approx 0,04 \rho V^2. \tag{4}$$

The resulting expressions allow us to reasonably determine the requirements to the sensitivity threshold, used the pressure pulsation transducers which installed behind

Table 9.1 Pulsations of pressure behind wedge-shaped pyramid for different altitudes and velocities of flight

V, km/h	30	50	100	200	400	800	1000
P_m, Pa							
$H = 0$	4.3	9.3	38	152	608.4	2433	3997
$H = 1000$ m	3.8	8.2	33.7	134	539	2158	3545
$H = 3000$ m	**2.9**	**6.4**	26.2	105	421	1683	2765
$H = 7000$ m	**1.7**	**3.7**	**15.4**	**61**	247	987	1622
$H = 11{,}000$ m	**1.5**	**3.3**	**13.8**	**55**	222	888	1458

streamlined body and in vortex path, into an electrical signal. We need this information for building of selection channels, transformation and information processing when building vortex sensor of aerodynamic angle and true airspeed.

Amplitude of pressure pulsations P_m calculated for different altitude and velocities of incoming air flow to wedge-shaped pyramid is shown in Table 9.1.

Bold combination altitude and velocities of flight for manned and unmanned subsonic aircraft plane are not working.

The use of the vortex method allowed the authors to develop an original design-functional diagram of the sensor system for controlling the altitude–speed parameters of an unmanned AP with a single integrated receiver of primary information, shown in Fig. 9.1 [7].

The integrated sensor system of altitude–velocity parameters of the unmanned AP contains two wedge-shaped pyramids 1, the bases of which located orthogonal to each other and counter the incoming air flow. The velocity vector of the incoming air flow $\boldsymbol{V}$ is equal in magnitude and has negative sign to the vector $\boldsymbol{V}_a$ of the true airspeed of unmanned AP. Angle of direction of flow is equal in magnitude and sign of measured aerodynamic angle, for example, the incidence angle α of unmanned AP.

Wedge-shaped pyramids are installed on board of the unmanned AP so that the axis of the wedge-shaped pyramids will be perpendicular to plane of changes of the measured incidence angle α.

When flying of the unmanned AP, the bases of the wedge-shaped pyramids will be located at different angles and $\phi_1 = \phi_0 + \alpha$ and $\phi_2 = \phi_0 - \alpha$ to the direction of the incoming air flow. This leads to different meaning of frequencies f_1 and f_2 of vortex formation behind the wedge-shaped pyramids. On one of the faces of the pyramids, the pneumo-electric converters 2 located, which perceive the pressure pulsations on the rear surfaces of the pyramids. The pneumo-electric converters 2 are connected to the recording devices 3 of frequency, which measure the frequencies f_1 and f_2 of vortex formation behind the wedge-shaped bodies. Measured frequency f_1 and f_2 fed to the input of the processing device 4, in which digital signals (codes) are determined and formed according to the developed algorithms according to the value of the true airspeed V_{B} and aerodynamic angle α.

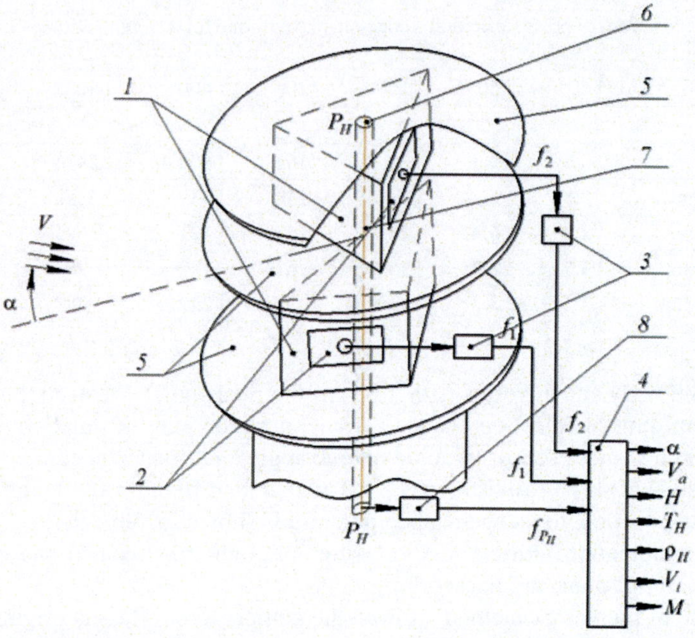

Fig. 9.1 Design-functional diagram of integrated sensor system for controlling the altitude–velocity parameters of unmanned AP based on the vortex method

When the constructive implementation of the integrated sensor system of the unmanned AP the wedge-shaped pyramids are installed coaxially above each other, as shown in Fig. 9.1. To ensure stable vortex formation and eliminate the influence of the skewing of the incoming air flow in a plane is perpendicular to the vertical axis of wedge-shaped pyramids on the lower and upper surfaces of pyramids the straightening vanes 5 are installed which made in the form of thin disks. These disks allocate zones with stable vortex formation behind the wedge-shaped pyramids in the incoming air flow and reduce errors caused by the skewing of the flow in the plane perpendicular to the plane of measurement.

If necessary to reduce the size of the receiver of primary information acting in the incoming flow, the wedge-shaped pyramids can be spaced and located in the measurement plane, while the number of straightening vanes is reduced to two.

The researches have shown [9] that the frequencies of vortex formation behind wedge-shaped pyramids, located by the angle $2\phi_0 = 90°$, are determined by the relations:

$$f_1 = \frac{\text{Sh}}{l}\frac{V_a}{\sin(\phi_0 + \alpha)} = \sqrt{2}\frac{\text{Sh}}{l}\frac{V_a}{\cos\alpha + \sin\alpha};$$
$$f_2 = \frac{\text{Sh}}{l}\frac{V_a}{\sin(\phi_0 - \alpha)} = \sqrt{2}\frac{\text{Sh}}{l}\frac{V_a}{\cos\alpha - \sin\alpha}, \tag{5}$$

where Sh—number Struhal's of wedge-shaped pyramid, l—size of base.

To calculate the true airspeed V_B and aerodynamic angle α with use of integrated sensor system for controlling the altitude–velocity parameters of unmanned AP, the authors developed original processing algorithms of frequencies of vortex formation of the form [8]:

$$V_a = \frac{l}{\sqrt{2}\mathrm{Sh}} \frac{f_1 f_2}{\sqrt{f_1^2 + f_2^2}}; \quad \alpha = \mathrm{arctg} \frac{f_2 - f_1}{f_1 + f_2} \tag{6}$$

To increase the functionality and measuring all altitude–velocity parameters of unmanned AP, the authors proposed [10] on the surface of the upper straightening vane 5 (Fig. 9.1) install the receiver 6 for the perception of static pressure P_H of incoming air flow. Receiver 6 through a pneumatic duct 7 is connected with pneumo-electric sensor 8 of absolute pressure with the frequency output signal. The output of the pneumo-electric sensor 8 in the form of the frequency f_{P_H} is proportional to the static pressure P_H of the incoming air flow connected to the input of the processing device 4. The processing device 4 is made in the form of a computer, which according with developed algorithms by the authors calculating and formatting output digital signals of integrated sensor system for controlling altitude–velocity parameters of unmanned AP.

9.3 Algorithms for Determining the Altitude–Velocity Parameters of the Unmanned Aircraft Plane

The theoretical base for building of algorithms for determining the altitude–velocity parameters of the integrated sensor system for controlling of unmanned AP based on vortex method used the identical bond of true airspeed V_{am}, measured at the output of the computing device, with calculated magnitude of true airspeed V_a which determined by the standard dependence according to GOST 5212-74 [11], using expression of form:

$$V_a = \sqrt{2gRT_H\left(\frac{k}{k-1}\right)\left[\left(\frac{P_d}{P_H}+1\right)^{\frac{k}{k-1}} - 1\right]} = \sqrt{2gRT_H\left(\frac{k}{k-1}\right)\left[\left(\frac{P_t}{P_H}\right)^{\frac{k-1}{k}} - 1\right]}, \tag{7}$$

where g—gravity acceleration; R—gas constant; k—ratio of specific heats for air; P_d—dynamic pressure (velocity head) incoming air flow; P_t—total pressure of incoming air flow; ρ_H—density of air at altitude of flight H, which is determines as [12]:

$$\rho_H = \rho_0 \frac{P_H T_0}{P_0 T_H}, \tag{8}$$

where ρ_0, P_0, and T_0—density of air, absolute pressure, and absolute temperature on altitude $H = 0$.

If the measured value of the true airspeed V_{am} by the integrated sensor system will be equated to the calculated value V_{B} in accordance with GOST 5212-74, we obtain an expression of the form:

$$V_{\text{am}} = \sqrt{2gR\left(\frac{k-1}{k}\right)T_H\left[\left(1 + \frac{\rho_0 T_0}{2P_0 T_H}V_{\text{am}}^2\right)^{\frac{k-1}{k}} - 1\right]}. \tag{9}$$

The resulting expression allows to establish an unambiguous implicit relationship between the measured value of the true airspeed V_{am} and absolute ambient temperature T_H outer air at the altitude of flight H at current flight altitude, determined by the aspect ratio:

$$T_H = T_0 - \tau H = \frac{V_{\text{am}}^2}{2gR\left(\frac{k}{k-1}\right)\left[\left(1 + \frac{\rho_0 T_0}{2P_0 T_H}V_{\text{am}}^2\right)^{\frac{k-1}{k}} - 1\right]}, \tag{10}$$

where τ—temperature gradient of air when changing the altitude H.

Using the ratio (8, 10), we can calculate the density of air ρ_H at altitude H of flight by formula:

$$\rho_H = \rho_0 \frac{P_H T_0}{P_0 V_{\text{am}}^2} \cdot 2gR\left(\frac{k}{k-1}\right)\left[\left(1 + \frac{\rho_0 T_0}{2P_0 T_H}\right)^{\frac{k-1}{k}} - 1\right] \tag{11}$$

Then, in accordance with GOST 5212-74, we can calculate the instrument speed of flight of an unmanned AP by formula:

$$V_{np} = \sqrt{2gRT_0\left(\frac{k}{k-1}\right)\left[\left(1 + \frac{\rho_0 T_0}{2P_0^2}\frac{P_H}{T_H}V_{\text{am}}^2\right)^{\frac{k-1}{k}} - 1\right]}. \tag{12}$$

Using the information from integrated sensor system, we can calculate the Mach number characterizing the ratio of the true airspeed of flight of the unmanned AP to the velocity of sound a_H at current altitude of flight by the form:

$$M = \frac{V_{\text{am}}}{a_H} = \sqrt{\frac{2}{k-1}\left[\left(1 + \frac{\rho_0 T_0}{2P_0 T_H}V_{\text{am}}^2\right)^{\frac{k-1}{k}} - 1\right]}. \tag{13}$$

By information about static pressure P_H perceived on the upper surface of streamlined, using standard dependence to GOST 4401-81, the current absolute barometric altitude of the unmanned AP is calculated according to the formulas [13].

– at -2000 m $< H <$ 11,000 m:

$$H = \frac{T_0}{\tau}\left[1 - \left(\frac{P_H}{P_0}\right)^{\tau R}\right]; \tag{14}$$

– at 11,000 m $\leq H <$ 15,000 m:

$$H = H_{11} + RT_{11}\ln\frac{P_{11}}{P_H}, \tag{15}$$

where P_{11}—absolute pressure at altitude $H_{11} = 11{,}000$ m.

Vertical velocity V_y of unmanned AP is determining by calculating of time derivative from absolute barometric altitude using the ratios:

$$V_y = \frac{dH}{dt} = \frac{H(t_i) - H(t_{i-1})}{t_i - t_{i-1}};$$

$$\text{or } V_y = \frac{1}{4\tau}[H(t_i) - H(t_i - 2\tau) + H(t_i - \tau) - H(t_i - 3\tau)], \tag{16}$$

where t_i, t_{i-1}—the current and previous time points at which the absolute barometric altitude is calculated; $\tau = t_i - t_{i-1}$—fixed time interval.

Analysis of instrumental and methodic errors of integrated sensor system for controlling of altitude–velocity parameters of unmanned aircraft plane on the basis of vortex method is necessary to carry out using the next methodic.

For barometric altitude channel of integrated sensor system for controlling of altitude–velocity parameters of unmanned aircraft plane on the basis of vortex method, ignoring the error of processing channel, excluding the calculation error, the instrumental error of channel will be determined by the instrumental error of used absolute pressure sensor.

Then expression to estimate of instrumental error of barometric altitude measuring channel of integrated sensor system for controlling of altitude–velocity parameters of unmanned aircraft plane on the basis of vortex method, using expression (14), will have form:

$$\Delta H_i = \frac{dH}{dP_H}\Delta P_{Hi} = \pm\frac{T_0 R}{P_0^{\tau R}}\frac{\Delta P_{Hi}}{P_H^{1-\tau R}}, \tag{17}$$

where ΔH_i and ΔP_{Hi}—instrumental errors of barometric altitude channel and absolute pressure sensor.

– At ΔP_{Hi}, P_0 and P_H, in Pa—will have:

$$\Delta H_i = \frac{941}{P_H^{0.80973825}} \Delta P_{Hi}.$$

– At ΔP_{Hi}, P_0, and P_H, in mm Hg:

$$\Delta H_i = \frac{2387,5}{P_H^{0.80973825}} \Delta P_{Hi}.$$

For domestic sensor of absolute pressure with frequency output-type PSFE with permissible relative error 0.01% below altitudes 11,000 m, the instrumental error of barometric altitude channel of integrated sensor system for controlling of altitude–velocity parameters of unmanned aircraft plane on the basis of vortex method does not exceed ±5–10 m.

To estimate the instrumental error ΔT_{Hi} determination of outdoor air temperature of integrated sensor system for controlling of altitude–velocity parameters of unmanned aircraft plane on the basis of vortex method, using the ratio [13] $T_H = T_0 - \tau H$, can use expression in form:

$$\Delta T_H \leq \tau \Delta H_i \leq \pm(0,0325 \ldots 0,065) K.$$

Instrumental error of vortex sensor of the aerodynamic angle and true airspeed, used in integrated sensor system for controlling of altitude–velocity parameters of unmanned aircraft plane on the basis of vortex method, does not exceed the value $\Delta V_{Ai} = \pm\ 0.8$ m/s—in channel of true airspeed; $\Delta \alpha_i = \pm\ 0.13°$—in aerodynamic angle channel [10].

Instrumental error ΔV_{ii} of indicated speed channel of integrated sensor system for controlling of altitude–velocity parameters of unmanned aircraft plane on the basis of vortex method can be estimated by expression:

$$\Delta V_{ii} = \sqrt{\Delta_H} \Delta V_{ai} = \sqrt{\frac{P_H T_0}{P_0 T_H}} \Delta V_{ai}. \quad (18)$$

Substituting the numerical values of parameters, the expression for error of indicated speed channel has a form:

$$\Delta V_{ii} = \sqrt{\frac{288.15 P_H}{101325 T_H}} \Delta V_{ai} = \sqrt{0.284 * 10^{-2} \frac{P_H}{T_H}} \Delta V_{ai}. \quad (19)$$

As shown by the calculations when value of instrumental error $\Delta V_{a\,i} = \pm\,2.8$ km/h (0.8 m/s) at altitude of $H = 0$—$\Delta V_{ii} \leq 0.8$ m/s (2.8 km/h), at altitude of $H = 11{,}000$ m $\Delta V_{ii} \leq \pm\ 0.24$ m/s (0.86 km/h).

Similarly, the instrumental error of Mach number channel of integrated sensor system for controlling of altitude–velocity parameters of unmanned aircraft plane on the basis of vortex method, according to expression $M = \frac{V_a}{\sqrt{2gRT_H}}$, can be estimated as:

$$\Delta M_i = \frac{dM}{dV_a}\Delta V_{a\,i} + \frac{dM}{dV_a}\Delta T_{Hi} = \frac{1}{\sqrt{2gRT_H}}\Delta V_{ai} + \frac{V_a}{\sqrt{2gRT_H}}\Delta T_{Hi} \quad (20)$$

As shown by the calculations when $V_a = 300$ m/s, $\Delta V_{ai} = 2.8$ km/h (0.8 m/s), $\Delta T_{Hi} = \pm\, 0.0325$–$0.065$ K for altitude $H = 0$ $\Delta M_i = \pm\, 0.002$, for altitude $H =$ 11,000 m, $\Delta M_i \leq \pm\, 0.0013$.

Therefore, instrumental errors of instrumentation channels of integrated sensor system for controlling of altitude–velocity parameters of unmanned aircraft plane on the basis of vortex method correspond to the instrumental errors of traditional air data systems built on basis aerometric, aerodynamic, and vane methods.

Besides, the reasons of appearing of methodic errors of instrumentation channels of integrated sensor system for controlling of altitude–velocity parameters of unmanned aircraft plane on the basis of vortex method are the same as for traditional systems. The methodic errors are determine by results of flight tests on a specific type of unmanned subsonic AP for a specific installation site of the vortex sensor of aerodynamic angle and true airspeed.

9.4 Conclusion

Thus, integrated sensor system for controlling altitude–velocity parameters of unmanned aircraft plane on the basis of vortex method according to the developed algorithms allows to determine all altitude–velocity parameters of the unmanned AP with the help of one fixed receiver of frequency–time primary informative signals.

If compared the considered integrated sensor system for controlling of altitude–velocity parameters of unmanned aircraft plane on the basis of vortex method with traditional systems for controlling altitude–velocity parameters which implementing aerometric, aerodynamic, and vane methods, the considered integrated sensor system has the following competitive advantages and dignities:

- Using vortex method, shown integrated sensor system for controlling of altitude–velocity parameters of unmanned aircraft plane simultaneous calculation all altitude–velocity parameters of movement of the unmanned AP.
- Using vortex method, shown integrated sensor system for controlling of altitude–velocity parameters of unmanned aircraft plane provides control of all altitude–velocity parameters of the unmanned AP using the one fixed small-sized receiver of primary information with virtually without distortion of the aerodynamics of the AP.

- Using vortex method shown integrated sensor system for controlling of altitude–velocity parameters of unmanned aircraft plane provides control of all altitude–velocity parameters of movement of the unmanned AP by the time–frequency primary informative signals, what allows to reduce errors and simplify the implementation of conversion and processing channels, to reduce the prime cost of the integrated sensor system for controlling the altitude–velocity parameters of the unmanned AP based on the vortex method.

These advantages and competitive advantages of the integrated sensor system for controlling of altitude–velocity parameters of unmanned aircraft plane determine the prospects for its application on unmanned aircraft planes of various classes and purposes.

Acknowledgements This work is supported by RFBR projects no. 18-08-00264 and no. 18-38-00094.

References

1. Yankevich, Y.: The use of unmanned aircraft systems for civilian purposes. Aerosp. Courier **6**, 55–57 (2006)
2. Moiseev, V.S., Guschina, D.S., Moiseev, G.V.: Fundamentals of the theory of creation and application of information unmanned aerial systems. Kazan: Publishing house Ministry of education and science Tatarstan Republic, 196 p. (2010)
3. Alekseev, N.V., et al.: The system of measuring air signals of a new. Aerosp. Instr. Mak. **8**, 31–36 (2003)
4. Novitskiy, P.V., Knoring V.G., Gutnikov, V.V.: Digital devices with frequency sensors. L: Energy, 423 p. (1970)
5. Kiyasbeili, A., Sh. Perel'shtein, M.E.: Vortex measuring instruments. M. Eng. 152 p. (1972)
6. Blohintsev, D.I.: Acoustics of inhomogeneous environment. M.: Gostehizdat, 168 p. (1946)
7. Soldatkin, V.M., Soldatkina, E.S.: Patent RF for invention №2506596, WPC G01P 5/00. Vortex sensor of aerodynamic angle and true airspeed. Declared 16.07.2012. Application №202130111/28. Patentee FSBEI HPE «Kazan National Research Technical University named after A.N. Tupolev–KAI» . Published 10.02.2014. Bulletin №4
8. Soldatkin, V.M., Soldatkina, E.S., et al.: Vortex sensor of aerodynamic angle and true airspeed. Russian Aeronaut. **55**(4), 402–407 (2012)
9. Soldatkina, E.S., et al.: System engineering design of a vortex aerodynamic angle and true airspeed sensor. Russian Aeronaut. **56**(3), 291–296 (2013)
10. Soldatkin, V.M., Soldatkina, E.S.: Patent RF for invention №2556760, WPC G01P 5/00. Vortex sensor of aerodynamic angle and true airspeed. Declared 21.04.2014. Application №20114116035/28. Patentee FSBEI HPE «Kazan National Research Technical University named after A.N. Tupolev–KAI» . Published 20.07.2015. Bulletin №20
11. GOST 5212-74. Aerodynamic table. Dynamic pressures and stagnation temperatures of air for flight velocities from 10 to 4000 km/h. Parameters. M.: Publishing house of Standards, 239 p. (1974)
12. Zalmanzon, L.A.: Flow elements of pneumatic devices of control and control. M.: Publishing house AS SSSR, 247 p. (1961)
13. Braslavskiy, D.A.: Instruments and sensors of aircraft plane. M. Eng. 392 p. (1970)

Chapter 10
Synthesis of SimMechanics Model of Quadcopter Using SolidWorks CAD Translator Function

Sergey Jatsun, Boris Lushnikov, Oksana Emelyanova, and Andres Santiago Martinez Leon

Abstract Currently, computer modeling is one of the most important scientific tools for investigating the behavior of complex dynamic systems. The choice of an algorithmic language depends on the simplicity of programming, the form of presentation of the simulation results, and different advantages provided by programs such as MATLAB, libraries of which include SimMechanics Visual Modeling Tool. This article discusses modern approaches to computer modeling of unmanned aerial vehicles (UAVs), describes the integration process of SolidWorks and MATLAB/Simulink environments by implementing a CAD model, created previously in SolidWorks and exported to MATLAB/Simulink. An algorithm for modeling a dynamical model of an UAV-type quadcopter based on PID control strategies has been implemented, and a software for modeling and testing control algorithms for a UAV-type quadcopter has been performed, creating automatic navigation systems and planning the trajectories of a quadcopter UAV.

10.1 Introduction

Nowadays, small-sized unmanned aerial vehicles (UAVs) represent a suitable alternative for monitoring, ground mapping, agricultural and environmental preservation practices, fire detection, transmission power line infrared supervision, etc. in large, dangerous, or remote areas. One of the most preferred configurations among multi-rotor aerial vehicles is a quadcopter due to their high maneuverability and ability to execute complex motion patterns [1–4].

S. Jatsun · B. Lushnikov · O. Emelyanova (✉) · A. S. M. Leon
Department of Mechanics, Mechatronics and Robotics, Southwest State University (SWSU), Kursk, Russian Federation
e-mail: newteormeh@inbox.ru

A. S. M. Leon
e-mail: asml1992@yandex.ru

A. Ronzhin and V. Shishlakov (eds.), *Proceedings of 15th International Conference on Electromechanics and Robotics "Zavalishin's Readings"*, Smart Innovation, Systems and Technologies 187, https://doi.org/10.1007/978-981-15-5580-0_10

One of the most promising fields of research on the development of UAV systems is devoted to the development of novel control techniques; this is a particularly discussed topic in several works of Russian and foreign researchers [5–8].

In order to solve tasks such as autonomous navigation, multi-agent control, real-time mapping, etc., first of all the UAV should achieve a good performance in terms of stability, hover, attitude control, etc. for a robust behavior of the system to external factor [9, 10]. There are leading companies around the world that have developed a variety of tools in the field of computer-aided design (CAD) software such as Simscape/MATLAB, V-Rep, Gazebo, SolidWorks Motion, and others [11–13]. This software is able to provide powerful results simulating any dynamical system, considering its geometry, mass, and inertia properties. Also, this software integrates a good database of elements such as electronic (sensors), electromechanical (motors) actuators, external forces (uncertainly factors), etc., which can be added to the model in use. However, these solutions require good theoretical skills and proficiency in terms of modeling, design, and control of dynamical systems.

In our study, we have used SolidWorks as CAD software for obtaining a 3D model of the UAV-type quadcopter, as well as SimMechanics library, which is a part of Simscape toolbox for model validation, simulation, and control development tasks. SimMechanics uses a block diagram programming method integrated with MATLAB and provides a multibody simulation environment for 3D mechanical systems [14, 15].

This paper is organized as follows: Section 10.2 presents a full description about SolidWorks and SimMechanics/MATLAB capabilities and interaction process; Section 10.3 shows our designed UAV-system-type quadcopter with all its constructive characteristics; Section 10.4 presents the simulation results obtained in this work; Section 10.5 concludes this purpose, such as some notes about future work.

10.2 Basic Concepts of 3D Model Export Process (Integration of SolidWorks and MATLAB/Simulink Environments)

SimMechanics is part of Simscape/MATLAB simulation software package and allows to model complex dynamical systems based on block diagram programming method. Also, SimMechanics is able to interact with other components of the Simulink library, increasing in this way the possibilities of modeling of mechatronic and robotic systems [16–19].

However, it is difficult to model a mechanical system in SimMechanics due to the need of determining some parameters such as the moment of inertia and coordinates of every linked element, etc. With the aim to simplify SimMehcanics modeling process, MathWorks has developed a CAD translator, which provides the creation of dynamical models in the SimMechanics environment based on their solid-state

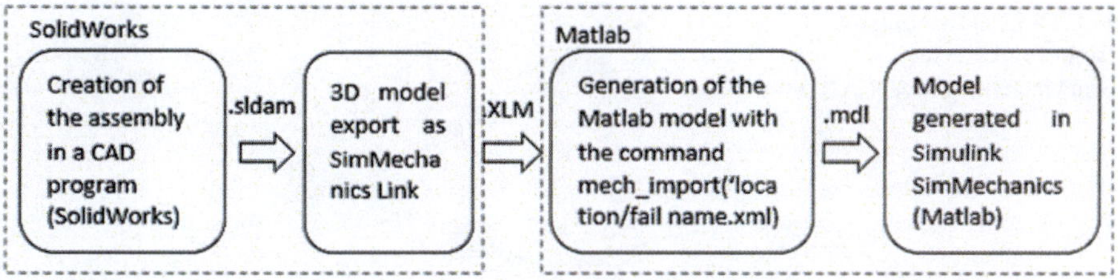

Fig. 10.1 Algorithm of generation of an exported model to SimMechanics/MATLAB

models developed in CAD systems such as SolidWorks using Simscape Multibody Link SolidWorks Plug-In [20–22].

The use of Simscape Multibody Link SolidWorks Plug-In allows us to export the 3D model from SolidWorks to SimMechanics including its properties in a whole form (moment of inertia, dimensions, etc.). At the same time, the performance of the exported model is validated by the CAD system establishing between the elements of the model the right links (kinematical pairs). This approach simplifies the process and extends the possibilities of simulation of any dynamical system [17, 19].

In Fig. 10.1, an algorithm of the export process of a 3D model designed in SolidWorks and its translation to SimMechanics is presented.

10.3 UAV Simulator Design

Simulink is a block diagram environment for multi-domain simulation and model-based design; also it supports a system-level design, simulation, automatic code generation, and continuous test and verification of embedded systems. Simulink provides a graphical editor, customizable block libraries, and solvers for modeling and simulating dynamic systems. Its integration with MATLAB allows to incorporate MATLAB algorithms into models and export simulation results to MATLAB for further analysis [23, 24].

One of the main advantages of using Simulink for the analysis of dynamic systems is related to allow the analysis of complex systems that may be difficult to analyze analytically. Simulink is able to solve any simulation task by using numerically approximate solutions to mathematical models that we are unable to describe or results complicated. Simscape automatically formulates the equations for a physical system and then uses symbolic manipulation and index reduction to identify the mathematical formulation that most efficiently represents the system in developing process [24, 25]. It results that mathematical equations representing a given system serve as the basis for a Simulink model and can be derived from physical laws.

In this paper, a concept of an UAV-type quadcopter has been designed using SolidWorks for a further validation of the developed model using MATLAB/Simulink tools (Fig. 10.2) and implementation of an algorithm of control related to the stabilization

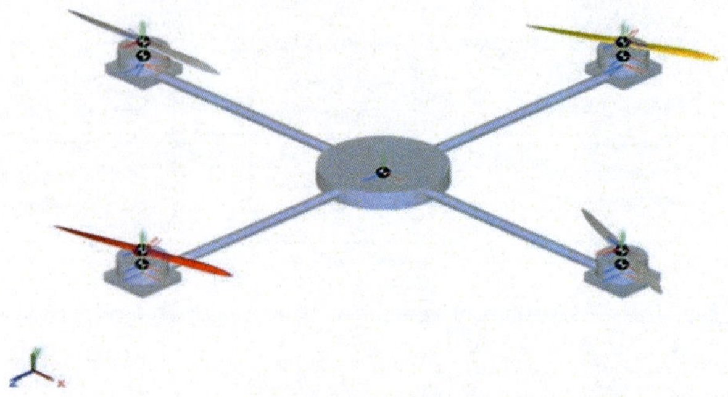

Fig. 10.2 Quadcopter isometric view in SimMechanics/MATLAB

of the quadcopter and its capability to hover at a given altitude to demonstrate the efficiency of the integration of SolidWorks and Simulink/MATLAB environments.

Any assembly consists of individual parts or subassemblies; it makes possible to introduce a control system and perform any motion analysis.

Due to approach the previously mentioned objective, the SimMechanics Link utility has to be installed and linked to SolidWorks. Then, it is necessary to use the SimMechanics Link exporter to create a physical model (XML file) that includes information related to the mass and inertia of each part of the assembly, definitions of constraints between parts, as well as a set of STL (stereo-lithographic) files for representing the geometry of the elements of the assembly. Further these files are received into MATLAB/Simulink and becomes possible the generation of a SimMechanics model [20, 22, 24–27].

The SimMechanics model, obtained after the implementation, includes a block scheme (Fig. 10.3) and allows visualizing the designed quadcopter model (Fig. 10.2).

SimMechanics is composed of a set of block libraries and special simulation interfaces (sensor and actuator blocks) for interconnection of the SimMechanics block diagram with the Simulink environment. The SimMechanics blocks present elements enabling to model mechanical systems consisting of rigid bodies connected by joints that represent translational and rotational degrees of freedom. SimMechanics automatically sets up a single absolute inertial reference frame and coordinate system (CS) called World, [20–23]. To understand the implemented blocks by SimMechanics after exporting the quadcopter model from SolidWorks, their function is detailed in Table 10.1.

The imported SimMechanics model of the quadcopter contains information related to the mass and geometry of the bodies, their center of mass and tensors of the inertia, etc. All these parameters are contained in the before explained blocks. The main properties of each body of the quadcopter model are presented in Table 10.2.

The SimMechanics imported model almost always requires modifications like removing unnecessary constraints between elements of the model or changing their types. Therefore, due to improve block diagram, some blocks can be transformed into subsystems. The modified SimMechanics block diagram after grouping and adding some control functions is shown in Fig. 10.4.

The first block represents the quadcopter airframe body. In Fig. 10.5, its internal

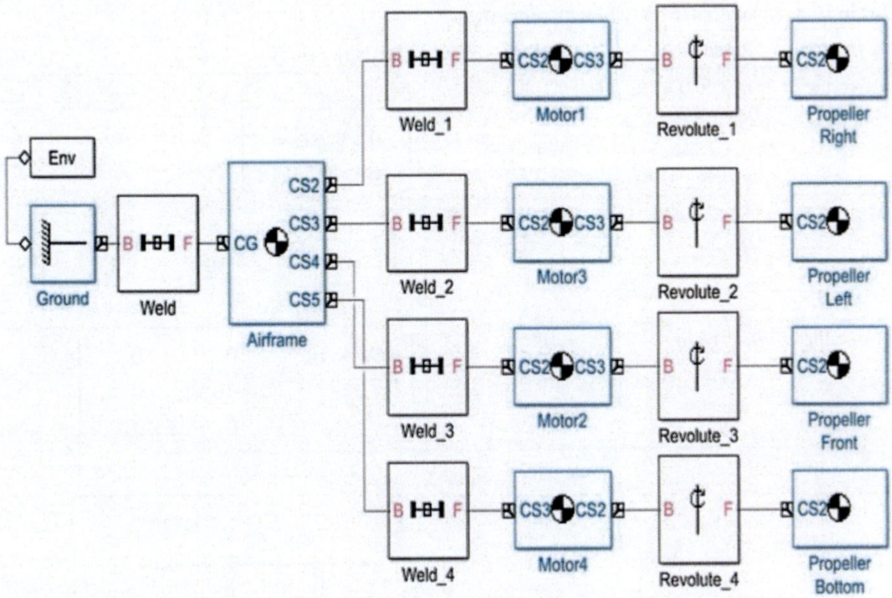

Fig. 10.3 SimMechanics diagram block model of the quadcopter

Table 10.1 Description of the used blocks in SimMechanics

Group	Block	Name	Description
Bodies	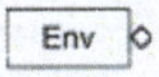	Env	Defines the environment for calculation of the scheme
	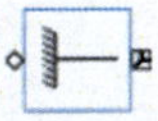	Ground	Represents a fixed point having infinite mass. At least one block ground connected with the machine environment must be involved
		Body	Replaces all fixed rigid bodies among which the degrees of freedom are added. The bodies are defined by their final and nonzero masses, inertia, positions, directions, and by coordinate systems that are connected to them
Joints		Weld	Describes a body without any degree of freedom
		Revolute	The Revolute block from the joints group represents one degree of freedom (rotation)

structure is presented. In this part of the study, it has been implemented a custom joint block between the ground and quadcopter airframe body blocks. A custom Joint block is a composite joint with a specified combination of primitives (prismatic, revolute, or spherical). Thus, the system is able to move in three dimensions, obtaining six degrees of freedom (DOF).

Table 10.2 Quadcopter body blocks properties

Element	Dimensions	Mass (kg)	Inertia (kg•m²)
Airframe	0.45 × 0.45 × 0.03 m	1.8	$\begin{bmatrix} 0.18 & 0 & 0 \\ 0 & 0.36 & 0 \\ 0 & 0 & 0.18 \end{bmatrix}$
Motor	Ø0.028 × 0.19 × 0.015 m	0.2	$\begin{bmatrix} 1.9 \times 10^{-4} & 0 & 0 \\ 0 & 1.9 \times 10^{-4} & 0 \\ 0 & 0 & 1.9 \times 10^{-4}8 \end{bmatrix}$
Propeller	9 × 14 in.	0.01	$\begin{bmatrix} 1.98 \times 10^{-5} & 0 & 0 \\ 0 & 2 \times 10^{-5} & 0 \\ 0 & 0 & 3 \times 10^{-7} \end{bmatrix}$

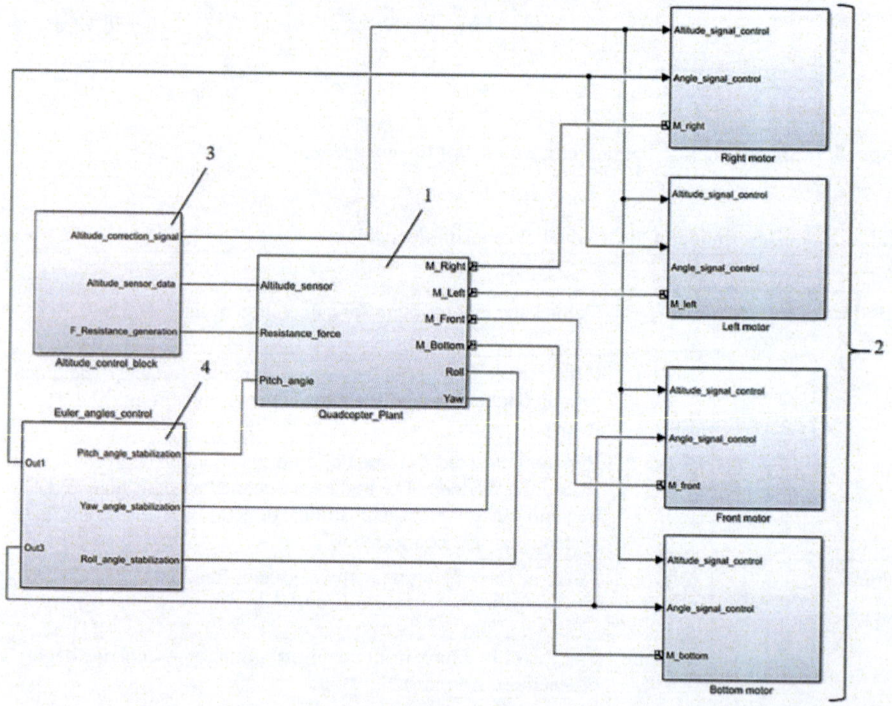

Fig. 10.4 Generalized SimMechanics block diagram of the quadcopter: 1—Quadcopter airframe block, 2—motor blocks, 3—altitude control block, 4—Euler angles stabilization block

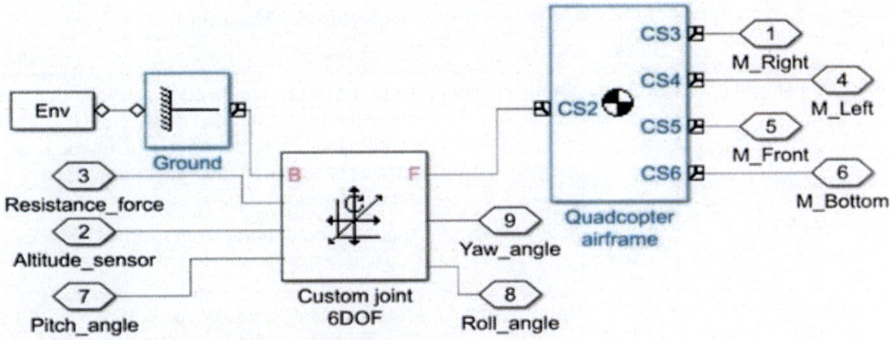

Fig. 10.5 Quadcopter airframe internal structure

Then, the next block is related to the modeling of the motor characteristics of the quadcopter (Fig. 10.6). In this case, our system is composed of four motors. The front and bottom motors rotate clockwise (CW), and the right and left motors rotate counterclockwise (CCW). In terms of their internal structure, they are all identical. With the aim to distinguish the direction of the rotation of the motors, CW motors are going to be represented by a positive value of the generated by them angular velocity, and the CCW motors will be represented by a negative value.

As we could see from Fig. 10.6, inside the motor blocks are located another two subsystems due to generating the necessary angular velocity and momentum on the motors.

Also, the SimMechanics library is equipped with a series of actuators (joint actuators) and sensors (joint sensors) that allow to apply a certain force/momentum or motion to a body, or to realize motion measurements, respectively [20–23]. As part of the elements used in Fig. 10.6, we have implemented some of the blocks already

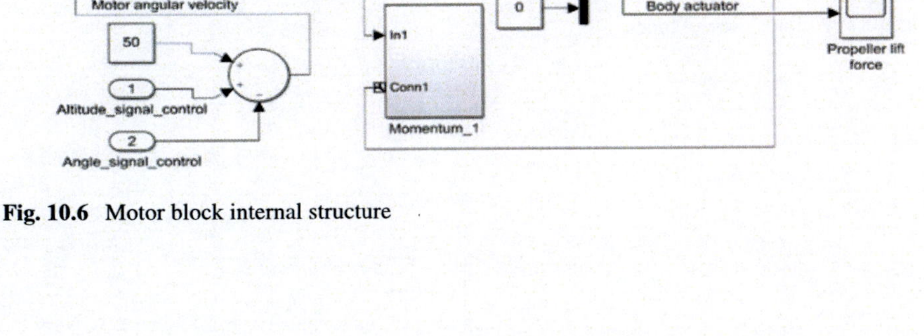

Fig. 10.6 Motor block internal structure

Table 10.3 Description of the use of joint actuators/sensors in SimMechanics

Group	Block	Name	Description
Sensors and actuators		Joint actuator	Applies a force along a prismatic joint or a momentum for rotational motion about a revolute joint, and motion to a prismatic joint in terms of linear/angular position, velocity, and acceleration
		Body actuator	Applies a force/momentum to a body block
		Joint sensor	Measures the linear/angular position, velocity, and acceleration of a joint, and also the reaction force/torque across the joint

above described. For better understanding of them, we present a brief description of them in Table 10.3.

One of the objectives of our study is related to develop an altitude control system, and in this way, the quadcopter must be able to hover at a certain altitude. In order to implement this algorithm of control, a subsystem called "altitude control block" has been implemented, its internal structure in shown in Fig. 10.7.

In order to control and minimize the Euler angles deviation, a stabilization subsystem has been implemented (not shown). In our study, we will consider zero values for the angle deviations around axes *x*, *y*, and z as the desired values for our simulation test.

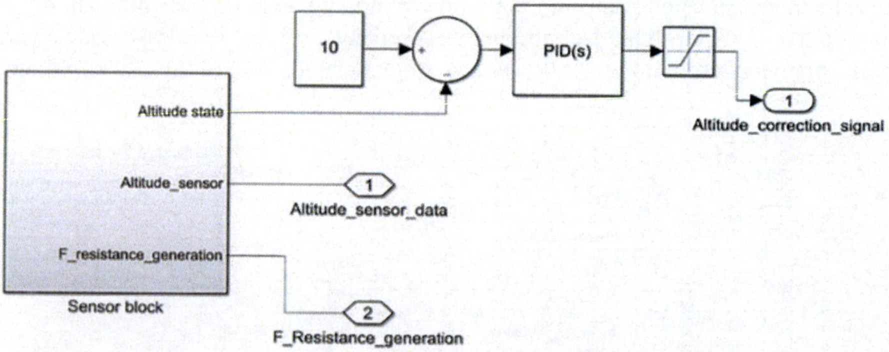

Fig. 10.7 Altitude control block

10.4 Simulation Results

As previously discussed, the aim of this study is related to demonstrate the efficiency of the integration of SolidWorks and Simulink/MATLAB environment, design a CAD model of a UAV-type quadcopter using SolidWorks, and validate the developed model using Simscape tools by implementing an algorithm of control in terms of stabilization and the capability of the quadcopter to hover at a given altitude.

The simulation was carried out taking into account the following boundary conditions as shown in Table 10.4.

Based on the established boundary conditions, the obtained results are presented in Figs. 10.8, 10.9, 10.10, 10.11, and 10.12. In Figs. 10.8, 10.9, the results of the generated angular velocity and lift force by the propellers of the CW motors (top and bottom motors) are presented. The results of the generated angular velocity and lift force by the propellers of the CCW motors (left and right motors) are identical.

In order to demonstrate the accuracy of the designed quadcopter model, the obtained result due to hover the quadcopter at 10 m has been carried out according to the established boundary conditions (Fig. 10.10).

Then, parameters like the linear velocity and acceleration of the center of mass were also obtained as shown in Figs. 10.11 and 10.12.

Table 10.4 Simulation boundary conditions

Parameter	Value	Units
Simulation time	10	s
Motor angular velocity	50	rad/s
Hover stabilization time	<5	s
Hover altitude	10	m
Overshoot	<10	%
Mathematical solver method	Bogacki–Shampine	–

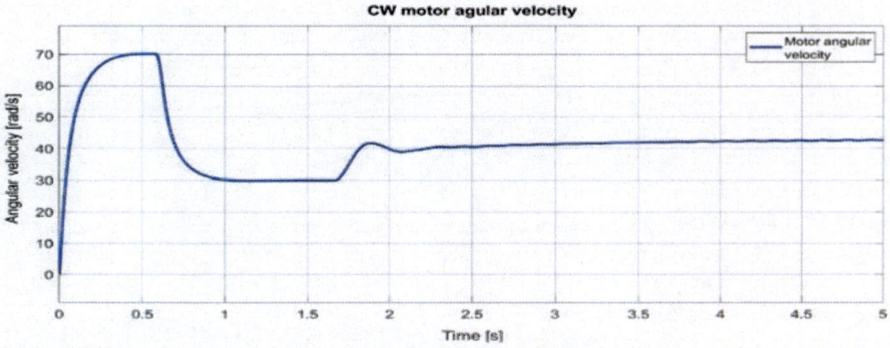

Fig. 10.8 Generated angular velocity by the propellers of the CW motors

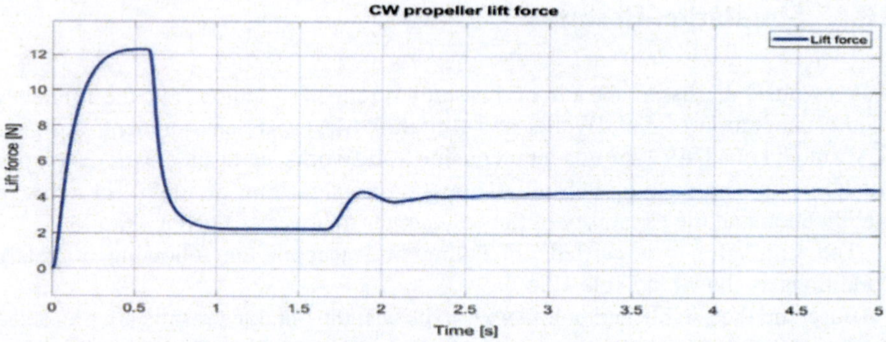

Fig. 10.9 Generated lift force by the propellers of the CW motors

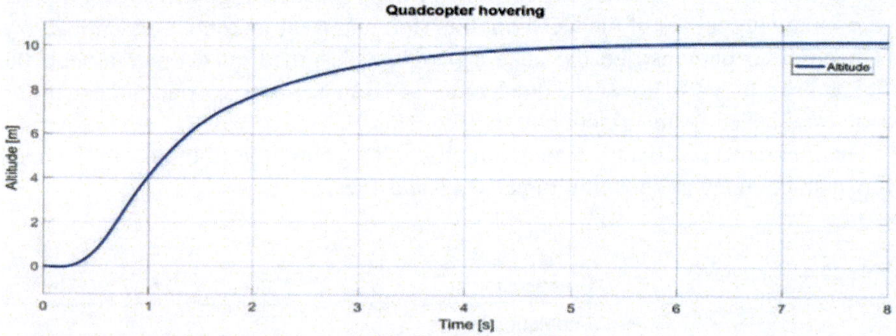

Fig. 10.10 Quadcopter hovering at given altitude

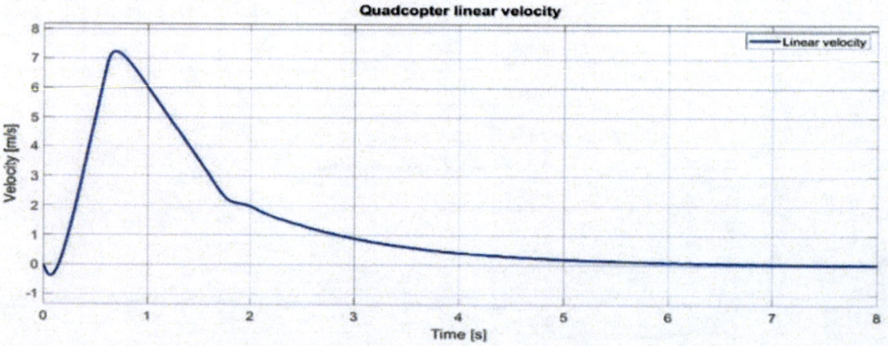

Fig. 10.11 Linear velocity of the center of mass of the quadcopter

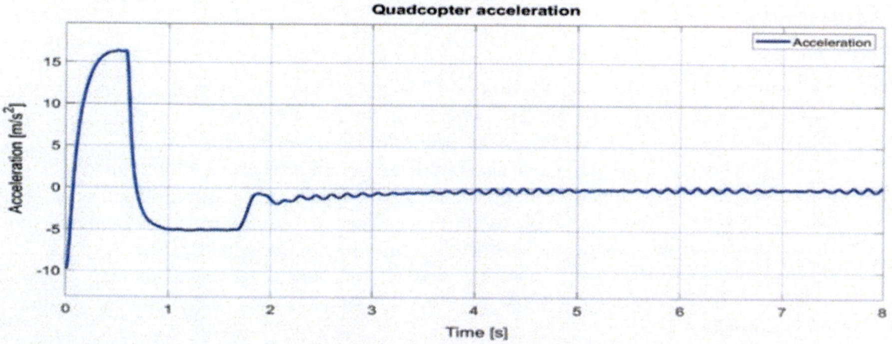

Fig. 10.12 Linear acceleration of the center of mass of the quadcopter

Once the results of the simulation of the UAV-type quadcopter have been carried out, we can conclude that developed system presents a very sensitive characteristic to external factors. In order to achieve the stabilization of the quadcopter, it is necessary to perform a high precise configuration of the PID regulators used in this model during the simulation process. An inaccurate configuration of the regulators can affect significantly to the stability of the quadcopter (roll over or crash).

10.5 Conclusions and Further Work

In this paper, we described the integration of SolidWorks and MATLAB/Simulink environments by implementing a CAD model, created previously in SolidWorks and exported to MATLAB/Simulink. The simulation of a UAV-type quadcopter has been performed in order to evaluate the efficiency of the SolidWorks and MATLAB/Simulink. The implemented algorithms of control based on PID controller method were carried out with the use of MATLAB PID Tuner Tool and Check Step Response block, obtaining high-precision coefficients in order to stabilize the quadcopter and reach the desired position even the presence of some generated perturbations random forces.

Further, there will be improved the model of the quadcopter, implementing different algorithms of control or a combination of them with the aim to obtain better results. Also, there will be extended dynamical model of the quadcopter with the aim to positioning it as a research tool, as well as a teaching tool for the development of control systems and design of UAV systems in general.

Acknowledgements The work was carried out with the financial support of the RFBR (project 18-08-0773A).

References

1. Emelyanova, O.V., et al.: The synthesis of electric drives characteristics of the UAV of "convertiplane-tricopter" type. MATEC Web Conf. **99**, 249–263 (2017)
2. Jatsun, S.F., Emelyanova, O.V., Martinez Leon, A.S.: Design of an experimental test bench for a UAV type convertiplane. IOP Conf. Series: Mater. Sci. Eng. **714**(1), 012009 (2020)
3. Jatsun, S.F. et al.: Investigation of oscillations of a quadcopter convertiplane in transient mode in the vertical longitudinal plane. In: Proceedings of 14th International Conference on Electromechanics and Robotics "Zavalishin's Readings", pp. 345–358 (2020)
4. Fetisov, V. et al.: Continuous monitoring of terrestrial objects by means of duty group of multicopters. In: Proceedings of XX IMEKO World Congress "Metrology for Green Grouth", pp. 9–14 (2012)
5. Zulu, A., John, S.: A review of control algorithms for autonomous quadrotors. Open J. Appl. Sci. **4**, 547–556 (2014)
6. Mellinger, D., Kumar, V.: Control and planning for vehicles with uncertainty in dynamics. In: Proceedings of the IEEE International Conference on Robotics and Automation (ICRA), pp. 960–965(2010)
7. Tang, S., Wüest, V., Kumar, V.: Aggressive flight with suspended payloads using vision-based control. IEEE Robot. Autom. Lett. **2**(3), 1152–1159 (2018)
8. Jatsun, S. et al.: Control flight of a UAV type tricopter with fuzzy logic controller. In: XIII International scientific and technical conference "Dynamics of Systems, Mechanisms and Machines", pp. 1–5 (2017)
9. Mustapa, Z. et al.: Altitude controller design for multi-copter UAV. In: 2014 International Conference on Computer, Communications, and Control Technology (I4CT), pp. 382–387
10. Koch, W., et al.: Reinforcement learning for UAV attitude control. ACM Trans. Cyber-Phys. Syst. **2**(3), 1–21 (2019)
11. Bernardeschi, C., et al.: ROS/gazebo based simulation of cooperative UAVs. In: International Conference on Modelling and Simulation for Autonomous Systems, pp. 321–334 (2018)
12. Olivares Mendez, M.A., Kannan, S., Voos, H.: Setting up a testbed for UAV vision-based control using V-REP & ROS: a case study on aerial visual inspection. In: 2014 International Conference on Unmanned Aircraft Systems (ICUAS), pp. 447–458 (2014)
13. Kuantama, E.E., Tarca, E.R.: Quadcopter attitude and thrust simulation based on Simulink platform. Robot. Manag. **2**(20), 40–44 (2015)
14. Cekus, D., Posiadala, B., Warys, P.: Integration of modeling in SolidWorks and Matlab/Simulink environments. Arch. Mech. Eng. **1**(61), 57–74 (2014)
15. Xing, Z., He, Y., Jian, C.: Design and implementation of UAV flight simulation based on Matlab/Simulink. In: 2015 International Conference on Advances in Mechanical Engineering and Industrial Informatics, pp. 190–193 (2015)
16. Yun, C., Li, X.M.: Aerodynamic model analysis and flight simulation research of UAV based on simulink. J. Softw. Eng. Appl. **2**(6), 2–43 (2013)
17. Lu, P., Geng, Q.: Real-time simulation system for UAV based on Matlab/Simulink. In: 2nd International Conference on Computing, Control and Industrial Engineering, vol. 1, pp. 399–404 (2011)
18. Schlotter, M.: Multibody system simulation with simmechanics. University of Canterbury (2003)
19. Frankovský, P., et al.: Modeling of dynamic systems in simulation environment MATLAB Simulink–SimMechanics. Am. J. Mech. Eng. **1**(7), 282–288 (2013)
20. Wood, G.D., Kennedy, D.C.: Simulating mechanical systems in Simulink with SimMechanics. The Mathworks Report (2003)
21. Simscape Multibody Link. https://www.mathworks.com/help/physmod/smlink/index.html, Last accessed 2019/12/11
22. Export a SolidWorks Robot Assembly Mode. https://www.mathworks.com/help/physmod/smlink/ug/export-robot-assembly-from-solidworks-software.html, last accessed 2019/12/11

23. Miller, S., Wendlandt, J.: Real-time simulation of physical systems using Simscape. MATLAB News and Notes, pp. 1–13 (2010)
24. Blinov, O.V., Kuznecov, V.B.: The study of mechanical systems in the environment of SimMechanics (MatLab) using the capabilities of three-dimensional modeling programs. Ivanovo State Polytechnic University (2012)
25. Tijonov, K.M., Tishkov, V.V.: SimMechanics Matlab as a dynamic modeling tool complex aviation robotic systems. J Trudy MAI **41**, 1–19 (2010)
26. Simscape Multibody. https://matlab.ru/products/simmechanics. Last accessed 2019/12/15
27. De Simone, M.C. et al.: Multibody model of a UAV in presence of wind fields. In: 2017 International Conference on Control, Artificial Intelligence, Robotics and Optimization (ICCAIRO), pp. 83–88 (2017)

Chapter 11
Technology for Constructing Multifunctional Controlling System of Motion's Parameters of Unmanned Single-Rotor Helicopter

Aleksandr Nikitin, Vyacheslav Soldatkin, and Vladimir Soldatkin

Abstract This chapter views the technology of construction of the controlling sensor system of motion's parameters of unmanned single-rotor helicopter based on the use of the information about position resulting velocity vector of incoming air flow of vortex column of rotor using original fixed multifunction receiver of primary information. The transformation of informative signals is proposed to provide according with two-channel scheme with ion-mark and aerometric channels. The algorithms of processing of informative signals and determination of air parameters of motion of an unmanned single-rotor helicopter in the channels of a multifunctional sensor system built according to the one fixed multifunctional receiver with ion-mark and aerometric channels are presented.

11.1 Introduction

In recent years, unmanned single-rotor helicopters of various classes and purposes have been used for civil and military purposes [1, 2]. The flight of a single-rotor unmanned helicopter takes place in the ground perturbed layer of the atmosphere. To ensure the operate and performance of flight tasks of unmanned single-rotor helicopters it is necessary to control altitude-velocity parameters of motion and state of surrounding air environment that influence to safety of helicopter's flight [3].

The reliability of control of air parameters and the quality of operate of motion of unmanned single-rotor helicopters are reduced, especially at starting and takeoff and landing modes with using traditional sensor systems built on the basis of aerometric, aerodynamic and vane methods of controlling the parameters of the incoming air flow due to significant perturbations from vortex column of rotor. The change of position

A. Nikitin (✉) · V. Soldatkin · V. Soldatkin
Kazan National Research Technical University named after A.N. Tupolev-KAI, 10 Karl Marks street, 420111 Kazan, Russian Federation
e-mail: nikitin.rf@mail.ru

V. Soldatkin
e-mail: w-soldatkin@mail.ru

A. Ronzhin and V. Shishlakov (eds.), *Proceedings of 15th International Conference on Electromechanics and Robotics "Zavalishin's Readings"*, Smart Innovation, Systems and Technologies 187, https://doi.org/10.1007/978-981-15-5580-0_11

of longitudinal axis of unmanned helicopter in azimuth in range ±180° during make three-dimensional flight also makes it difficult to control [3, 4]. This determines the importance of use of new technology of construction of sensor systems for controlling parameters of motion taking into account of influence of aerodynamics field of vortex column of rotor and features flight of the unmanned single-rotor helicopter.

11.2 Theoretical Bases of Construction of System

For ensuring noise immunity of controlling of motion's parameters of unmanned single-rotor helicopter the unconventional approach proposed for purposes of measuring to use parameters of disturbances from aerodynamic field of vortex column of rotor, which are error for traditional sensor systems [5, 6].

When implementing of unconventional approach to control the air parameters of motion on board an unmanned single-rotor helicopter, we propose to use as informative parameters of aerodynamic field of vortex column the parameters of velocity vector $\mathbf{V}_\Sigma$ of resulting air flow from vortex column in the form of the velocity value and angular position of the velocity vector $\mathbf{V}_\Sigma$ relative to the axes of the coordinate system associated with the helicopter.

Assuming the symmetry of the inductive flow of the vortex column of a single-rotor helicopter relative to the axis of rotation of rotor in hovering operation when absence of translational motion of helicopter relative the earth's surface, the velocity vector $\mathbf{V}_\Sigma$ of air flow of vortex column of rotor can be represented as the following mathematical model:

$$\mathbf{V}_\Sigma = \mathbf{V} + \mathbf{V}_i + \mathbf{V}_f, \tag{11.1}$$

where $\mathbf{V}$—stationary component of velocity vector of translational motion of unmanned helicopter relative to air environment; $\mathbf{V}_i$—stationary component of velocity vector of inductive flow of rotor and the air flow due to the thrust of rotor; $\mathbf{V}_f$—fluctuation component caused by the flapping motion of the blades, and work of the swashplate and other factors.

For unmanned single-rotor helicopters, the circular pulsation frequencies of the velocity vector $\mathbf{V}_f$ are multiples to the angular rotation velocity of rotor, which allows them to be significantly reduced using digital filtration technologies [7].

Then, as informative signals of sensor system for controlling of motion's parameters of unmanned single-rotor helicopter, the vector $\mathbf{V}_\Sigma$ filtered from the fluctuation component of the form can be used

$$\mathbf{V}_\Sigma = \mathbf{V} + \mathbf{V}_i = \mathbf{V}_i - \mathbf{V}_a, \tag{11.2}$$

where $\mathbf{V}_a = -\mathbf{V}$—vector of true airspeed of motion of unmanned single-rotor helicopter relative to the surrounding air environment.

11.3 Variants of Construction of Sensor System

An important aspect of technology for controlling air parameters on board unmanned single-rotor helicopter is the building of the sensor controlling system based on a single (integrated) fixed receiver of primary information.

The original fixed aerometric receiver of informative parameters of the aerodynamic field of vortex column of rotor of single-rotor helicopter is proposed in works [6, 8] (Fig. 11.1).

The basis of construction of receiver is multi-channel flow aerometric receiver 1 installed on fuselage of single-rotor helicopter. In flow channel between screen disks 2 and 3, the total pressure tubes 4 are installed which perceive pressures P_i with use of which according to developed algorithms the value of velocity V and the angle ψ of the direction of the incoming air flow in azimuthal plane are determined [9]. The angle α of incoming air flow in the vertical plane in the parking before starting the power plant of single-rotor helicopter determines according to pressures $P_{\alpha i}$, $P_{\alpha i-1}$ and $P_{st.d}$ perceived of multi-channel flow receiver.

When the fixed multi-channel flow receiver is located in zone of vortex column from rotor of a single-rotor helicopter, we proposed to install in outer surface of upper screen disk 3 the additional fixed receiver 7 in the form of hemisphere with diameter is equaled to the diameter of the upper screen disk [10]. Total pressure $P_{t\Sigma}$, static pressure $P_{st\Sigma}$ of resulting air flow of vortex column of rotor and pressure P_1, P_2 and P_3, P_4 are perceived with help of holes 8, 9, 10, 11 on the upper surface of the

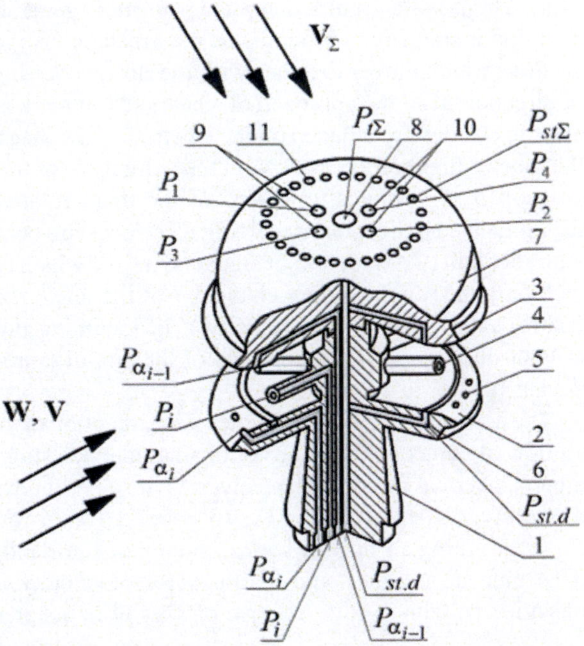

Fig. 11.1 Design scheme of fixed space-distributed combined aerometric receiver

hemisphere. The air parameters are determined by perceived pressures and developed algorithms [10] while fixed combined aerometric receiver is located in area of vortex column of rotor when starting of power plant, when taxing and maneuvering on the earth's surface (the starting modes), during takeoff and climbing, descending, hovering and landing (takeoff and landing modes) of single-rotor helicopter.

As the analysis has shown, the presence of a large number of total pressure tubes installed counter to incoming air flow causes the need to protect them from icing, dust and moisture in the real conditions of operation of an unmanned single-rotor helicopter that reduce reliability of working system. The large number of conversion channel of perceived pressures defines the strict requirements to identity and stability of characteristics and complicates the design, reduces the manufacturability, increases the cost and constrains application of shown sensor system for controlling motion's parameters on board of unmanned single-rotor helicopter.

As studies have shown [11], one of the ways to eliminate the shortcomings of the sensor system for controlling air parameters with the use of a fixed combined aerometric receiver is the construction of a sensor system of unmanned single-rotor helicopter according to a two-channel scheme with ion-mark and aerometric channels as shown in Fig. 11.2.

To ensure control of air parameters of unmanned single-rotor helicopter while changing the position of its longitudinal axis in azimuth in the range $\pm 180°$ the ion-mark instrumentation channel of sensor system is proposed [12] to perform based on the original combined ion-mark sensor of aerodynamic angle and true airspeed [13]. The sensor provides registration of parameters of velocity vector $\mathbf{V}$ of incoming air flow and consists the board 1 in center of which a spark gap switch 2 is installed. When a high-voltage pulse is applied from mark generator (MG) 4 to spark gap switch 2, an ion marks with a pronounced electrostatic charge are created in the incoming air flow, which moves together with the flow, acquiring its velocity V and the angle of direction β. At the approach of a charged ion mark above receiving electrodes the electric charges are induced on the electrodes, the magnitude of which depends from the time of flight of ion mark with the velocity V of the distance R and from angular position β of motion's trajectory of the mark relative to the longitudinal axis of the unmanned helicopter. Receiving electrodes are connected to the inputs of coarse amplifier unit (CAU) 5, output signals $U(\alpha_i)$ of which are registered by the scheme 6, including the rough reading channel 7 of the angle the direction of incoming flow, control value channel 8 of velocity V of incoming flow, precise reading channel 9 of controlled angle β. The outputs of the circuit 6 are connected to the calculating device 10.

The aerometric channel provides the perception of velocity vector $\mathbf{V}_\Sigma$ of resulting air flow of vortex column of rotor of an unmanned single-rotor helicopter. It includes a hemispherical aerometric receiver 11 installed above board 1. On the surface of the receiver, there are holes 12, 13, 14 and 15, 16 for perception of total pressure $P_{t\Sigma}$ of resulting air flow of vortex column of rotor and pressures P_1, P_2 and P_3, P_4, determine the angles φ_1 and φ_2 of position of velocity vector $\mathbf{V}_\Sigma$ of resulting air flow of vortex column of rotor relative of axes of helicopter. On hemispherical receiver,

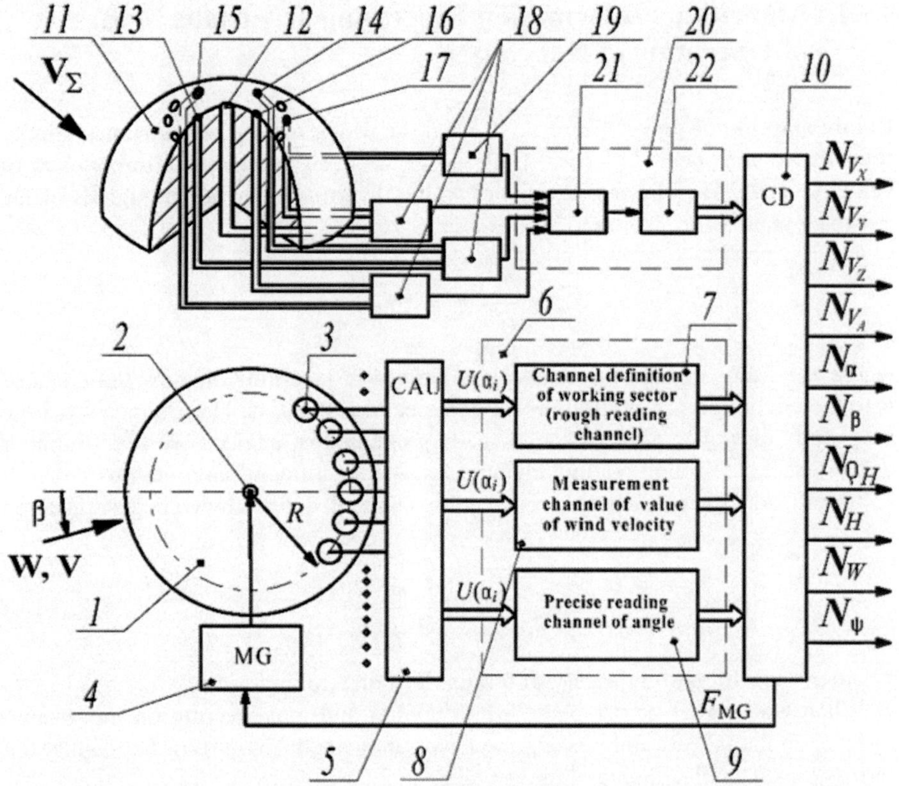

Fig. 11.2 Functional scheme of sensor controlling system of air parameters on board unmanned single-rotor helicopter, designed by two-channel scheme with ion-mark and aerometric channels

there are also holes 17 for perception of averaged value of static pressure $P_{st\Sigma}$ of resulting air flow of vortex column of rotor.

The pressures $P_{t\Sigma}$, P_1, P_2 and P_3, P_4, $P_{st\Sigma}$ are converted into electrical signals by means of pneumoelectric converters of differential pressure 18 and absolute pressure 19, which are fed to input of computing device 10 through the channel of analog-to-digital conversion including the multiplexer 21 and analog-to-digital converter (ADC) 22. Computing device calculates and gives out digital signals according to developed algorithms about motion's controlled air parameters of unmanned single-rotor helicopter in operating modes when sensor system located in area of vortex column of rotor.

11.4 Algorithms Information Processing at Various Modes of Operating of Helicopter

Before starting of power plant, to prevent the tilting motion of unmanned single-rotor helicopters, necessary measuring the velocity W_h and angle of direction ψ of velocity vector $\mathbf{W}_h$ of horizontal wind which determined by output signals of the ion-mark channel according to algorithms:

$$W_h = \frac{R}{\tau_W};\ \psi = i\alpha_0 + \alpha_w, \tag{11.3}$$

where τ_W—time of motion of ion mark carried in incoming air flow the distance R to the circumference with the receiving electrodes; α_0—angle, characterized the magnitude of work sector of rough reading channel of angle; $i = \overline{1,4}$—number of work sector of rough reading channel; α_{wi}—magnitude of angle in range of i-th sector of work channel of the precise reading channel, defined by voltages registered CAU:

$$\alpha_{wi} = \mathrm{arctg}\frac{U \sin\alpha_i}{U \cos\alpha_i}, \tag{11.4}$$

where α_i—position of trajectory of ion mark in ith working sector.

When working of power plant, when taxiing and maneuvering on the ground surface, on the takeoff and landing modes of unmanned single-rotor helicopter, the aerometric channel is included in the work.

The magnitude of velocity V_Σ of resulting air flow vector of vortex column of rotor is determined by the pressure $P_{t\Sigma}$ and $P_{st\Sigma}$ perceived by the hemispherical receiver according to the expression of the form [10]:

$$V_\Sigma = 44{,}826\sqrt{\frac{T_{b\Sigma}\left[\left(\frac{P_{t\Sigma}-P_{st\Sigma}}{P_{st\Sigma}}+1\right)^{0{,}2857143}-1\right]}{\left(\frac{P_{t\Sigma}-P_{st\Sigma}}{P_{st\Sigma}}+1\right)^{0{,}2857143}}}, \tag{11.5}$$

where $T_{b\Sigma}$—braking temperature of resulting air flow of vortex column of rotor perceived by the receiver of braking temperature installed on fuselage of unmanned single-rotor helicopter in zone of vortex column of rotor.

According to pressures $P_{b\Sigma}$, P_1, P_2, P_3, P_4, $P_{st\Sigma}$, the motion's air parameters of unmanned single-rotor helicopter are determined using the formulas [10]

$$\begin{aligned} V_x = {} & V_\Sigma \sin\left[\frac{1}{2}\arcsin\left(\frac{4}{9\sin 2\varphi_{01}}\frac{P_1 - P_2}{P_{t\Sigma} - P_{st\Sigma}}\right)\right] \\ & \times \cos\left[\frac{1}{2}\arcsin\left(\frac{4}{9\sin 2\varphi_{02}}\frac{P_3 - P_4}{P_{t\Sigma} - P_{st\Sigma}}\right)\right] - K_{ix}V_{i0} - (\omega_y z - \omega_x y); \end{aligned}$$

$$V_z = V_\Sigma \sin\left[\frac{1}{2}\arcsin\left(\frac{4}{9\sin 2\varphi_{02}}\frac{P_3 - P_4}{P_{t\Sigma} - P_{st\Sigma}}\right)\right] - K_{iz}V_{i0} - (\omega_x y - \omega_y x); \tag{11.6}$$

$$V_a = V_i = \sqrt{V_x^2 + V_y^2 + V_z^2}; \quad \alpha = \text{arctg}\frac{V_y}{V_x}; \quad \beta = \text{arsin}\frac{V_z}{V_A}; \tag{11.7}$$

$$P_H = P_{st\Sigma} - K_p\frac{\rho_\Sigma V_\Sigma^2}{2}; \tag{11.8}$$

$$H = \frac{T_0}{\tau}\left[1 - \left(\frac{P_H}{P_0}\right)^{\tau R}\right], \tag{11.9}$$

where $V_a = V_i$—true airspeed and indicated speed of flight; K_p, K_{ix}, K_{iy} and K_{iz}—nondimensional coefficient are determined during flight tests for a particular type of unmanned single-rotor helicopter and installation site on fuselage of fixed multifunctional receiver; P_0 and T_0—absolute pressure and temperature of outer air at altitude $H = 0$; R—specific gas constant of air; τ—temperature gradient of change the temperature of outer air during changing of altitude H.

The components of vector $\mathbf{W}_h$ of horizontal wind and angle of direction ψ on takeoff and landing modes of unmanned single-rotor helicopter will be determined as [14]:

$$\begin{aligned} &W_x = |V_x| - V_g\cos\psi_c; \; W_z = |V_z| - V_g\sin\psi_c; \\ &\psi = \beta + \psi_c; \; W_h = \sqrt{W_x^2 + W_z^2}, \end{aligned} \tag{11.10}$$

where ground-speed V_g and crab angle ψ_c measured by the doppler velocity and drift angle gauge (DVDAG).

When the fixed multifunctional receiver of sensor system leaves from zone of vortex column of rotor, the air parameters of motion of unmanned single-rotor helicopter are determined by information of ion-mark channel, in particular the indicated V_i and true airspeed V_a and the gliding angle β—by the formulas:

$$V_i = V_a = \frac{R}{\tau_V}; \quad \beta = i\alpha_0 + \alpha_{wi}; \quad \alpha_{wi} = \text{arctg}\frac{U\sin\alpha_i}{U\cos\alpha_i}, \tag{11.11}$$

where τ_V—motion time of ion mark to the circumference with receiving electrodes of ion-mark channel.

To determine the barometric altitude H in flight modes, when a fixed multifunctional receiver of sensor system leaves from zone of vortex column of rotor of unmanned single-rotor helicopter, the hole-receiver to perceive the static pressure P_H of the incoming air flow is installed on the outer surface of board with receiving electrodes of ion-mark sensor, and the barometric altitude H will be determined using the ratio (11.8).

11.5 Analysis of Instrumental Errors of the System

Instrumental errors of multifunctional controlling system of motion's parameters of unmanned single-rotor helicopter, like and other devices and system, are caused by errors of sensors of primary information.

That is why instrumental error Δy multifunctional controlling system of motion's parameters of unmanned single-rotor helicopter can determine by form:

$$\Delta y = \frac{\partial f}{\partial x_1}\Delta x_1 + \frac{\partial f}{\partial x_2}\Delta x_2 + \cdots + \frac{\partial f}{\partial x_n}\Delta x_n + \Delta_{\text{pro}}, \tag{11.12}$$

where Δx_i—errors of sensors of primary information; Δ_{pro}—error of information processing channel.

By analogy with air data system of manned single-rotor helicopter, the analysis of instrumental errors of multifunctional controlling system of motion's parameters of unmanned single-rotor helicopter will be shown on example of control channel of parameters of wind vector on various modes of operation of unmanned single-rotor helicopter.

Neglecting of conversion error and processing of controlling parameters the errors of instrumentation channel of velocity and direction angle of horizontal wind vector before starting of power plant will be defined by form:

$$\Delta W_h = \frac{1}{\tau_W}\Delta R - \frac{R}{\tau_W}\Delta\tau_W;\ \Delta\psi = \Delta\alpha_0 + \Delta\alpha_W, \tag{11.13}$$

where ΔR—manufacturing tolerance on the radius R of installing of receiver of primary information; $\Delta\tau_W$—error of registration of time-of-flight ion mark from gap switch to receivers; $\Delta\alpha_0$ and $\Delta\alpha_W$—errors of rough count channel of angular position of ion mark and registration the angular position of ion mark inside the work sector.

Parameters of horizontal wind vector $\mathbf{W}_h$ when starting of power plant are determined in accordance with the algorithms (11.10).

By analogy with air data system of manned single-rotor helicopter the instrumental of measuring the velocity W_h and angle of direction ψ of horizontal wind vector of multifunctional controlling system of motion's parameters of unmanned single-rotor helicopter in the parking when starting the power plant and rotating the rotor without taking into account the error of calculation will be determined by relations:

$$\Delta W_x = \frac{4}{9\sin 2\varphi_0}\left[\begin{array}{l}\frac{P_1 - P_2}{P_{t\Sigma} - P_{st\Sigma}}\Delta V_\Sigma \\ +V_\Sigma \frac{(P_{t\Sigma} - P_{st\Sigma})\Delta(\Delta P_1) - (P_1 - P_2)\Delta(\Delta P_\Sigma)}{(P_{t\Sigma} - P_{st\Sigma})^2}\end{array}\right];$$

$$\Delta W_z = \frac{4}{9\sin 2\varphi_0}\left[\begin{array}{l}\frac{-P_3 - P_4}{P_{t\Sigma} - P_{st\Sigma}}\Delta V_\Sigma \\ +V_\Sigma \frac{(P_{t\Sigma} - P_{st\Sigma})\Delta(\Delta P_2) - (P_3 - P_4)\Delta(\Delta P_\Sigma)}{(P_{t\Sigma} - P_{st\Sigma})^2}\end{array}\right];$$

$$\Delta V_{\Sigma} = \frac{V_{\Sigma}}{2T_{b\Sigma}} + \frac{a^2\sqrt{T_{b\Sigma}}}{2V_{\Sigma}} b\left(\frac{\Delta P_{\Sigma}}{\Delta P_{st\Sigma}} + 1\right)^{b-1} \cdot \left[\frac{1}{P_{st\Sigma}}\Delta(\Delta P_{\Sigma}) - \frac{\Delta P_{\Sigma}}{P_{st}^2}\Delta P_{st\Sigma}\right];$$

$$\Delta W_h = \frac{W_x \Delta W_x + W_z \Delta W_z}{W_h}; \quad \Delta\psi = \frac{1}{\cos^2 \frac{W_z}{W_x}} \cdot \frac{W_x \Delta W_z + W_z \Delta W_x}{W_x^2}, \tag{11.14}$$

where $\Delta(\Delta P_1)$, $\Delta(\Delta P_2)$, $\Delta(\Delta P_{\Sigma})$—errors of using sensors of absolute pressure; $T_{b\Sigma}$—braking temperature of resulting air flow.

Longitudinal W_x and lateral W_z components of vector W_h and angle of direction ψ of horizontal wind during taxiing and maneuvering the helicopter on the earth's surface will be determined by relations [9]:

$$W_x = V_{\Sigma}\frac{4}{9\sin 2\varphi_{01}}\frac{P_1 - P_2}{P_{t\Sigma} - P_{st\Sigma}} - K_{ix}V_{i0} - V_{tx};$$

$$W_z = V_{\Sigma}\frac{4}{9\sin 2\varphi_{02}}\frac{P_3 - P_4}{P_{t\Sigma} - P_{st\Sigma}} - K_{iz}V_{I0} - V_{tz};$$

$$W_h = \sqrt{W_x^2 + W_z^2}; \quad \psi = \operatorname{arctg}\frac{W_z}{W_x}, \tag{11.15}$$

where V_{tx} and V_{tz}—components of velocity vector $\mathbf{V}_t$ taxiing and maneuvering of unmanned single-rotor helicopter on earth's surface, determined by signals of satellite navigation system.

In mode of taxiing and maneuvering on the earth's surface of unmanned single-rotor helicopter, the instrumental errors of velocity channel and angular position of horizontal wind of multifunctional controlling system of motion's parameters of unmanned single-rotor helicopter can be estimated using the relations (11.14).

Parameters of the horizontal wind vector $\mathbf{W}_h$ on the takeoff and landing modes of unmanned single-rotor helicopter, using information about magnitude of V_g and crab angle ψ_c of vector ground-speed $\mathbf{V}_g$ from the doppler velocity and drift angle gauge, will be determined by relations [14]:

$$W_x = |V_x| - V_g\cos\psi_c; \; W_z = |V_z| - V_g\sin\psi_c; \; \psi = \beta + \psi_c; \; W_h = \sqrt{W_x^2 + W_z^2},$$

$$V_x = V_{\Sigma}\frac{4}{9\sin 2\varphi_0}\frac{P_1 - P_2}{P_{t\Sigma} - P_{st\Sigma}} - K_{ix}V_{i0};$$

$$V_z = V_{\Sigma}\frac{4}{9\sin 2\varphi_0}\frac{P_3 - P_4}{P_{t\Sigma} - P_{st\Sigma}} - K_{iz}V_{i0}. \tag{11.16}$$

Then, without taking into account the error of the calculate channel the instrumental error of channels of multifunctional system the parameters of horizontal vector of velocity of wind will be determined as

$$
\begin{aligned}
\Delta W_x &= \frac{4}{9\sin 2\varphi_0}\begin{bmatrix} \frac{P_1-P_2}{P_{t\Sigma}-P_{st\Sigma}}\Delta V_\Sigma \\ +V_\Sigma \frac{(P_{t\Sigma}-P_{st\Sigma})\Delta(\Delta P_1)-(P_1-P_2)\Delta(\Delta P_\Sigma)}{(P_{t\Sigma}-P_{st\Sigma})^2} \end{bmatrix} \\
&\quad - \cos\varphi_c \Delta V_g + \sin\varphi_c \Delta\varphi_c - z\Delta\omega_y + y\Delta\omega_x; \\
\Delta W_z &= \frac{4}{9\sin 2\varphi_0}\begin{bmatrix} \frac{P_3-P_4}{P_{t\Sigma}-P_{st\Sigma}}\Delta V_\Sigma \\ +V_\Sigma \frac{(P_{t\Sigma}-P_{st\Sigma})\Delta(\Delta P_2)-(P_3-P_4)\Delta(\Delta P_\Sigma)}{(P_{t\Sigma}-P_{st\Sigma})^2} \end{bmatrix} \\
&\quad - \sin\varphi_c \Delta V_g - V_g\cos\varphi_c - y\Delta\omega_x + x\Delta\omega_y; \\
\Delta V_\Sigma &= \frac{V_\Sigma}{2T_{b\Sigma}}\Delta T_{b\Sigma} + \frac{a^2\sqrt{T_{b\Sigma}}}{2V_\Sigma} b\left(\frac{\Delta P_\Sigma}{\Delta P_{st\Sigma}}+1\right)^{b-1} \\
&\quad \cdot\left[\frac{1}{P_{st\Sigma}}\Delta(\Delta P_{st\Sigma}) - \frac{\Delta P_\Sigma}{P_{st\Sigma}}\Delta P_{st\Sigma}\right]; \\
\Delta W_h &= \frac{W_x\Delta W_x + W_z\Delta W_z}{W_h}; \quad \Delta\psi = \Delta\beta - \Delta\varphi_c,
\end{aligned} \tag{11.17}
$$

where ΔV_g and $\Delta\varphi_c$—errors of doppler velocity and drift angle gauge; $\Delta\omega_x$, $\Delta\omega_y$, $\Delta\omega_z$—measuring errors of angular velocities of rotation of unmanned single-rotor helicopter relative to axes of associated coordinate system; *x*, *y*, *z*—coordinates of installation location of multifunctional fixed receiver on board of unmanned single-rotor helicopter.

The results of study of experimental samples in wind tunnel have shown [10, 14], instrumental errors of sensor system for controlling air parameters on board unmanned single-rotor helicopter with one fixed multifunctional receiver, ion-mark and aerometric channels can be estimated by next values: Root-mean-square errors of the wind velocity channel do not exceed 0.55 … 0.83 m/s, angle of wind direction—1.5 … 2 degree of arc; root-mean-square errors of channel of indicated speed and true airspeed are the interval 3.6 ± 1.1 km/h, channel of gliding angle in the range from 0 to 360 degree of arc—0.25 ± 1 degree of arc; channel of absolute barometric altitude—3.5 ± 0.15 m, which meet modern requirements for the accuracy of sensor systems for controlling air parameters of unmanned single-rotor helicopter.

11.6 Conclusion

In conclusion, it should be noted that the proposed building of sensor system based on one fixed multifunctional receiver with ion-mark and aerometric channels and the developed algorithms provide reliable control of air parameters on board an unmanned single-rotor helicopter in all operating modes. The use of sensor system for controlling and operating of motion with one fixed multifunctional receiver and two-channel measuring scheme allows to provide safe operate of unmanned single-rotor helicopters at parking, starting, takeoff and landing, flight modes in conditions of significant disturbances of vortex column of rotor.

Acknowledgements This work is supported by RFBR project no. 18-08-00264.

References

1. Moiseev, V.S., Gushchina, D.S., Saleev, A.V.: Unmanned aircraft systems. Classification, basic principles of creation and application. Russian Aeronaut. **3**, 3–5 (2006)
2. Moiseev, V.S., Gushchina, D.S., Shafigullin, R.R.: Rational placement and required number of information unmanned aerial systems for on-line monitoring of large territories. Russian Aeronaut. **55**(3), 223–229 (2012)
3. Kozitsin, V.K., Makarov, N.N., Porunov, A.A., Solatkin, V.M.: Analysis of the principles of construction of systems for measuring air signals of helicopter. Aerosp. Instrum. Making **10**, 2–13 (2003)
4. Nikitin, A.V., Soldatkin, V.V.: A starting system for measuring low airspeeds of a single-rotor helicopter. Russian Aeronaut. **55**(1), 68–75 (2012)
5. Soldatkin, V.V.: Aerometric system for measuring low helicopter airspeeds based on the data of the vortex column position. Russian Aeronaut. **52**(4), 455–462 (2009)
6. Soldatkin, V.V.: Methodological foundations for helicopter altitude-speed parameter measurement using aerometric information from the main rotor vortex column. Russian Aeronaut. **53**(3), 320–327 (2012)
7. Emaletdinova, L.Y., Lyasheva, S.A., Shleimovich, M.P.: Wavelet filtering of measurement information in unmanned aerial vehicle control system. Russian Aeronaut. **55**(3), 158–163 (2012)
8. Soldatkin, V.V., et al.: Patent RF for invention №2427844, WPC G01P5/14. Air data system of helicopter. Declare 09.03.2010. Published 27.08.2011. Bulletin No. 24
9. Porunov, A.A., Soldatkin, V.V., Soldatkin, V.M.: Methodology of construction and models of informative signals of fixed multichannel flow aerometric receiver of air data system helicopter's. Russian Aeronaut. **4**, 58–63 (2010)
10. Soldatkin, V.V.: Air data system of helicopter on bases fixed aerometric receiver and information about aerodynamic field of vortex column of main rotor: monograph. Kazan: Publishing house Kazan government technical university, 264 p. (2012)
11. Soldatkin, V.M.: Measuring system of air parameters of helicopter with fixed receiver, ion-mark and aerometric instrumentation channels. Mekhatronika Avtomatizatsiya Upravlenie **19**(11), 744–752 (2018)
12. Soldatkin, V.M., et al.: Patent RF for invention №2587389, WPC G01P5/14. Onboard measuring system of parameters of wind velocity vector at parking, starting and takeoff and landing modes. Declare 10.12.2014. Published 15.04.2016. Bulletin No. 17
13. Ganeev, F.A., Soldatkin, V.M.: Ion-marking aerodynamic angle and airspeed sensor with logometric informative signals and interpolation processing scheme. Russian Aeronaut. **53**(3), 312–319 (2010)
14. Nikitin, A.V., Soldatkin, V.M.: Measuring system of parameters of wind and true airspeed on board helicopter. Datchiki Systemi (Sens. Syst.) **4**, 48–54 (2015)

Chapter 12
Mathematical Modeling of Stable Position of Manipulator Mounted on Unmanned Aerial Vehicle

Vinh Nguyen, Quyen Vu, and Andrey Ronzhin

Abstract The stability of the manipulator is important issue especially in the domain of unmanned aerial vehicle (UAV). Any movements of manipulator greatly affects UAV stability. In particular, the horizontal shift of the center of gravity requires of the UAV the powerful controller, that change forced the propellers of quadrotor to bring the UAV to a stable state. In this paper, we have studied the movements of manipulator and its center of gravity at least horizontally when UAV is in hover mode. As a result, we developed mathematical model and software for calculation of deflection angles between successive links of the manipulator, which provides deviation of the center of mass (COM) of the UAV horizontally less than 1 mm.

12.1 Introduction

In recent years, UAV, drones, quadrotor, or quadcopter are the keyword that has been searched a lot on websites. There is a lot of research on UAV for different purposes. However, [1] indicates that only 2% of the studies refer to unmanned aircraft fitted with manipulating devices. This figure is very small compared to the total research on UAV in general. In the last few years, research in this area has increased rapidly. The results obtained from those studies are so large and also the basis for future research.

There are many proposals for the aerial manipulation system and a lot of different approaches. In [2], the grippers attached directly to the bottom of each quadrotor enable them to pick up, transport, and assemble the structural elements. In [3] using a helicopter with a gripper attached to it to pick different object up, the stability

V. Nguyen · Q. Vu
Department of Electromechanics and Robotics, SUAI, Bolshaya Morskaya Str., 67, 190000 Saint-Petersburg, Russia

A. Ronzhin (✉)
Laboratory of Autonomous Robotic Systems, SPIIRAS, 14-th Line of V.I., 39, 199178 Saint-Petersburg, Russia
e-mail: ronzhin@iias.spb.su

A. Ronzhin and V. Shishlakov (eds.), *Proceedings of 15th International Conference on Electromechanics and Robotics "Zavalishin's Readings"*, Smart Innovation, Systems and Technologies 187, https://doi.org/10.1007/978-981-15-5580-0_12

of system was proved when working with PID controller. In [4], using self-sealing suction technology for grasping a wide range of objects, the self-sealing suction cup attached to each gripper and the gripper attached to bottom of quadrotor.

The next approach is to suspend a payload with cables. In [5], a quadrotor with a cable-suspended load with eight degrees of freedom and four degrees of underactuation is described. A nonlinear geometric control design is presented, that enables tracking of either the quadrotor attitude, the load attitude, or the position of the load. Almost global exponential stability for the load attitude tracking, and almost global exponential attractivity of the load position tracking, is demonstrated. In [6], the proposed system is a hexarotor with slung load, length of the tether is decided to be 2 m. This paper presents a practical swing-angle estimation method for slung load operations of UAV, which is essential to maintain the safety during the operation.

The other approach is a manipulator attached to UAV to capture object. In [7], the aerial manipulation system consists of two-link manipulator attached to the bottom of a quadrotor. By deriving the inverse kinematics, the authors design the controller such that the desired end-effector position and orientation can be tracked. The design in [8] is aerial manipulation system consists of the four-DOF robotic arm attached to bottom of tricopter. The simulation scenario can be summarized as follows: (a) UAV hovers with the folded arm for the first 10 s; (b) the robotic arm is unfolded in the next 10 s followed by the folding of the arm for another 10 s; (c) the above two steps are repeated one more time; (d) finally, UAV hovers with the folded arm for the final 10 s. The strengths of this study are in combination of PD and T2FNN controller (PD + T2FNN) results in a superior performance. In [9], the authors present aerial manipulation using a quadrotor with a two-DOF robot arm. The objective of this research is picking up an object, moving, and releasing it inside of a shelf. The difference in this study is the robustness of the controller, to control the combined system, an adaptive sliding mode controller is used. In research's results, during the pickup phase, root-mean-square error of the quadrotor position is only 2.08 cm. In the release phase, root-mean-square error of the quadrotor is 2.56 cm.

Also there is a research of quadrotor team for cooperative transport of common payload. Such quadrotors are rigidly attached to a payload. The main contribution of this study is to propose a cooperative control algorithm based on a decentralized algorithm [10]. In [11], the authors present a framework to enable a fleet of rigidly attached quadrotor aerial robots to transport heavy objects along a known reference trajectory without inter-robot communication or centralized coordination.

In the first category using a gripper, it has the following disadvantages owing to its mechanical structure: limited workspace and limited grasping ability of mass and volume. In this situation, payloads are rigidly connected to the body of an UAV. Accordingly, not only the attitude of the payload is restricted to the attitude of the UAV, but also the accessible range of the end-effector is confined because of the UAV body and blades. In the second type using cables, the movement of the payload cannot be always regulated directly because manipulation is achieved using a cable which cannot always drive the motion of the payload as desired. To cope with these limitations, one alternative approach is to equip an aerial vehicle with a robotic manipulator that can actively interact with the environment. By combining the mobility of

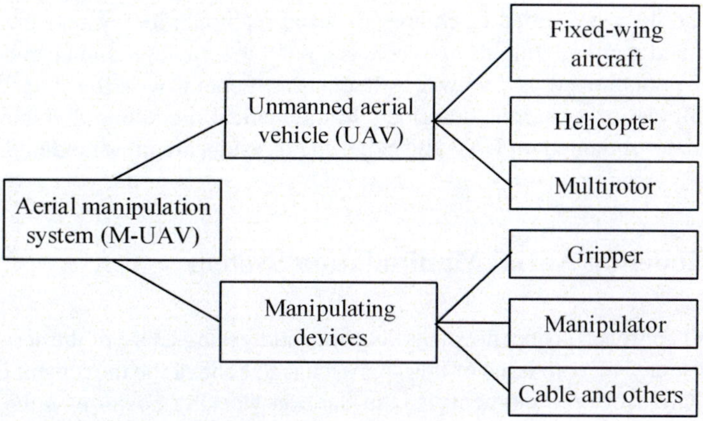

Fig. 12.1 Component classification of aerial manipulation system

the aerial vehicle with the versatility of a robotic manipulator, the utility of mobile manipulation can be maximized [9].

UAVs are subdivided into two general categories, fixed wing and rotary wing. Rotary wing UAVs divided into two types: helicopter and multirotor. Multirotor has several advantages over fixed-wing airplanes. They can move in any direction and are capable of hovering and fly at low speeds [12]. So multirotor is an optimal choice for simple tasks. A component classification of aerial manipulation systems is shown in Fig. 12.1.

There are two approaches to address modeling and control problems. The first independent approach divides a system into two independent parts and considers the modeling methods of each part, respectively. This approach considers the motion and dynamics of manipulating devices as external disturbances of UAVs; hence, it is easier to implement than an alternative. The second approach considers the complete system as an overall system, addressing challenges that the center of mass (COM) is changing constantly and the internal dynamics are coupled. Increasingly researchers are adopting the overall approach for high accuracy. The independent approach faces the challenge that the system model is not sufficiently precise. Although the overall approach should solve the dynamic decoupling problem, model and COM changes during manipulating continue. Meanwhile, controllers for UAVs and manipulating devices have been developed from early-time PID to the present situation. Now adaptive sliding mode control and back-stepping control are used for UAVs, while impedance control and visual servoing control are used for manipulating devices. A controller should address the following challenges: (i) variable mass, inertia, and model, (ii) unknown features and model of a target, (iii) balanced performance of flight and manipulation, attempting to achieve agile flight with a heavy payload at the same time as providing accurate and stable manipulation, and (iv) continuous position control of the end-effector is difficult to achieve because an underactuated multirotor cannot control the position and attitude simultaneously [1].

In our study, we decided to choose an aerial manipulation system including of multirotor and manipulating devices consisted of a manipulator and an end-effector. The paper is organized as follows. In Sect. 12.2, general working diagram of the aerial manipulation system is described. Mathematical modeling of stable position of manipulator mounted on UAV and experiment results are presented in Sect. 12.3.

12.2 Modeling Aerial Manipulation System

The overall approach is one that dives deep into addressing COM of the aerial manipulation system. The core issue of this approach is to control the movement of manipulator's COM from the moment; it is in the non-working position to the working one, especially the change of manipulator's and the entire system's COM at the time the robotic arm performs object capture. The goal of the study is to analyze to find the pose of the robotic arm when it is in the non-working and working position. And then we can determine the angle of deviation between the links to control the robotic arm so that the COM is least changed during work.

There are many ways to design a robotic arm to stabilize the COM of the aerial manipulation system. In [13], the proposed aerial manipulation system consisted of two symmetrical manipulators. In [14], the authors designed a manipulator have two end-effectors: one for the payload and another for the counterbalance weight. The study shows that UAVs are highly affected by COM horizontal shifting. Therefore, quadrotors must overcome any COM horizontal shifting in order to stabilize itself. COM vertical shifting, on the other hand, causes change in magnitude of mass moment of inertia. COM location shifts downwards in most cases, leading into an increased UAV stability. Another remarkable result of the study is that the aerodynamic drag does not have a significant effect on the stability of the quadrotor since the proposed manipulator was designed so the manipulator movement is only done in the hover state. In our study, a robotic arm is proposed with the first two links of that robotic arm plays a role as counterbalance weight instead of using another end-effector is a counterbalance weight like in [14].

In Fig. 12.2 shown that with the pose of the robotic arm in Fig. 12.2a, the COM of the robotic arm is kept closer to vertical axis of UAV. In Fig. 12.2b, the COM is quite

Fig. 12.2 Influence of manipulator's pose to the COM of UAV

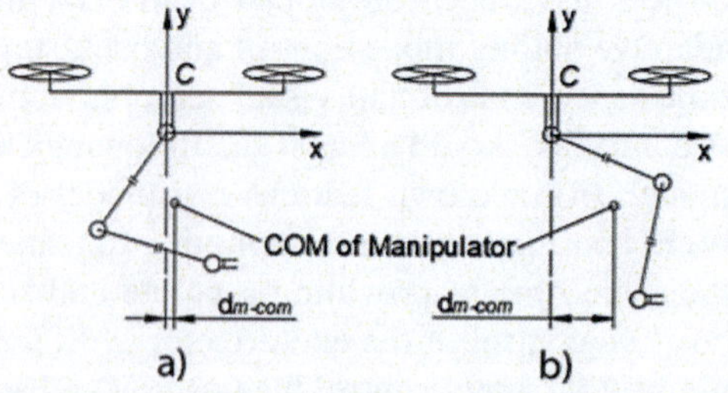

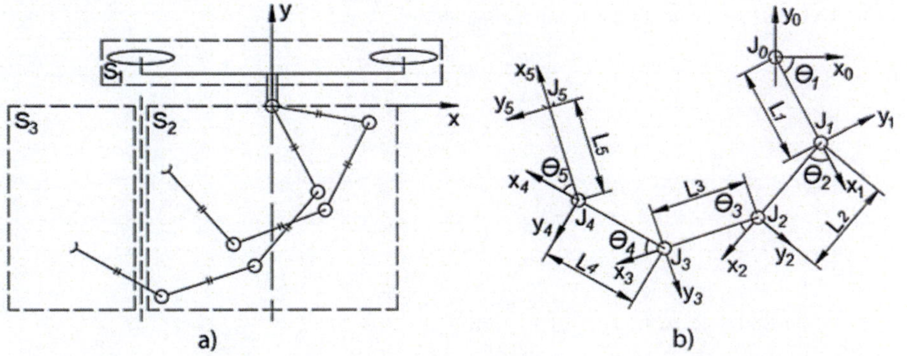

Fig. 12.3 **a** General working diagram of the aerial manipulation system. **b** Geometric parameters of manipulator

far from the vertical axis of UAV and we will not consider the pose of the robotic arm as stable.

We offer the general working diagram of the aerial manipulation system presented in Fig. 12.3a: S_1 is a UAV, S_2 is the space where the robotic arm is placed when it is in the non-working state (parking and takeoff stages), S_3 is the workspace of robotic arm. S_3 is the space containing objects to manipulate.

In this study, we will analyze the theta deflection angle between the links of the arm (Fig. 12.3b) when the robotic arm moves from S_2 to S_3. Through that we will find out the optimal theta angle set, so that the robotic arm moves but still keeps its the change of COM is minimal along the horizontal axis X. The objective is always to keep the COM of the robotic arm moving on the close space of vertical axis Y.

12.3 Mathematical Modeling of Stable Position of Manipulator Mounted on UAV

Suppose that there is a manipulator consists of five links mounted on UAV. The kind of joint between two consecutive links is only revolute joints such as Fig. 12.3b. So we have geometric parameters table of the serial chain manipulator in Table 12.1. The purpose of this forward kinematic is determining the coordinates of the joints J_i ($i = 1 \div 5$) so that we can calculate the COM coordinates of the manipulator. Pay attention that J_5 is like the end-effector, the coordinates of J_0 are ($x_0 = 0$; $y_0 = 0$).

a_i is the distance from z_{i-1} to z_i along x_{i-1},
α_i is the angle from z_{i-1} to z_i about x_{i-1},

Table 12.1 DH parameter of the chain manipulator

i	α_i	a_i	d_i	θ_i
1	0	L_1	0	θ_1
2	0	L_2	0	θ_2
3	0	L_3	0	θ_3
4	0	L_4	0	θ_4
5	0	L_5	0	θ_5

d_i is the distance from x_{i-1} to x_i along z_i,
θ_i is the angle from x_{i-1} to x_i about z_i.
Here: z coincides with the rotation of axes of the joints.

In [15], the type of joint is revolute, so joint rotation matrix is:

$$ {}^{i-1}R_i = \begin{bmatrix} c_i & -s_i & 0 \\ s_i & c_i & 0 \\ 0 & 0 & 1 \end{bmatrix} \tag{12.1} $$

where $c_i = \cos\theta_i$; $s_i = \sin\theta_i$;

$$ s_{ij} = \sin(\theta_i + \theta_j) = \cos\theta_i \sin\theta_j + \cos\theta_j \sin\theta_i; $$
$$ c_{ij} = \cos(\theta_i + \theta_j) = \cos\theta_i \cos\theta_j + \sin\theta i \sin\theta_j; $$

Coordinate frame i can be located relative to coordinate frame $i-1$ by executing a rotation through an angle. ${}^{i-1}T_i$ is a matrix to determine the relative position coordinates of link i relative to link $i-1$.

$$ {}^{i-1}T_i = \begin{bmatrix} c_i & -s_i & 0 & a_i c_i \\ s_i & c_i & 0 & a_i s_i \\ 0 & 0 & 1 & 0 \\ 0 & 0 & 0 & 1 \end{bmatrix}, \tag{12.2} $$

The coordinates of the link i $(x_i; y_i)$ relative to the original coordinates are determined through Forward Kinematics (12.3):

$$ {}^{0}T_i = {}^{0}T_1 {}^{1}T_2 \ldots {}^{i-1}T_i = \begin{bmatrix} r_{11} & r_{12} & r_{13} & x_i \\ r_{21} & r_{22} & r_{23} & y_i \\ r_{31} & r_{32} & r_{33} & 0 \\ 0 & 0 & 0 & 1 \end{bmatrix}, \tag{12.3} $$

From (12.2) and (12.3), we have coordinates of i-joint:

Table 12.2 Parameters of manipulator links

Link i	Mass, m (kg)	Length, L (m)
1	0.45	0.2
2	0.25	0.2
3	0.15	0.2
4	0.15	0.2
5	0.15	0.2

$$\begin{cases} x_i = \sum_{k=1}^{i} L_k c_{1\ldots k}, \\ y_i = \sum_{k=1}^{i} L_k s_{1\ldots k} \end{cases} \tag{12.4}$$

Suppose that COM of each link i is the midpoint of that link. The mass and the length of link i are shown in Table 12.2:

COM of link i:

$$\begin{cases} x^i_{\text{com}} = \dfrac{x_{i-1} + x_i}{2} \\ y^i_{\text{com}} = \dfrac{y_{i-1} + y_i}{2} \end{cases} \tag{12.5}$$

Manipulators' COM coordinate:

$$\begin{cases} x_{\text{com}} = \dfrac{\sum_{i=1}^{5} m_i x^i_{\text{com}}}{\sum_i^n m_i}, \\ x_{\text{com}} = \dfrac{\sum_{i=1}^{5} m_i x^i_{\text{com}}}{\sum_i^n m_i} \end{cases}, \tag{12.6}$$

From (12.4), (12.5), and (12.6), we have the manipulator's COM coordinates:

$$x_{\text{com}} = 0.185_{C_1} + 0.115_{C_{12}} + 0.115_{C_{123}} + 0.045_{C_{1234}} + 0.115_{C_{12345}},$$

$$y_{\text{com}} = 0.185_{S_1} + 0.115_{S_{12}} + 0.115_{S_{123}} + 0.045_{S_{1234}} + 0.115_{S_{12345}}.$$

In the previous studies [16, 17], the tasks in the implementation of manipulation of UAVs with ground objects were given (Fig. 12.4).

Based on the diagram in Fig. 12.4 we will calculate the position of the robot arm when the system is in parking, takeoff, and object grasp stages. The calculation process is shown in Fig. 12.5, where: 1—the position of manipulator' end-effector when the system is in parking and takeoff stages; 2, 3, 4—the position of manipulators' end-effector when the system moves to point to the workspace; 5—the position of manipulators' end-effector when the system is at workspace. The end-effector has

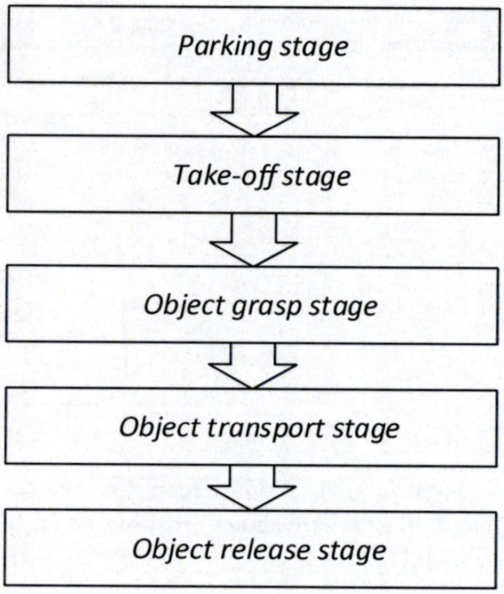

Fig. 12.4 Working diagram of aerial manipulation system

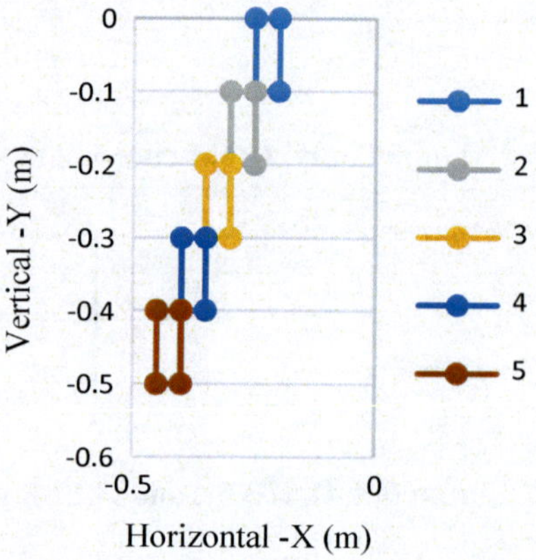

Fig. 12.5 Diagram calculating the position of the end-effector in space $\mathbf{S}_3$ according to Fig. 12.3a

the coordinates $(x_5; y_5)$.

$$\text{In 1-position:} -0.25\,\text{m} \le x_5 \le -0.20\,\text{m and} -0.1\,\text{m} \le y_5 \le 0\,\text{m}; \tag{12.7}$$

$$\text{In 2-position:} -0.30\,\text{m} \le x_5 \le -0.25\,\text{m and} -0.2\,\text{m} \le y_5 \le -0.1\,\text{m}; \tag{12.8}$$

$$\text{In 3-position:} -0.35\,\text{m} \le x_5 \le -0.30\,\text{m and} -0.3\,\text{m} \le y_5 \le -0.2\,\text{m}; \tag{12.9}$$

$$\text{In 4-position:} -0.40\,\text{m} \le x_5 \le -0.35\,\text{m and} -0.4\,\text{m} \le y_5 \le -0.3\,\text{m}; \tag{12.10}$$

$$\text{In 5-position:} -0.45\,\text{m} \le x_5 \le -0.40\,\text{m and} -0.5\,\text{m} \le y_5 \le -0.4\,\text{m}; \quad (12.11)$$

With each COM coordinate of the manipulator ($x_{\text{com}}; y_{\text{com}}$) we calculate:

$$\begin{cases} -0.001\,\text{m} \le x_{\text{com}} \le 0.001\,\text{m}, \\ y_{\text{com}} = -0.4\,\text{m} : d_y : 0, \\ d_y = 0.002\,\text{m}, \end{cases}$$

and consider a set of angles ($\theta_1, \theta_2, \theta_3, \theta_4, \theta_5$):

$$\begin{cases} \theta_i = -180^\circ : \Delta\theta : 0^\circ, i = 1 \div 5, \\ \Delta\theta = 10^\circ. \end{cases}$$

If there is available a set of angles ($\theta_1, \theta_2, \theta_3, \theta_4, \theta_5$), which satisfies the conditions (12.7), (12.8), (12.9), (12.10) and (12.11), then that is the angle set we need to find.

The results of calculation of the theta deflection angle between two consecutive links and coordinates of the joints are shown in Tables 12.3 and 12.4. Table 12.3 shows the theta deflection angle of the manipulators' links; Table 12.4 shows the coordinates of joints when the manipulator moves from parking and takeoff stages to workspace, and Table 12.5 is COM coordinates of manipulator. In Fig. 12.6, we selected a typical set of angles to describe the arm movement from parking position to workspace. The calculation results show that it is possible to determine the COM of the aerial manipulation system so that the system is most stable when the COM changes horizontally less than 1 mm. From Tables 12.4, 12.5 and Fig. 12.6 we also have determined the position of the robotic arm when it is in parking and takeoff stages with the theta angle set between successive links [−0.872664626; −0.872664626; −1.221730476; −1.221730476; −0.34906585] (radian) and the minimum length of the quadrotor's leg is approximately 0.40 m when it is in the parking stage.

Through the calculation results above, we propose the working plan of the M-UAV as following: According to the diagram in Fig. 12.4, the M-UAV will fly from parking position to the workspace. The robotic arm will pick up the object while UAV is in hover mode. In order to control the robotic arm from parking position to workspace while ensuring the least moving of COM on the horizontal axis, we propose a control plan for the robotic arm is control each link of the manipulator in Fig. 12.6.

Table 12.3 Theta deflection angle of the manipulators' links

Angle	θ_1 (Radian)	θ_2 (Radian)	θ_3 (Radian)	θ_4 (Radian)	θ_5 (Radian)
1	−0.872664626	−0.872664626	−1.047197551	−1.221730476	−0.698131701
2	−0.872664626	−0.872664626	−1.221730476	−1.221730476	−0.34906585
3	−0.698131701	−1.570796327	0	−2.094395102	0
4	−0.523598776	−1.745329252	−0.174532925	−1.570796327	−0.698131701
5	−0.872664626	−0.872664626	−0.698131701	−1.221730476	−0.872664626
6	−0.698131701	−1.221730476	−0.523598776	−1.047197551	−1.047197551
7	−0.872664626	−0.872664626	−0.872664626	−1.396263402	−0.34906585
8	−0.523598776	−1.570796327	−0.34906585	−0.872664626	−1.396263402
9	−0.523598776	−1.745329252	0	−1.396263402	−1.047197551
10	−0.34906585	−1.745329252	−0.698131701	−0.34906585	−1.570796327
11	−0.872664626	−0.872664626	−0.523598776	−1.221730476	−0.698131701
12	−0.698131701	−1.396263402	0	−1.396263402	−0.872664626
13	−0.698131701	−1.396263402	0	−1.570796327	−0.523598776
14	−0.523598776	−1.570796327	−0.34906585	−0.34906585	−1.745329252
15	−0.523598776	−1.396263402	−0.872664626	−0.174532925	−1.396263402
16	−0.34906585	−1.919862177	−0.174532925	−0.174532925	−1.919862177
17	−0.174532925	−2.094395102	−0.34906585	0	−1.745329252
18	−0.698131701	−1.221730476	−0.34906585	−0.698131701	−1.047197551
19	−0.698131701	−1.221730476	−0.34906585	−0.872664626	−0.872664626
20	−0.698131701	−1.047197551	−1.047197551	0	−1.221730476
21	−0.523598776	−1.570796327	−0.174532925	−0.523598776	−1.047197551
22	−0.523598776	−1.570796327	−0.174532925	−0.698131701	−1.047197551
23	−0.698131701	−1.047197551	−0.872664626	−0.174532925	−0.34906585
24	−0.698131701	−1.047197551	−0.872664626	−0.174532925	−0.523598776
25	−0.698131701	−1.047197551	−0.872664626	−0.174532925	−0.698131701
26	−0.523598776	−1.570796327	−0.174532925	−0.34906585	−0.698131701

12.4 Conclusion

We studied a UAV robotic manipulator consisted of 5 degrees of freedom and assigned it as a specific weight and size in which first two links of the manipulator play a role of a counterbalance weight. From the results of the calculation, it is possible to control the position of the manipulator's COM when the quadrotor is in hover mode, so that the aerial manipulation system is as stable as possible. However, in [18, 19] it is shown that there is a big difference between theoretical calculations and actual models. In order to ensure the practicality of the research, we will improve the developed model and mount a manipulator to test UAV stability in future work.

Table 12.4 Coordinate of the manipulators' joints

Coordinate of joints	Joint 1		Joint 2		Joint 3		Joint 4		Joint 5	
	x_1	y_1	x_2	y_2	x_3	x_1	y_1	x_2	y_2	y5
1	0.1286	−0.1532	0.0938	−0.3502	−0.0941	−0.4186	−0.2227	−0.2654	−0.2227	−0.0654
2	0.1286	−0.1532	0.0938	−0.3502	−0.1031	−0.3849	−0.2031	−0.2117	−0.2379	−0.0147
3	0.1532	−0.1286	0.0247	−0.2818	−0.1039	−0.435	−0.1723	−0.247	−0.2407	−0.0591
4	0.1732	−0.1	0.0446	−0.2532	−0.1086	−0.3818	−0.2371	−0.2286	−0.2371	−0.0286
5	0.1286	−0.1532	0.0938	−0.3502	−0.0594	−0.4787	−0.2326	−0.3787	−0.2673	−0.1818
6	0.1532	−0.1286	0.0848	−0.3165	−0.0684	−0.4451	−0.2563	−0.3766	−0.2911	−0.1797
7	0.1286	−0.1532	0.0938	−0.3502	−0.0794	−0.4502	−0.2079	−0.297	−0.2763	−0.109
8	0.1732	−0.1	0.0732	−0.2732	−0.08	−0.4018	−0.277	−0.367	−0.277	−0.167
9	0.1732	−0.1	0.0446	−0.2532	−0.0839	−0.4064	−0.2571	−0.3064	−0.2571	−0.1064
10	0.1879	−0.0684	0.0879	−0.2416	−0.1	−0.31	−0.3	−0.31	−0.3	−0.11
11	0.1286	−0.1532	0.0938	−0.3502	−0.0347	−0.5034	−0.2227	−0.435	−0.3227	−0.2618
12	0.1532	−0.1286	0.0532	−0.3018	−0.0468	−0.475	−0.2347	−0.4066	−0.3031	−0.2186
13	0.1532	−0.1286	0.0532	−0.3018	−0.0468	−0.475	−0.22	−0.375	−0.32	−0.2018
14	0.1732	−0.1	0.0732	−0.2732	−0.08	−0.4018	−0.2679	−0.4702	−0.3027	−0.2732
15	0.1732	−0.1	0.1048	−0.2879	−0.0831	−0.3563	−0.2801	−0.3911	−0.3485	−0.2031
16	0.1879	−0.0684	0.0594	−0.2216	−0.0938	−0.3502	−0.267	−0.4502	−0.3018	−0.2532
17	0.197	−0.0347	0.0684	−0.1879	−0.1048	−0.2879	−0.278	−0.3879	−0.3464	−0.2
18	0.1532	−0.1286	0.0848	−0.3165	−0.0438	−0.4697	−0.2407	−0.5044	−0.3693	−0.3512
19	0.1532	−0.1286	0.0848	−0.3165	−0.0438	−0.4697	−0.2438	−0.4697	−0.3723	−0.3165
20	0.1532	−0.1286	0.1185	−0.3255	−0.0695	−0.3939	−0.2574	−0.4623	−0.386	−0.3091

(continued)

Table 12.4 (continued)

Coordinate of joints	Joint 1		Joint 2		Joint 3		Joint 4		Joint 5	
	x_1	y_1	x_2	y_2	x_3	x_1	y_1	x_2	y_2	y5
21	0.1732	−0.1	0.0732	−0.2732	−0.0554	−0.4264	−0.2433	−0.4948	−0.3965	−0.3663
22	0.1732	−0.1	0.0732	−0.2732	−0.0554	−0.4264	−0.2523	−0.4611	−0.3809	−0.3079
23	0.1532	−0.1286	0.1185	−0.3255	−0.0547	−0.4255	−0.2427	−0.4939	−0.4427	−0.4939
24	0.1532	−0.1286	0.1185	−0.3255	−0.0547	−0.4255	−0.2427	−0.4939	−0.4396	−0.4592
25	0.1532	−0.1286	0.1185	−0.3255	−0.0547	−0.4255	−0.2427	−0.4939	−0.4306	−0.4255
26	0.1732	−0.1	0.0732	−0.2732	−0.0554	−0.4264	−0.2286	−0.5264	−0.4255	−0.4917

Table 12.5 COM coordinates of manipulator

Coordinate of COM	Xcom	Ycom	Coordinate of COM	X_{com}	Y_{com}
1	−0.0005	−0.2312	14	0.0004	−0.2409
2	0	−0.2143	15	0.001	−0.2199
3	−0.0009	−0.2081	16	0.0009	−0.2073
4	−0.0001	−0.1793	17	−0.0008	−0.1661
5	−0.0001	−0.2659	18	0.0002	−0.2808
6	0	−0.245	19	−0.0005	−0.2729
7	−0.0001	−0.2439	20	−0.0007	−0.2617
8	0.0009	−0.2175	21	0.0007	−0.2553
9	−0.0009	−0.2005	22	0.0005	−0.2459
10	0.0009	−0.1735	23	−0.0005	−0.2851
11	0.001	−0.284	24	−0.0003	−0.2825
12	−0.0007	−0.254	25	0.0004	−0.2799
13	0.0002	−0.248	26	0.0008	−0.2694

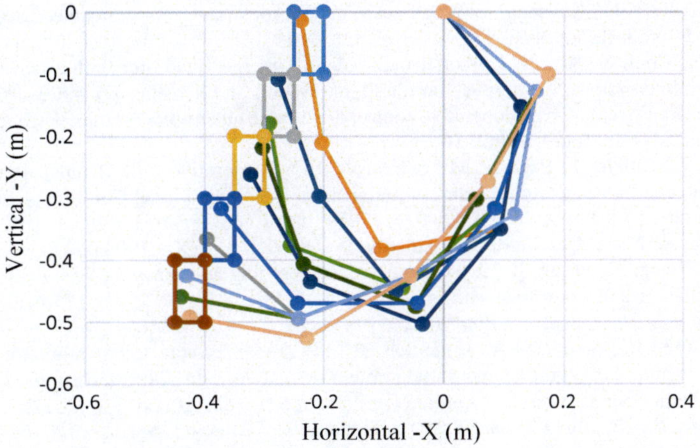

Fig. 12.6 Pose of manipulator from parking to working stage

Acknowledgements This work is supported by the RFBR (grant No. 18-58-76001_ERA.Net) in the framework of the ERA.Net Plus Project 99-HARMONIC.

References

1. Xilun, D.I, Pin, G.U.O., Kun, X.U., Yushu, Y.U.: A review of aerial manipulation of small-scale rotorcraft unmanned robotic systems. Chin. J. Aeronaut. **32**(1), 200–214 (2019). https://doi.org/10.1016/j.cja.2018.05.012
2. Lindsey, Q., Mellinger, D., Kumar, V.: Construction of cubic structures with quadrotor teams. In: Proceedings of Robotics, Science & Systems VII (2011)
3. Pounds, P.E., Bersak, D.R., Dollar, A.M.: Grasping from the air: hovering capture and load stability. In: IEEE International Conference on Robotics and Automation, pp. 2491–2498 (2011)
4. Kessens, C.C., Thomas, J., Desai, J.P., Kumar, V.: Versatile aerial grasping using self-sealing suction. In: International Conference on Robotics and Automation (ICRA), pp. 3249–3254 (2016)
5. Sreenath, K., Lee, T., Kumar, V.: Geometric control and differential flatness of a quadrotor UAV with a cable-suspended load. In: 52nd IEEE Conference on Decision and Control, pp. 2269–2274 (2013)
6. Lee, S.J., Kim, H.J.: Autonomous swing-angle estimation for stable slung-load flight of multi-rotor UAVs. In: International Conference on Robotics and Autom. (ICRA), pp. 4576–4581 (2017)
7. Khalifa, A., Fanni, M., Ramadan, A., Abo-Ismail, A.: New quadrotor manipulation system: Inverse kinematics, identification and ric-based control. Int. J. Recent Adv. Mech. Eng. (IJMECH) **4**(3), 39–58 (2015)
8. Imanberdiyev, N., Kayacan, E.: A fast learning control strategy for unmanned aerial manipulators. J. Intell. Robot. Syst. 94(3–4), 805–824 (2019)
9. Kim, S., Choi, S., Kim, H.J.: Aerial manipulation using a quadrotor with a two dof robotic arm. In: International Conference on Intelligent Robots and Systems, pp. 4990–4995 (2013)
10. Babaie, R., Ehyaei, A.F.: Cooperative control of multiple quadrotors for transporting a common payload. AUT J. Model. Simul. **50**(2), 147–156 (2018)
11. Wang, Z., Singh, S., Pavone, M., Schwager, M.: Cooperative object transport in 3d with multiple quadrotors using no peer communication. In: International Conference on Robotics and Automation (ICRA), pp. 1064–1071 (2018). https://doi.org/10.1109/icra.2018.8460742
12. Khalifa, A.: Novel Quadrotor Manipulation System (2019). arXiv:1904.05090
13. Yu, P., Wang, Z., Wong, K.C.: Exploring aerial perching and grasping with dual symmetric manipulators and compliant end-effectors. Int. J. Micro Air Veh. **11**, 1756829319877416 (2019). https://doi.org/10.1177/1756829319877416
14. Hafez, O.M.A., Jaradat, M.A., Hatamleh, K.S.: Stable under-actuated manipulator design for mobile manipulating unmanned aerial vehicle (MM-UAV). In: 7th International Conference on Modeling, Simulation, and Applied Optimization (ICMSAO), pp. 1–6 (2017)
15. Siciliano, B., Khatib, O. (Eds.): Springer Handbook of Robotics. Springer (2016). http://link.springer.com/book/10.1007/978-3-319-32552-1
16. Nguyen, V., Solenaya, O., Smirnov, P.: Issues of physical interaction of unmanned aircraft manipulators with ground objects. In: MATEC Web of Conference, vol. 161, p. 03021 (2018)
17. Ronzhin A.L., Nguen V.V., Solenaya O.Y.: Analysis of the problems of unmanned flying manipulators development and UAV physical interaction with ground objects. Trudy MAI 98, (2018). http://trudymai.ru/published.php?ID=90439
18. Mironov, V.I., Mironov, Y.V., Khegay, D.K.: Optimal determination of space objects orbit by angular measurements of ground-based optoelectronic stations. SPIIRAS Proc. **18**, 1239–1263 (2019). https://doi.org/10.15622/sp.2019.18.5.1239-1263
19. Muscolo, G.G., Caldwell, D., Cannella, F.: Calculation of the center of mass position of each link of multibody biped robots. Appl. Sci. **7**(7), 724 (2017). https://doi.org/10.3390/app7070724

Chapter 13
Active Phased Antenna Arrays for Long-Range Mobile Radars Based on Quadcopters

Denis Milyakov, Vladimir Verba, Vladimir Merkulov, and Andrew Plyashechnik

Abstract Strengthening the role of unmanned aerial vehicles (UAVs) for various purposes due to their group use allows obtaining a number of advantages in solving a wide class of tasks. At the same time, the new advantages of group use of UAVs are due to: the difficulty of separate observation of the group members and, accordingly, the difficulties of tracking and target distribution; the inability to serve the entire large group with the number of participants exceeding the capacity of the information control system of the opposing side; the increase in the behavioral complexity of the UAV in solving various problems through the use of artificial intelligence; random change of the spatial position of individual UAVs within the group, preventing their detection and selection of virtually all types of information systems. The noted advantages of the UAVs groups are especially pronounced in the implementation of such a new task as the formation and use of temporary active phased antenna arrays (APAA) of large sizes based on multicopter for the implementation of long-range radar systems. In this regard, the purpose of the paper is to present a variant of the algorithm for the formation and functioning of such a APAA. On the example of solving the task of maintaining an air object, the features of the operation of a radar with a APAA based on a group of UAVs are illustrated.

13.1 Introduction

The relevance of the group use of unmanned aerial vehicles (UAVs) is due to the capabilities and advantages that it gives when solving problems of both military-technical confrontation and the civilian sector. The influence of the group use of UAVs is clearly manifested in the formation of new opportunities for the construction of radio-electronic complexes (REC) for various purposes, performing the functions

---

D. Milyakov (✉) · V. Verba · V. Merkulov · A. Plyashechnik
Joint-Stock Company "Radio-Engineering Corporation "Vega"", Kutuzov Avenue 34, 121170 Moscow, Russia
e-mail: from_fn@mail.ru

A. Ronzhin and V. Shishlakov (eds.), *Proceedings of 15th International Conference on Electromechanics and Robotics "Zavalishin's Readings"*, Smart Innovation, Systems and Technologies 187, https://doi.org/10.1007/978-981-15-5580-0_13

of radar systems (RS), radio communications, radio navigation, etc. in solving many problems, among which, first of all, the following should be highlighted:

- Radar control of air, ground, and surface space for strategic important objects and dangerous goods in a limited range;
- Ensuring precision mapping of natural and artificial objects of various types;
- Control of outer space and the study of pollution of near-Earth orbits by space debris;
- An operational search for places of air crashes and shipwrecks;
- Providing high-speed communication lines (up to several Gbps) "air-space" and "air-surface" at the most important construction sites, in emergency situations and in man-made disasters;
- Use as an alternative to permanent satellite communications channels for data networks with a small coverage area (within a radius of 500–1000 km);
- Monitoring and traffic control on critical sections of roads and highways of municipalities, including in emergency mode;
- Comprehensive geophysical studies of the Earth's surface and near-surface layers;
- Sounding of the troposphere to determine the characteristics of fast-moving atmospheric processes and their trajectories.

At the same time, the research and development of antenna systems implemented on the basis of UAVs are an important task, which largely determines the functionality of such a REC.

The noted advantages of UAV groups are especially pronounced when implementing a new technique, which consists in providing requirements for the locator in solving the problem (in range, resolution, etc.) due to the formation and use of temporary active phased antenna arrays (APAA) of large sizes based on the UAV multicopter groups for the implementation of long-range radars. Such an application of UAV groups in solving radiolocation problems instead of stationary large-sized systems, which are potentially an easily accessible object for defeat by the opposing side, becomes especially relevant.

This technique potentially allows to simultaneously increase all the radar system performance indicators (overall effectiveness, survivability, dynamism, information content, sensitivity).

An example of the radar problem considered in this work, which is solved using a group of UAV quadcopters, is the construction of a radar with such an APAA and tracking of a maneuvering air object with an illustration of the features of the functioning of such a radar.

Such APAA is formed by a group of UAVs spaced in space and stabilized at points (nodes). Each UAV is a carrier of a unified set of equipment.

Viewing of the surrounding space is ensured by controlling the generated uniform radiation pattern (RP).

13.2 Literature Review

The formation of coordinated antenna arrays for airborne and ground-based groupings of means aimed at solving various communication and relay tasks in various modes ("air-to-air," "air-to-surface," "surface-to-air") has been the subject of a number of works in Russian, and foreign science. Among open publications, one can note, for example, [1–4]. At the same time, it is necessary to note the practical absence by the publication, and, accordingly, of proven solutions for various problems of radar.

Theory of control of single UAV multicopter has been the subject of many works, among which the sources significant for the project should be highlighted [5, 6].

One of the important problems necessary for building an APAA based on a UAV is the implementation of a reflective APAA screen. The presence of a reflective screen in the APAA allows for a single-beam mode, as well as to increase its gain [7]. Traditionally, reflective screens (ground plane) are a solid or mesh material with a high conductivity value (metal sheet). The disadvantages of such screens in the framework of the proposed project are their high weight, structural complexity, and significant wind loads.

An alternative way to implement a reflective screen is to use frequency-selective structures (FSS) [8–10], which are resonant structures consisting of metal elements, dielectric layers, and also absorption layers and are used as spatial filters. Such structures:

- Possess spatially selective and polarizing properties and are used as spatial filters;
- Possess frequency selectivity, are characterized by their own working band of radio frequencies, in the general case, different from the working band of frequencies of the antenna array;
- Have a frequency response such as a low-pass filter (LPF), a high-pass filter (HPF), a bandpass filter (BPF), a band-stop filter (BSF), etc.

A reflective screen based on heart rate can be realized by using heart rate with frequency-blocking spatial-frequency characteristic. The bandwidth-blocking heart rate is based on discrete conductive elements and, as a rule, is used to develop absorbers, mirrors, and reflective antennas [11]. Some implementation options for single-layer and multi-layered heart rate barriers were considered in [12–14]. In [15], the use of band-blocking heart rates as a conductive base of an antenna to reduce its effective scattering surface (ESS) was considered. However, in this work, heart rate is made using printing technology and its elements are located on a dielectric screen, which has the drawbacks noted above.

A feature of the proposed work is the ability to change the APAA period due to reconfiguration of the spatial position of the UAV during the transition from the search mode of an air object to the mode of tracking it. Changes in the APAA period or the transition from an equidistant to a non-equidistant lattice leads to a change in the mutual influence between the elements [16], which, in turn, will lead to a deterioration in the characteristics of the APAA. Therefore, when implementing a reconfigurable

APAA based on a UAV, it is necessary to use emitters with the possibility of adjusting their geometric and electrophysical parameters in the process of reconfiguring the lattice period. At the same time, each necessary period for the location of the emitters must correspond to their parameters of the emitter APAA and the element of the heart rate screen.

Reconfigurable antennas with the ability to change the operating frequency, polarization, beam pattern (BP) by introducing diode and MEMS switches, mechanical drives, etc. into their design were considered in [17, 18].

A feature of the use of reconfigurable emitters in the proposed project is the need to preserve, and not change, their radio technical characteristics (primarily the standing wave coefficient and the shape of the partial radiation pattern) when changing the APAA period.

13.3 Problem Solution

The problem of the formation and functioning of an APAA based on UAVs was solved in two stages.

The first was devoted directly to the removal of all UAVs from the initial points of their ground location to points with coordinates set for each UAV in the air and the formation of a stabilized fixed UAV group in this way.

This stage in the work was presented by the results of modeling an algorithm consisting of the following steps:

1. Construction of a reference trajectory for each UAV;
2. Calculation of derivatives of the trajectory up to 4th order inclusive;
3. Substitution of the derived derivatives into the laws of control of each UAV;
4. Recalculation of the status of each UAV.

At the first step, when constructing the UAV reference trajectory, mathematical relations were used for models of its movement in the horizontal plane [5]:

$$\begin{cases} \dot{\mathbf{p}}_h = \mathbf{v}_h, \\ \dot{\mathbf{v}}_h = -g\mathbf{A}_\psi \mathbf{\Theta}_h, \end{cases}$$

$$\mathbf{p}_h = \begin{bmatrix} p_x \\ p_y \end{bmatrix}, \ \mathbf{A}_\psi = \mathbf{R}_\psi \begin{bmatrix} 0 & 1 \\ -1 & 0 \end{bmatrix}, \ \mathbf{R}_\psi = \begin{bmatrix} \cos\psi & -\sin\psi \\ \sin\psi & \cos\psi \end{bmatrix}, \ \mathbf{\Theta}_h = \begin{bmatrix} \varphi \\ \theta \end{bmatrix},$$

where g is the acceleration of gravity, φ is the roll angle, θ is the pitch angle, Ψ is the yaw angle, in the vertical plane [5]:

$$\begin{cases} \dot{p}_z = v_z; \\ \dot{v}_z = g - \dfrac{f}{m}, \end{cases}$$

where f is the traction force of the engines, m is the mass of the UAV, and angular motion [5]:

$$\begin{cases} \dot{\Theta} = \boldsymbol{\omega}; \\ \mathbf{J}\dot{\omega} = \boldsymbol{\tau}, \end{cases}$$

where $\boldsymbol{\Theta}$ is the vector of Euler angles, $\boldsymbol{\omega}$ is the vector of angular velocity of rotation of the UAV, $\mathbf{J}$ is the matrix of inertia moments of the UAV, $\boldsymbol{\tau}$ is the total torque of the UAV screws.

The second step associated with the calculation of derivatives can be represented by the following relationships [6]:

$$\mathbf{R}_d = \left[\mathbf{b}_{1d}\ \mathbf{b}_{3d} \times \mathbf{b}_{1d}\ \mathbf{b}_{3d}\right] \in \mathrm{SO}(3);$$

$$\mathbf{b}_{1d} \equiv \Psi_d;\ \mathbf{b}_{3d} = \mathbf{R}_d\mathbf{e}_3 = -\frac{-k_x\mathbf{e}_x - k_v\mathbf{e}_v - mg\mathbf{e}_3 + m\ddot{\mathbf{x}}_d}{\|-k_x\mathbf{e}_x - k_v\mathbf{e}_v - mg\mathbf{e}_3 + m\ddot{\mathbf{x}}_d\|};$$

$$\boldsymbol{\Omega} \equiv \boldsymbol{\omega};\ \mathbf{e}_x = \mathbf{x} - \mathbf{x}_d;\ \mathbf{e}_v = \mathbf{v} - \mathbf{v}_d;\ \mathbf{e}_R = \frac{1}{2}\left(\mathbf{R}_d^T\mathbf{R} - \mathbf{R}^T\mathbf{R}_d\right)^{\vee};$$

$$\mathbf{e}_{\Omega} = \Omega - \mathbf{R}^T\mathbf{R}_d\Omega_d;$$

$$\dot{\mathbf{R}} = \mathbf{R}\cdot\hat{\Omega} \Rightarrow \dot{\mathbf{R}}_d = \mathbf{R}_d\hat{\Omega}_d \rightarrow \hat{\Omega}_d = \mathbf{R}_d^T\dot{\mathbf{R}}_d \rightarrow \hat{\dot{\Omega}}_d = \dot{\mathbf{R}}_d^T\dot{\mathbf{R}}_d + \mathbf{R}_d^T\ddot{\mathbf{R}}_d \rightarrow$$

$$\rightarrow \begin{bmatrix} \ddot{\Psi}_d = \ldots \\ \ddddot{x}_d = \ldots \end{bmatrix},$$

$$\hat{\Omega}_d = \mathbf{R}_d^T\dot{\mathbf{R}}_d \rightarrow *^{\vee} \rightarrow \boldsymbol{\Omega}_d = \left(\mathbf{R}_d^T\dot{\mathbf{R}}_d\right)^{\vee}, \quad (13.1)$$

where $\mathbf{R}$ and $\mathbf{R}_d$ are the current and desired UAV orientation matrix in space, respectively SO(3) is group of orthogonal rotation matrices; $(\mathbf{b}_1, \mathbf{b}_2, \mathbf{b}_3)$ is triple of vectors corresponding to the direction cosines associated with the UAV, in this case $\mathbf{b}_{1d}$ is determined by the desired UAV orientation around its axis OZ when moving along the trajectory as the law of change in time of rotation of the axis OX around the axis OZ of the UAV; $\mathbf{b}_{3d} \times \mathbf{b}_{1d}$ is determined by the vector product of the vectors $\mathbf{b}_{3d}$ and $\mathbf{b}_{1d}$; $\mathbf{e}_3$ is the unit vector of the OZ axis; k_x, k_v are some positive constants; $\mathbf{e}_x$, $\mathbf{e}_v$, $\mathbf{e}_R$, $\mathbf{e}_{\Omega}$ are UAV following errors along the desired trajectory; $\boldsymbol{\Omega}$ and $\boldsymbol{\Omega}_d$ are the vector and its desired value of the angular velocity of rotation in the coordinate system associated with the UAV; $*^{\vee}$ is the sign of the operation of converting the matrix into a vector.

As a result of performing the second step from (13.1), we obtain derivatives $\ddot{\Psi}_d$ and $\ddddot{x}$ of following along the desired trajectory.

At the third step, the propeller thrust $\mathbf{f}$ and torque $\boldsymbol{\tau}$ as UAV control signals were generated according to the laws:

$$\begin{bmatrix} f \\ \tau_x \\ \tau_y \\ \tau_z \end{bmatrix} = \begin{bmatrix} c_T & c_T & c_T & c_T \\ 0 & -dc_T & 0 & dc_T \\ dc_T & 0 & -dc_T & 0 \\ c_M & -c_M & c_M & -c_M \end{bmatrix} \begin{bmatrix} \omega_1^2 \\ \omega_2^2 \\ \omega_3^2 \\ \omega_4^2 \end{bmatrix};$$

$$\mathbf{f} = -(-k_x\mathbf{e}_x - k_v\mathbf{e}_v - mg\mathbf{e}_3 + m\ddot{\mathbf{x}}_d)\mathbf{R}\mathbf{e}_3;$$

$$\boldsymbol{\tau} = -k_R\mathbf{e}_R - k_\Omega\mathbf{e}_\Omega + \boldsymbol{\Omega} \times \mathbf{J}\Omega - \mathbf{J}\left(\hat{\Omega}R^T\mathbf{R}_d\boldsymbol{\Omega}_d - \mathbf{R}^T\mathbf{R}_d\dot{\Omega}_d\right);$$

$$\boldsymbol{\tau} = \begin{bmatrix} \tau_x & \tau_y & \tau_z \end{bmatrix}^T. \tag{13.2}$$

Here $\mathbf{f}$ and f is the thrust vector directed along the Z-axis of the UAV, and its module; $\boldsymbol{\tau}$ is torque of UAV screws with projections (τ_x, τ_y, τ_z) on the coordinate axis; c_T and c_M are dimensionless coefficients corresponding to the thrust and torque of the screws, respectively; d is the distance between the center of the UAV and the center of its ith engine; ω_i is angular rotation speed of the screws of the ith UAV engine; k_R, k_Ω are some positive constants.

For the UAV configuration considered in this work in the form of an X-shaped quadrocopter with perpendicular guides, the control signals in (13.2) can be converted into the following form [5]:

$$f = c_T(\omega_1^2 + \omega_2^2 + \omega_3^2 + \omega_4^2)$$

$$\tau_x = dc_T\left(\frac{\sqrt{2}}{2}\omega_1^2 - \frac{\sqrt{2}}{2}\omega_2^2 - \frac{\sqrt{2}}{2}\omega_3^2 + \frac{\sqrt{2}}{2}\omega_4^2\right);$$

$$\tau_y = dc_T\left(\frac{\sqrt{2}}{2}\omega_1^2 + \frac{\sqrt{2}}{2}\omega_2^2 - \frac{\sqrt{2}}{2}\omega_3^2 - \frac{\sqrt{2}}{2}\omega_4^2\right);$$

$$\tau_z = c_M(\omega_1^2 + \omega_2^2 + \omega_3^2 + \omega_4^2).$$

At the fourth step, the UAV state was calculated according to the model [5]:

$$\begin{cases} {}^e\dot{\mathbf{p}} = {}^e\mathbf{v}; \\ {}^e\dot{\mathbf{v}} = g\mathbf{e}_3 - \dfrac{f}{m}\mathbf{R}\mathbf{e}_3; \\ \dot{\mathbf{R}} = \mathbf{R}\left[{}^b\boldsymbol{\omega}\right]_\times; \\ J \cdot {}^b\dot{\omega} = -{}^b\boldsymbol{\omega} \times \left(J \cdot {}^b\boldsymbol{\omega}\right) + \mathbf{G}_a + \boldsymbol{\tau}. \end{cases} \tag{13.3}$$

$$\left[{}^b\boldsymbol{\omega}\right]_\times = \begin{bmatrix} 0 & -{}^b\omega_z & {}^b\omega_y \\ {}^b\omega_z & 0 & -{}^b\omega_x \\ -{}^b\omega_y & {}^b\omega_x & 0 \end{bmatrix} \tag{13.4}$$

In (13.3) ${}^e*$ means the indication of $*$ in the Earth's coordinate system, and ${}^b*$—in the onboard coordinate system associated with the UAV; $\mathbf{p}$ and $\dot{\mathbf{p}}$ are vectors of coordinates and rates of change for the center of gravity of the UAV, respectively; the operation $[*]_\times$ in (13.4) means constructing a matrix by a vector *.

The UAV trajectory was drawn according to the model [5]:

$$\begin{cases} {}^e\dot{\mathbf{p}} = {}^e\mathbf{v}; \\ {}^e\dot{\mathbf{v}} = g\mathbf{e}_3 - \frac{f}{m}\mathbf{R}\mathbf{e}_3; \\ \dot{\theta} = \mathbf{W} \cdot {}^b\boldsymbol{\omega}; \\ J \cdot {}^b\dot{\omega} = -{}^b\boldsymbol{\omega} \times \left(J \cdot {}^b\boldsymbol{\omega}\right) + \mathbf{G}_a + \boldsymbol{\tau}, \end{cases}$$

whose kinematics is associated with the representation of the rotation matrix $\mathbf{R}$ from (13.3) in the form of a matrix $\mathbf{W}$ of Euler angles.

Thus, we obtained the results of modeling the output of a group of 9 UAVs at given points in space, illustrated in Fig. 13.1.

The second stage of solving the problem was devoted to the formation of the RP of obtained APAA based on the UAV and the tracking of the air object (AO) during its maneuverable movement.

The research results of this step are illustrated in Figs. 13.2–13.4.

Also, studies were carried out on reconfiguration of the APAA in the case of nodes diverging from 0.25λ to 0.5λ and from 0.25λ to 0.9λ (here λ is the wavelength of the emitted signal), the results of which are illustrated in Figs. 13.5 and 13.6, respectively.

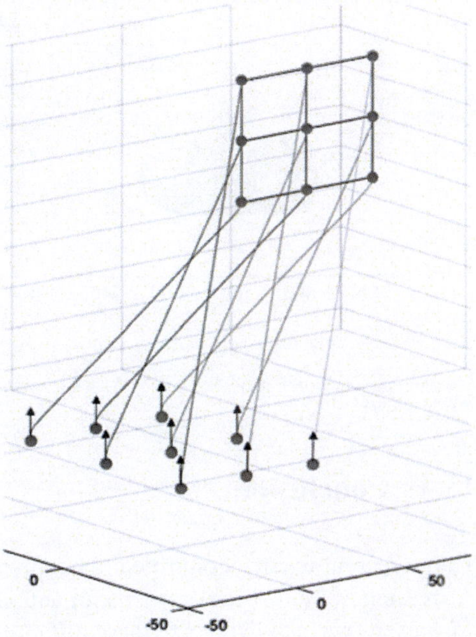

Fig. 13.1 Example of the formation of phased array UAV

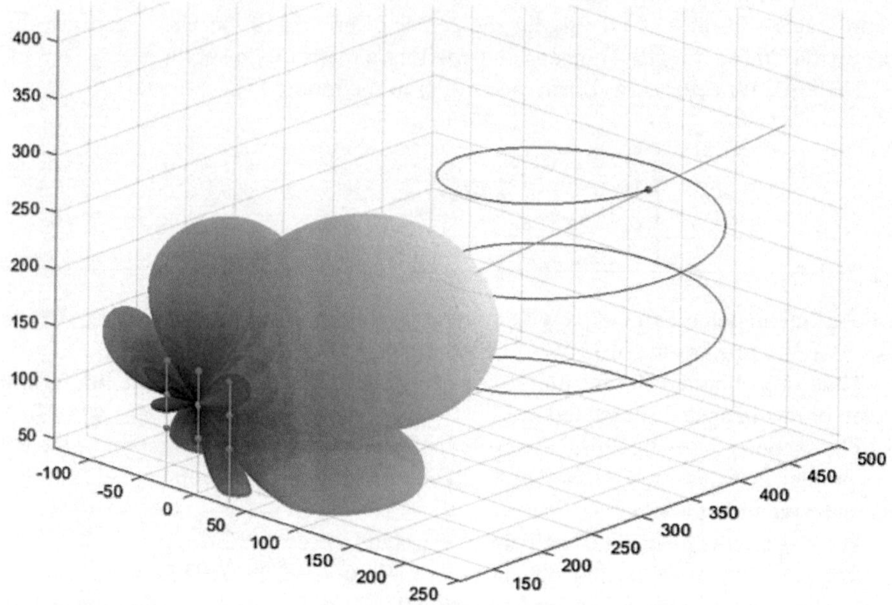

Fig. 13.2 Maintenance of maneuvering AO using APAA based on the UAV group

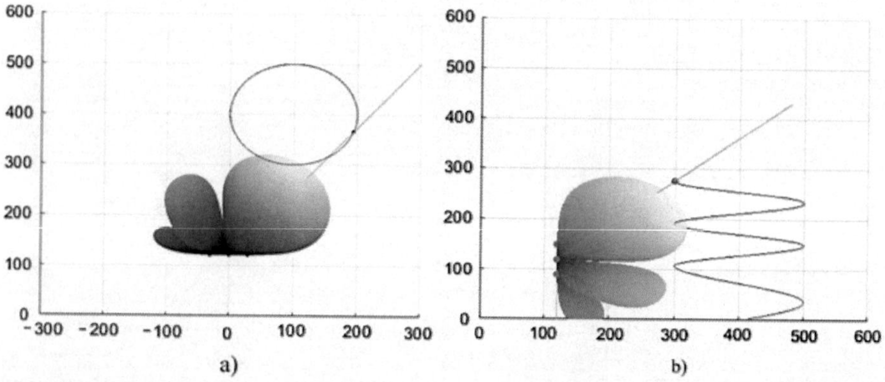

Fig. 13.3 Maintenance of maneuvering AO using APAA based on the UAV group (**a** top view, **b** side view)

13.4 Conclusion

The research results confirmed the emergence of standard difficulties for the processing of multi-beam patterns for antenna systems. It should be noted that there are various ways to overcome these difficulties that determine the various properties of the resulting systems.

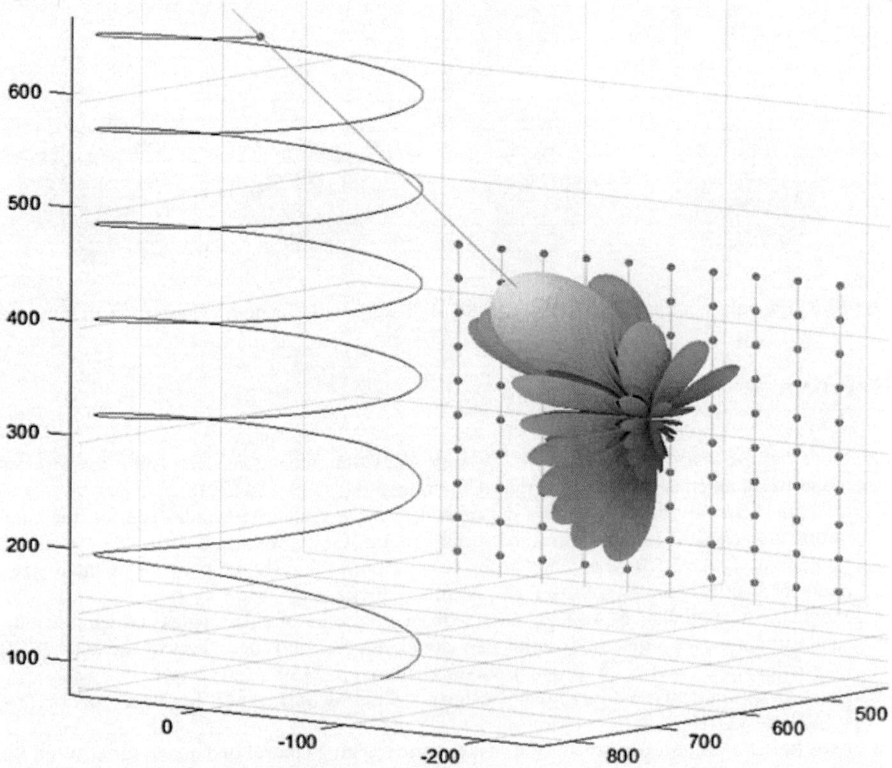

Fig. 13.4 Appearance of the radiation pattern of the APAA when changing its dimension

Fig. 13.5 Reconfiguration of the APAA in case of divergence of its nodes from 0.25λ to 0.5λ

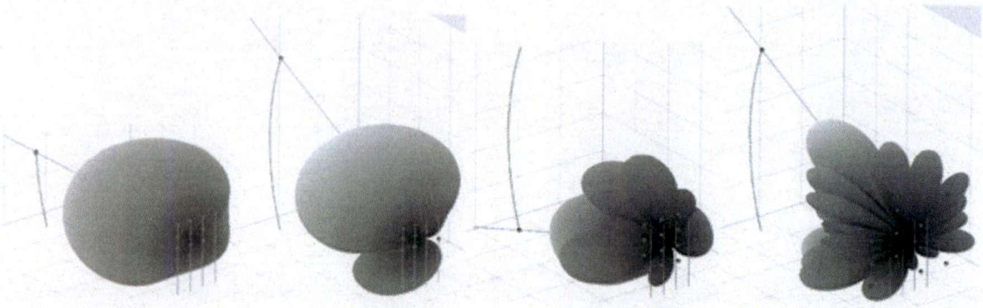

Fig. 13.6 Reconfiguration of the APAA in case of divergence of its nodes from 0.25λ to 0.9λ

References

1. Mozaffari, M., Saad, W., Bennis, M., Debbah, M.: Communications and control for wireless drone-based antenna array. IEEE Trans. Commun. **67**(1), 820–834 (2018)
2. Luchkin, S.A.: Terrestrial spatially distributed antenna systems for radio control lines and information transmission of unmanned aerial systems. Ph.D. Abstract (2010)
3. Ved'en'kin, D.A.: Focused antenna arrays as a part of radio-electronic means of small-sized unmanned aerial vehicles groups. Ph.D. Abstract (2012)
4. Mokhtari, A., Benallegue, A.: Dynamic feedback controller of euler angles and wind parameters estimation for a quadrotor unmanned aerial vehicle. In: Proceedings of the 2004 IEEE International Conference on Robotics & Automation, pp. 2359–2366 (2004)
5. Quan, Q.: Introduction to Multicopter Design and Control. Springer (2017). https://doi.org/10.1007/978-981-10-3382-7
6. Lee, T., Leok, M., McClamroch, N.H.: Geometric tracking control of a quadrotor UAV on SE (3). In: 49th IEEE Conference on Decision and Control (CDC), pp. 5420–5425 (2010)
7. Voskresenskij, D.I., Gos'uhin, V.L., Maksimov, V.M., Ponomar'ov, L.I.: Microwave Devices and Antennas. Radiotechnika, 376 p. (2008)
8. Wu, T.K.: Frequency-Selective Surface and Grid Array. Wiley, N.Y. (1995)
9. Munk, B.A.: Frequency-Selective Surfaces: Theory and Design. Wiley, N.Y. (2000)
10. Cwik, T.: Frequency-selective screens. Modern Antenna Handbook, pp. 779–828. (2008)
11. Narayan, S., Jha, R.M.: Electromagnetic techniques and design strategies for FSS structure applications. IEEE Antennas Propag. Mag. **57**(5), 135–158 (2015)
12. Yang, Y., Zhou, H., Wang, X.H., Mi., Y.: Low-pass frequency selective surface with wideband high-stop response for shipboard radar. J. Electromag. Waves Appl. **27**(1), 117–122 (2013)
13. Da Silva, M.R., Nóbrega, C.D.L., Silva, P.D.F., D'Assunção, A.G.: Dual-polarized band-stop FSS spatial filters using vicsek fractial geometry. Microw. Opt. Technol. Lett. **55**(1), 31–34 (2013)
14. Kashanianfard, M., Sarabandi, K.: Metamaterial inspired optically transparent band-selective ground planes for antenna application. IEEE Trans. Antennas Propag. **61**(9), 4624–4631 (2013)
15. Wang, W.-T., Gong, S.-X., Wang, X., Yuan, H.-W., Ling, J.: RCS reduction of array antenna by using bandstop FSS reflector. J. Electromagn. Waves Appl. **23**, 1505–1514 (2009)
16. Amitej, N, Galindo, V., Vu, Ch.: Theory and analysis of phased array antennas. Moscow. Mir (1974)
17. Bernhard, J.T.: Reconfigurable antennas. Modern Antenna Handbook, pp. 983–1003 (2008)
18. Constantine, J., Tawk, Y., Barbine, S.E., Christodoulou, C.G.: Reconfigurable antennas: design and application. Proc. IEEE **103**(3), 424–438 (2015)

Chapter 14
Collaborative Robots: Development of Robotic Perception System, Safety Issues, and Integration of AI to Imitate Human Behavior

Rinat Galin and Roman Meshcheryakov

Abstract The development of collaborative robotics as a research area is based on the study of safety and machine vision issues. The process of integrating artificial intelligence into robotic systems is gradually taking place. The process of intelligent robotic automation is based on a combination of machine learning of robots and high vision technologies on the way to interactive intelligent collaborative robotics. Given the lack of barriers for modern robots that work with humans, the issues of safety interaction remain the basic basis, which is considered an integral part of any implementation of new technological solutions. Such intelligent robotic solutions aimed to provide complementing and augmenting human capabilities, not replacing them. To take full advantage of this collaboration between robots and humans, we must understand how humans can most effectively augment robots and how robots can enhance what humans do best.

14.1 Introduction

The progress of modern robotics has allowed us to explore a new area of research—human–robot interaction. This process of interaction is called collaborative. Compared to industrial robots, collaborative robots have a number of advantages and include technological innovations. Due to their flexibility, safety in the workplace, and high productivity, such work will become popular in the next decade and will be massively introduced into people's lives and production processes.

The continuous process of implementing new technologies in collaborative robots in terms of vision systems and deep machine learning will be the basis for the evolution of the future robotics industry. This is primarily related to the safety of

R. Galin (✉) · R. Meshcheryakov
V.A. Trapeznikov Institute of Control Sciences of Russian Academy of Sciences, 65, st. Profsoyuznaya, 117997 Moscow, Russia
e-mail: rinat.r.galin@yandex.ru

R. Meshcheryakov
e-mail: mrv@ieee.org

A. Ronzhin and V. Shishlakov (eds.), *Proceedings of 15th International Conference on Electromechanics and Robotics "Zavalishin's Readings"*, Smart Innovation, Systems and Technologies 187, https://doi.org/10.1007/978-981-15-5580-0_14

human–machine interaction, since robots began working in a shared space with a human.

Robotic automation is growing every year, with smarter and more compact robots automating tasks that industrial robots cannot. Due to this, the market for collaborative robots will grow and will occupy a large part of the robot market in the future.

Nowadays, machine learning and deep learning are the latter has been one of the most used keywords in many conferences and reports in robotics community. The introduction of machine learning as a form of artificial intelligence is becoming an integral part of modern collaborative robots. This allows you to train the robot without processing large databases and many different applications. The capabilities of modern intelligent robots allow them to be multitasking on the place.

In this paper, we consider the concept of a collaborative robot, its characteristics, and security standards. We reveal the essence of the development of robots with artificial intelligence, or to be more precise, a form of artificial intelligence—machine learning. We tell you about one of the most popular collaborative robot YuMi and its parameters.

14.2 Collaborative Robots

A collaborative robot is capable of performing multiple tasks, as assisted by the humans. In contrast, autonomous robots are built to perform same task repeatedly and work independently. Modern robots no longer need enclosing perimeters, they become intelligent and flexible. There is a transition from cooperation and coexistence to collaboration. A human and a robot become partners, mutually reinforcing when working in a single workspace. Working in workplace is not only sequential, but also parallel.

To realize harmonious coexistence, new human–robot cooperation robot should be guaranteed in safety, comfortability, adaptation, and easy programming. Safety means to protect people from possible injuries caused by robots during interaction. Comfortability means actions of robot have to conform with the cognitive habits of people and people can recognize robot's intentions. Adaptation means that robot can understand people's demand and adapt to people's motion and different missions accurately. Easy programming means that it should be easy for people to program, to learn the operation methods, and to easily control the robot.

The main differences between collaborative robots and traditional robotic solutions in context of robot manipulators are present in Table 14.1. For a common understanding of the difference, four characteristics of a collaborative robot are described [1–6].

Collaborative robots have additional features [1, 5–7, 8] that are regulated in the standards for robotics solutions and special for safety issues:

Table 14.1 Differences between collaborative robots and traditional robotic solutions

Difference item	Description
Ability to interact safely with a human	Traditional robots perform operations in accordance with a given program without taking into account the people working next to them. Collaborative robots no need protective barriers. For example, collaborative robots pass parts to workers who then perform assembly or quality control operations
Risk reduction in the implementation of hazardous tasks.	Collaborative robots perform operations that pose a risk to humans. These risks include a safety transportation of sharp and sharpened or hot items, tightening of bolts, etc. At this time, specialists can focus on less complex aspects of production
Flexibility and learning	Control of traditional robots requires special programming skills. Programming collaborative robots is easier. Some robots are capable of self-learning. For example, it is necessary to show such robot any movement and it is accurate to repeat it
Possibility of wide use and quick adjustment	Collaborative robots are easy to reprogram and easy to move them to another points in the production chain. Many models of robots can be installing on any surface—horizontal, vertical, and even on the ceiling

- "Safety-rated monitored stop"—robot stops when human-operator enters the collaborative workspace and continues when workspace is free (allows for direct operator interaction with the robot system under specific circumstances);
- "Hand guiding"—robot movements are controlled by human-operator (operator uses a hand-operated device to transmit motion commands);
- "Speed and separation monitoring"—operator and robot system may move concurrently in the collaborative workspace;
- "Power and force limiting"—contact forces between human-operator and moving robot are technically limited to safe level (physical contact between the robot system and a human-operator can occur either intentionally or unintentionally).

The purpose of the standards [1, 7, 8] is to increase the interoperability of robots and their components and reduce the cost of their development and maintenance through standardization and harmonization of processes, interfaces, and parameters. In particular, the standards can be classified in three categories [5] that are shown in Fig. 14.1. This classification is not a complete description of safety standards, only common review.

Today there are already many robotic solutions based on the standards. Collaborative robots are being used in multiple verticals, which include supply chain management, manufacturing, and health care. Global collaborative robot market is projected to register a CAGR of 55.2% in the forecast period of 2019–2026 [9].

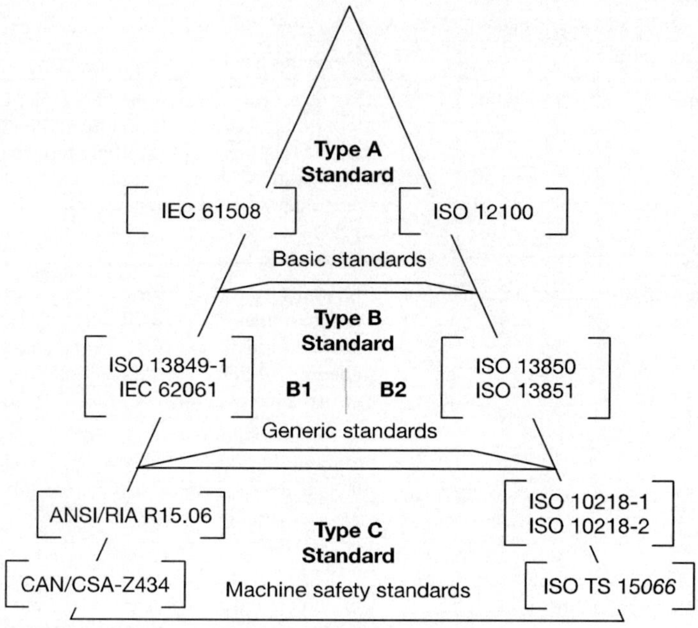

Fig. 14.1 Categories for safety standards and specification of types

Global collaborative robot market is segmented into six notable segments which are component, solution, safety method, payload capacity, application, and end user. Segmentations are as follows:

- On the basis of component, the market is segmented into hardware and software;
- On the basis of solution, the market is segmented into safety light curtain, safety laser scanner, camera-based 3D space monitoring system, and safe gripping;
- On the basis of safety method, the market is segmented into speed and separation monitoring, power and force limiting, safety-rated monitored stop, and hand guiding;
- On the basis of payload capacity, the market is segmented into up to 5, 5–10 kg, and above 15 kg;
- On the basis of application, the market is segmented into material handling, quality inspection, pick and place/drop, packaging and pelletizing, machine tending, molding operations, gluing and welding, test and inspection, assembly, polishing, laboratory analysis, entertainment, and others;
- On the basis of end user, the market is segmented into electrical and electronics, furniture and equipment, automotive, aerospace and defense, pharmaceuticals, plastics and polymers, metals and machining, food and beverages, cosmetics, and others.

Main key market players are Universal Robots A/S (Denmark), Rethink Robotics (US), ABB (Switzerland), KUKA AG (Germany), FANUC Corporation (Japan). There are more than 50 companies that develop robots including collaborative robots.

For example, Universal Robots is a leader in the collaborative robot market. The company plans to invest aggressively to strengthen its development, production, and sales capabilities for the future, including enhancing overseas production and sales networks in growth areas and accelerating the development of next-generation products. To boost the use of robotics, the company has set up robot centers, along with its distribution network, that offer the know-how required for introducing robotics to system integrators dealing in automation in various fields [9].

14.3 Development of Intelligent Robotic Perception System

As time goes on, robots are becoming smarter and more autonomous but they still cannot think, show emotions, and make decisions as a human. They do not have the ability to cognitive activity. The main hurdle to the cognitive ability is attributable to highly unpredictable nature of human behavior and thought process. This created a challenge to create a truly collaborative environment. As the cognitive behavior of humans is culturally specific, evolving continuously, developing a perfect AI system is a major challenge faced by the providers of AI technology for the robotics market.

The integration of AI into collaborative robots might create more challenges for robot programmers with rising complexities in tasks. However, the level of intelligence of robotic systems is determined by various machine learning technologies. Perhaps this is why it is believed that there is no unanimous acceptance of *strong* AI understanding, but rather machine learning and deep learning [9, 10].

The collaborative robots can automate high-volume, repeatable tasks and free humans to perform more intelligent assignments. In robotics, intelligent robotic perception system is understood as a system that endows the collaborative robot with the ability to perceive, comprehend, and reason about the surrounding environment.

The key modules of intelligent robotic perception system [11] are represented by sensors, environment modeling, and machine learning algorithms. These components are illustrated in Fig. 14.2. Agreeing with the authors [11, 12] that artificial intelligent is not yet developed at a high level (strong AI) in robotics, so this section of the paper will focus on machine learning.

Intelligent robotics perception system is very important for a collaborative robot to make decisions, plan, and operate in real-world environments, by means of numerous functionalities and operations from occupancy grid mapping to object detection.

Most of robotic perception systems use machine learning techniques, ranging from classical to deep learning approaches. Machine learning for robotic perception can be in the form of unsupervised learning, or supervised classifiers using handcrafted features, or deep learning neural networks, or even a combination of multiple methods [13].

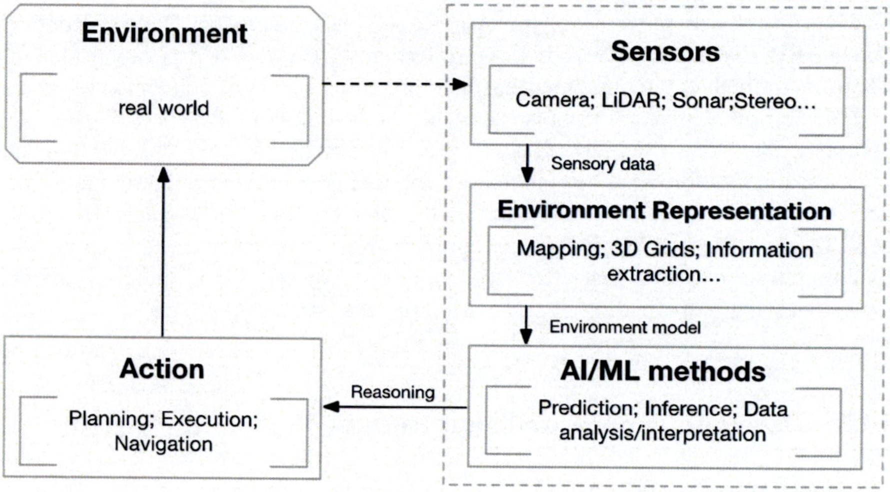

Fig. 14.2 Components of intelligent robotic perception system. Based on information from Premebida et al. [11]

The key object of machine learning research is data that is transmitted from the sensors of a robot or a group of robots. This data can carry information that needs to be combined for processing using machine learning. Data processing is necessary depending on the task in question for the application of machine learning.

Sensor representation of the environment is an important part of an intelligent robotic perception system. The display here covers the metric model and its semantic interpretation, which allows you to represent the environment. Machine learning is used at various levels in this process.

The increase in data volume and robustness combined with the deep learning approach made the development of perception systems easier and more accessible. Often, the result can be obtained directly from raw data by providing a large number of training examples. Choosing a method often comes down to getting the latest pre-prepared network from an online repository and fine-tuning it to the current problem. But it is worth noting that at the moment there is no ready-made deep learning solution for each problem. When processing big data, the need to use a pre-trained network for huge data becomes obvious. The recent surge and interest in deep learning methods for perception has greatly improved performance in a variety of tasks. These tasks are as follows: object detection, recognition, semantic segmentation, etc. One of the main reasons that allowed us to develop this direction is the availability of experimental experiments on publicly available data sets, as well as their comparison with other methods using standard criteria [11–15].

Perception is a very important part of a complex, embodied, active, and goal-driven intelligent robotic system.

14.4 Safety Zone of a Collaborative Robot in a Shared Space

First, safety zone in this context means common workplace for a collaborative robot and a human where the result and safety of human–robot collaboration could maximize and the harm with costs minimize. Second, this interaction between human and robot usually calls human–robot interaction. Considering human–robot interaction, you should understand the taxonomy of human–robot interaction. Taxonomies of human–robot interaction, cooperation, and collaboration proposed in these works [5, 6, 16, 17]. Taxonomic classification for the level of shared interaction among agents can have one of eight following values [17]: one human–one robot; one human–robot team; one human–multiple robots; human team–one robot; multiple humans–one robot; human team–robot team; human team–multiple robots; and multiple humans–robot team.

Human–robot interaction during collaboration is rapidly growing field of research in robotics and AI technologies. Safety issues become very important point in human–robot collaboration when developing robotic solutions.

Basis for all safety regulations and recommendations is through the three "Laws of Robotics" by science fiction writer Asimov [18]. Based on these laws, we can conclude that there are two distinct ways, in which a robot could inflict harm on a human being. It is a direct physical contact and a psychological harm. The main causes of harm considered at list of potential risks of hazardous from human–robot interaction during collaboration are described in follow block.

The analysis of publications with a higher impact factor allowed us to make the following conclusions on the main themes of ensuring the safety of human–robot interaction. These themes explain aspects of safety interaction through control, planning, prediction, and consideration of psychological factors [19, 20].

- *Control method.* The main factors are pre- and post-collision control methods for safety interaction between robot and human. Pre-collision control method deals prior to contact between a human and robot. Key points of this method are to limitless main robot parameters such as velocity or energy, or preventing collisions from occurring through the use of methods including defining safety regions, tracking separation distance, and guiding robot motion away from humans. Post-collision methods include different ways to minimize injury by switching between various control methods when a collision is detected. Also, for effective collaboration distinguishing between intentional and non-intentional contact, and allowing for safe physical contact.
- *Motion planning.* All works focus on planning safer robot paths and motions in order to avoid potential collisions. During this work we are taking in account various human-related parameters. Such parameters as separation distance between robot and human or gaze direction when forming motion plans.
- *Prediction.* This is the way to predict different solutions for safety interaction. This involves predicting human actions and motions through a variety of methods,

including sequence matching, probabilistic plan recognition, and motion characteristic analysis. We have to consider the predictability of the robot, because of multi-agent interaction when the robot role as assistant also important.

- *Psychological factors*. This sphere of safety methods includes different metrics through physiological sensing, questionnaires, and behavioral metrics and identifying. It is very important to understand the psychophysiological state of a human during interaction with a robot that could prevent possible harm. There are factors such as fear of working with a robot, fatigue, poor interaction experience, and others.

The development of collaborative robots will be expressed by the close interaction between human and robot. In such conditions, the working space of the robot will intersect with the personal space of a human. To ensure a safe and effective interaction between a human-operator and a robot(s), it is necessary to consider a complex task with a multitude of factors influencing the goal.

For example, collaborative robot YuMi-IRB 14050. The single-arm YuMi is ABB's most agile and compact collaborative robot. This robot was introduced in 2015 as truly the first collaborative robot. The technical characteristics of this robot meet all the requirements of the standards. Collaborative robot YuMi is compact and lightweight (9.5 kg) and supports mounting in any direction including ceiling, table, and wall mounting for rapid and flexible installation to fit into existing production lines. The ultralight magnesium arm rotates on seven axes to mimic human-like movements with greater agility than 6-axis robots. By ABB inform center this robot was specifically designed to meet the flexible production needs required by small parts assembly processes including consumer electronics, consumer goods, and small and medium enterprises [21]. The robot is presented in Fig. 14.3, and also its working range parameters are presented.

This collaborative robot has intuitive and easy-to-use lead through programming that workers can teach the robot motions and positions easily and quickly, greatly

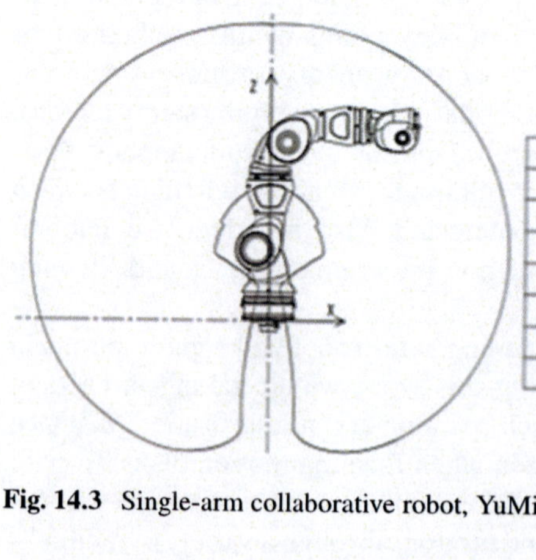

	Working range	
Ax.1	-168.5°	+168.5°
Ax.2	-143.5°	+43.5°
Ax.7	-168.5°	+168.5°
Ax.3	-123.5°	+80°
Ax.4	-290°	+290°
Ax.5	-88°	+138°
Ax.6	-229°	+229°

Fig. 14.3 Single-arm collaborative robot, YuMi-IBR 14050 with working range [21]

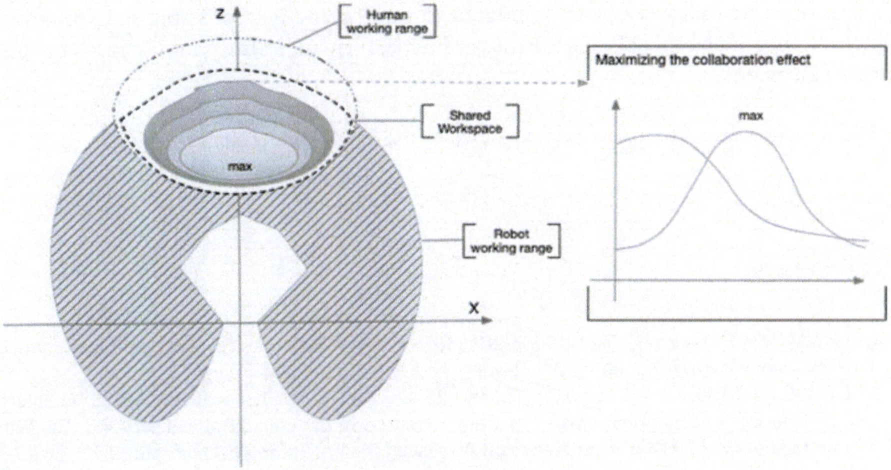

Fig. 14.4 Modeling of HRI in a shared workspace and areas of intersection of actions

simplifying robot programming. Combining this simplicity with the robot's deployment flexibility, it will help manufacturers in many industries to offset shortages of skilled workers and help reduce entry barriers for new potential robot users, particularly small and medium enterprises [21–23].

Maximizing the effectiveness of human–robot interaction is achieved at the intersection of the robot's work area and the human's work area during the task execution process. Figure 14.4 presented human–robot interaction (HRI) during collaboration in a shared workspace.

Achieving the maximum effect is possible in a certain area where the direct interaction of a human and a robot on the task is carried out in a shared workspace. Further results will contribute to the development of an interactive approach to HRI.

14.5 Conclusion

Making conclusions, we would like to emphasize that due to collaborative robots, production processes will bring cooperation and productivity to a new level. And it will be not so much about security and the latest machine vision technologies, but about interactivity. It will be possible to handle more complex tasks in full interactive human–robot mode, when the robot is not limited in actions or scenarios, and the person does not think about what is working with the robot. The robot will be able to make decisions by analyzing the state, words, and behavior of the person, and the person will perceive the robot as a full-fledged partner. There will be no need to develop software code, it will be enough to just work, and work on gestures and voice commands will be trained from the first shot.

Further research work will be related to finding the best possible human–robot interaction, which will achieve the maximum performance effect, taking into account safety measures.

Acknowledgements This research is supported by the RFBR Project No. 19-08-00331.

References

1. ISO/TC 299 Robotics: ISO/TS 15066:2016 Robots and robotic devices—collaborative robots. https://www.iso.org/standard/62996.html. Accessed 28 Jan 2020
2. Khalid, A., Kirisci, P., Ghrairi, Z., Thoben, K.-D., Pannek, J.: Towards implementing safety and security concepts for human-robot collaboration in the context of industry 4.0. In: 39th International MATADOR Conference on Advanced Manufacturing (Manchester, UK), pp. 0–7 (2017)
3. Robla-Gomez, S., et al.: Working together: a review on safe human-robot collaboration in industrial environments. IEEE Access **5**, 26754–26773 (2017)
4. Djuric, A., Urbanic, R., Rickli, J.: A framework for collaborative robot (cobot) integration in advanced manufacturing systems. SAE Int. J. Mater. Manuf. **9**(2), 457–464 (2016)
5. Villani, V., Pini, F., Leali, F., Secchi, C.: Survey on human–robot collaboration in industrial settings: safety, intuitive interfaces and applications. Mechatronics **55**, 248–266 (2018)
6. Galin, R., Meshcheryakov, R.: Review on human–robot interaction during collaboration in a shared workspace. In: International Conference on Interactive Collaborative Robotics, pp. 63–74. Springer, Cham (2019)
7. ISO 10218-1, 2:2011: Robots and robotic devices—safety requirements for industrial robots—part 1, 2: robot systems and integration, Geneva (2011)
8. Galin, R.R., Meshcheryakov, R.V.: Human-robot interaction efficiency and human-robot collaboration. In: Robotics: Industry 4.0 Issues & New Intelligent Control Paradigms, vol. 272, pp. 55–63. Springer, Cham (2020)
9. Global Collaborative Robot Market—Industry Trends and Forecast to 2026. Data Bridge Market Research. Report. https://www.databridgemarketresearch.com/reports/global-collaborative-robot-market. Accessed 27 Jan 2020
10. Ivanov, D.A., Ivanova, M.A., Sokolov, B.V.: Analysis of trends in enterprise management principles in the context of the development of Industry 4.0 technologies. SPIIRAS Proc. **5**, 97–127 (2018)
11. Premebida, C., Ambrus, R., Marton, Z.C.: Intelligent robotic perception systems. In: Applications of Mobile Robots. IntechOpen (2018).
12. Russell, S.J., Norvig, P.: Artificial Intelligence: A Modern Approach, 2nd edn. Prentice Hall, Upper Saddle River, New Jersey (2003)
13. Sünderhauf, N., et al.: The limits and potentials of deep learning for robotics. Int. J. Robot. Res. **37**(4–5), 405–420 (2018)
14. Bore, N., Jensfelt, P., Folkesson, J.: Multiple object detection, tracking and long-term dynamics learning in large 3D maps. https://arxiv.org/abs/1801.09292 (2018)
15. Handa, A., Patraucean, V., Badrinarayanan, V., Stent, S., Cipolla, R.: Understanding real world indoor scenes with synthetic data. In: 2016 IEEE Conference on Computer Vision and Pattern Recognition (CVPR), pp. 4077–4085 (2016)
16. Geravand, M., Flacco, F., DeLuca, A.: Human–robot physical interaction and collaboration using an industrial robot with a closed control architecture. In: Proceedings of the IEEE International Conference Robotics and Automation (ICRA), pp. 4000–4007. IEEE (2013)

17. Yanco, H., Drury, J.: Classifying human-robot interaction: an updated taxonomy. In: Proceedings of the IEEE International Conference Systems, Man and Cybernetics (SMC), pp. 2841–2846. IEEE (2004)
18. Asimov, I.: Runaround. Astounding Science Fiction. Street and Smith Pub, New York (1942)
19. Lasota, P., Shah, J.: Analyzing the effects of human-aware motion planning on close-proximity human-robot collaboration. Human Fact. **57**(1), 21–33 (2015)
20. De Luca, A., Flacco, F.: Integrated control for pHRI: Collision avoidance, detection, reaction and collaboration. In: 2012 4th IEEE RAS & EMBS International Conference on Biomedical Robotics and Biomechatronics (BioRob), pp. 288–295. IEEE (2012)
21. IRB 14050 Single-arm YuMi® Collaborative Robot. Presentation. https://new.abb.com/products/robotics/industrial-robots/irb-14050-single-arm-yumi. Accessed 30 Jan 2020
22. Kirschner, D., Velik, R., Yahyanejad, S., Brandstötter, M., Hofbaur, M.: YuMi, come and play with Me! A collaborative robot for piecing together a tangram puzzle. In: International Conference on Interactive Collaborative Robotics, pp. 243–251. Springer, Cham (2016)
23. Spacek, L., Vojtesek, J.: Ball & Plate Model on ABB YuMi Robot. In: Computer Science On-line Conference, pp. 283–291. Springer, Cham (2019)

Chapter 15
Hand Gestures Recognition Model for Augmented Reality Robotic Applications

Youshaa Murhij and Vladimir Serebrenny

Abstract Augmented reality (AR) is a research promising field. Its main idea is to integrate and merge the virtual world with the real world. Augmented reality could improve or enhance our perception of the real world by integrating virtual objects. The existing hand gesture applications related to augmented reality can detect hand motion or track it in addition to the ability to construct a 3D model of tracked hand using markers and motion sensing devices like Kinect, LeapMotion, and AR/VR instruments. In this paper, a hand gesture recognition model based on deep convolutional neural network is proposed to be used in 3D virtual environments for robotic teleportation. This model is tested on HTC VIVE Pro AR/VR instruments using the VIVE eye and on a Kinect v2 to control an industrial manipulator in real time using only the hand movements in both online and offline control modes.

15.1 Introduction

In the field of robotics, augmented and virtual reality applications focus on teleportation missions. So, the robot or the manipulator will follow the movements of the user in real time to achieve a remote-control task. Which means, the more accurate the tracking and recognizing system is, the more reliable teleportation system we get. Based on the researches in the last few years, we can say the neural networks and artificial intelligent took the lead from old image processing techniques which encouraged us to integrate them in our project to build recognition and tracking system to construct three-dimensional (3D) hand models to be used in 3D virtual environment.

Y. Murhij (✉) · V. Serebrenny
Robotic Systems and Mechatronics Department, Bauman Moscow State Technical University, Moscow, Russia
e-mail: murkhizhyu@student.bmstu.ru

V. Serebrenny
e-mail: vsereb@bmstu.ru

A. Ronzhin and V. Shishlakov (eds.), *Proceedings of 15th International Conference on Electromechanics and Robotics "Zavalishin's Readings"*, Smart Innovation, Systems and Technologies 187, https://doi.org/10.1007/978-981-15-5580-0_15

15.1.1 Related Work

In [1], they proposed a graph convolutional neural network (Graph CNN)-based method to reconstruct a full 3D mesh of hand surface that contains rich information of both 3D hand shape and pose. To train networks with full supervision, they create a large-scale synthetic dataset containing both ground truth 3D meshes and 3D poses. When fine-tuning the networks on real-world datasets without 3D ground truth, they proposed a weakly supervised approach by leveraging the depth map as a weak supervision in training. Their proposed method can produce accurate and reasonable 3D hand mesh and can achieve superior 3D hand pose estimation accuracy.

In [2], they introduced a novel convolutional network architecture for the task of human pose estimation. Features are processed across all scales and consolidated to best capture the various spatial relationships associated with the body. They showed how repeated bottom-up, top-down processing used in conjunction with intermediate supervision is necessary to improve the performance of the network.

In [3], they presented a formulation of CNNs in the context of spectral graph theory, which provides the necessary mathematical background and efficient numerical schemes to design fast localized convolutional filters on graphs.

In [4], they proposed a weakly supervised method, adapting from fully annotated synthetic dataset to weakly labeled real-world dataset with the aid of a depth regularizer, which generates depth maps from predicted 3D pose and serves as weak supervision for 3D pose regression.

In [5], they tested for potential ways to control KUKA KR10 industrial arm manipulator using virtual reality technology and checked for the advantages of applying this control method. They aimed to achieve this goal by establishing an interaction between the user and the manipulator inside a virtual environment developed using the game engine Unity3D and the HTC VIVE Pro headset for the virtual visualization.

In [6], they presented a residual learning framework to ease the training of networks that are substantially deeper than those used previously. They explicitly reformulate the layers as learning residual functions with reference to the layer inputs, instead of learning unreferenced functions. They provided comprehensive empirical evidence showing that these residual networks are easier to optimize, and can gain accuracy from considerably increased depth.

In [7], they proposed a novel graph-convolution operator to establish correspondences between filter weights and graph neighborhoods with arbitrary connectivity. The key novelty of their approach is that these correspondences are dynamically computed from features learned by the network, rather than relying on predefined static coordinates over the graph as in previous works.

Despite the development of more powerful hardware and software, the usage of AR systems is mainly limited to research and gaming applications. Main applications comprise installation and maintenance tasks [8, 9]. In industrial practice, for example, Boeing successfully uses Google Glass to visualize assembly instructions [10]. Funk [11] developed various systems, with which employees are supported in assembling and commissioning tasks by virtual information and instructions that are visualized via a projector. In addition, the user's actions are detected with a camera. Sand et al. [12] developed a similar system to provide assembly and picking instructions, which is based on a projector as well.

In [13], they made a survey about augmented reality applications and advanced robotic technologies that have been deployed in a camera shooting system to create a robotic cameraman so that the performance of AR broadcasting could be further improved.

15.1.2 *Hardware and Software*

This research is based on HTC VIVE Pro AR/VR instruments including its eye cameras in addition to KUKA KR3 R540 manipulator with the integration of Kinect-V2 camera and depth sensor. The mainly used platforms and software are Unity3D, MATLAB, and Autodesk MAYA in addition to Python PyTorch machine learning framework to train the deep neural network model.

15.2 Methodology and Main Procedure

15.2.1 *Overview*

The general idea of this research is to construct a full hand 3D model and determine its position information based on an input RGB picture from the headset of an AR/VR instrument or Kinect sensor so the 3D generated model could be visualized in AR/VR application and its position data could be used to achieve the teleportation mission of a manipulator in the 3D environment which is connected to a real robot in the laboratory.

The research focuses on using convolutional neural network (CNN) to generate the 3D mesh of the hand from the input image and using neural regressor to determine the 3D position of the hand details and its skeleton from the output mesh. We did not aim to discuss the performance and the accuracy of using such structures of deep networks as they are already discussed before [1, 2, 4].

15.2.2 *General Neural Network Structure*

First, the RGB picture that we capture using the eye cameras of the headset or the Kinect camera will be converted to a tensor as an input to a stack network which will detect and recognize the hand and isolate it in the picture by generating the heat map of this picture in addition to extract the hand features. The output of the two stacked network (the features and the heat map) will be an input to a residual network from which we get a latent feature vector that will be the input to a graph convolutional neural network CNN model which will work as a generator to the 3D hand mesh. The generated 3D mesh of the hand will be an input to a 3D pose regressor that determines the 3D location of the skeleton of the hand in 3D coordinates system [1].

We will start with the stacked hourglass network which's name is based on the successive steps of pooling and upsampling that are done to produce a final set of predictions. This network includes a repeated bottom-up, top-down processing used in conjunction with intermediate supervision. We chose to use the hourglass structure because of the need to capture information at every scale. While local evidence is essential for identifying features like hands and fingers, a final pose estimate requires a coherent understanding of the full picture. The hand's orientation, the arrangement of fingers, and the relationships of adjacent joints are among the many cues that are best recognized at different scales in the image. The hourglass is a simple, minimal design that has the capacity to capture all of these features and bring them together to output pixel-wise predictions for our model as shown in Fig. 15.1 [2].

Residual learning part. In this neural network with residual blocks, each layer feeds into the next layer and directly into the layers about 2–3 hops away. The convolutional layers mostly have 3×3 filters and follow two design rules: (i) for the same output feature map size, the layers have the same number of filters, and (ii) if the feature map size is halved, the number of filters is doubled so as to preserve the time complexity per layer. We perform down sampling directly by convolutional layers that have a stride of 2 (Fig. 15.2). The network ends with a global average pooling layer and a fully connected layer with SoftMax. The total number of weighted layers is 34 [6].

Implemented Graph CNN model. Chebyshev spectral graph CNN as shown in Fig. 15.3 was applied to generate 3D coordinates of vertices in the hand mesh and

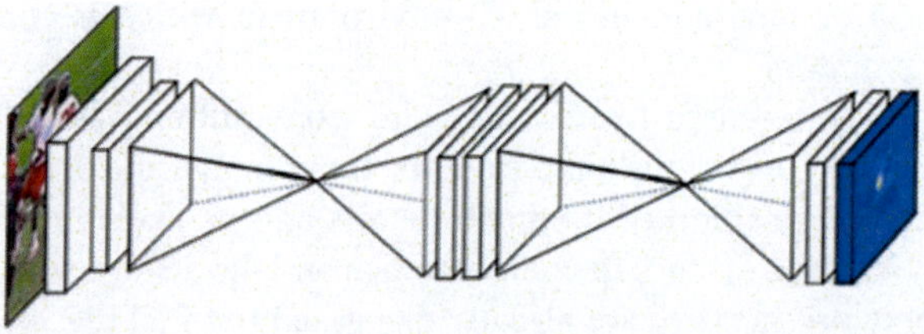

Fig. 15.1 Two-stacked hourglass network

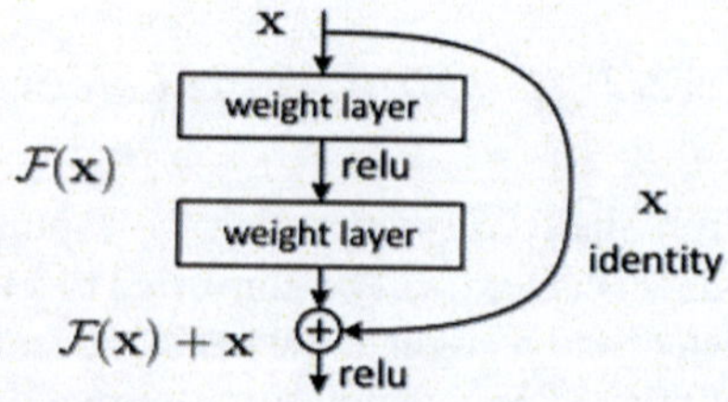

Fig. 15.2 Residual learning: a building block

Fig. 15.3 Our implemented graph CNN structure

$$L_{fully} = \gamma_H L_H + \gamma_M L_M + \gamma_J L_J .$$

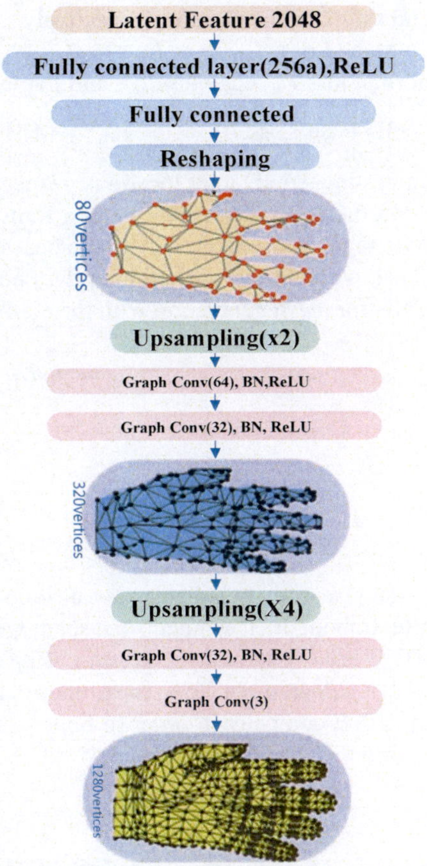

estimate 3D hand pose from the generated mesh. A 3D mesh can be represented by an undirected graph $M = (V, E, W)$, where $V = \{v_i\}_{i=1}^{N}$ is a set of N vertices in the mesh, $E = \{e_i\}_{i=1}^{E}$ is a set of E edges in the mesh, $W = (w_{ij})_{N \times N}$ is the adjacency matrix, where $w_{ij} = 0$ if $(i, j) \notin E$, and $w_{ij} = 1$ if $(i, j) \in E$ [3].

15.2.3 Applied Loss Functions

We first train the networks on a hand shape and pose dataset in a supervised manner. The networks are supervised by heat-map loss L_H, mesh loss L_M, and 3D pose loss L_j.

Heat-map Loss. $L_H = \sum_{j=1}^{J} \left\| H_j - \hat{H}_j \right\|_2^2$, where H_j and $\hat{H}_j$ are the ground truth and estimated heat-maps, respectively.

Mesh Loss. $L_M = \gamma_v L_v + \gamma_n L_n + \gamma_e L_e + \gamma_l L_l$, is composed of vertex loss L_v, normal loss L_n, edge loss L_e, and Laplacian loss L_l.

3D Pose Loss.$L_J = \sum_{j=1}^{J} \left\| \emptyset_j^{3D} - \hat{\emptyset}_j^{3D} \right\|_2^2$, where $\emptyset_j^{3D}$and $\hat{\emptyset}_j^{3D}$ are the ground truth and estimated 3D joint locations, respectively.

In the implementation, we first train the stacked hourglass network and the 3D pose regressor separately with the heat-map loss and the 3D pose loss, respectively. Then, we train the stacked hourglass network, the residual network, and the graph CNN for mesh generation with the combined loss L_{fully}:

$$L_{\text{fully}} = \gamma_H L_H + \gamma_M L_M + \gamma_J L_J$$

15.2.4 Custom Dataset

In our research, we built our own custom hand tracking dataset to fine-tune the model after training it on synthetic dataset as we used real hand photos extracted from HTC VIVE Pro eye with their skeleton using the VIVE hand tracking technology, and as shown below in Fig. 15.4, the proposed hand skeleton is similar to the one provided by VIVE company in order to build our own dataset for training the network and testing the proposed neural network model on their device.

15.2.5 Augmented Reality in Robotics

There are systems that not only display information and instructions, but also support the user while programming a robot. In this case, the use of AR systems is focused on path planning and the simulation of the created robot program in order to detect and avoid collisions.

When considering the presented systems in the field of robot programming, it is noticeable that they are characterized by an intricate structure, which generally consists of a camera system and an AR display. In addition, tablets or laptops are often used to display AR contents; however, such systems are impractical in application and do not allow for free-handed operations. In contrast, the HTC VIVE Pro connects a camera display with an AR display inside an HMD, which yields a simple system setup in total. In addition, unlike many other HMDs, holograms can be displayed in 3D and a permanently scanning process of the environment allows the user to place

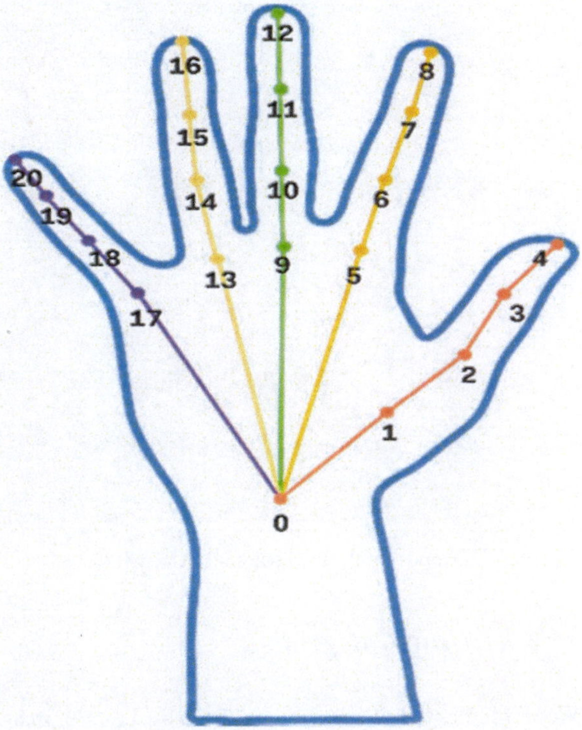

Fig. 15.4 Proposed hand skeleton

and fix holograms in the workspace. Furthermore, the gesture and voice control provide an intuitive control functionality.

15.2.6 Concept of Programming

The goal of this work is to develop an intuitive programming tool for robots. As shown in the previous section, AR is a method to allow human–machine interactions. In this context, the VIVE Pro offers new possibilities to realize communication and information processing, which are used to develop an application that converts virtual robotic movements into a real robot movement in real time. In this way, the first step will be the generation of virtual robot motion based on the previously explained deep neural network model in order to locate 3D position of the robot's end effector (Fig. 15.5).

Fig. 15.5 3D model of the manipulator in Unity3D

15.2.7 Unity Integration

We implemented the previously discussed model in a custom *U*nity*3D* package that is able to take live pictures from HTC VIVE Pro eye cameras and build the 3D hand model in *U*nity*3D* environment in addition to constructing the skeleton of the hand in order to take the position of the palm in each frame. Also we built a 3D model of the controlled robot which is KUKA KR3 R540 in the same environment with a complete forward/inverse kinematics and direct connectivity with the real robot in the laboratory using Ethernet connection in order to control the position of the end effector of a virtual robot in the same environment as shown in Fig. 15.5.

And we compared the output skeleton of the network with the constructed model from hand tracking VIVE technology and the actual and virtual trajectories, later.

15.3 Results

By comparing between the actual generated 3D trajectory which is computed using this model and the trajectory that is done by user's hand, we can compute the reliability of this model. In Fig. 15.6, where the red trajectory represents the 3D hand position and the blue one represents the trajectory of the manipulator, we can notice that the movement of the robot's end effector is a scaled version of the trajectory generated by the model, which gives the system enough reliability to be used to accomplish different tasks.

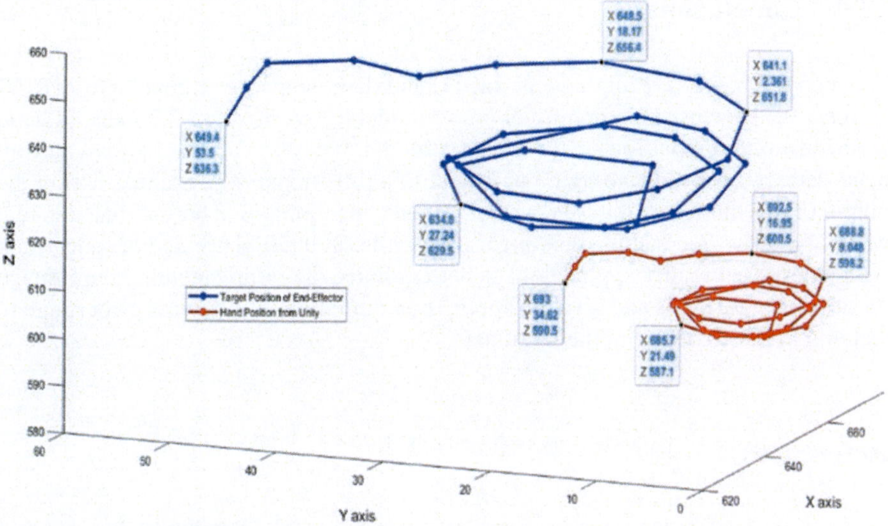

Fig. 15.6 3D coordinates of Palm's center of mass and the end effector of the manipulator in the same system coordinates

This method is able to not only estimate 2D/3D hand joint locations, but also recover a full 3D mesh of hand surface from RGB image. We show our model results on real-world hands examples took from the eye camera of HTC VIVE instrument. Figure 15.7 shows the 3D mesh of the captured hands as seen from the headset of AR instrument and their skeleton (the skeleton and the mesh are two separated models). We should mention here that when the hand model position is out of the range or cannot be reached by the robot, no control data would be sent to the robot.

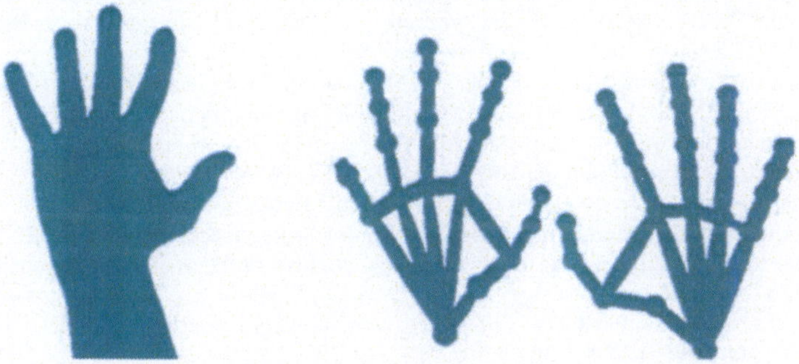

Fig. 15.7 Model and skeleton for a real hand in Unity3D

15.4 Conclusions

In our research, we have implemented a 3D hand shape and pose estimation from RGB images. We have used a graph CNN-based model to reconstruct a 3D mesh of hand surface from the input image. To train the model, we have created our own custom hand dataset, on which we train our model in a fully supervised manner. Using the output of this model in Unity3D, we controlled a manipulator in both virtual and real world. Finally, we conclude that integrating artificial intelligence and deep learning with augmented reality applications is better choice than implementing pure image processing algorithms and give the application more flexibility or less error range to respond to the changing in the input data.

References

1. Ge, L., et al.: 3D hand shape and pose estimation from a single RGB image. In: Proceedings of the IEEE Conference on Computer Vision and Pattern Recognition, pp. 10833–10842 (2019)
2. Newell, A., Yang, K., Deng J.: Stacked hourglass networks for human pose estimation. In: European Conference on Computer Vision, pp. 483–499 (2016)
3. Defferrard, M., Bresson, X., Vandergheynst, P.: Convolutional neural networks on graphs with fast localized spectral filtering. Adv. Neural Inf. Process. Syst. 3844–3852 (2016)
4. Cai, Y., Ge, L., Cai, J., Yuan, J.: Weakly-supervised 3d hand pose estimation from monocular RGB images. In: Proceedings of the European Conference on Computer Vision (ECCV), pp. 666–682 (2018)
5. Murhij, Y., Serebrenny, V.: An application to simulate and control industrial robot in virtual reality environment integrated with IR stereo camera sensor. IFAC-PapersOnLine **52**(25), 203–207 (2019). https://doi.org/10.1016/j.ifacol.2019.12.473
6. He, K., Zhang, X., Ren, S., Sun, J.: Deep residual learning for image recognition. In: 2016 IEEE Conference on Computer Vision and Pattern Recognition (CVPR), pp. 770–778 (2016). https://doi.org/10.1109/cvpr.2016.90
7. Verma, N., Boyer, E., Verbeek, J.: Feastnet: Feature steered graph convolutions for 3d shape analysis. In: Proceedings of the IEEE Conference on Computer Vision and Pattern Recognition, pp. 2598–2606 (2018)
8. Fraunhofer-Anwendungszentrum Industrial Automation, smart factory-design space (2017)
9. Elia, V., Gnoni, M.G., Lanzilotto, A.: Evaluating the application of augmented reality devices in manufacturing from a process point of view: an AHP based model. Exp. Syst. Appl. **63**, 187–197 (2016). https://doi.org/10.1016/j.eswa.2016.07.006
10. Sacco, A.: Google glass takes flight at Boeing. 2016. http://www.cio.com/article/3095132/wearable-technology/google-glass-takes-flight-at-boeing.html. Accessed 26 Nov 2021
11. Funk, M.: Augmented reality at the workplace: a context-aware assistive system using in-situ projection (2016)
12. Sand, O., Sebastian, B., Paelke, V., Carsten, R.: Virtual. Augmented Mixed Real. (2016). https://doi.org/10.1007/978-3-319-39907-2
13. Yan, D., Hu, H.: Application of augmented reality and robotic technology in broadcasting: a survey. Robotics **6**(3), 18 (2017)

Chapter 16
An Experimental Analysis of Different Approaches to Audio–Visual Speech Recognition and Lip-Reading

Denis Ivanko, Dmitry Ryumin, and Alexey Karpov

Abstract In this paper, we have analyzed different approaches to audio–visual speech recognition. We mainly focused on testing different modalities fusion techniques, rather than other parts of AVSR (e.g., feature extraction methods). Tree audio–visual modalities integration methods were under consideration, namely GMM-CHMM, DNN-HMM and end-to-end approaches, defined as the most promising and commonly found in scientific literature. The testing was performed on two different datasets: on GRID corpus for the English language and on HAVRUS corpus for the Russian. Obtained results once again confirms the superiority of neural network approaches compared to the others in conditions when we have enough data to effectively train NN models, which was demonstrated by our experiments on the GRID dataset. On a more compact in size HAVRUS database, the best recognition results were demonstrated by the traditional GMM-CHMM approach. This paper presents our vision on current state of audio–visual speech recognition field and possible directions for the further research.

16.1 Introduction

Significant success has been achieved in the field of automatic speech recognition over the past decades—there are many commercial applications that make investments in this area profitable. Automatic speech recognition systems are widely used in smart virtual assistants (e.g., Siri from Apple, Alexa from Amazon, Cortana from Microsoft, Alisa from Yandex, etc.), including voice control, voice search, voice text

---

D. Ivanko (✉) · D. Ryumin · A. Karpov
St. Petersburg Institute for Informatics and Automation of the Russian Academy of Sciences, St. Petersburg, Russia
e-mail: denis.ivanko11@gmail.com

D. Ryumin
e-mail: dl_03.03.1991@mail.ru

A. Karpov
e-mail: karpov@iias.spb.su

A. Ronzhin and V. Shishlakov (eds.), *Proceedings of 15th International Conference on Electromechanics and Robotics "Zavalishin's Readings"*, Smart Innovation, Systems and Technologies 187, https://doi.org/10.1007/978-981-15-5580-0_16

input, etc. Speech recognition technologies are widely used in various areas of business, for example, in telephony by creating voice self-service systems and automating the processing of incoming and outgoing calls, which allow receiving reference information and consultations, ordering goods and services, conducting surveys, questionnaires, informing, and any other scenarios. Automatic speech recognition systems are embedded in the voice interfaces of smart home systems, voice interfaces of electronic robots, household appliances, and even car navigation systems. With the help of automatic speech recognition technologies, social services to organize contactless interaction are created for people with disabilities.

Nevertheless, despite significant successes, the main goal of research, which was originally implied in ASR field—the free communication of man and the "machine"—has not yet been achieved. The development of the direction has revealed new difficulties that challenge researchers at the present stage.

For voice control tasks, automatic recognition systems have already reached a certain level and are used in a number of applications, however, the quality and reliability of computer-based speech analysis in real-world applications remain insufficient. One of the most difficult tasks in the field of ASR is the recognition of continuous speech. The complexity of the speech recognition problem is mainly related to the variability of its main parameters, which are influenced by many factors. First of all, this is a random component of the process of speech formation, which leads to a variety of descriptions of the same word spoken by the same speaker. More significant variability is associated with individual differences in the speech apparatus of different speakers due to the influence of the speaker's gender, age, accents, and the emotional and physical state of the speaker. In addition, the acoustic aspect, i.e., microphone change, its location relative to the mouth, surrounding acoustic environment also has a strong influence on recognition accuracy. It should be noted that the existing systems and models of automatic speech recognition are still significantly inferior to human speech abilities, especially in real conditions of functioning, which indicates their insufficient adequacy and makes the use of speech technologies in industry and everyday life ineffective. In many operating conditions (in particular, with a low sound signal quality, the presence of external noise or extraneous conversations), automatic recognition systems cannot provide the required quality of operation even when using various filtering and noise reduction methods.

At the same time, we must not forget that oral speech is the result of the interconnected work of the articulatory organs of the human vocal apparatus: vocal cords, larynx, lungs, tongue, lips, etc. Accordingly, oral speech is transmitted from the speaker to the listener both through sound/auditory, and through visual channels (communication modalities).

Natural speech is generated multimodally and transmitted simultaneously through several channels in the form of audio and video information, which duplicates and complements each other, and helps to correctly perceive speech in many difficult situations, e.g., when exposed to dynamic noise or simultaneous speaking of several people. In order to increase the accuracy and reliability of automatic speech recognition systems, in the late 90s, in addition to the analysis of acoustic information, the researchers also began to use visual information about speech [1]. The results of world studies on the creation of speech recognition systems that use audio–visual information show that they work better than uni-modal systems that use only audio information. When combining information flows from audio and video recognizers, the result of joint recognition may exceed the total recognition accuracy for each of the modalities. This fact is also known in human speech recognition as the McGurk effect [2].

There are several reasons for the lack of effectiveness of existing recognition systems for audio–visual Russian speech. Firstly, there are no publicly available audio–visual Russian speech corpora and generally accepted bases for assessing the quality of recognition systems for audio–visual Russian speech. Secondly, the Russian language, related to inflectional languages, has a significantly larger number of word forms compared to analytic languages. Thirdly, the task is complicated by the phonetic features of Russian speech, namely sluggish articulation, the phenomena of assimilation (combination of sounds), and reduction (reduction of sounds). An effective recognition system for Russian speech should be resistant to the acoustic variability of the speech signal caused by these phonetic features.

Given the above, we can conclude that it is necessary to develop methods, algorithms, and software that provide increased recognition accuracy of audio–visual Russian speech.

16.2 Background

In general, researchers divide the solution of the audio–visual speech recognition problem into two parts. First, in extracting the most informative features from each modality and second, in the most successful way of fusion both modalities [3]. A typical structure of bimodal speech recognition is shown in Fig. 16.1.

Thus, the first important step in the design of AV recognition systems is to create the best possible uni-modal recognizers. And the second step is to determine how to integrate knowledge from both modalities (in our case, audio and video) in order to preserve the useful information of each modality, and at the same time get rid of the shortcomings of both.

Representing modalities (audio and visual) in an appropriate and efficient feature space is an important step before their fusion.

A common metric for evaluating the quality of extracted features is the speech recognition accuracy. To ensure this, the constructed information features should contain the valuable characteristic of the speech signal. One of the most commonly used features in modern speech recognition systems is Mel-frequency cepstral coefficients (MFCC) [4]. Some other well-known representative features that have been used widely by speech recognition community are: perceptual linear prediction (PLP) [5] coefficients, linear predictive coding (LPC) [6], phoneme posterior features [7], and prosodic features [8].

To date, there are several basic types of visual features, which can be found in the literature. The most frequently used of them are: pixel-based features [9, 10]—raw

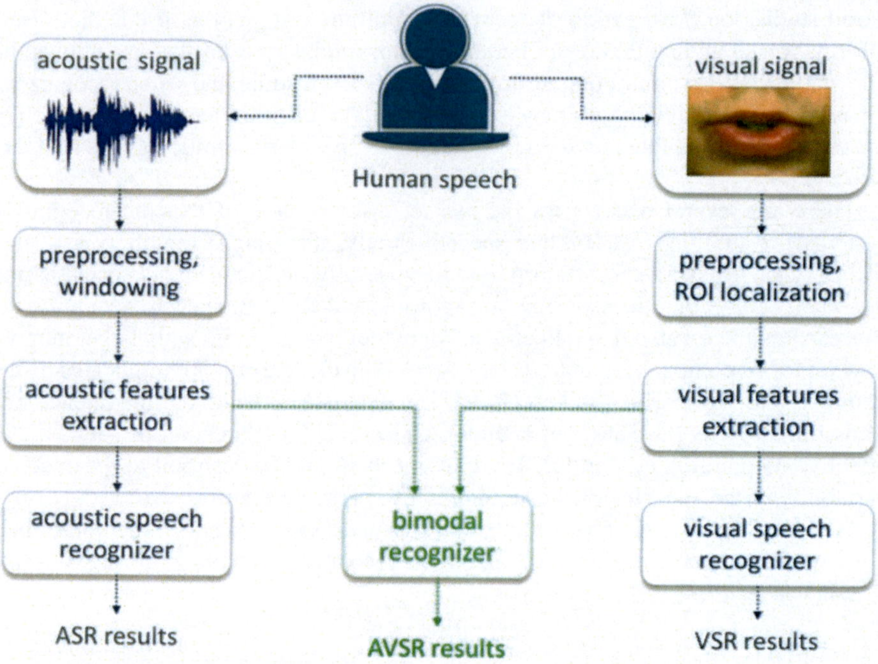

Fig. 16.1 Basic structure of audio-visual speech recognizer

pixel data used directly or after some image transformation; geometry-based features [11, 12]—geometric information of the talking lips is extracted as features.

There are several different approaches to fuse audio and visual information: early integration (feature-level fusion, as was done in the works [13, 14]), late integration (decision level fusion [15]), and hybrid fusion, which combines both methods (e.g., [16]).

Modalities fusion can be performed at several different levels. Combining at the feature vectors level is done before the start of the modeling process by combining features from all modalities. This method is called early integration. On the other hand, a way of combining modalities at the decision level is also possible. In this case, the modeling of each stream is performed separately, and then the outputs or solutions of the models are integrated to make the final decision [17]. This method is known as late integration. In addition, there is another way, which is called intermediate integration. Also, it is possible to combine two methods of integration, performing the fusion at two levels simultaneously, which is called a hybrid approach [18].

Initially, the most widespread methods were based on the use of Hidden Markov Models (HMM) for lip-reading and their coupled or multistream versions for audio–visual speech recognition [19]. However, at present, the approaches based on the use of neural networks of different architectures have become increasingly popular [20]. Some methods based on SVM can also be found in the literature [21].

16.3 Analysis and Implementation of State-of-the-Art Approaches

16.3.1 GMM-CHMM Model

To determine the Coupled Hidden Markov Model (CHMM) of a certain audio–visual unit of speech, the following parameters must be set $\lambda = \langle L, D, B, \gamma \rangle$:

1. The number of hidden states of the model L.
2. The distribution (matrix) of transition probabilities between states—$D = \{d_{ij}\}, 1 \le i \le L, 1 \le j \le L$.
3. The probability distribution of the appearance of observation symbols (feature vectors of audio–visual speech) in the states—$B = \{b_j(O)\}$. Usually, mixtures of normal (Gaussian) probability distributions are used:

$$b_j(O) = \sum_{m=1}^{M} c_{jm} N(O, \mu_{jm}, \sigma_{jm}), \sum_{m=1}^{M} c_{jm} = 1, \tag{16.1}$$

where O is the modeled vector of parameters (audio or video signal), c_{jm} is the weight coefficient of the mth component in state j, N is the probability density (Gaussian distribution) with the mean value and the standard deviation (variance) for the mth component mixtures in state j; M is the number of mixtures of normal distributions in the model.

4. Weights of informativeness (significance) $\gamma = \{\gamma^A, \gamma^V\}$ of speech modalities (audio and video streams), which can be adjusted during model training or adaptation to environmental conditions and the voice channel.

Figure 16.2 shows the topology of the model of the audio–visual unit of speech (phoneme/viseme pair) with several states for each stream of feature vectors. The circles indicate the CHMM states that are hidden for observation, and the squares indicate the mixture of normal distributions of the observation vectors in the states. The coupled hidden Markov model is a set of parallel HMMs, one for each information stream (modality). The model states at some point in time t for each HMM depend on the hidden states at time $t-1$ of all parallel HMMs. Thus, the general state of the CHMM is determined by the set of states of two parallel HMM. The advantage of such a topology is that it allows several streams of feature vectors to independently pass through the states of the model, which makes it possible to simulate natural asynchrony in audio and video data.

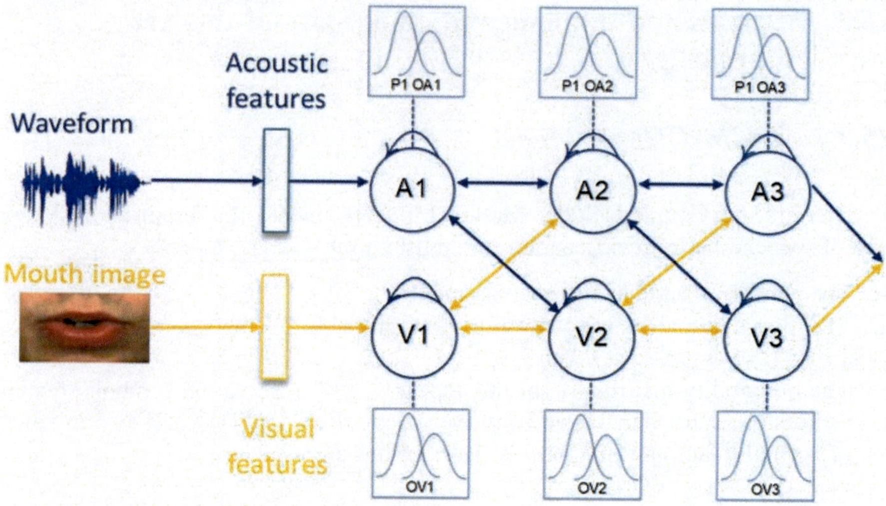

Fig. 16.2 CHMM audio–visual unit topology

16.3.2 DNN-HMM Model

An alternative way to calculate the probabilities of emissions is to use deep neural networks (DNN). The use of DNN as models for speech recognition has been the subject of many studies, in particular [22].

DNN training is the adjustment of the $\Theta = \{W, b\}$ parameters according to the available training examples $S = \{\mathrm{o^m}, \mathrm{y^m} | 0 \ll m < M\}$ where M is the number of examples, $\mathrm{o^m}$ and $\mathrm{y^m}$ are the observation vectors and the desired output vector for the mth example. The learning process is characterized by a learning criterion and a learning algorithm.

The learning criterion should strongly correlate with the ultimate goal of the task, so that the improvement of the learning criterion leads to an improvement in the final result. In classification tasks, y represents the probability distribution of acoustic classes, and the criterion for minimizing cross-entropy is often used.

Given the training criteria, the parameters of the model $\{W, b\}$ can be trained using the well-known error backpropagation algorithm, meant to use the differentiation rules of a complex function to calculate the gradient.

In its simplest form, model parameters are updated according to the formulas:

$$W_{t+1}^{l} = W_{t}^{l} - \varepsilon \Delta W_{t}^{l} p; \tag{16.2}$$

$$b_{t+1}^{l} = b_{t}^{l} - \varepsilon \Delta b_{t}^{l}, \tag{16.3}$$

where W_t^l and b_t^l are the weight matrix and the bias vector for the layer l after the tth update:

$$\Delta W_t^l = \frac{1}{M_b} \sum_{m=1}^{M_b} \nabla_{W_t^l} J(W_t, b_t; o^m, y^m), \tag{16.4}$$

$$\Delta b_t^l = \frac{1}{M_b} \sum_{m=1}^{M_b} \nabla_{b_t^l} J(W_t, b_t; o^m, y^m), \tag{16.5}$$

respectively: the average gradient of the weight matrix and the average gradient of the bias vector at iteration t, calculated from the training portion (batch) containing M_b examples, ε is the learning speed, and $\nabla_x J$ is the gradient of the function J with respect to x.

DNN cannot be directly used to model a speech signal, because the speech signal is a time series, while a deep neural network requires an input signal of a fixed length. The combination of artificial neural networks and hidden Markov models is an alternative paradigm for automatic speech recognition.

In our research, we used structure of a hybrid system combining deep neural networks and hidden Markov models (DNN-HMM) which is shown in Fig. 16.3. In this system, the dynamics of a speech signal is modeled using HMM, and DNN is trained to classify the states of hidden Markov models and is used to calculate the probabilities of emissions.

16.3.3 End-to-End Model

Due to the shortcomings of the HMM-based models, coupled with the development of deep learning technology, more and more researchers began to study end-to-end speech recognition and lip-reading. The end-to-end model is a system that directly maps input data sequence to sequence of words or other graphemes. Unlike the HMM-based model that is composed of several parts, the end-to-end model replaces multiple modules with one DNN, which makes the direct mapping of input signals (audio or video) into label sequences without hand-crafted intermediate states. Following parts are already included into end-to-end model: encoder (maps input sequence to feature sequence), aligner (makes the alignment between feature sequence and language), and decoder (decodes the final recognition results).

In our research, we used LipNet architecture developed by Oxford University, Google, and CIFAR [23] for visual speech recognition task. It is meant to map variable-length sequences of video frames to text sequences and is trained end-to-end. Figure 16.4 illustrates out implementation of this architecture.

The network main building blocks are spatiotemporal convolutions, gated recurrent units, and connectionist temporal classification.

Spaciotemporal Convolutions. A basic 2D convolution layer (without a bias and with unit stride):

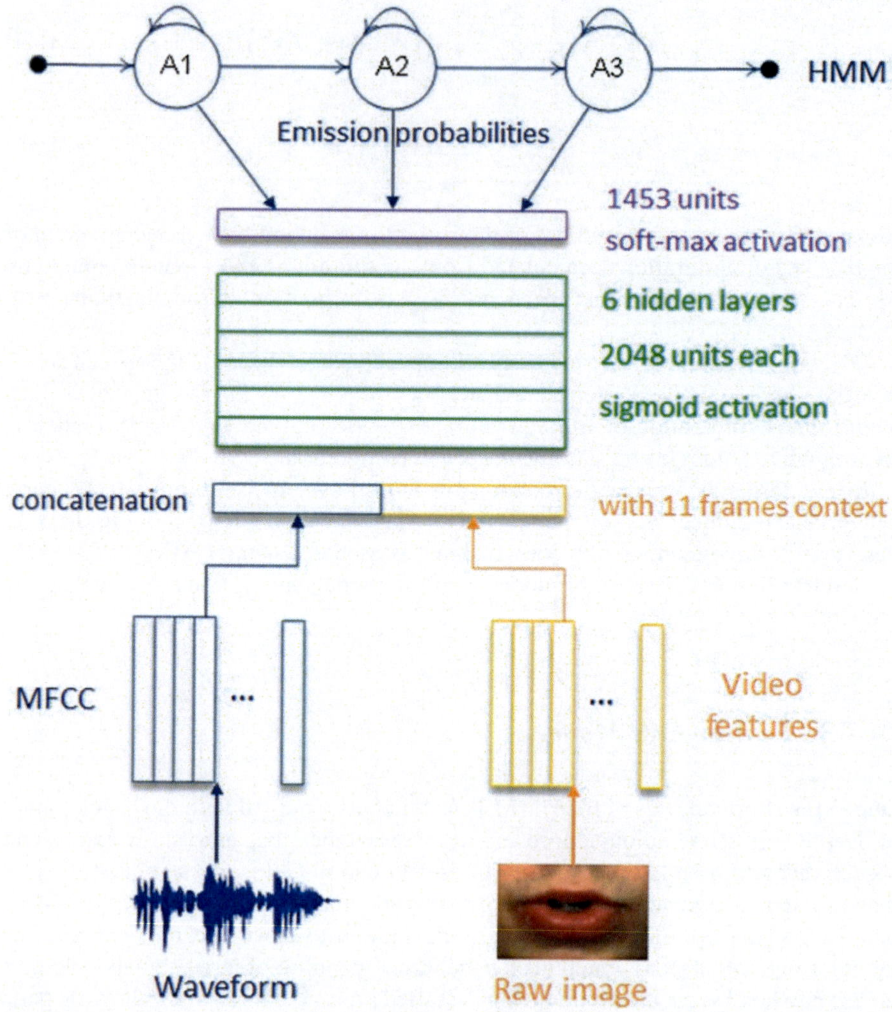

Fig. 16.3 Architecture of the used DNN-HMM model

$$[\mathrm{conv}(x, w)]_{c'ij} = \sum_{c=1}^{C} \sum_{i'=1}^{k_w} \sum_{j'=1}^{k_h} w_{c'ci'j'} x_{c,i+i',j+j'} \tag{16.6}$$

For input x and weights $w \in R^{C' \times C \times k_w \times k_h}$, where we define $x_{c,i} = 0$ for i, j out of bounds.

Hence, spaciotemporal convolutions:

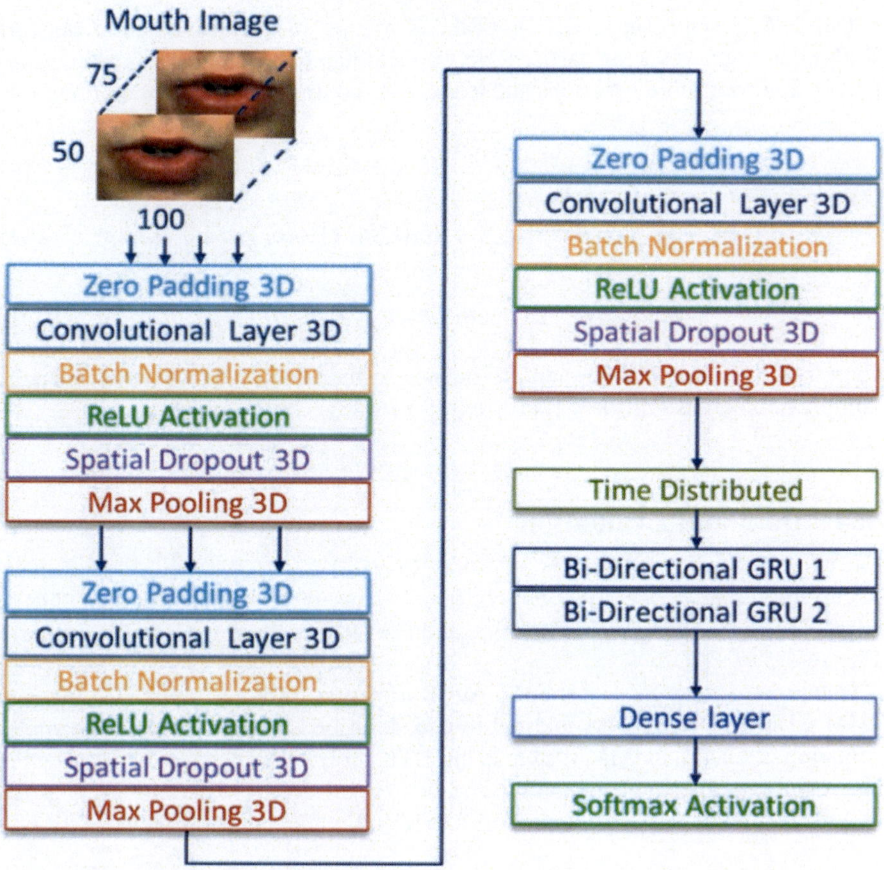

Fig. 16.4 LipNet network architecture

$$[\text{st conv}(x, w)]_{c'tij} = \sum_{c=1}^{C} \sum_{t'=1}^{k_t} \sum_{i'=1}^{k_w} \sum_{j'=1}^{k_h} w_{c'ct'i'j'} x_{c,t+t',i+i',j+j'}. \tag{16.7}$$

Connectionist Temporal Classification (CTC) *loss.* CTC is widely used in modern speech recognition as it eliminates the need for training data that aligns inputs to target outputs.

Let V denote the set of tokens that the model classifies at a single time-step of its output (vocabulary), and the blank-augmented vocabulary $\tilde{V} = V \cup \{\_\}$, where _ denotes the CTC blank symbol.

Define the function $\beta : \tilde{V}^* \to V^*$, that, given a string over $\tilde{V}$ deletes adjacent duplicate characters and removes blank tokens. For a label sequence, CTC defines $p(y|x) = \sum u \in \beta^{-1}(y) \text{s.t.} |u| = T^{p((u_1,\ldots,u_T|x)}$, where T is the number of time steps in the sequence model.

Gated Recurrent Unit (GRU). GRU is a type of recurrent neural network (RNN) that improves upon earlier RNNs by adding cells and gates for propagating information over more timesteps and learning to control this information flow:

$$[u_t, r_t]^T = \mathrm{sigm}\left(W_z z_t + W_h h_{t-1} + b_g\right); \tag{16.8}$$

$$\tilde{h}_t = \tanh\left(U_z z_t + U_h\left(r_t^\circ h_{t-1}\right) + b_h\right); \tag{16.9}$$

$$h_t = (1 - u_t)^\circ h_{t-1} + u_t^\circ \tilde{h}_t, \tag{16.10}$$

where $z \in \{z_1, ..., z_T\}$ is the input sequence to the RNN. $^\circ$ denotes element-wise multiplication and $\mathrm{sigm}(r) = 1/(1 + \exp(-r))$ [23].

16.4 Data and Evaluations

In current research, we used two different datasets for model training. GRID dataset for English language (Fig. 16.5) and HAVRUS dataset for Russian language (Fig. 16.6).

Using these datasets, we trained AVSR's of tree different topologies, namely GMM-CHMM, DNN-HMM, and end-to-end, described in Sects. 3.1–3.3. The implementation of GMM-CHMM system using HTK toolkit was previously described in the work [26]. The implementation of DNN-HMM AVSR was done using Kaldi toolkit and similar to approach proposed in [27]. Finally, the end-to-end AVSR was

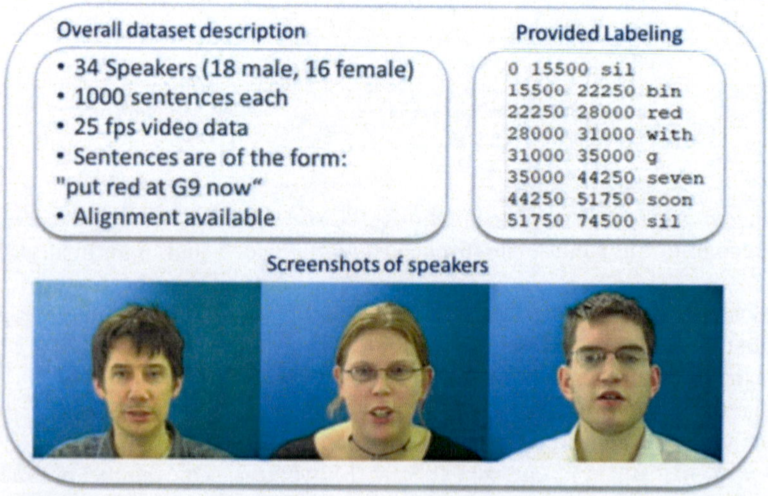

Fig. 16.5 GRID dataset description [24]

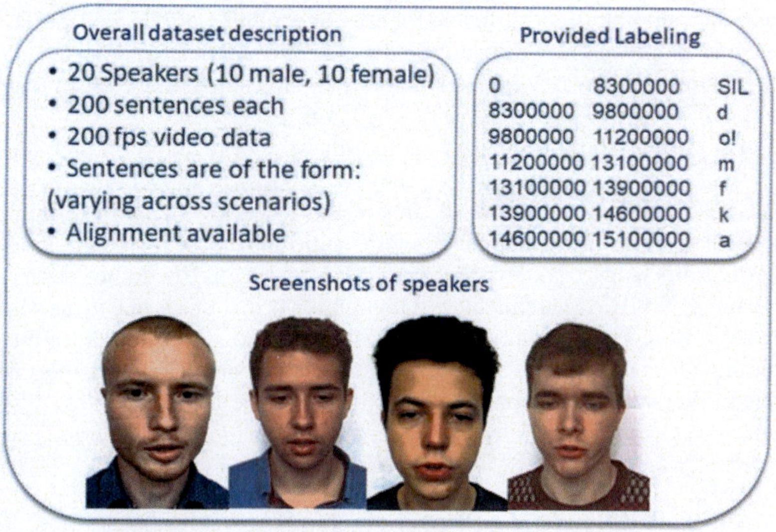

Fig. 16.6 HAVRUS dataset description [25]

implemented using tensorflow deep learning framework in a way of [28], on the base of the LipNet architecture proposed in [23]. Overall results are presented in Table 16.1.

As we can observe from obtained preliminary results on the video part of the GRID corpus, end-to-end approach (84.3%) outperforms GMM-CHMM (55.12%) and DNN-HMM (71.34%) ones. This result is quite expected and once again confirms the superiority of neural network approaches compared to the others in conditions when we have enough data to effectively train NN models. However, the situation drastically changes when we train same models on the video part of the Russian HAVRUS corpus.

In this case, the best recognition results were obtained using traditional GMM-CHMM approach (35.17%), and the results on NN topologies were much worse. These recognition results can be explained be the fact that HAVRUS corpus is simply much smaller (4000 phrases in total) than the GRID one (34,000 phrases in total)

Table 16.1 Word recognition rate (WRR, %) comparison of different AVSR systems

	GMM-CHMM	DNN-HMM	End-to-end
GRID video	55.12	71.34	84.3
HAVRUS audio	89.51	67.86	–
HAVRUS video	35.17	15.34	–
HAVRUS combined	91.28	64.72	–

and these data amount is not sufficient for effective training of NN models. The most illustrative example of this we observe in the case of end-to-end approach. Due to the lack of data the model was completely undertrained, resulted in the final recognition accuracy on HAVRUS corpus below one percent.

We also present some results on audio-only speech recognition and combined audio–visual speech recognition on HAVRUS dataset. It is demonstrated that combined use of audio and visual streams results in improvement of speech recognition accuracy (91.28%), compared with audio-only (89.51%) and video-only (35.17%) results using traditional GMM-HMM approach. We do not observe such results using DNN-HMM architecture. Our opinion is that this is due to the small size of the HAVRUS corpus, the amount of data of which was not enough for the effective training of models (only 15.34% WRR for visual part). However, this certainly deserves further study and experimental confirmation.

16.5 Conclusions

In this paper, we have analyzed the effect of different audio–visual fusion techniques on word recognition accuracy rate for English and Russian languages (on GRID and HAVRUS corpora, respectively). Specifically, we have compared GMM-CHMM, DNN-HMM, and end-to-end methods to AV speech recognition. All experiments on GRID dataset have clearly shown that NN-based methods are superior to traditional GMM-CHMM approach to audio–visual speech recognition. However, when it comes to small-sized corpora, such as HAVRUS (4000 phrases in comparison of 34,000 of GRID), the situation becomes different. For the HAVRUS dataset, the traditional GMM-CHMM approach demonstrated the best recognition results.

Acknowledgements This research is supported by the RFBR projects No. 19-29-09081 (Section 3) and No. 18-37-00306 (Sections 2 and 4), as well as state research No. 0073-2019-0005.

References

1. Katsaggelos, K., Bahaadini, S., Molina, R.: Audiovisual fusion: challenges and new approaches. Proc. IEEE **103**(9), 1635–1653 (2015)
2. McGurk, H., MacDonald, J.: Hearing lips and seeing voices. Nature **264**, 746–748 (1976)
3. Ivanko, D., et al.: Multimodal speech recognition: increasing accuracy using high speed video data. J. Multimodal User Interfaces **12**(4), 319–328 (2018)
4. Davis, S.: Comparison of parametric representations for monosyllabic word recognition in continuously spoken sentences [Text]. Proces.s IEEE Trans. Acoust. Speech Signal **28**(4), 357–366 (1980)
5. Hermansky, H.: Perceptual linear predictive (PLP) analysis of speech [Text]. J. Acoust. Soc. Am. **87**(4), 1738–1752 (1990)
6. Bundy, A., Wallen, L.: Linear predictive coding. In: Catalogue of Artificial Intelligence. Springer-Verlag (1984)

7. Bahaadini, S., Asaei, A., Imseng, D., Bourlard, H.: Posterior-based sparse representation for automatic speech recognition. In: Proceeding of Interspeech, pp. 2454–2458 (2014)
8. Konar, A., Chakraborty, A.: Emotion Recognition: A Pattern Analysis Approach. Wiley (2015)
9. Hong, X., Yao, H., Wan, Y., Chen, R.: A PCA Based visual DCT feature extraction method for lip-reading. In: Proceedings International Conference Intelligent Information Hiding Multimedia, Signal Process, pp. 321–326 (2006)
10. Ivanko, D., Ryumin, D., Kipyatkova, I., Axyonov, A., Karpov, A.: Lip-Reading Using pixel-based and geometry-based features for multimodal human–robot interfaces. In: Proceedings of 14th International Conference on Electromechanics and Robotics "Zavalishin's Readings", pp. 477–486 (2020)
11. Cetingul, H., Yemez, Y., Erzin, E., Tekalp, A.: Discriminative analysis of lip motion features for speaker identification and speech-reading. Proc. IEEE Trans. Image Process. **15**(10), 2879–2891 (2006)
12. Ivanko, D., Ryumin, D., Axyonov, A., Zelezny, M.: Designing advanced geometric features for automatic Russian visual speech recognition. In: Proceedings of International Conference on Speech and Computer, pp. 245–254 (2018)
13. Ngiam, J. et al.: Multimodal deep learning. In: Proceedings of 28th international Conference of Machine Learning, pp. 689–696 (2011)
14. Chetty, G., Wagner., M.: Audio-visual multimodal fusion for biometric person authentication and liveness verification. In: Proceedings NICTA-HCSNet Multimodal user interaction workshop, vol. 57, pp. 17–24 (2006)
15. Atrey, P.K., Hossain, M.A., Saddik, E., Kankanhalli, M.S.: Multimodal fusion for multimedia analysis: A survey. Multimedia Syst. **16**(6), 345–379 (2010)
16. Xu, H., Chua, T.S.: Fusion of AV features and external information sources for event detection in team sport video. ACM Trans. Multimedia Comput. Commun. Appl. **2**(1), 44–67 (2006)
17. Snoek, C.G., Worring, M., Smeulders, A.W.: Early versus late fusion in semantic video analysis. In: Proceedings 13th Annual ACM International Conference on Multimedia, pp. 399–402 (2005)
18. Wu, Z., Cai, L., Meng, H.: Multi-level fusion of audio and visual features for speaker identification. In: Advances in Biometrics, pp. 493–499 (2005)
19. Ivanko, D., et al.: Using a high-speed video Camera for robust audio-visual speech recognition in acoustically noisy conditions. SPECOM **2017**, 757–766 (2017)
20. Chung, J.S., Zisserman, A.: Lip reading in the wild. In: Computer Vision—ACCV 2016. pp. 87–103 (2016)
21. He, J., Zhang, H.: Lipreading recognition based on svm and dtak. In: 2010 4th International Conference on Bioinformatics and Biomedical Engineering, pp. 1–4 (2010)
22. Hinton, G., et al.: IEEE signal process. Mag. **29**(6), 82–97 (2012)
23. Assael, Y.M., Shillingford, B., Whiteson, S., Freitas, N. LipNet: End-to-End Sentence-level Lipreading. In: arXiv:1611.01599, pp. 1–13 (2016)
24. Cooke, M., Barker, J., Cunningham, S., Shao, X.: An audio-visual corpus for speech perception and automatic speech recognition. J. Acoust. Soc. Am. **120**(5), 2421–2424 (2006)
25. Verkhodanova, V., Ronzhin, A., Kipyatkova, I., Ivanko, D., Karpov, A., Zelezny, M.: HAVRUS corpus: high-Speed recordings of audio-visual Russian speech. In: International Conference on Speech and Computer, pp. 338–345 (2016)
26. Ivanko, D. et al.: Using a high-speed video Camera for robust audio-visual speech recognition in acoustically noisy conditions. In: International Conference on Speech and Computer, pp. 757–766 (2017)
27. Meutzner, H., Ma, N., Nickel, R., Schymura, C., Kolossa, D.: Improving audio-visual speech recognition using deep neural networks with dynamic stream reliability estimates. In: 2017 IEEE International Conference on Acoustics, Speech and Signal Processing (ICASSP), pp. 5320–5324 (2017)
28. Markovnikov, N., Kipyatkova, I.: An analytic survey of end-to-end speech recognition systems. SPIIRAS Proc. **58**(3), 77–110 (2018)

Chapter 17
The Concept of Robotics Complex for Transporting Special Equipment to Emergency Zones and Evacuating Wounded People

Mark Mamchenko, Pavel Ananyev, Alexander Kontsevoy, Anna Plotnikova, and Yuri Gromov

Abstract Ministry of the Russian Federation for Civil Defense, Emergencies and Disaster Management (EMERCOM of Russia, EMERCOM) devotes special attention to equipping subordinate units with advanced robotics. Despite this, EMERCOM of Russia lacks robotics complex capable of transporting special rescue equipment to emergency zone, bringing medicines to the wounded, and evacuating them to safe areas. This fact reduces efficiency and increases duration of emergency operations: if special vehicles are not able to reach the emergency zone, Russian EMERCOM employees will have to deliver rescue equipment and tools manually, and evacuate the wounded using stretchers on their own. This article proposes a concept of advanced robotics complex capable of carrying out the aforementioned tasks. We do not seek to provide full and detailed technical description of the whole complex. Only its composition, requirements, and peculiarities of power/communications systems, cargo delivery and safe evacuation mechanism, and possible electronic components shall be described.

17.1 Introduction

The widespread robotization of complex technological processes allows to exclude human activity in difficult environment, including life-threatening actions (e.g.,

M. Mamchenko (✉)
Institute of Control Sciences V. a. Trapeznikov Academy of Sciences, 65, st. Profsoyuznaya, Moscow, Russia 117342
e-mail: markmamcha@gmail.com; cigt@mail.ru; iait_tstu@mail.ru

P. Ananyev · A. Kontsevoy · A. Plotnikova
Non-Commercial Partnership "Scientifically-Educational Centre "Innovation Mining Technologies, 4A, Leninsky Prospekt, Moscow, Russia 119991

Y. Gromov
Institute of Automation and Information Technologies, 112, st. Michurinskaya, Tambov, Russia 392032

A. Ronzhin and V. Shishlakov (eds.), *Proceedings of 15th International Conference on Electromechanics and Robotics "Zavalishin's Readings"*, Smart Innovation, Systems and Technologies 187, https://doi.org/10.1007/978-981-15-5580-0_17

during emergency management) [1]. Therefore, it is logical, that robotic systems and complexes play an important role in technical equipment of the formations of EMERCOM of Russia. These means are indispensable in emergency management at explosive and dangerous objects, technological accidents, fires and natural disasters, while maintaining situational awareness using remotely controlled devices.

EMERCOM of Russia officials was fully aware of the need to adopt and exploit modern special robotics, and in 2016, the EMERCOM of Russia Special Robotic Systems Development Concept 2030 has been approved. Currently, the robotics "forces" of EMERCOM of Russia are mainly represented by robots designed to:

- conduct reconnaissance and surveillance operations;
- deal with accidents at dangerous objects;
- detect and neutralize the explosive objects;
- fight fires in high temperature areas;
- conduct above-water and underwater operations.

Despite the constant improvement of special robotic systems, it is not yet possible to exclude people (operators) from robots' management processes, since the existing robotic systems are still not fully autonomous. Besides, currently operating robotic models are not able to perform the full range of tasks during emergency management (fire-fighting, evacuation of the wounded, dismantling of the rubble, surrounding area monitoring, delivery of medicines, etc.), so they are not all-service. As the result EMERCOM of Russia formations may lack special-purpose robotic systems or suffer from insufficient effectiveness of engaged ones during rescue operations and other urgent works. In some cases, there is no ready-made robotics complex to carry out special missions, taking into account strict requirements, set by EMERCOM of Russia.

Our goal is to:

- review the current situation in equipping the EMERCOM of Russia units with robots that are able to transport the emergency-and-rescue equipment to emergency zones through impassable areas, deliver medicines to victims, and evacuate wounded to safe areas;
- in case of absence of such robotic systems—offer the concept of a promising robotics complex, capable to perform the aforementioned tasks.

The article will mainly describe electronic systems and components of the concept of the proposed robotics complex.

17.2 Analysis of the Current State in the Field of EMERCOM of Russia Robotics

Currently, there is a dual situation with equipping Russian EMERCOM units with robotics.

On the one hand, within the second stage (2018–2020) of robotic systems and respective technologies development, EMERCOM units are being equipped with robotics and unmanned aerial vehicles (UAV); devices and equipment modernization is in process; and improvement of relevant legal framework for robotics management, as well as operators and relevant maintenance personnel training are being held. In-service robotic vehicles are also intensively used. In addition, improvement of EMERCOM UAV and robotics application methods as well as optimization of the structure of respective administrative and operational units is scheduled to be complete until 2030 [2]. The focus is on the development, promotion, adoption, and engagement of domestic robotics in order to mitigate the dependence on foreign technologies and analogues.

On the other hand, EMERCOM of Russia current in-service robots are designed primarily for emergency area reconnaissance (“MRK-RP”) and emergency management, including in conditions of powerful thermal radiation and high temperatures, at explosive objects (“MPSK-RGWT”, “El-4”, “El-10”), and in case of chemical and radiation contamination (“El-4”, “El-10”).

Special mention should be made to the rescue and firefighting robot “ABR-ROBOT”. It is designed for emergency-and-rescue operations, and delivering rescue crew (5 persons), instruments, and respective equipment to the emergency zone. This complex includes the “MRK-RP” mobile robot, capable of remotely carrying out emergency-and-rescue operations, and extinguishing fires in dangerous conditions, in settlements and at industrial facilities. However, judging by the appearance of the robot, “MRK-RP” is not designed for evacuation of the wounded and the delivery of medicines. It is assumed that the “ABR-ROBOT”, being constructed on the basis of KA-MAZ-4326 chassis, will deliver the crew and equipment immediately to the emergency zone.

Similarly, the following robots are not intended for the wounded evacuation: TELEMAX and TEODOR (from the Tel-630 robotic complex), RTK-10, BROKK Mini-Cut, BROKK-180 (from the MAKORR robotic complex), BROKK-110D, and RTS-S (from the robotic system SHIELD) [3]. It should be noted that the lack of ready-made samples of special robotics designed for safe evacuation of wounded from the emergency zones is inherent not only in Russian EMERCOM but also in similar structures of other countries [4].

Due to the lack of special robotics, automobiles on the chassis of heavy trucks with large full weight and load capacity, including fire trucks, rescue and first-aid vehicles, tankers, and ladder trucks are mainly used to deliver crews and equipment to emergency zones [5].

Moreover, it is assumed that the evacuation of wounded from the emergency zone and providing them immediate care are carried out by fire-fighting units on stretchers or by first-aid vehicles crews (if these vehicles are possible to reach the emergency zone). It should be noted that the provision of first aid to the wounded and their evacuation to medical facilities are regarded as rescue-and-emergency operations (according to the current legislation) [6].

Thus, it is possible to distinguish three main negative scenarios (relating to the theme of this article) for EMERCOM units during emergency management:

- The impossibility of delivering the EMERCOM equipment and cargo directly to the emergency zone using special vehicles (due to impassable roads, as well as rough terrain), simultaneously with the impossibility and/or inexpediency of using heavy remote-controlled robotics (e.g., "El-4", "El-10", and "MPSK-RGWT");
- The increase of units to be engaged and time required for manual transporting of equipment and medicines to the emergency zone (e.g., in the case of an impassable area), which actually increases the duration and reduces the effectiveness of the emergency management;
- The necessity of by EMERCOM of Russia crews engagement for manual evacuation of wounded (e.g., using stretchers) to a safe zone (in case of lack of special equipment, or absence of organic in-service search-and-rescue robotics).

Thus, currently, there are no available Russian EMERCOM robotic assets capable of transporting cargo to the emergency zone and evacuating the wounded. In this regard, the challenge is to develop a concept of promising robotics complex in order to fulfill the above-mentioned tasks with no risk of death or injury for the EMERCOM crews, reduce the time of emergency response, as well as provide rapid evacuation of wounded victims (as soon as possible). The solution of this problem will significantly improve the efficiency of rescue and other urgent operations by eliminating the need for performing such time-consuming labor-intensive tasks, as manual deliver of cargo and equipment to emergency zone through rough terrain or evacuation of wounded to a safe zone (using stretchers).

17.3 The Purpose and Composition of the Promising Robotics Complex, as Well as the Imposed Requirements

17.3.1 The Purpose and Composition of the Promising Robotics Complex

In proposing this robot, we realize that it is both a heterogeneous robotic system and a multi-agent socio-cyber-physical system, and it has the characteristics and the limitations of both systems [7]. The promising robotics complex should be designed for transportation of rescue tools and equipment through impassable areas, delivery of medicines to the wounded and their evacuation to safe areas.

The following composition of the robotic system is proposed: robotic platform (RP)—1, portable remote control console (PRCC)—1, operator's automated workstation (AW)—1, reconnaissance UAV—2 (1 operational, 1 reserve), and communication aerial drone—2 (1 operational, 1 reserve).

17.3.2 Requirements for the Promising Robotics Complex

The proposed robot should be intelligent, multi-sensor, as autonomous as possible, and cost-effective (with minimal production costs due to widespread availability of ready-made cheap components in its architecture) [8].

One RP implementation option has been proposed in article [9]. However, this armored robot system is a ready-made robotics complex, and it will be difficult to adapt it according to the requirements of Russian EMERCOM. In particular, it is questionable whether it can be transposed via a rescue vehicle or unloaded by a crew of five persons. An interesting concept of robotic system is presented in the article [10]: using a swarm of reconnaissance, robots may enable the maximum flow of emergency situational data. However, the question of evacuating the wounded remains open: it is not clear how the swarm can achieve this. Thus, designing a swarm scheme is not appropriate for the main purpose of the robot. Paper [11] explains the key requirement for such heterogeneous robotic system: during design of communication/interaction between robots, they should act as an integrated system. Other important requirements are the need to manage the emergency as soon as possible, minimize the damage, and exclude the risk of death (injury) of EMERCOM personnel. While analyzing different components of the robotics complex, it is useful to understand that the RP should be able to operate both as an indoor robot and as an armored one. Thus, its design should be explosion-proof and fireproof [12].

To summarize, the following requirements are imposed:

- time of autonomous operation—not less than 6 h;
- range of RP operation (remote control mode; from the PRCC)—not less than 3 km;
- range of RP operation ("follow the leader" mode)—not less than 30 m;
- time for the deployment from the transportation to the operational position—not more than 7 min;
- RP speed of movement—2–7 km per hour; with the option of selectable constant speed of movement within the set range;
- time to change the power supply after returning to the rescue machine—not more than 5 min;
- standard deviation error from specified route (should not exceed the error of the navigation equipment used)—2–4 m (for GPS satellite constellation), 3–6 m (for GLONASS satellite constellation);
- tracking accuracy for a set route in remote control, beacon-based movement, and "follow the leader" modes—not more than 0,5 m;
- maximum height of the obstacle to be overcome by the RP—not less than 0,25 m;
- RP turn radius—not more than 2 m;
- RP weight—not more than 150 kg;
- RP width—not more than 0,71 m (allows to move freely inside the rooms, taking into account the standard width of the door panel of 0,8 m);
- payload—250–300 kg;

- modular design of the RP with the possibility of adding additional engine modules and forming a robotic "train";
- RP operating modes:
  - manual remote control from the PRCC;
  - semi-automatic mode (following the operator's route at a set speed or following the beacon);
  - automatic mode (following the pre-set route with GLONASS/GPS navigations systems' data path correction);
  - automatic mode (following the pre-set route to the operation starting point in case of communications disruption, when signals from the PRCC and GLONASS/GPS navigations systems are not available).

In addition, the robotics complex should provide:

- the ability to evacuate one person through rough terrain, minimizing possible negative impact on the wounded;
- the ability to overcome obstacles (e.g., stairs and curbs) and be operational in the buildings;
- day and night operational engagement;
- secure radio channel for the RP control from the PRCC, transmitting the coordinates of the RP to the operator, and saving received information on the AW for further analysis;
- ability to switch between RP's different operational modes at any time and at any stage of the operation;
- protection against loss of control, communication diagnostics and automatic return to the starting point after a pre-determined waiting time (when signal is lost);
- RP transportability (on the emergency-rescue machine).

17.4 Promising Robotics Complex Engagement Concept

17.4.1 Robotic System Deployment and Moving to the Emergency Zone

The deployment of robotics complex involves the usage of additional 3–5 small mobile autonomous reconnaissance robots (recon robot). Their task is to assess the passability of various routes and the current conditions in the emergency area before engaging the RP. The route, which the recon robot was able to pass, is regarded as available one. If the recon robot is not able to pass the route, it reports to the operator (sends information to the AW) and moves along the available route. In case of absence of available routes, any other route (which the recon robot will consider optimal in terms of time to reach the target) will be selected. Among all available routes to reach the target, the optimal one is chosen, along which the RP starts to

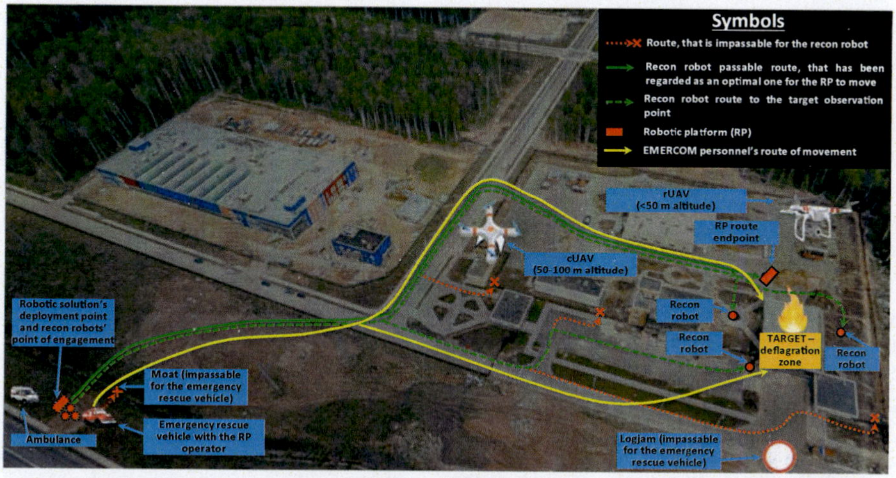

Fig. 17.1 Robotics complex deployment and operation scheme

move. All routes of all robot system's components with geolocation data are to be sent to the operator's AW. Proposed in [13] thin reconnaissance robot is not suitable for this purpose because of its questionable passability through rough terrain and operability in emergency conditions.

The navigation and autonomous action systems should not allow the RP to make deliberately false decisions or move toward a wrong direction, which may lead to catastrophic consequences [14]. In particular, additional video data and data from recon robots and rUAV are transmitted not only to the operator's AW but also to the RP. Information exchange between recon robots, the RP, and the operator's AW is conducted via the communication UAV (cUAV). The reconnaissance UAV (rUAV) is launched simultaneously with the recon robots' engagement in order to obtain video data from the emergency zone and transmit data to the operator's AW. It allows to rapidly provide and maintain situational awareness. The deployment and operation scheme of the proposed robotics complex are shown in Fig. 17.1.

17.4.2 *Safe Evacuation of the Wounded*

Using mobile telemedicine complex (UAV with a medical manipulator) as a means of medicines delivery (as presented in [15]) is not advisable, as the drones are unable to evacuate at least one person, and the quantity and composition of medicines being transported may not be sufficient (due to small payload capacity). Therefore, we believe that the best option is to install a special hermetic evacuation capsule on the RP. It can be attached above the platform or behind it (with a flexible coupling). The wounded or the rescue equipment can be transported in this capsule. The first variant of attachment is preferable, since it allows to optimize the center of the RP's gravity, as well as increase its productivity, reduce the probability of its overturning, while ensuring maximum safety of the wounded/transported equipment.

The capsule contains the module with medicines to render the first medical aid. During evacuation, the main concern is to ensure that the health of the evacuee does not deteriorate. To this end, life support and evacuation systems must meet strict safety standards [16]. In addition, the capsule should be equipped with finger pulse oximeters (e.g., NSI20303) and oxygen masks to prevent death of the wounded from suffocation and provide first aid in case of loss of consciousness as a result of inhalation of carbon dioxide and carbon monoxide, as well as track his health status during the transportation. Real-time data should be transmitted from pulse oximeters to the operator's AW.

In addition, the RP is to be equipped with a rollover prevention and control system, which receives data from position sensors (triaxial acceleration gauges and gyroscopes) to continually monitor its position. This system will consist of three levels of rollover prevention and:

1. The first level is the "rollover threat"—the RP, having received a signal from the position sensors about a possible rollover, stops and attempts to change its position/movement direction.
2. The second level (when the capsule is attached behind the RP)—"return to initial position"—assumes that the RP has turned over. In this case, the motors at the ends of the coupling elements rotate around the axis to move the overturned module to its operational position (the optimal direction of rotation is chosen according to the data from the triaxial acceleration gauges and gyroscopes).
3. The third level—"distress call"—assumes that the RP is stuck or is not able to return to its operational position on its own. In this case, a "distress signal" will be sent to the operator's AW, and red LED light on the RP starts to blink for rapid detection of the RP.

While designing the models of electromechanical elements, it is advisable to use the developments, specified in the paper [17].

17.5 Electronic Components of the Robot

RP power system. Each RP module has two commutator motors with a built-in 350 W reduction gear and 24 V voltage. The module's power system is based on two 12 V batteries connected in series. The main task is to use batteries that can be connected in series to achieve the value of capacity of 100 Ampere-hours. A robotics complex of three connected modules and one evacuation/cargo capsule can be operated for at least 6 h on one charge.

Hitches are used to connect modules with each other, as well as with the capsule for transportation of equipment/wounded people. Each module is equipped with two additional electric motors, which are designed to perform axial rotation of the module to return the RP to the normal position after overturning. Resetting the RP to normal position is only possible when the capsule is attached behind the last module, but not above the RP.

Electronic components of the RP module include two STM32 microcontrollers, four VNH3SP30 integrated drivers, one MCP2515 CAN bus controller, one 50 A and one 100 A current sensors, two pulse voltage stabilizers, a set of terminals and connectors, three three-axis accelerometers, and an LED strip with at least 96 LEDs per meter and a power of 7.7 W.

RP control system. The main component is a single-board microcomputer (e.g., Raspberry Pi Zero); it is also advisable to include a 4-channel 16-bit analog-to-digital converter, as well as four US-015 ultrasonic ranging sensors.

Module's remote control and "follow the leader" subsystems use the Arduino PRO micro microcontroller; commands from the operator are transferred using a key fob with a radio transceiver to the SS1101-SPI receiving module. ATmega-based Arduino modules are widely used in robotic systems, including rescue ones, due to the large amount of digital input/output pins and analog inputs [18, 19]. RP's synthetic vision system is implemented via high-detail cameras and infrared sensors, which allow the robot to operate at any time of the day. Streaming video data is transmitted to the operator and saved on the AW. It is proposed to use binocular vision sensors in the front of the RP as a basis of day vision system [20].

Communication system of the robotics complex. Among the considered options (IEEE 802.15.4 ZigBee [21], Bluetooth [22], etc.), the choice of Wi-Fi mesh technology to build a communication architecture, has been regarded as optimal in terms of speed and distance of data transmission. Mesh network concept involves creating a distributed, peer-to-peer, self-organizing connection, which improves the management of the robotics complex and its fault tolerance. In order to ensure and maintain robot's operation with lack of cellular communication infrastructure, recon robots, rUAV, and the RP are to be included in the mesh Wi-Fi network, where the cUAV provides overall connection with each component of the network. As described in [23], cUAV should be able to function not only as part of the mesh network but also as a wireless base station. It is suggested to use asymmetric encryption scheme to ensure secure communication and data transmission [24]. Another promising option for secure and sustainable communication is applying digital watermark technology to "conceal" control signals, transmitted from operator to robot scouts and RP [25, 26].

As a backup option, we propose to equip all robotic components with GSM modules; this will allow using the existing cellular operators' infrastructure and reducing the deployment time of the W-Fi mesh network. In addition, the simultaneous availability of W-Fi and GSM modules enables further application of "Internet of Things" solutions [27, 28]. As a "last chance" additional option, the possibility of direct control of the RP and recon robots via the radio channel (within line-of-sight distance) should be provided. It is proposed to equip the RP and recon robots with Arduino PRO micro microcontrollers and engage their SS1101-SPI radio channel modules.

Recon robots electronic equipment. Besides the battery and engines, each recon robot should be equipped with a Arduino PRO micro microcontroller, high resolution video camera (day mode), infrared video sensor (night mode), and a Wi-Fi mesh module.

rUAV electronic equipment. It should be noted that the use of small unmanned aerial systems for reconnaissance missions during emergency management operations was a success [29]. For carrying out autonomous flight missions, rUAV should be equipped in accordance with recommendations specified in the article [30]:

- Pixhawk 4 flight controller;
- 900 MHz manual control telemetry system;
- 2.4/5 GHz digital video transmission system;
- GPS/GLONASS transceiver module;
- T-motor engine propulsion system;
- depth camera sensor (for the collision avoidance system);
- reserve 1.2 GHz video transmission system;
- neural network recognition module;
- autonomous navigation system module with collision avoidance algorithm;
- ducted fan with a fairing, designed for protecting the propeller.

cUAV electronic equipment. The equipment of the communication UAV differs from the rUAV by the absence of digital video transmission system, depth camera sensor, reserve 1.2 GHz video transmission system, neural network recognition module, autonomous navigation system module with collision avoidance algorithm, and the presence of Wi-Fi mesh transceiver module [31].

The general composition of the robotics complex radio electronic equipment is shown in Fig. 17.2.

17.6 Conclusion and Future Work

This paper justifies the need for the development of a robotics complex, designed for transportation of EMERCOM equipment and cargo directly to the emergency zone, delivery of medicines to the wounded, as well as their evacuation to safe zones to provide the wounded with qualified medical care. The concept of such robotics complex has been proposed and the possible composition of its electronic components has been described.

It should be noted that such a robot can be exploited not only by Russian EMERCOM units but also by other organizations (including private enterprises), where people have to operate in life-threatening conditions (e.g., at the mines or mining and processing integrated plants, etc.). In particular, these robots may be engaged in order to carry out complex operations, which do not require the involvement of Russian EMERCOM formations or other special units (e.g., cargo delivery, situation monitoring in complex conditions, etc.).

This article is a starting point. The following articles will provide detailed information on every component and subsystem of the proposed robotics complex.

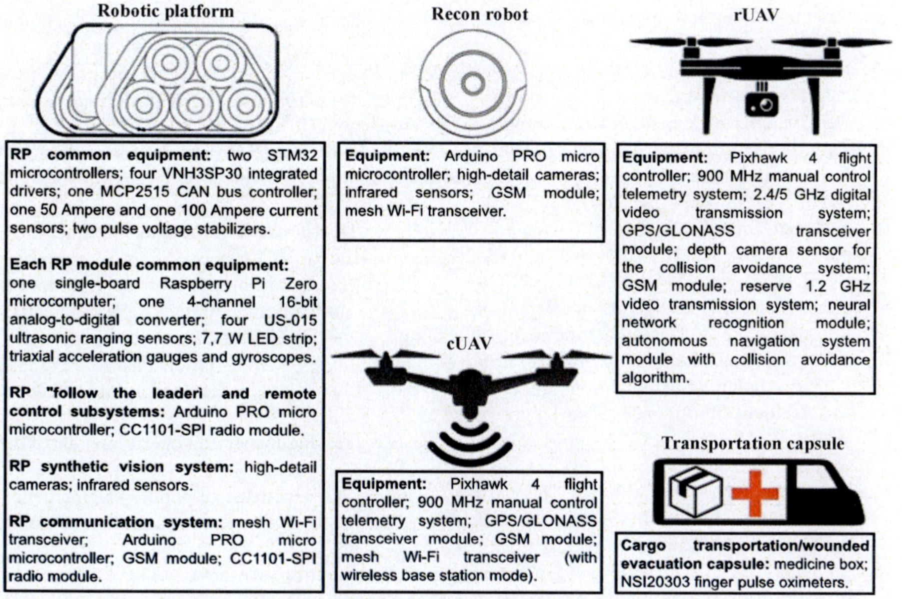

Fig. 17.2 Radio electronic equipment of the robotics complex

Acknowledgements This research is supported by the RFBR Project No. 19-08-00331.

References

1. Tadokoro, S.: Challenge of disaster robotics. In: 34th Chinese Control Conference (CCC), pp. 21–22. IEEE (2015)
2. Conclusion of the Panel of Ministry of the Russian Federation for Civil Defense, Emergencies and Disaster Management № 14-I (dated 2014/11/12) "Concerning the exploitation of robotic complexes, unmanned aerial vehicles in Russian EMERCOM units, and further development of robotics and respective technologies of its application", https://www.mchs.gov.ru/dokumenty/1826. Last Accessed 13 Jan 2020
3. Datskov, I.: Special-purpose robotic complexes deployed in Special Risk Rescue Operations Center "LEADER". In: Proceedings of 27th International Scientific and Practical Conference "Prevention. Rescue. Assistance", pp. 20–30 (2017)
4. Schneider, F., Wildermuth D.: Using robots for firefighters and first responders: Scenario specification and exemplary system description. In: 2017 18th International Carpathian Control Conference (ICCC), pp. 216–221. IEEE (2017)
5. Developed samples of fire and technical equipment for the Ministry of the Russian Federation for Civil Defense, Emergencies and Disaster Management, https://www.mchs.gov.ru/uploads/resource/05.09.2019/acf46516063db27a1181336939d0f2ca.pdf. Last Accessed 13 Jan 2020
6. Motienko, A., Ronzhin, A., Pavlyuk, N.: Modern developments of rescue robots: possibilities and principles of their application. Sci. Bull. NSTU **3**(60), 147–165 (2015)
7. Galin, R., Trefilov, P.: Technology of multi-agent systems in human and robot interaction modelling In: 5th International Scientific-Practical Conference (Seminar School) of Young

Scientists "Applied mathematics and Informatics: modern Research in Natural and Technical Sciences", pp. 155–158, (2019)
8. Maddukuri, S., Renduchintala, U., Visvakumar, A., Pang, C., Mittapally, S.: A low cost sensor based autonomous and semi-autonomous fire-fighting squad robot. In: 2016 Sixth International Symposium on Embedded Computing and System Design (ISED), pp. 279–283. IEEE (2016)
9. Jang, J., Suh, J., Kong, H., Lee, S., Kim, D.: Design of a Moving Substructure for an Armored Robot for Complex Disasters. In: 2019 16th International Conference on Ubiquitous Robots (UR), pp. 478–482. IEEE (2019)
10. Kuswadi, S., Adji, S., Sigit, R., Tamara, M., Nuh, M.: Disaster swarm robot development: On going project. In: 2017 International Conference on Electrical Engineering and Informatics (ICELTICs), pp. 45–50. IEEE (2017)
11. Zhao, Y., et al.: Design of intelligent algorithms for multi-mobile robot systems. In: 2015 International SoC Design Conference (ISOCC), pp. 177–178. IEEE (2015)
12. SeungSub, O., Jehun, H., Hyunjung, J., Soyeon, L., Jinho, S.: A study on the disaster response scenarios using robot technology. In: 2017 14th International Conference on Ubiquitous Robots and Ambient Intelligence (URAI), pp. 520–523. IEEE (2017)
13. Walker, I., University, C.: Use of continuum robots for remote inspection operations. In: 2017 Computing Conference, pp. 1382–1385. IEEE (2017)
14. Robinette, P., Li, W., Allen, R., Howard A., Wagner, A.: Overtrust of robots in emergency evacuation scenarios. In: 2016 11th ACM/IEEE International Conference on Human-Robot Interaction (HRI), pp. 101–108. IEEE (2016)
15. Bayanbay, N., et al.: The Use of Unmanned Aerial Vehicle for Emergency Medical Assistance. In: 2019 20th International Conference of Young Specialists on Micro/Nanotechnologies and Electron Devices (EDM), pp. 597–600. IEEE (2019)
16. Nallathambi, D.: Comprehensive evaluation of the performance of rescue robots using victim robots. 2018 4th International Conference on Control. Automation and Robotics (ICCAR), pp. 60–64. IEEE, Auckland (2018)
17. Osipov, O., Meshcheryakov, R., Shepelenko, M.: Designing digital models of elements of the electromashine part of electromechatronic modules of robotic systems. Extreme Robotics **1**, 160–164 (2017)
18. Mittal, S., Rana, M., Bhardwaj, M., Mataray, M., Mittal, S.: CeaseFire: The Fire Fighting Robot. 2018 International Conference on Advances in Computing. Communication Control and Networking (ICACCCN), pp. 1143–1146. IEEE, Greater Noida (UP), India (2018)
19. Suresh, J.: Fire-fighting robot. In: 2017 International Conference on Computational Intelligence in Data Science (ICCIDS), pp. 1–4. IEEE, Chennai (2017)
20. Li, H., Li B., Xu, W.: Development of a remote-controlled mobile robot with binocular vision for environment monitoring. In: 2015 IEEE International Conference on Information and Automation, pp. 737–742. IEEE, Lijiang (2015)
21. Brodeur, T., Regis, P., Feil-Seifer, D., Sengupta, S.: Search and Rescue Operations with Mesh Networked Robots. 2018 9th IEEE Annual Ubiquitous Computing. Electronics & Mobile Communication Conference (UEMCON), pp. 6–12. IEEE, New York City, NY, USA (2018)
22. Ren, Q., Man, K., Lim, E., Lee, J., Kim, K.: Cooperation of multi robots for disaster rescue. In: 2017 International SoC Design Conference (ISOCC), pp. 133–134. IEEE, Seoul (2017)
23. Lin, Y., Wang, T., Wang, S.: UAV-assisted emergency communications: an extended multi-armed bandit perspective. IEEE Commun. Lett. **23**(5), 938–941 (2019)
24. Dolgiy, E., Shokina, K., Aleshko, R.: Development of secure long-range channel scheme for ground control station of unmanned aerial vehicle. Molodoy ucheniy **13–1**(93), 10–12 (2015)
25. Shumskaya, O., Iskhakova, A.: Problems of masking of operating signals of agents of mobile robotic groups. VSPU 2019, 3086–3091. ICS RAS, Moscow (2019)
26. Shumskaya, O., Iskhakova, A., Iskhakov, A.: Masking of operating signals of agents in mobile robotic groups with network-centric management. Extreme Robotics. 303–306 (2019)
27. Anantha Raj P., Srivani, M.: Internet of Robotic Things Based Autonomous Fire Fighting Mobile Robot. In: 2018 IEEE International Conference on Computational Intelligence and Computing Research (ICCIC), pp. 1–4. IEEE, Madurai, India (2018)

28. Kanwar, M., Agilandeeswari, L.: IOT Based Fire Fighting Robot. 2018 7th International Conference on Reliability. Infocom Technologies and Optimization (Trends and Future Directions) (ICRITO), pp. 718–723. IEEE, Noida, India (2018)
29. Murphy, R., et al.: Two case studies and gaps analysis of flood assessment for emergency management with small unmanned aerial systems. 2016 IEEE International Symposium on Safety. Security, and Rescue Robotics (SSRR), pp. 54–61. IEEE, Lausanne (2016)
30. Meshcheryakov, R., et al.: An application of swarm of quadcopters for searching operations. 19th IFAC Conference on Technology. Culture and International Stability TECIS 2019, pp. 14–18. IFAC, Sozopol, Bulgaria (2019)
31. Trefilov, P., Meshcheryakov, R., Chekhov, A.: Development of multicopter UAV for searching for people in forests. In: 5th All-Russian scientific-practical seminar "Unmanned vehicles with artificial intelligence 2019", pp. 173–180 (2019)

Chapter 18
Implementation of Robot–Human Control Bio-Interface When Highlighting Visual-Evoked Potentials Based on Multivariate Synchronization Index

Sergey Kharchenko, Roman Meshcheryakov, Yaroslav Turovsky, and Daniyar Volf

Abstract The introduction part, the main ways of integration of human–machine control systems integration of systems as well as the ways of visual-evoked potentials extraction in electroencephalograms (EEGs) are described. The article includes researches of electroencephalogram (EEG) signals with steady-state visual-evoked potentials for various photostimulation frequencies basing on the method of the multivariate synchronization index. Influence of length of window being processed on recognition accuracy of frequency of the signal being studied is explored. Ratio of the correctly recognized states is considered in the function of accuracy metric. Examined necessity of original signal preprocessing by way of the signal band-pass filtering. In addition, application of the multivariate synchronization index in multi-channel mode is reviewed. The result of the authors' investigation is a number of recommendations on parameters used for extraction of steady-state visual-evoked potentials in the method of the multivariate synchronization index. Results obtained are of considerable practical importance as they can be used for brain-computer interface producing on the basis of visual-evoked potentials and later can be taken for building of control theory of robot systems of different application and for implementation of decisions on man–machine interoperability within narrow practical tasks.

S. Kharchenko (✉)
Tomsk State University of Control Systems and Radioelectronics, 40, ave. Lenina, Tomsk, Russia 634050
e-mail: ss.kharchenko@gmail.com

R. Meshcheryakov · Y. Turovsky · D. Volf
Institute of Control Sciences Academician VA Trapeznikov, 65, st. Profsoyuznaya, Moscow, Russia 117997
e-mail: mrv@ieee.org

Y. Turovsky
Voronezh State University, 1, sq. Universitetskaya, Voronezh, Russia 394018

A. Ronzhin and V. Shishlakov (eds.), *Proceedings of 15th International Conference on Electromechanics and Robotics "Zavalishin's Readings"*, Smart Innovation, Systems and Technologies 187, https://doi.org/10.1007/978-981-15-5580-0_18

18.1 Introduction

Building the man–machine interaction system goes back to the beginning of the twentieth century, when in 1924, Hans Berger for the first time registered a brain activity on EEG. Starting form 1970s, this concept started to be developed intensively when in 1977, Jacques Vidal performed an experiment, in which on the basis of visual-evoked potentials, on-screen cursor control was implemented [1]. Co-use of neuroprosthetics together with brain-computer interface even today helps to make up for vision and hearing loss, muscle-skeleton disorders (MSDs) [2]. The brain-computer interface is a bridge that is able to connect the neural structures of the motor cortex with robotic arms or exoskeletons. To control robotic devices, bidirectional brain-computer systems are used (see Fig. 18.1) with feedback. EEG got widespread use due to noninvasiveness but still there are a number of problems such as low space resolution, unability to use all-pass signals [3–6]. Besides, EEG includes not only signal of interest but also noises—brain spontaneous activity, which make it difficult to find signals [7–10]. One of realization ways of such systems based on EEG is usage of visual-evoked potentials (VEP). Registration of the evoked brain potentials (EP) serves as an objective and noninvasive method of testing functions of the central nervous system. Evoked potentials are electrical brain responses on visual, auditory

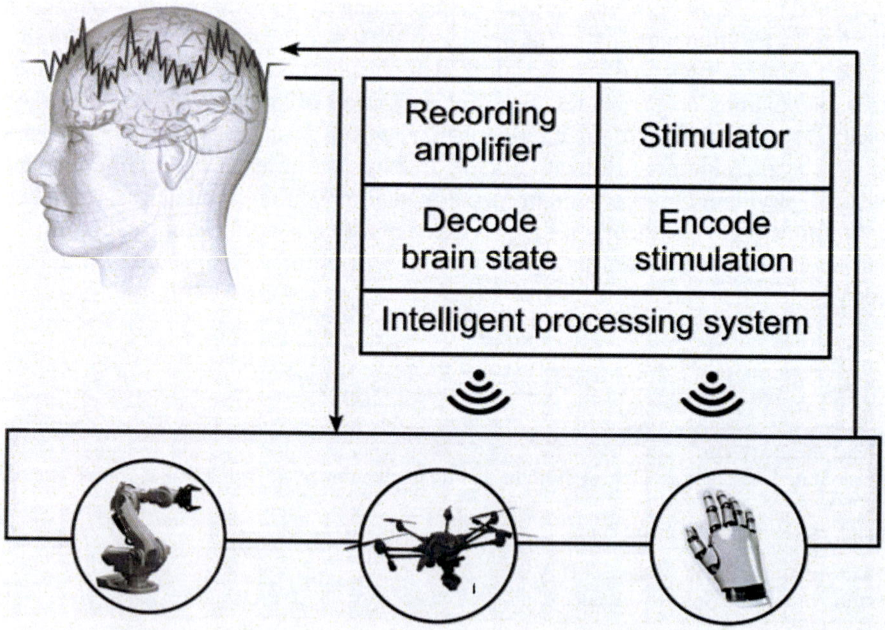

Fig. 18.1 Graphical visualizing bidirectional interfaces of computer-brain type by the example of management of hand-type robot unit

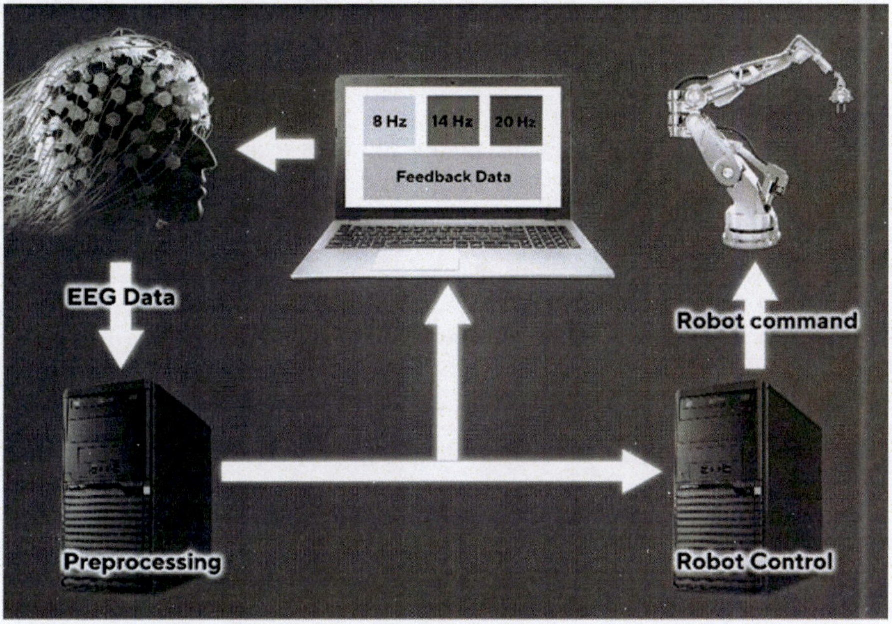

Fig. 18.2 Brain-computer system based on steady-state visual-evoked potentials

or sensory stimuli. The important fact is that evoked potential signals on some kind of irritators are similar to signals generated in case when a patient imagines an irritator.

Year after year, a number of articles on current subject and items implementing man–machine interaction, featured in exhibitions, are increasing [11–17]. Such kind of systems is of broad use starting from entertainment AR experience to medical products, military industry items, and robotic devices. All of that shows how high the relevance is speaking about brain-computer control system integration basing on visual-evoked potentials with accuracy increase of their extraction in EEG (see Fig. 18.2).

18.2 Overview

There are different ways for VEP extraction in EEG signal and the most popular among them at the present moment are the following: frequency analysis on the basis of Fourier transformation and wavelets, canonical correlation analysis, methods of synchronous coherent integration and/or averaging, and multivariate synchronization index.

The methods of coherent accumulation and averaging are aimed at increasing the signal-to-noise ratio, and in a general case, the number of required summations is found by Eq. (18.1):

$$N = (2 \cdot (1/K)^2)^2, \tag{18.1}$$

where K is the original signal-to-noise ratio, N is the number of summations, required for selection evoked potentials in a signal.

However, if initial signal-to-noise ratio is low, it is required a huge amount of summations. As a rule, the methods of coherent accumulation and averaging are realized at the hardware level in devices of performing EEG. To determine the states of synchronization and desynchronization of sensorimotor rhythms, EEGs in realizations of human–machine interaction usually use transition from the representation of signals as a function of time to the representation of a signal in the frequency domain. Transition to the frequency domain is carried out by signal decomposition to harmonic components by discrete Fourier transform Eq. (18.2):

$$F(n) = \sum_{t=0}^{N-1} x(t) e^{-\frac{2\pi j}{N} nt}, n = 0, 1 \ldots, N-1, \tag{18.2}$$

where $x(t)$ is measured values of the EEG signal at the moment of time t.

The wavelet transform is used for transition to the frequency domain, which is different from the discrete Fourier transform that considers signal as stationary. The wavelet transform allows to track the dynamic of the harmonic components of the signal over time Eq. (18.3):

$$W(\tau, a) = \int_{-\infty}^{\infty} x(t) \psi^*_{\tau,a}(t) \mathrm{d}t, \tau \in R, a \in R^+, \tag{18.3}$$

where $W(\tau, a)$ is the coefficients of the wavelet decomposition; τ, a is parameters of the time shift and scale, respectively.

Canonical correlation analysis is a multi-parameter statistical method for finding linear combinations that maximize correlation between two data sets Eq. (18.4). The EEG signal is used as the first data set, and as the second data set are used signals that satisfying Eq. (18.5):

$$\max_{W_x, W_y} p(x, y) = \frac{W_x^T X Y^T W_y}{\sqrt{W_x^T X X^T W_x W_y^T Y Y^T W_y}}, \tag{18.4}$$

$$Y(t) = \begin{bmatrix} \sin(2\pi \mathrm{ft}) \\ \cos(2\pi \mathrm{ft}) \\ \ldots \\ \sin(2\pi N \mathrm{ft}) \\ \cos(2\pi N \mathrm{ft}) \end{bmatrix}, \quad t = \frac{1}{F_s}, \frac{2}{F_s}, \ldots \frac{K}{F_s}, \tag{18.5}$$

where N is the harmonic sequence number, K is the number of samples, Fs is the sampling frequency, $Y(t)$ is the reference signal, and $X(t)$ is the original EEG signal.

18.3 Methods and Results

18.3.1 Multivariate Synchronization Index Method

This article looks at the method which shows the highest results from the perspective of various researchers—multivariate synchronization index method. Synchronization principle of multi-channel EEG and reference signals, meant to be in line with visual stimulus frequency (photostimulation frequency), is a core of the multivariate synchronization index method. Selection of reference signals is based on Eq. (18.5). Initially, the correlation matrix between two signals is calculated using Eqs. (18.6)–(18.8).

$$D_{11} = \left(\frac{1}{M}\right) X X^{\mathrm{T}} \tag{18.6}$$

$$D_{22} = \left(\frac{1}{M}\right) Y Y^{\mathrm{T}}, \tag{18.7}$$

$$D_{21} = D_{12} = \left(\frac{1}{M}\right) X Y^{\mathrm{T}}. \tag{18.8}$$

A linear transformation is calculated to eliminate the effect of autocorrelation on the measure of synchronization Eq. (18.9):

$$T = \begin{bmatrix} \frac{1}{\sqrt{D_{11}}} & 0 \\ 0 & \frac{1}{\sqrt{D_{22}}} \end{bmatrix}. \tag{18.9}$$

After that, the correlation matrix is calculated by Eq. (18.10):

$$S = \begin{bmatrix} I_{1\times 1} & \frac{1}{\sqrt{D_{11}}} D_{12} \frac{1}{\sqrt{D_{22}}} \\ \frac{1}{\sqrt{D_{22}}} D_{21} \frac{1}{\sqrt{D_{11}}} & I_{2N_h \times 2N_h} \end{bmatrix}. \tag{18.10}$$

Then multivariate synchronization index is determined by Eq. (18.11):

$$R_i = 1 + \frac{\sum \lambda_i' \log(\lambda_i')}{\log(P)}, \tag{18.11}$$

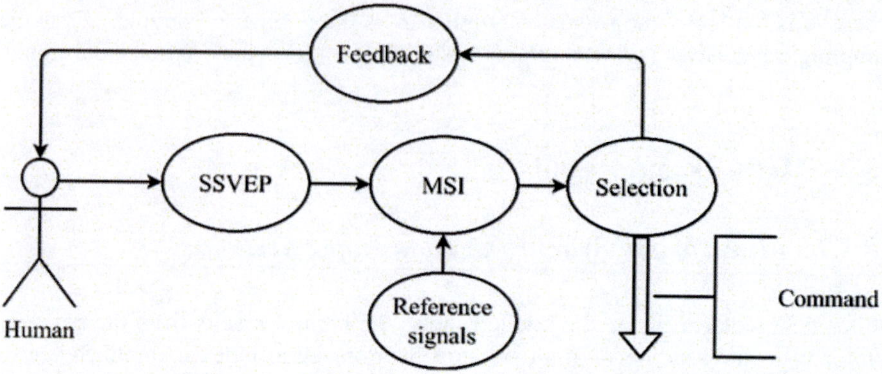

Fig. 18.3 Use of MSI to allocate SSVEP in human–machine interaction

where λ'_l is normalized eigenvalues S, $P = N + Nh$ и Nh note the number of lines in the reference signal $B(t)$. The frequency of photostimulation is defined as the frequency corresponding to the maximum value of R.

Nowadays, this method pretends to be the most advantageous as it allows extraction of visual-evoked potentials on the basis of 1–4 channels, losing not too much of recognition rate, which in the future contributes to a better brain-computer system creation [18–23] to control robotic devices (see Fig. 18.3).

Software implementation MATLAB script for determining multivariate synchronization index is shown below.

```
function S = ind(x, y)
[N, M]=size(x);
[NH, ~] = size(y);
P = N + NH;
c12=(1/M)*x*y';
c21=c12';
c11=(1/M)*(x*x');
c22=(1/M)*(y*y');
C=[c11 c12;c21 c22];
C11=c11^-0.5;
C22=c22^-0.5;
[a, b]=size(C22);
U=zeros(b+N,b+N);
U(1:N,1:N)=C11;
U(N+1:b+N,N+1:b+N)=C22;
R=U*C*U';
e=eig(R);
E=e/sum(e);
S=1+(sum(E.*log(E)))/log(P);
```

18.3.2 Single-Channel Mode

An algorithm was designed on the basis of the multivariate synchronization index method [21] and realized in MATLAB application program package. As the first part of the research 60 EEG signals with steady-state visual-evoked potentials of 30 adult people, men and women, without contraindications to EEG and photostimulation reading were studied. Photostimulation frequency amounted to 8 and 14 Hz. Sampling frequency of signal release amounted to 5000 Hz. Each signal was divided into windows of 500, 750, and 1000 samples with 250-sampling overlap which is equivalent of analysis epoch of 100, 150, and 200 ms, respectively. Multivariate synchronization index method for each analysis epoch was applied with prefiltration basing on a band-pass Butterworth filter [24] and without it. Edges of a filter were set as 4 and 30 Hz. Recognition rate was defined as a number of epochs within which a frequency to be determined was recognized in the presence of competing frequencies of 6, 8, 10, 12, 14, 16, 18, 20, 22, 24, and 26 Hz. The results of each set of signals with a photostimulation frequency of 14 Hz for all situations described are presented in Fig. 18.4.

As it is revealed from the figure, a number of correctly recognized states are increasing due to a window size increase, besides when using prefiltration on small window size (500 samples) correctly recognized states amounted to 2 which is equal to 0.05 for one signal, and for the whole sample collection, it is on the level of statistical error. In order to define directivity of proportion changes of correctly recognized states and their intensity when using signal prefiltration, Wilcoxon t-test [25] was conducted. The first step was the following: absolute values of proportion changes of correctly recognized states were calculated both with filtration and without it. The second step referred to hypothesizing:

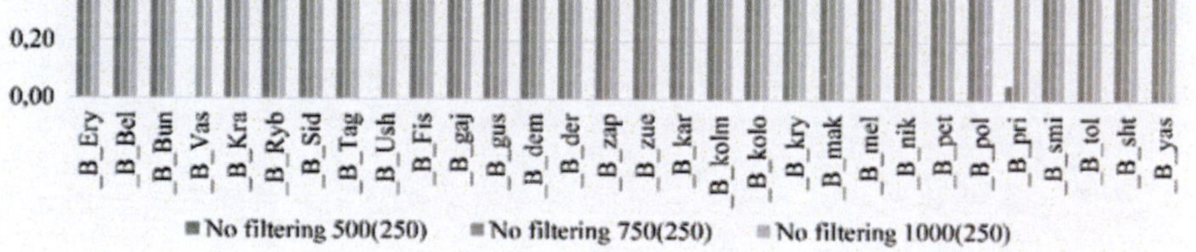

Fig. 18.4 Research results for signal set with a frequency of 14 Hz without prefiltration

- H0: Values of multivariate synchronization index used without filtration exceed values of multivariate synchronization index used with prefiltration of a signal under study;
- H1: Values of multivariate synchronization index used with prefiltration of a signal under study are lower than values of multivariate synchronization index used without filtration (reverse H0).

At the third step, ranks were assigned to each indicator difference with prefiltration and without it. Zero shifts were preliminary excluded from calculation as it is one of obligatory requirements of Wilcoxon test. Validity check of rank matrix compiling was run at the fourth step.

The column sum and check sum amounted to 55 and were equal with each other, which means ranking was carried out properly. At the fifth step, nontypical directions—positive in this case—were defined. Rank sum of these directions amounts to empirical value of test $T = 0$; Basing of critical value table of Wilcoxon t-test, the following critical values were found for $n = 10$: T-test $= 5$ for $p \leq 0.01$ and T-test $= 10$ for $p \leq 0.05$. In that case, significance area spreads to the left, i.e., empirical value T falls within significance area Temp < T-test (0.01).

Hypothesis H0 is accepted. Values of multivariate synchronization index used without filtration exceed values of multivariate synchronization index used with prefiltration of a signal under study. Influence of window size and prefiltration occurrence under analysis are presented in more details in Figs. 18.4 and 18.5.

The results for the second set of EEG signals taken from the same group of people on the same hardware but with SSVEP frequency of 8 Hz on similar window sizes with and without filtration are presented in Figs. 18.6 and 18.7 . For signals with SSVEP 8 Hz, we have the same situation, the bigger a window size is the more states are recognized correctly and at the same time better results for the same analysis epochs can be found from the algorithm without signal prefiltration. But as

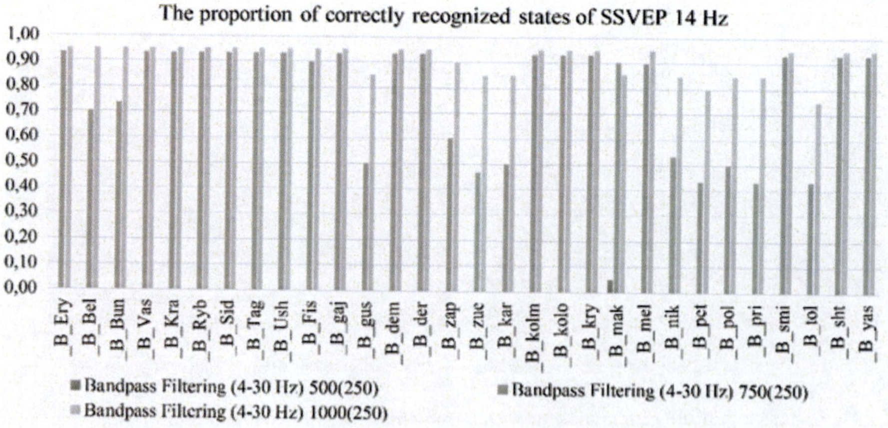

Fig. 18.5 Research results for signal set with a frequency of 14 Hz with prefiltration

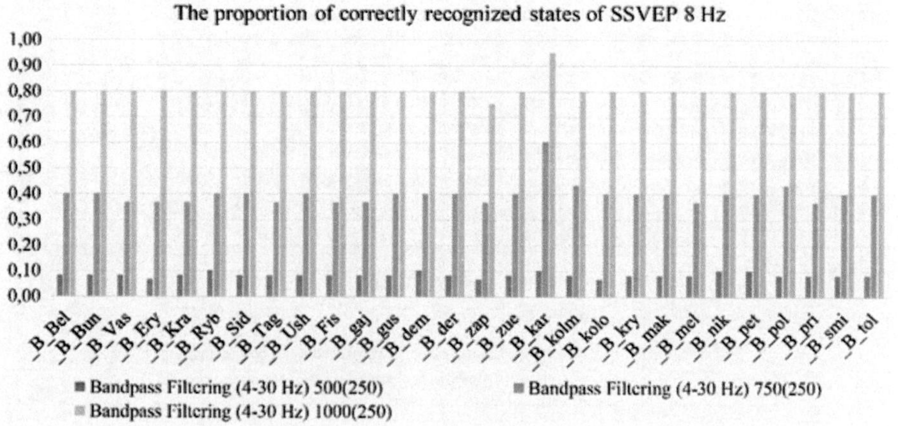

Fig. 18.6 Research results for signal set with a frequency of 8 Hz without prefiltration

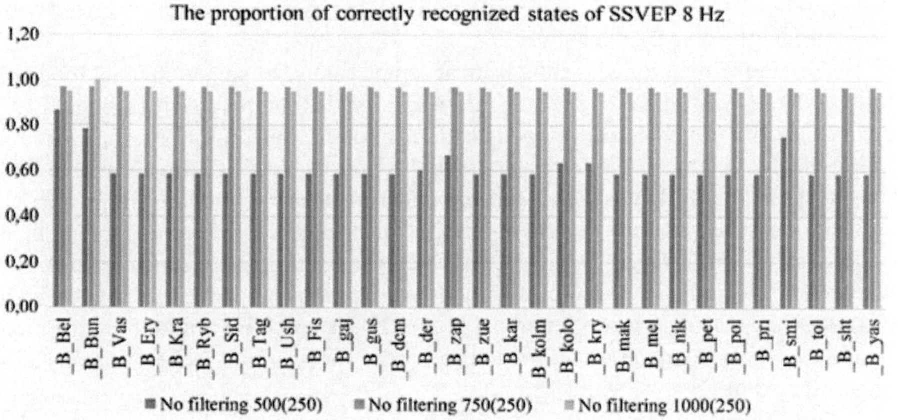

Fig. 18.7 Research results for signal set with a frequency of 8 Hz with prefiltration

for algorithm without filtration, a proportion of correctly recognized states with an average value for a window size of 500 samples is higher by 0.02 than that with a window size of 750 samples. Nevertheless, after further enlarging of a window size, a proportion of correctly recognized states increases and amounts to 1 when having a window size of 1500 samples (300 ms); such results remain the same up to a window size of 5000 samples (1 s).

In a similar manner to a signal set with 14 Hz photostimulation frequency, there were two hypotheses, H0 and H1, developed. For this data set, T equaled 0; Basing of critical value table of Wilcoxon t-test, the following critical values were found for $n =$ 29: T-test $= 110$ for $p \leq 0.01$ and T-test $= 140$ for $p \leq 0.05$. In this case, significance area spreads to the left, i.e., empirical value T falls within significance area Temp < T-test (0.01). Hypothesis H0 is accepted. Values of multivariate synchronization index

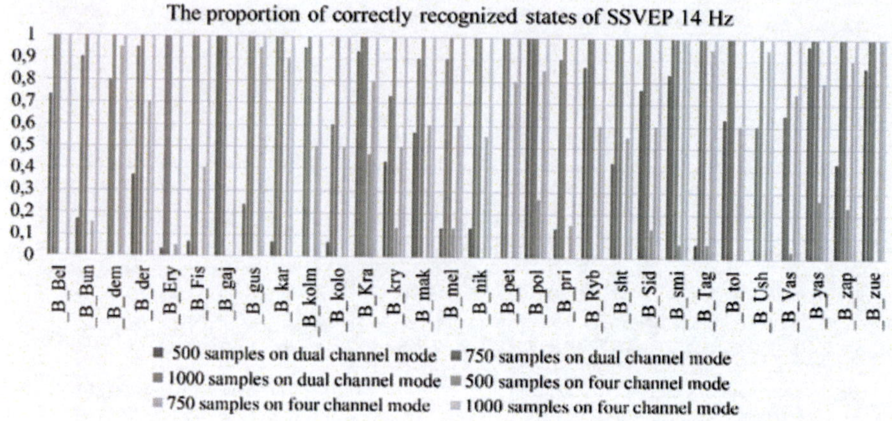

Fig. 18.8 Research results for signal set with a frequency of 14 Hz in multi-channel mode

used without filtration exceed values of multivariate synchronization index used with prefiltration of a signal under study.

All the above-mentioned results were accepted for use of single-channel mode of multivariate synchronization index, a reference signal is compared with a single EEG channel, and this algorithm developed on the basis of the multivariate synchronization index enables comparison of reference signals with several EEG signals at once. As long as studies show that use of a band-pass filtering does not produce proper results, further researches on influence of a number of channels used were implemented without signal prefiltration. A proportion of correctly recognized states for window sizes of 500, 750, and 1000 samples without overlapping was defined for the research of multi-channel usage mode for signal sets of 14 Hz and 8 Hz without prefiltration within dual-channel mode and four-channel mode. The results of recognition of states are presented in Figs. 18.8 and 18.9.

As can be seen from Figs. 18.8 to 18.9, the use of multi-channel mode has a positive effect on the quality of the allocation of established visual-evoked potentials. However, if the analysis window length of 500 samples and 750, that is 100 and 150 ms, respectively, in some cases, the proportion of true, the recognized states are less than or equal to 0.5. With increasing analysis epoch to 1000 samples (200 ms) for both sets of signals (8 Hz and 14 Hz), the initial rate was correctly detected the algorithm in all cases. With further increase of analysis window, the results of dual-channel mode and four-channel mode are the same.

18.4 Conclusion

This research was made to discover influence of analysis epoch duration and usability of pre-band-pass filtration of a signal while using of the multivariate synchronization

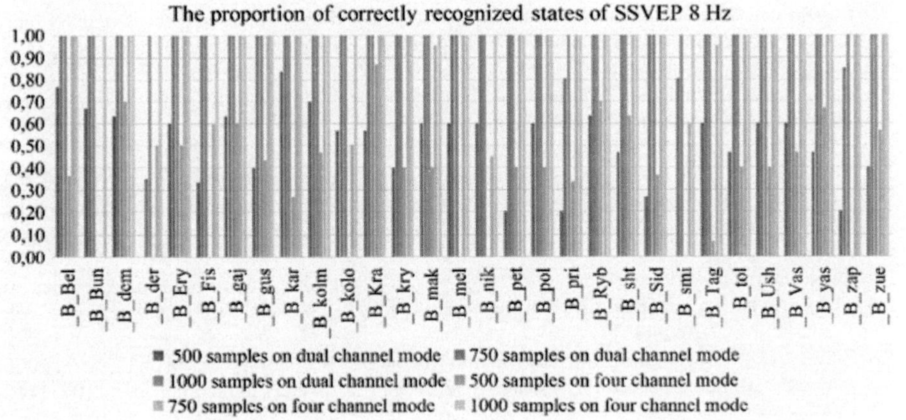

Fig. 18.9 Research results for signal set with a frequency of 8 Hz in multi-channel mode

index for extraction of visual-evoked potentials. It is shown that use of prefiltration within a single-channel mode cuts a proportion of correctly recognized states while on the contrary increase of analysis window enhances a proportion of correctly recognized states. The positive effect of analysis epoch increase within multi-channel mode is also considered.

Acknowledgements This research is supported by the RFBR projects No. 19-08-00331 and No. 19-29-01156.

References

1. Nam, C.S., Nijholt, A., Lotte, F.: Brain–Computer Interfaces Handbook: Technological and Theoretical Advances, CRC Press (2018)
2. Pfurtscheller, G., et al.: Current trends in Graz brain-computer interface (BCI) research. IEEE Trans Rehabil Eng **8**(2), 216–219 (2000)
3. Tyagi, A., Semwal, S., Shah, G.: A review of eeg sensors used for data acquisition. J. Comput. Appl. (IJCA). 13–17 (2012)
4. Malmivuo, J., Plonsey, R.: Bioelectromagnetism—Principles and Applications of Bioelectric and Biomagnetic Fields. Oxford University Press, Oxford (1995)
5. Ma, W. et al.: Using EEG artifacts for BCI applications. In: 2014 International Joint Conference on Neural Networks (IJCNN), pp. 3628–3635 (2014)
6. Han, J.: Recent progress of non-invasive optical modality to brain computer interface: a review study. In: The 3rd International Winter Conference on Brain-Computer Interface, pp. 1–2 (2015)
7. Gnezditsky, V.V.: Evoked Brain Potentials in Clinical Practice. Publishing House of TSURE (1997)
8. Maksimenko, V.A., Kuc, A.K., Pisarchik, A.N., Hramov, A.E.: Mathematical modeling of visual sensory information processing using multi-channel EEG signals. Inf. Control Syst. **6**, 58–65 (2019)

9. Stankevich, L.A., et al.: Classification of electroencephalographic patterns of imaginary one-hand finger movements for brain-computer interface development. SPIIRAS Proc. **3**(40), 163–182 (2015)
10. Kurkin, S.A., Pitsik, E.H., Khramov, A.E.: Classifying electrical activity of the brain during imaginary movements of untrained subjects using artificial neural networks. Inf. Control Syst. **6**, 77–84 (2020)
11. Chueshev, A., Melekhova, O., Meshcheryakov, R.: Cloud robotic platform on basis of fog computing approach. In: Lecture Notes in Computer Science (including subseries Lecture Notes in Artificial Intelligence and Lecture Notes in Bioinformatics), vol. 11097, pp. 34–43 (2018)
12. Shepelenko, M., Meshcheryakov, R.: Algorithm for automated calculation of a segmented electromechatronic module of robotic system motion. In: MATEC Web of Conferences, vol. 113, article no. 02005 (2017)
13. Zhu, Y. et al.: The Research of Rehabilitation Robots BCI Technology Based on EEG. In: 2008 International Conference on Computer Science and Software Engineering, pp. 1107–1110 (2008)
14. Cietto, V. et al.: Evaluating commercial BCIs for moving robots. In: Proceedings of the 13th Biannual Conference of the Italian SIGCHI Chapter: Designing the next interaction, pp. 1–4 (2019)
15. Voznenko, T.I., Chepin, E.V., Urvanov, G.A.: The control system based on extended BCI for arobotic wheelchair. Procedia Comput. Sci. **123**, 522–527 (2018)
16. Gridnev, A.A., Voznenko, T.I., Chepin, E.V.: The decision-making system for a multichannel-robotic device control. Procedia Comput. Sci. **123**, 149–154 (2018)
17. Kobozeva, I.M., Sidorov, G.O., Zimmerling, A.V.: Module for dialog management in the interaction system between user and mobile robotic guide. SPIIRAS Proc. **2**(33), 186–206 (2014)
18. Friman, O., Volosyak, I., Graser, A.: Multiple channel detection of steady-state visual evoked potentials for brain–computer interfaces. IEEE Trans. Biomed. Eng. **54**, 742–750 (2007)
19. Lin, Z., Zhang, C., Wu, W., Gao, X.: Frequency recognition based on canonical correlation analysis for SSVEP-based BCIs. IEEE Trans. Biomed. Eng. **54**, 1172–1176 (2006)
20. Friman, O., et al.: Detection of neural activity in functional MRI using canonical correlation analysis. Magn. Reson. Med. **45**, 323–330 (2001)
21. Zhang, Y., Xu, P., Cheng, K.: Multivariate synchronization index for frequency recognition of SSVEP-based brain-computer interface. J. Neurosci. Methods **221**, 32–40 (2014)
22. Nan, W. et al.: A comparison of minimum energy combination and canonical correlation analysis for SSVEP detection. In: Proceedings of the 5th International IEEE EMBS Conference on Neural Engineering Cancun, Mexico, pp. 469–472 (2011)
23. Purushothaman, G., Prakash, P., Kothari, S.: Investigation of multiple frequency recognition from single-channel steady-state visual evoked potential for efficient brain–computer interfaces application. IET Signal Proc. **12**(3), 255–259 (2018)
24. Miroslav D.: Filter Design for Signal Processing using MATLABand Mathematica. Prentice Hall, New Jersey, USA (2001)
25. Wilcoxon, F.: Individual comparisons by ranking methods. Biometrics **1**, 80–83 (1945)

Chapter 19
Human–Machine Interface of Rehabilitation Exoskeletons with Redundant Electromyographic Channels

Andrey Trifonov, Sergey Filist, Sergey Degtyarev, Vadim Serebrovsky, and Olga Shatalova

Abstract A method for controlling an exoskeleton by means of instructions obtained by decoding an electromyography (EMG) signal is considered. The method allows to minimize errors when positioning the exoskeleton in the verticalization mode. The EMG signals are segmented into intersecting or non-intersecting windows and for each segment obtained in the previous step, they receive many signs of the EMG signal (vector of informative signs). The vector of informative features is fed to the neural network classifier, which controls the controller of the exoskeleton servomotors. The vector of informative features is obtained through a multilevel comparator, and the number of levels of which determines the dimension of the vector of informative features. The EMG classifier includes a comparator unit, a multiplexer, an informative feature calculation unit, a first neural network, a memory unit and a second neural network, the outputs of which control the servo motor controller. In order to adapt the exoskeleton control system to the patient, additional channels for classifying EMG signals are introduced into the human–machine interface. Each channel of the EMG signal is associated with a specific muscle or group of muscles that control the movement of the same limb joint. The servo motor controller uses a third neural network to aggregate these signals into a single control signal. The neural network control method with redundant EMG channels has been tested on the exoskeleton at the moment of controlling the verticalization of the patient.

19.1 Introduction

The clinical use of verticalization has proven to be highly effective in conducting rehabilitation therapy for a wide variety of neurological disorders. Verticalization can effectively improve and stabilize the performance of the cardiovascular system, normalize the breathing process, improve muscle mobility, activate work,

A. Trifonov · S. Filist (✉) · S. Degtyarev · V. Serebrovsky · O. Shatalova
Southwest State University, 94, st. 50 years of October, 305040 Kursk, Russia
e-mail: SFilist@gmail.com

A. Ronzhin and V. Shishlakov (eds.), *Proceedings of 15th International Conference on Electromechanics and Robotics "Zavalishin's Readings"*, Smart Innovation, Systems and Technologies 187, https://doi.org/10.1007/978-981-15-5580-0_19

and improve the innervation of the musculoskeletal system. Moreover, verticalization not only accelerates the rehabilitation process but also significantly reduces the risk of secondary complications associated with prolonged immobility of the patient. Verticalizers also contribute to the development of motor actions and the development of the functions of the upper limbs.

The systematic use of training devices has a passive and active effect on the musculoskeletal system (muscles, joints), stimulates or normalizes the activity of brain structures (hypothalamus, motor centers of the cerebral cortex and other departments), the activation of which helps to maintain a vertical posture, locomotor acts, manipulating objects, etc.

One of the methods of rehabilitation is verticalization using an exoskeleton, that is, the transition from the "sitting" to the "standing" position [1, 2]. Despite the numerous exoskeleton rehabilitation in the market of medical services, today, there are no effective algorithms that ensure the patient's stable movement in the exoskeleton during verticalization; questions of mathematical modeling of patient behavior in the exoskeleton remain open. The main complaint against medical exoskeletons on the part of consumers is the inability to adapt the exoskeleton to a particular patient, therefore, very often, specialists prefer cheaper and more effective means of rehabilitation using patient verticalization technology.

19.2 Biotechnological Rehabilitation System with Medical Exoskeleton

A generalized diagram of a biotechnological system with a medical exoskeleton is shown in Fig. 19.1. In this system, we can select a patient who performs information interaction with the environment using vision and tactile sensors, a human–machine interface (HMI) that controls the power and information interaction of a person with an exoskeleton. Man and the exoskeleton carry out cooperative interaction with the environment, the role of which is played by the supporting surface, obstacles, and technological load.

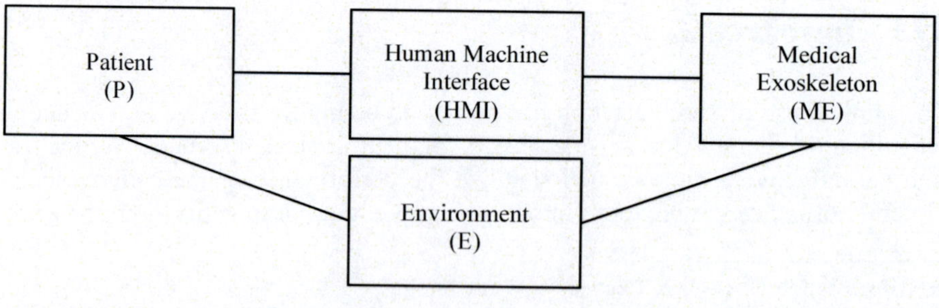

Fig. 19.1 Generalized scheme of a biotechnological system with a medical exoskeleton

A person constantly evaluates the environment and affects the technological load, estimated by the moments created by the muscular system. The interaction of a person and an exoskeleton is carried out using an intelligent power HMI in known technical solutions. A measure of force interaction is forces that are of the nature of both point and distributed. This distributed interface, which controls both the power and kinematic parameters of the system, consists of modules of the back, hip, foot, and arm. The system of sensors installed at predetermined points of the exoskeleton allows you to control human–machine interaction in the biotechnical system. The HMI also has information channels through which the patient acts on the exoskeleton, setting the nature of the desired movements of the exoskeleton links.

19.3 Non-invasive Electromyography in Implementation of Human–Machine Interfaces

Non-invasive electromyography (EMG) is an effective means of implementing HMI in modern mechatronic systems. The most promising field of application of bioinformation systems based on EMG data is the creation of exoskeleton devices with biocontrol [3–6].

In order to distinguish and classify muscle movements, the most significant parts of EMG (signs) must be extracted, which represent the current signal in the form of a code that carries information about the specific movement of the limbs. This information is decoded in the decoder (classifier).

Various coding methods are used to classify EMG signals: spectral coefficients, autoregression coefficients, and wavelet coefficients. However, given the time limitations on decoding the command, preference is given to methods for encoding an EMG signal in the time domain based on an analysis of signal amplitudes. Such features can be easily identified, and are highly stable for pattern recognition using neural networks. The code (feature vector) must contain enough information to reflect the essential properties of the EMG signal to achieve the best results in command recognition. Therefore, the main requirement for the code is its ease of receipt and its decoding speed (team classification). The following values in the time domain, measured as functions of time, are used as features: integral EMG; arithmetic mean; average value of the module; finite differences; sum of elementary areas; variance; standard deviation; signal length; and maximum value of the EMG signal. More information about the selection of features for classification is given in [7].

Most often, EMG signals are used to identify gestures [8]. The system for identifying gestures includes a sensor that responds to user-made gestures, and a processor whose software is designed to assess the likelihood of gestures presented in gesture libraries based on the classification of EMG signals.

A drawback of the gesture identification system is the data classification technology, which involves correlating one property of a data window with a particular class, which leads to the binding of several classes to one window and the need to calculate the most likely class for the current window.

The patient's EMG signal is processed by non-overlapping segmentation in an intelligent system for controlling a bionic limb [9]. Moreover, for each segment obtained in the previous step, a set of features of the EMG signal is formed based on the amplitude of the EMG signal for classifying gestures; transmit a set of signs of the EMG signal of each segment in the control system of the intellectual bionic limb; determine the type of gesture based on a set of signs of the EMG signal by using an artificial neural network; generating a control signal based on a certain type of gesture; transmit the generated control signal to the engines, driving the fingers of the bionic limb; receive feedback from the control system of the intellectual bionic limb by receiving information from external sensors. The system allows to increase the accuracy of positioning and making decisions about the capture of the subject.

The main drawback of the methods for encoding/decoding an EMG signal in the above HMI is that when generating a vector of informative features based on the analysis of the EMG signal in the window, only the amplitude characteristics of the EMG samples are used, which leads to a loss of information content of the EMG signal and a high level of errors in the classification of commands. The accuracy of the classification of commands can be improved by duplicating EMG signals over several channels and decrypting the vector of informative features thus formed by means of a trained neural network. However, such a technical solution can only be used when taking signals from the upper or lower extremities and is unacceptable when decoding the back muscle muscles EMG, the interpretation of which is necessary when controlling rehabilitation exoskeletons in the verticalization mode or industrial exoskeletons in the load lifting mode. In addition, the use of duplicate channels does not lead to an increase in the information content of the signal, but is most likely aimed at increasing its noise immunity. This circumstance entails a decrease in the reliability of the neural network classifier, the inputs of which are highly correlated, obtained on the basis of the analysis of signals that depend not only on the EMG signal but also on the position of the electrode in the current experiment or on the different dynamics of changes in skin resistance under the electrodes in the process of the experiment, which cannot be taken into account in the process of training a neural network.

19.4 Method for Classifying Electromyography Signals

The proposed method allows to minimize errors in the positioning of the exoskeleton by means of servomotors controlled by commands obtained from the results of decoding an electromiosignal. The essence of the method is to segment the current EMG signal into intersecting or non-intersecting windows with a width of *TW*. An informative sign *FD* is formed from the samples of each window.

Figure 19.2 shows EMG signal plots illustrating the process of segmentation of an EMG signal into intersecting and non-intersecting windows.

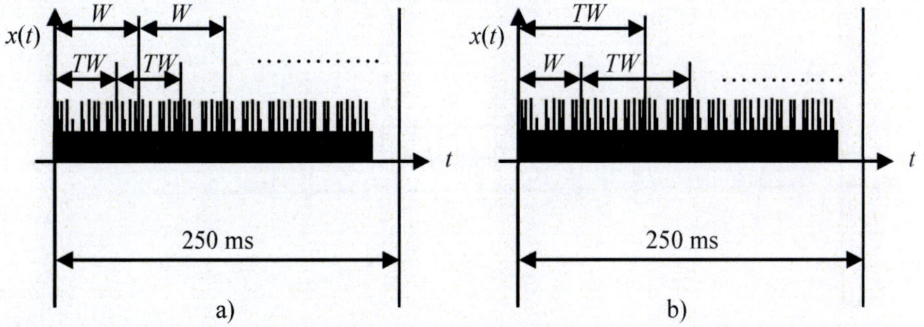

Fig. 19.2 Determining the number of analysis windows: for windows without overlapping (**a**), for overlapping windows (**b**)

A transition is made from the current discrete EMG sample x_τ to the sample y_τ obtained by comparing the current sample with threshold Θ and calculated according to the expression:

$$y_\tau = \begin{cases} 1, & \text{if} \quad |x_\tau| \geq \Theta; \\ 0, & \text{if} \quad |x_\tau| < \Theta. \end{cases} \tag{19.1}$$

The transition from the EMG signal $x(t)$ to the signal $y(t)$ is illustrated by the diagrams in Fig. 19.3.

Based on samples (see 19.1), the informative feature in the TW window is calculated as:

$$\text{FD} = \frac{1}{\text{TW}} \sum_{1}^{\text{TW}} y_\tau. \tag{19.2}$$

On the width of the window, you can use many thresholds Θ and as a result, get a lot of informative signs FD.

The block diagram of the EMG signal classifier, which is implemented in the on-board exoskeleton processor at the hardware–software level is shown on Fig. 19.4. The procedure for transforming the samples of the EMG signal according to (see 19.1) is carried out by the block of comparators 1. Thus, one sample x_τ is associated with N samples y_τ, where N is the number of comparation levels Θ_i. This procedure can be carried out both at the hardware and software levels. In the calculators of the computing unit 2, the procedure for computing the informative feature (see Fig. 19.2) is implemented.

Through the multiplexer 3, the obtained N-dimensional vector of informative features is fed to the input of the first neural network NET1. After classifying it by means of NET1, the corresponding solution is entered in the corresponding memory cell of the memory block 6. After this, it is necessary to skip W samples and perform a similar procedure. The number of analyzed MW windows necessary for making

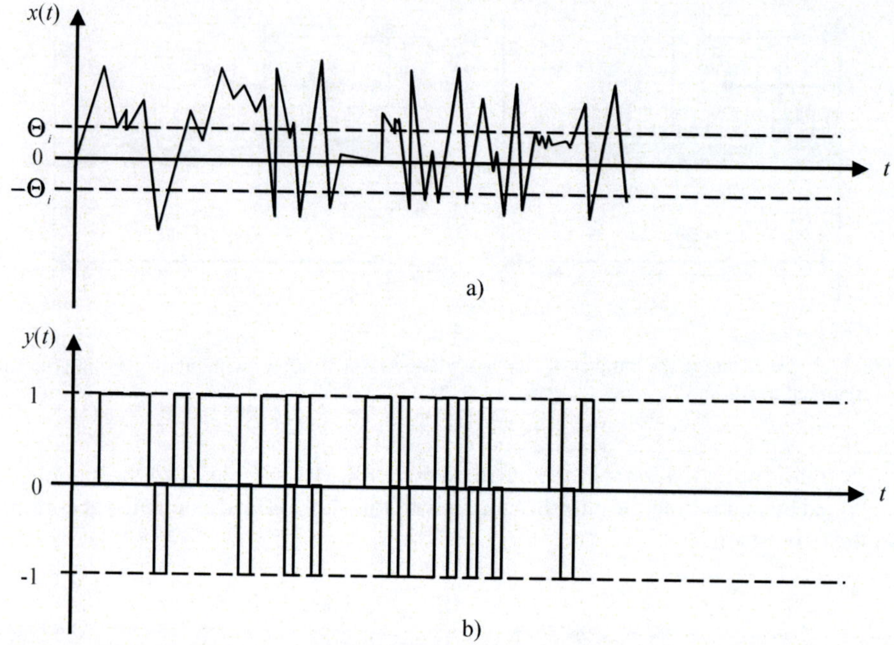

Fig. 19.3 Illustration of the transition from the EMG signal $x(t)$ (**a**) to the signal $y(t)$ (**b**)

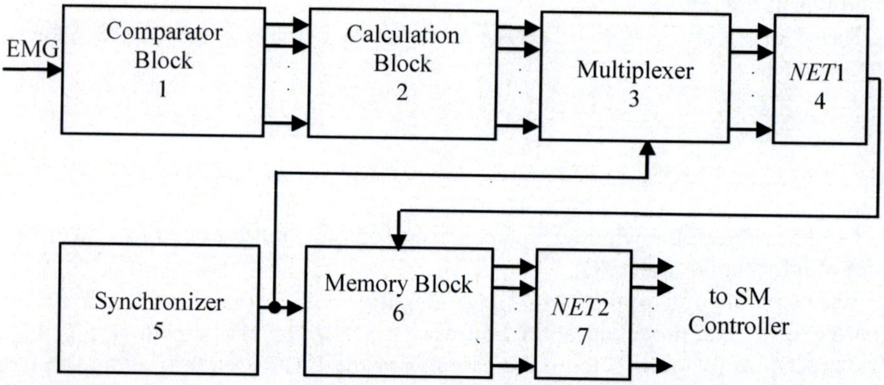

Fig. 19.4 Block diagram of the EMG signal classifier

decisions is calculated according to Fig. 19.2. If $W > \text{TW}$, then non-overlapping windows are used. Otherwise, windows with overlap are used. These informative features are inputs of the *NET*1 learning network, which is used as an EMG signal classifier (see block 4 in Fig. 19.4).

Given the EMG aperture, on which the decision is made to turn on the corresponding servomotor, we can select a lot of windows on it, and, therefore, many

NET1 outputs. Since the outputs of NET1 are separated in time, for making a decision on the totality of the results of these outputs, storage devices are required (see block 6 of Fig. 19.4), in which these results are stored. After analyzing the appropriate number of windows, the results of this analysis are sent to the trained neural network NET2 (see block 7 of Fig. 19.4), the outputs of which are connected to the servo motor (SM) controller.

The number of analysis windows, i.e., NET2 inputs, depends on the decryption time of the command encrypted in the EMG signal and on the minimum quantum of information carried by the EMG signal. According to published sources, the temporal aperture of decoding the EMG signal is 250 ms, and the minimum information quantum, that is, the minimum length of the EMG signal that carries relevant information, is 25 ms. Thus, when using windows without overlapping, MW = int(250/W). If windows with overlapping are used, then MW = int((250–TW)/W). The process of determining the number of analysis windows is illustrated in Fig. 19.2. It shows the process of forming windows without overlapping, and in Fig. 19.2b, the process of forming windows with overlapping.

Synchronizer 5 (see Fig. 19.4) synchronizes the operation of the classifier in such a way that a sequence of vectors of informative features characterizing time windows is formed, so that on the observation aperture of the EMG signal corresponding to MW analysis windows, a vector of informative features $\{\mathrm{NET1}_j\}$ is formed at the input NET2, where $j = \overline{1, \mathrm{MW}}$. Since the formation of this vector of informative features is separated in time, a gate signal should be applied to NET2, signaling the completion of the formation of this vector.

After completing the procedure for turning on the corresponding servomotor, the user can choose two options for further work. If the first option is selected, the following MW windows are analyzed. Therefore, the next command to the servomotor can be received only through TW × MW number of discrete samples. This slows down the decision-making process. Therefore, when deciding on the second option, the NET2 neural network only changes the input corresponding to the first result of the analysis of the NET1 neural network. The remaining NET2 inputs are shifted by one. Thus, the last discrete readout is replaced by the first discrete readout of the next window, and the first discrete readout of the previous window is replaced by the second discrete readout of the previous window.

19.5 Results of Technical Solution

In order to adapt the patient's functional state and verticalization processes, the classifier uses many duplicate channels of EMG signals associated with a muscle or group of muscles that control the movement of the same limb joint. As a result, at the output of the classifier of each channel, we obtain a number corresponding to the confidence in the command to rotate the exoskeleton servomotor. To aggregate decisions on the channels of classifiers, all outputs of the channel classifiers go to a

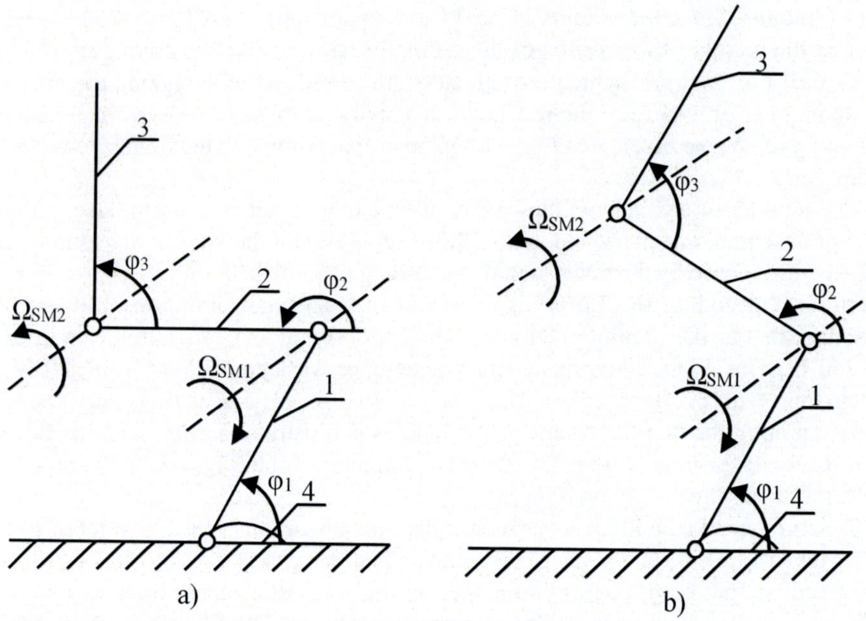

Fig. 19.5 Kinematic diagram of the exoskeleton: initial state of verticalization (**a**), intermediate state of verticalization (**b**). 1 lower leg, 2 thigh, 3 spine, 4 foot

fuzzy neural network; the defusifier of which generates a control signal to the servo motor controller. As a result of the analysis of this signal, the controller determines the speed and direction of rotation of the servomotor.

A simplified kinematic diagram of an exoskeleton for a verticalization mode is shown in Fig. 19.5. For simplicity of calculations, we assume that $\varphi1 = 90°$ and does not change during the verticalization process, and then the verticalization process can be carried out using two servomotors SM1 and SM2 controlling angles $\varphi2$ and $\varphi3$, respectively.

We use five EMG channels that read the EMG signal from the gluteus maximus muscle, biceps femoris muscle, semi-membranous muscle, semi-tendon muscle, and major adductor muscle to control servomotors in the verticalization process (see Fig. 19.6).

The verticalization process is controlled by decoding the EMG in the channels and the rate of verticalization depends on the intensity of the EMG signals in the channels. With certain combinations of EMG intensities in the channels, the servomotors can stop in any phase of verticalization. In this case, feeling that he "lacks the strength to stand up," the patient will want to sit down. In this case, it is possible generation of corresponding EMG signals, which can be used to reverse the servomotors.

To implement the "sit down" mode, EMG channels from the iliopsoas muscle are used (see Fig. 19.6). Thus, for a verticalizer, at least six EMG channels and one more neural network structure used as an aggregator of channel EMG classifiers should

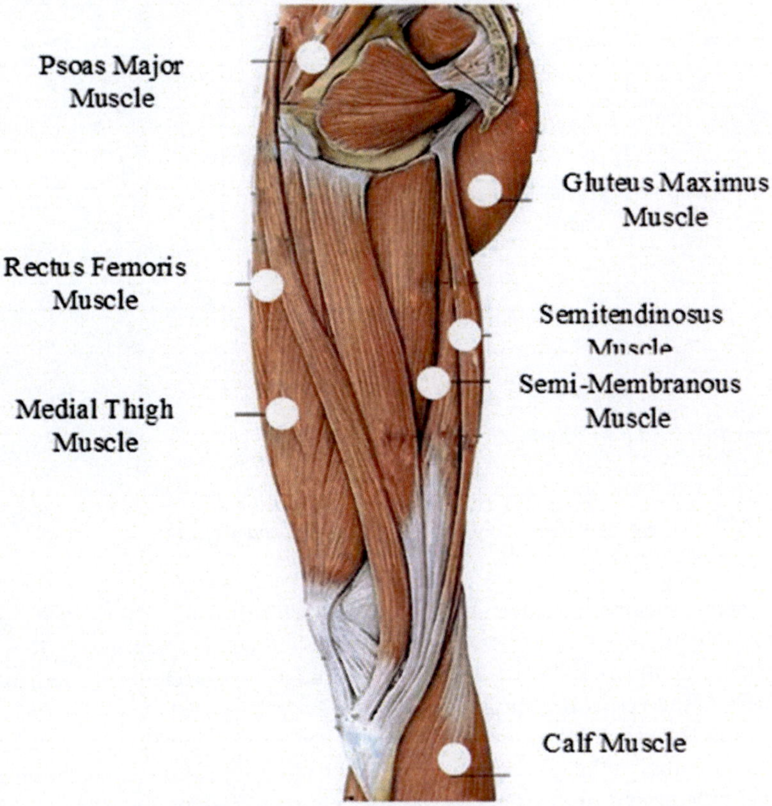

Fig. 19.6 Topology of installing sensors on the muscles forming the five EMG channels

be used as a decoder of EMG signals. The structural diagram of such a classifier is shown in Fig. 19.7.

Neural network NET3 is a fuzzy neural network. Its output is calculated by defusing six membership functions. The second servomotor is controlled by a servo motor controller based on the exoskeleton motion model in the "stand up—sit down" mode. The parameters of the model are adapted to the physical condition of the patient, which eliminates the main disadvantage of the known exoskeleton rehabilitation.

19.6 Conclusion

Method is proposed for the implementation of HMI, which allows minimizing errors in positioning the exoskeleton by means of servomotors controlled by commands obtained by decoding the surface electromyography signal.

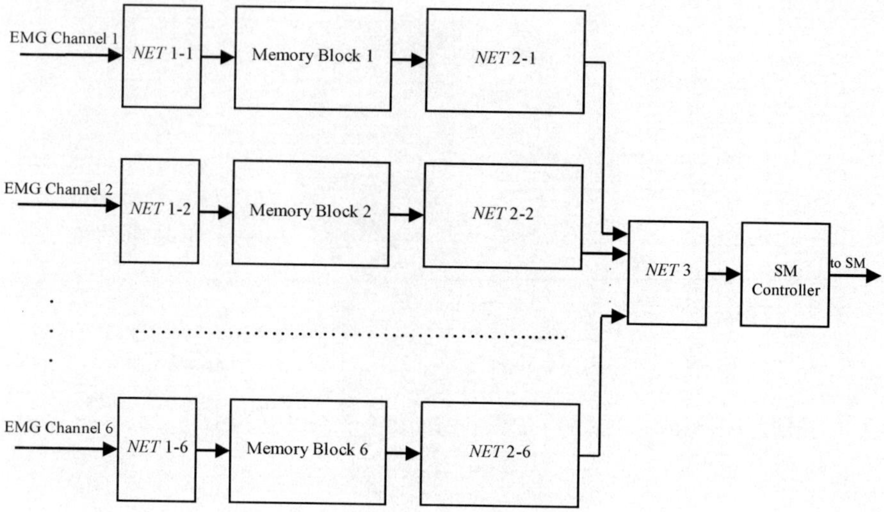

Fig. 19.7 Block diagram of classic EMG using duplicate channels

A human–machine interface has been developed that uses surface EMG signals as control channels, which allows controlling verticalization parameters depending on the patient's functional state based on the study of exoskeleton movement with the patient in the verticalization mode.

Experimental graphs of the movement of the center of mass of the exoskeleton and patient depending on the number of duplicate channels are obtained. A comparative analysis of the graphs of the trajectory of the center of mass of the exoskeleton with the HMI, based on control with neural network and with duplicating channels of the surface EMG, has been carried out. An analysis of the results showed that when using duplicate channels and a control with neural network with a fuzzy neural network at the output of the classifier of control EMG signals, the accuracy of positioning the center of mass of the exoskeleton increases. A comparative analysis of the quality of exoskeleton verticalization control with a PI controller and a neurocontroller was performed, which showed a smaller deviation of the exoskeleton's center of mass from the model trajectory when verticalization was controlled by a neurocontroller, constructed according to the developed method for classifying EMG signals and duplication of verticalization control channels. The obtained method, algorithms and models for the classification of surface EMG signals can be used in the construction of HMI for both medical and industrial exoskeletons.

Acknowledgements This research is supported by the RFBR Project No. 19-38-90112.

References

1. Al-Bared, A.Y.S., Brezhnev, A.N., Tomakova, R.A.: Algorithms for the synthesis of optimal control in biotechnological systems of a rehabilitation type based on neural network technologies. Syst. Anal. Manag. Biomed. Syst. **17**(3), 750–754 (2018)
2. Lin W.-B., Hsieh F.-H.: Lower limb rehabilitation method and apparatus using the method. Patent US, No. 2017007489 A1 (2017)
3. Filist, S.A., Petrunina, E.V., Trifonov, A.A., Serebrovsky, A.V.: Code images of electroencephalogram signals for controlling robotic devices through the brain-computer interface. Modeling, optimization and information technology. Sci. J. **7**(1), 67–79 (2019)
4. Gavrilov, A.I., So So Tav, U.: Application of electromyography data in control systems of exoskeletal devices. Eng. J. Sc. Inn. **6** (2017)
5. Fan Y., Yin Y.: Active and progressive exoskeleton rehabilitation using multisource information fusion from EMG and force-position EPP. IEEE Trans. Biomed. Eng. **60**(12), 3314–3321 (2013)
6. Jayanth D., Sai E.S., Sangeetha Y.: Surface EMG based intelligent signal processing for exoskeleton. In: 2017 International Conference on Inventive Communication and Computational Technologies (ICICCT), pp. 76–82. IEEE (2017)
7. Budko, RYu., Starchenko, I.B.: Creation of a classifier of mimic movements based on the analysis of an electromyogram. Proc. SPIIRAS **46**, 76–89 (2016)
8. Ataee Pedram, Aleem Idris S.: Systems, devices, and methods for gesture identification. Patent US, No. 20150370333 A1 (2015)
9. Ivanyuk, N.M., et al.: Method and system of intellectual bionic limb control. Patent RF **2635632**, C1 (2017)

Chapter 20
Neuro Sliding Mode Control for Exoskeletons with 7 DoF

Haci Mehmet Güzey

Abstract In this paper, a novel neuro-sliding-mode controller (SMC) is developed for a 7-degree-of-freedom (DoF) upper exoskeleton. RBF-like neural network control is used to estimate the exoskeleton dynamics in the configuration of the sliding mode control. Stability of the neuro-SMC is derived on the basis of Lyapunov stability criteria. To validate our theoretical claims, simulation results are given at the end of the paper.

20.1 Introduction

Cerebrovascular accident or stroke is one of five main chronic diseases which seriously affect patients' daily lives [1]. As many societies enter into rapid population aging, incidences of stroke will increase worldwide. More and more cases of motor function or disabilities of the upper limb will arise. Research shows that therapeutic treatment consisting of intensive repetitive movements of the impaired limbs is one of the most effective approaches to partly restore the motorability [2]. However, this treatment is labor-intensive and costly, posing a potentially unacceptable burden to national healthcare services and patients' families' time and energy. In view of this, a significant demand exists for rehabilitation robots that can be used to improve the motor ability of upper limbs and streamline the rehabilitation process [7].

In [3], the proposed control method comprises three main layers: an active impedance control which estimates the user's arm motion intention and guarantees an intuitive response of the suit to the wearer's motion; a mid-level controller which compensates for the backlash in the transmission and converts the reference arm motion to the desired position of the actuator; and a low-level controller which is responsible for driving the actuation stage by compensating for the nonlinear

H. M. Güzey (✉)
Department of Electrical-Electronics Engineering, Erzurum Technical University, Erzurum, Turkey
e-mail: hmgb79@mst.edu

A. Ronzhin and V. Shishlakov (eds.), *Proceedings of 15th International Conference on Electromechanics and Robotics "Zavalishin's Readings"*, Smart Innovation, Systems and Technologies 187, https://doi.org/10.1007/978-981-15-5580-0_20

dynamics occurring in the bowden cable to provide the desired assistive torque at the joint.

In [4], human–machine interaction is dealed. Intentional reaching direction (IRD) is estimated. A human–robot interface comprised of force-sensing resistors (FSRs) is designed to estimate the motion intention of the wearer's upper limb in real time.

The exoskeletons developed in [3, 4] control only arm, exclude the hand and fingers. However, in [5], authors focused on the design and control of a 7-degree-of-freedom cable-driven arm exoskeleton (CAREX-7) used for dexterous motion (including translation and rotation) training or assistance of the whole-arm, including hand and fingers.

In all aforementioned works [3–5], nonlinear system dynamics are assumed to be known. However, the dynamics model of the exoskeletons may not be accurate and changes for each patient. Therefore, uncertain exoskeleton dynamics should be approximated through neural network algorithms. Recently, an adaptive control algorithm was developed for the upper-limb exoskeleton in [6] by using neural networks (NN).

In the meanwhile, one of nonlinear control schemes which does not require system's dynamic model is the sliding mode control (SMC) scheme. SMC is a simple and powerful since it has been used in various motion control applications. One of the advantages of SMC is that stability of the controlled system is guaranteed through Lyapunov's stability theorem for either model-based or non-model-based schemes [6–10]. The performance of non-model-based SMC depends on the appropriate selection of nonlinear gains, which is time consuming. The nonlinear gain has to deal with the entire exoskeleton dynamics which is usually uncertain and time-varying/patient-varying. To deal with the uncertainty problem, neural network is used [9–11], and fuzzy logic is used in the framework of the sliding model control [12].

In this paper, a RBF-like neural network control is used to estimate the exoskeleton dynamics in the configuration of the sliding mode control. Combination of SMC and neural network forms the neuro-SMC structure. Stability of the neuro-SMC is derived on the basis of Lyapunov stability criteria.

Next, the dynamics of the exoskeleton shown in Fig. 20.1 and saturated controller scheme are provided.

20.2 Exoskeleton Dynamics and Saturated Controller Design

The general robotic dynamics with saturation is considered as [6]:

$$M(q)\ddot{q} + C(q, \dot{q})\dot{q} + G(q) = S(\tau) - f_{\text{dis}}, \tag{20.1}$$

where $q \in \Re^n$ is the vector of joint variables, $S(\tau)$ is the saturation function of control torque τ. $C(q, \dot{q}) \in \Re^{n \times n}$, $M(q) \in \Re^{n \times n}$ and $G(q) \in \Re^n$ are the dynamics functions

Fig. 20.1 Myomo mPower100

of $q, \dot{q}$. The matrix $C(q, \dot{q})$ defined by $\dot{M}(q) - 2C(q, \dot{q})$ is skew-symmetric. $f_{dis} \in \Re^n$ represents the time-varying disturbance, and there exists a nonnegative constant f_M^* such that $\| f_{dis} \| \leq f_M^*$.

According to (20.1), define the state variables as $x_1 = q$, $x_2 = \dot{q}$, then the exoskeleton dynamics can be re-written as:

$$\dot{x}_1 = x_2; \tag{20.2}$$

$$\dot{x}_2 = M(q)^{-1}(S(\tau) - f_{\text{dis}} - C(x_1, x_2)x_2 + G(x_1)). \tag{20.3}$$

Define the error signals as:

$$\begin{aligned} z_1 &= x_1 - x_d, \\ z_2 &= x_2 - \alpha_1 \end{aligned} \tag{20.4}$$

with x_d is the reference angles of each joints of exoskeleton, α_1 is the virtual controller being $\alpha_1 = \dot{x}_d - K_1 z_1$, $K_1 = K_1^T > 0$. To eliminate the effect of saturation nonlinearity, the auxilary design system is given similar to [6] as:

$$\dot{\beta} = \begin{cases} -K_\beta \beta - \frac{|z_2^T \Delta\tau| + 0.5\Delta\tau^T \Delta\tau}{\|\beta\|^2} \beta + \Delta\tau, & \|\beta\| \geq \chi, \\ 0, & \|\beta\| < \chi, \end{cases} \tag{20.5}$$

where $\Delta\tau = S(\tau) - \tau$, $K_\beta = K_\beta^T > 0$, $\chi > 0$ is a small positive value, and $\beta \in \Re^n$ is the state of auxiliary design system. Then, the controller is given in [6] as:

$$\tau = -z_1 - K_2(z_2 + \beta) + f_{\text{dis}} + C(x_1, x_2)\alpha_1 + G(x_1) + M(x_1)\dot{\alpha}_1 \tag{20.6}$$

with $K_2 = K_2^T > 0$.

Realize that the control torque in (20.6) requires all the exoskeleton dynamics, accurately. In this work, neuro-sliding-mode controllers in the presence of unknown exoskeleton dynamics are investigated.

Assumption 1 In this work, the saturation function $S(\tau)$ is assumed to be zero and the control torque in the dynamics (20.1) is denoted as τ.

20.3 Neuro-Sliding-Mode Controller

In this part, the neuro-sliding-mode control of an exoskeleton (20.1) with 7 degree of freedom (DoF) is proposed. The main contribution of this work is an upgraded second-order sliding mode controller designed to denote the robustness of the system against both the unknown exoskeleton dynamics (Fig. 20.2).

First, define the sliding surface as:

$$s = k_1 z_1 + k_2 \dot{z}_1, \tag{20.7}$$

where $k_1 > 0, k_2 > 0$ are positive design parameters.

Define the Lyapunov function based on the sliding surface (20.7) as:

$$L = \frac{1}{2} s^T s. \tag{20.8}$$

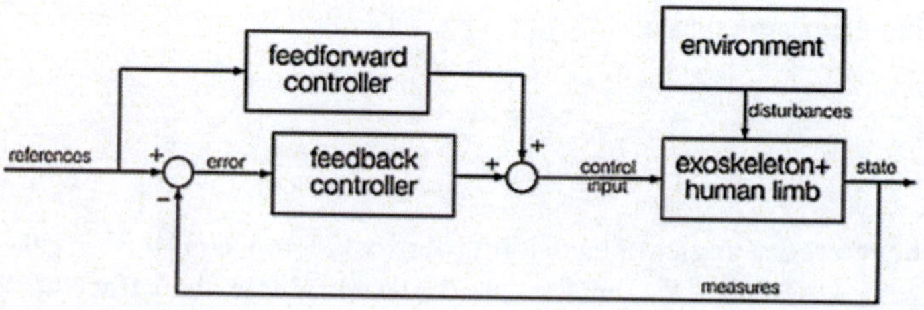

Fig. 20.2 Exoskeleton controller scheme [13]

Take derivative of (20.8) and get:

$$\begin{aligned} L &= s^T \dot{s} \\ &= s^T (k_1 \dot{z}_1 + k_2 \ddot{z}_1) \\ &= s^T (k_1 (x_2 - \dot{x}_d) + k_2 (\dot{x}_2 - \ddot{x}_d)) \\ &= s^T \left(\begin{array}{l} k_1 (x_2 - \dot{x}_d) \\ + k_2 \left(\begin{array}{l} M(q)^{-1} (\tau - f_{dis} - C(x_1, x_2) x_2 + G(x_1)) \\ - \ddot{x}_d \end{array} \right) \end{array} \right). \end{aligned} \quad (20.9)$$

Define the control torque in (20.9) as:

$$\tau = \frac{1}{k_2} \left(\begin{array}{l} M(q)(f_{\text{dis}} + C(x_1, x_2) x_2 - G(x_1)) \\ + \ddot{x}_d - k_1 (x_2 - \dot{x}_d) - k_s s \end{array} \right) \quad (20.10)$$

with $k_s > 0$ is a design parameter, then (20.9) becomes:

$$L = -k_s \|s\|^2. \quad (20.11)$$

From (20.11), it can be concluded that the system exoskeleton (20.1) tracks its desired trajectory, x_d asymptotically.

However, when the dynamics equation used in controller (20.11) has uncertainties, the controller (20.10) is not feasible. Therefore, combine the dynamics equations and define:

$$f_{\text{unc}} = (f_{dis} + C(x_1, x_2) x_2 - G(x_1)). \quad (20.12)$$

The uncertain nonlinear dynamics (20.12) are represented as:

$$f_{\text{unc}}(\bar{z}) = \Theta^T \psi\left(H^T \bar{z}\right) + \varepsilon, \quad (20.13)$$

where $\Theta \in \Re^{2 \times h}$ is the desired NN weights with h being the number of hidden-layer neurons, $\psi\left(H^T \bar{z}\right)$ is the basis function with $H^T \in \Re^{h \times n}$ is the mapping between the inputs and the hidden-layer neurons, n is the number of inputs to the NN, ε is the bounded NN reconstruction error satisfying $\|\varepsilon\| \leq \varepsilon_M$, with ε_M being a positive constant, $\bar{z}$ is the NN inputs to estimate the exoskeleton dynamics. The unknown NN weights can be estimated as $\hat{\Theta}$ and estimated uncertain dynamics can be given by:

$$\hat{f}_{\text{unc}}(\bar{z}) = \hat{\Theta}^T \psi(\bar{z}). \quad (20.14)$$

Now, the NN weight estimation error is defined as $\tilde{\Theta} = \Theta - \hat{\Theta}$ and the estimation error dynamics can be obtained as $\dot{\tilde{\Theta}} = -\dot{\hat{\Theta}}$. The control torque, using (20.14), is obtained as:

$$\tau = \frac{1}{k_2}\Big(\hat{f}_{\text{unc}} + \ddot{x}_d - k_1(x_2 - \dot{x}_d) - k_s s\Big). \tag{20.15}$$

Assumption 2 The target NN weights are bounded by positive values, such that $\|\Theta\| \le \Theta_M$ with Θ_M being a positive-bounded constant.

Theorem Consider the exoskeleton system (20.1) along with the control torque (20.15) using the NN estimate (20.14) of the uncertain dynamics by using the following NN weight update rule:

$$\dot{\hat{\Theta}} = \psi(z)s - k_\Theta \hat{\Theta}, \tag{20.16}$$

where $k_\Theta > 0$ is the small positive learning rate. Consider assumption 1 and assumption 2 holds. Then, the tracking errors and the NN weight estimation errors are semi-globally uniformly ultimately bounded (SGUUB).

Proof Define the NN estimation error on the uncertain dynamics as:

$$\tilde{f}_{\text{unc}}(\bar{z}) = \tilde{\Theta}^T \psi(z) + \varepsilon \tag{20.17}$$

and the Lyapunov function based on the sliding surface as well as the NN estimation error as:

$$L_u = \frac{1}{2}s^T s + \frac{1}{2}tr\Big\{\tilde{\Theta}^T\tilde{\Theta}\Big\}. \tag{20.18}$$

Take derivative of (20.18) and get:

$$\begin{aligned}\dot{L}_u &= s^T\dot{s} + tr\Big\{\tilde{\Theta}^T\dot{\tilde{\Theta}}\Big\}\\ &= s^T(k_1\dot{z}_1 + k_2\ddot{z}_1) + tr\Big\{\tilde{\Theta}^T\dot{\tilde{\Theta}}\Big\}\\ &= s^T(k_1(x_2 - \dot{x}_d) + k_2(\dot{x}_2 - \ddot{x}_d)) + tr\Big\{\tilde{\Theta}^T\dot{\tilde{\Theta}}\Big\}\\ &= s^T\left(k_1(x_2 - \dot{x}_d) + k_2\left(\begin{matrix}M(q)^{-1}(\tau - f_{\text{unc}})\\ -\ddot{x}_d\end{matrix}\right)\right)\\ &\quad + tr\Big\{\tilde{\Theta}^T\dot{\tilde{\Theta}}\Big\}.\end{aligned} \tag{20.19}$$

Use the controller (20.15) in (20.19) and obtain:

$$\begin{aligned}\dot{L}_u &= s^T\Big(\tilde{f}_{\text{unc}} - k_s s\Big) + tr\Big\{\tilde{\Theta}^T\dot{\tilde{\Theta}}\Big\}\\ &= -k_s\|s\|^2 + s^T\tilde{\Theta}^T\psi(z) + s^T\varepsilon - tr\Big\{\tilde{\Theta}^T\dot{\hat{\Theta}}\Big\}.\end{aligned} \tag{20.20}$$

By using adaptation law (20.16) in (20.20) yields:

$$\begin{aligned}\dot{L}_u &= -k_s\|s\|^2 + s^T\varepsilon + k_\Theta tr\left\{\tilde{\Theta}^T\hat{\Theta}\right\}\\ &= -k_s\|s\|^2 + s^T\varepsilon + k_\Theta tr\left\{\tilde{\Theta}^T\left(\Theta - \tilde{\Theta}\right)\right\} = .\\ &= -k_s\|s\|^2 + s^T\varepsilon - k_\Theta\left\|\tilde{\Theta}\right\|^2 + k_\Theta tr\left\{\tilde{\Theta}^T\Theta\right\}\end{aligned} \tag{20.21}$$

Using upper bounds on NN approximation error and NN weight estimation error, ε_M, Θ_M, respectively, in (20.21) yields:

$$\dot{L}_u \le -k_s\|s\|^2 + s^T\varepsilon_M - k_\Theta\left\|\tilde{\Theta}\right\|^2 + k_\Theta\left\|\tilde{\Theta}\right\|\Theta_M.$$

Now, use the Young's inequality to get:

$$\begin{aligned}\dot{L}_u \le &-k_s\|s\|^2 + \frac{1}{2}\|s\|^2 + \frac{1}{2}\varepsilon_M^2 - k_\Theta\left\|\tilde{\Theta}\right\|^2\\ &+ \frac{k_\Theta}{2}\left\|\tilde{\Theta}\right\|^2 + \frac{k_\Theta}{2}\Theta_M^2.\end{aligned} \tag{20.22}$$

Combine the similar terms in (20.22) and obtain:

$$\dot{L}_u \le -\left(k_s - \frac{1}{2}\right)\|s\|^2 - \frac{k_\Theta}{2}\left\|\tilde{\Theta}\right\|^2 + \frac{1}{2}\varepsilon_M^2 + \frac{k_\Theta}{2}\Theta_M^2. \tag{20.23}$$

Assume that we choose $k_s > \frac{1}{2}$ and define sum of the positive terms in (20.23) as $Pt = \frac{1}{2}\varepsilon_M^2 + \frac{k_\Theta}{2}\Theta_M^2$, the bounds on NN estimation error and sliding surface can be calculated as:

$$\begin{aligned}0 &\ge -\left(k_s - \frac{1}{2}\right)\|s\|^2 + Pt,\\ \left(k_s - \frac{1}{2}\right)&\|s\|^2 \ge Pt;\\ \|s\| &\ge \sqrt{\frac{Pt}{\left(k_s - \frac{1}{2}\right)}}.\end{aligned} \tag{20.24}$$

The system is asymptotically stable when the system moves away from the sliding surface with the altitude of (20.24). The bound can be reduced by choosing k_s as large as possible. Or, second bound can be found as:

$$0 \ge -\frac{k_\Theta}{2}\left\|\tilde{\Theta}\right\|^2 + Pt,$$

$$\frac{k_\Theta}{2}\left\|\tilde{\Theta}\right\|^2 \geq Pt;$$
$$\left\|\tilde{\Theta}\right\| \geq \sqrt{\frac{2Pt}{k_\Theta}}. \tag{20.25}$$

Finally, if one of the bounds (20.24) or (20.25) satisfies, the system becomes asymptotically stable. Therefore, the system is SGUUB.

Next, simulation results are provided to validate our theoretical claims.

20.4 Simulation Results

In this section, an exoskeleton system with 7 degree of freedom is controlled through neuro-sliding-mode controller. Each link is forced to track a sinusoidal trajectory given as:

$$x_d = \begin{bmatrix} \sin(t)+1, \cos(t), 2\sin(t)+1, 2\cos(t)+2, \\ \cos(t)+3, \sin(t)+3, 3\cos(t) \end{bmatrix}.$$

The controller gains are selected as $k_1 = 10.2, k_2 = 10.1, K_1 = 1.3, k_s = 120.3$ while the initial conditions are chosen as:

$$q(0) = \begin{bmatrix} 15\ 14\ 13\ 12\ 11\ 10\ 9 \end{bmatrix};$$
$$\dot{q}(0) = \begin{bmatrix} -1\ -1\ -1\ -1\ -1\ -1\ -1 \end{bmatrix}.$$

As shown in Fig. 20.3, the actual trajectory of the first link of the exoskeleton

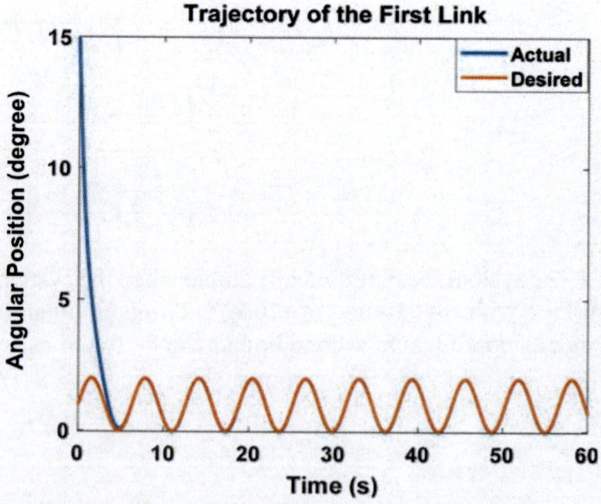

Fig. 20.3 Desired and actual trajectory of the first link of the exoskeleton

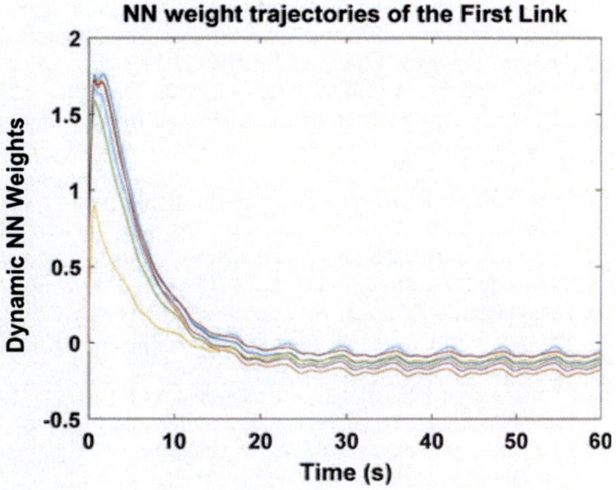

Fig. 20.4 Trajectories of the dynamic NN weights of the first link of exoskeleton

converge actual trajectory around 5 s.

Figure 20.4 illustrates the NN weights trajectories. It is obvious that the NN weights converge after some time which means the dynamics of the exoskeleton estimated through our two-layer NN design.

Next, some concluding results are provided.

20.5 Conclusion

In this paper, a neuro-sliding-mode controller was developed for an exoskeleton with 7 degree of freedom. Simulation results showed that the proposed controller provides highly efficient outcomes. In the future work, the focus will be considering saturation on the control torque.

Acknowledgements The work described in this paper is supported by TUBITAK (The Scientific and Technological Research Council of Turkey) under project number 217E138.

References

1. TheWorldHealthReport. http://www.who.int/whr/2008/en. Last Accessed 21 Nov 2019
2. Barreca, S., Wolf, L., Fasoli, S., Bohannon, R.: Treatment interventions for the paretic upper limb of stroke survivors: a critical review. Neurorehabilitation Neural Repair **17**(4), 220–226 (2003)
3. Dinh, B.K., Xiloyannis, M., Antuvan, C.W., Cappello, L., Masia, L.: Hierarchical cascade controller for assistance modulation in a soft wearable arm exoskeleton. IEEE Robotics Autom. Lett. **2**(3), 1786–1793 (2017)

4. Huang, J., Huo, W., Xu, W., Mohammed, S., Amirat, Y.: Control of upper-limb power-assist exoskeleton using a human-robot interface based on motion intention recognition. IEEE Trans. Autom. Sci. Eng. **12**(4), 1257–1270 (2015)
5. Cui, X., Chen, W., Jin, X., Agrawal, S.K.: Design of a 7-DOF cable-driven arm exoskeleton (CAREX-7) and a controller for dexterous motion training or assistance. IEEE/ASME Trans. Mechatron. **22**(1), 161–172 (2017)
6. He, W., Li, Z., Dong, Y., Zhao, T.: Design and adaptive control for an upper limb robotic exoskeleton in presence of input saturation. IEEE Trans. Neural Netw. Learn. Syst. **30**(1), 97–108 (2019)
7. Hashimoto, H.: A microprocessor-based robot manipulator control with sliding mode. IEEE Trans. Ind. Electron. **34**(1), 11–18 (1987)
8. Istefanopulos, Y., Jafarov, E.M., Parlakci, M.N.A.: A new robust continuous sliding mode control for robot manipulators with parameter perturbation. In: American Control Conference, pp. 3202–3205 (2002)
9. Soltanpour, M., Otadolajam, P., Khoban, M.H.: Robust control strategy for electrically driven robot manipulators: adaptive fuzzy sliding mode. IET Sci. Meas. Technol. 322–334 (2015)
10. Chaouch, D.E., Ahmed-Foitih, Z., Khelfi, M.F.: A sliding mode based control of 2dof robot manipulator using neural network. In: International conference on Sciences of Electronics Technologies of Information and Telecommunications, pp. 906–911 (2012)
11. Chen, Z., Zhang, J., Wang, Z., Zeng, J. Sliding mode control of robot manipulators based on neural network reaching law. In: IEEE Conference on Control and Automation, pp. 370–373 (2007)
12. Salen, M., Khelfi, M.F.: Online RBF and fuzzy based sliding mode control of robot manipulator. In: International Conference on Sciences of Electronics Technologies of Information and Telecommunications, pp. 896–901 (2012)
13. Jarrasse, N., et al.: Robotic exoskeletons: a perspective for the rehabilitation of arm coordination in stroke patients. Frontiers Hum. Neurosci. **8**, 947 (2014)

Chapter 21
Modeling of the Exoskeletal Human-Machine System Movement Lifting a Load

Andrey Karlov, Ekaterina Saveleva, Andrey Yatsun, and Aleksey Postolny

Abstract The mechanization and automation are introduced in various industries for loading and unloading operations. However, the manual labor continues to be widespread in a number of industries. One of the ways to improve the quality of working conditions due to the transition from extreme to comfortable conditions is using of exoskeletons. The technological operations of lifting and transfer of goods implementation by a person can be considered in two aspects: from the point of view of the workload performed by a person with this type of work, and on the other hand, the functional stress of the body as an integral response to the load. The study of the human-machine system (HMS) elements interaction allows us to identify new opportunities for providing a cooperative solution to technological problems by obtaining the exoskeleton's links movement given accuracy. The article is devoted to the HMS motion kinematic model development, as well as the determination of the dependence of the linear gravity compensator (LGC) length change on the magnitude of the exoskeleton's back rotation angle for various geometric dimensions that determine the position of the LGC. The determination of this dependence allows us to construct an algorithm for the HMI, including the LGC functioning. For this, the vector method of mathematical modeling is used.

21.1 Introduction

One of the ways to improve the quality of working conditions due to the transition from extreme to comfortable conditions is using of exoskeletons, which significantly increase a person's the strength while the body functional stress level decreases

A. Karlov · E. Saveleva (✉) · A. Yatsun · A. Postolny
Department of Mechanics, Mechatronics and Robotics, Southwest State University (SWSU), Kursk, Russian Federation
e-mail: katyasavelkursk@yandex.ru

A. Yatsun
e-mail: teormeh@inbox.ru

A. Ronzhin and V. Shishlakov (eds.), *Proceedings of 15th International Conference on Electromechanics and Robotics "Zavalishin's Readings"*, Smart Innovation, Systems and Technologies 187, https://doi.org/10.1007/978-981-15-5580-0_21

[1, 2]. This effect is achieved when the operator and exoskeleton form a human-machine system, the efficiency of which is determined by the degree of the system elements actions coordination. To ensure this, the system is complemented by another important element—the human-machine interface (HMI) [2, 3], which is a technical tool providing interaction between an operator and the exoskeleton. Such a system, which consists of an operator, an exoskeleton, and a HMI, is called a biotechnical walking system (BTWS) [4, 5].

The development of robotic systems, including human-machine systems began recently. The term "gravitational compensation" and its English analog, "gravitational compensation" are used to denote the system individual parts' properties that have statically stable positions independent of vertical movement of the links. Such an approach finds application in exoskeletons, as it allows expand person's capabilities significantly in performing tasks that a person in the normal state cannot perform. In the first place, this applies to the implementation of rigging works.

There are such terms as: passive gravitational compensation and active gravitational compensation [6].

(1) Gravity compensation in passive exoskeletons is achieved by usage of three main solutions:
(2) Additional masses for changing links the centers of mass position,
(3) Elastic elements creating forces compensating the gravity force,
(4) Devices using self-braking electric drives due to friction in articulated joints and transmissions,
(5) Controllable drives without self-braking.

Currently, there are a large number of publications devoted to researching and development of human-machine systems [6–12].

The HMI of the exoskeletal human-machine system provides control and the system elements movement management. In such a system, the operator occupies a dominant position, and the power element of the system, the exoskeleton, carries out controlled movements specified by the operator, necessary for performing a particular technological operation [7, 8, 12–15].

21.2 BTWS Structure

The structural diagram of the BTWS interaction with the environment is shown in Fig. 21.1. The main elements of the system are operator (O), industrial exoskeleton (IE), and human-machine interface (HMI) environment (E). A person constantly evaluates the environment, setting the movement of the lower and upper limbs, through the HMI that acts on the exoskeleton, which helps to perform technological operations, reducing the moments created by the muscular system [2, 16].

The exoskeleton is a system of variable-length links interconnected by hinges, which can be equipped with various rotary or linear drives, as well as potential energy storage devices that create the necessary moments in various hinges, which helps

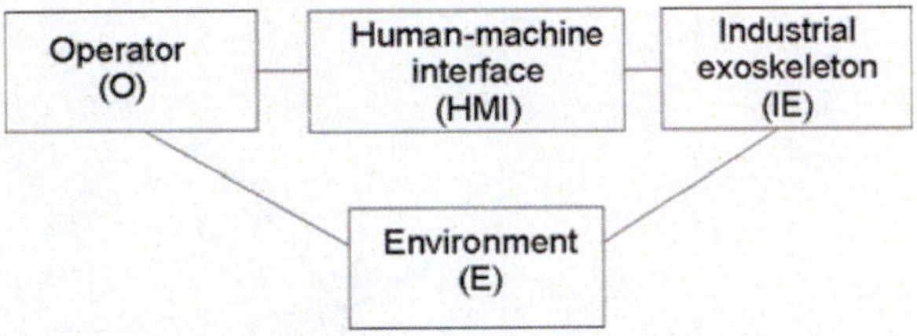

Fig. 21.1 Structural scheme of BTWS

operator in performing a particular technological operation, significantly reducing the muscle system load level, and in some cases completely unloading the person [4, 5, 17–20]. Personification is achieved due to the possibility of the parameters of the exoskeleton and HMI individual adjustment [3, 6]. A person and an exoskeleton carry out cooperative interaction with the environment which is represented by the supporting surface, obstacles, and technological load. The criterion of the interaction effectiveness between the operator and the exoskeleton is the accuracy and speed of the reproduction of the operator's legs and back movements by the exoskeleton.

The maximum of this criterion can be achieved by using the HMI. HMI allows to use the sensor system and signal processing algorithms to evaluate the instantaneous position of the person's elements (legs, back) and use this information for the control driver of voltages supplied to the corresponding exoskeleton electric drives and generating the necessary forces and moments. For this, it is important to have a dependence of the change in the length of the LGC on the magnitude of the generalized coordinates.

21.3 Mathematical Model of BTWS

At this stage, a BTWS providing an assistive movement of the operator's back is considered, due to the installation of an active hip joint (HJ) in the exoskeleton based on a combined LGC [4–6].

Such a biotechnological system, called BTWS, allows us to unload the lumbar spine muscle system, while the human-machine interface, which provides a control system for the interaction of the person and the exoskeleton, is characterized by relative simplicity, a minimal number of sensors, and the relative simplicity of the control algorithm. The human model consists of the thigh 3, the exoskeleton's back 4, and the operator's back 5. The movement of BTWS is considered in Fig. 21.2.

There are two kinds of compensation systems, which can be distinguished: active, having a controlled electric drive and passive, without a controlled drive. Next, we dwell on a combined type of a linear gravity compensator (LGC), which uses the positive properties of passive and active LGCs [6].

The LGC scheme is shown in Fig. 21.3. LGC consists of four main elements: 1—the rod, 2—the body, 3—the elastic element, 4—the electric drive (not shown in the diagram).

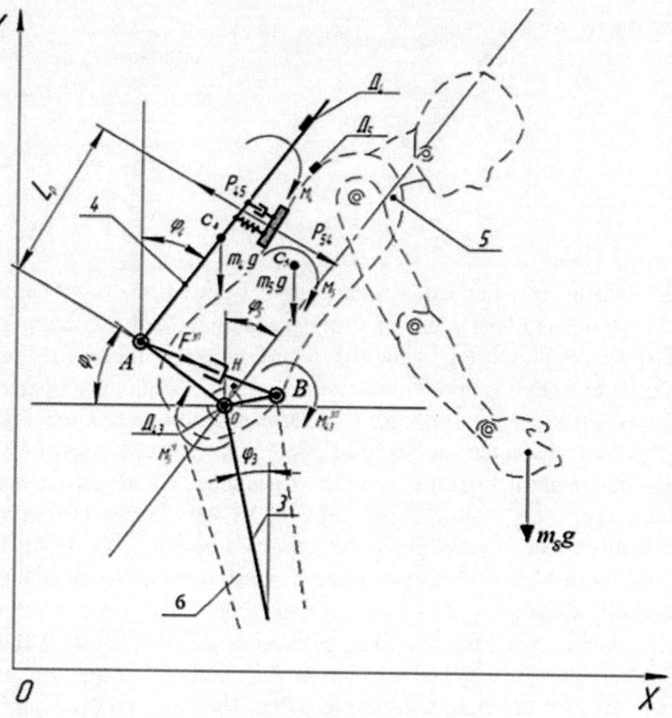

Fig. 21.2 Interaction scheme of man and the exoskeleton as the BTWS

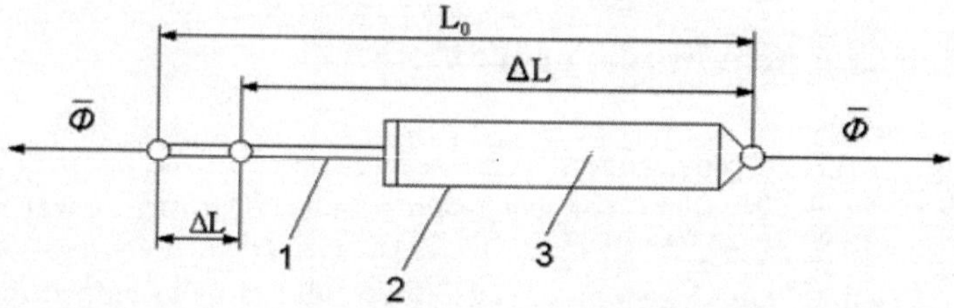

Fig. 21.3 Scheme of the LGC (1 the elastic element, 2 the body, 3 the rod)

The combined LGC operation principle is that, the potential energy of the elastic element accumulates during its' compression, and then, when it is stretched, the energy is transmitted to the exoskeleton in the form of a moment $M_{O_{43}}(\Phi_1)$ applied by the HJ. In parallel with this, the electric drive installed in LGC creates a moment $M_{O_{43}}(\Phi_2)$. Thus, we get an assisting moment acting on the exoskeleton's back 4 relative to the hip 3 and helping the person to perform the necessary movements.

$$M_{O_{43}}(\Phi) = M_{O_{43}}(\Phi_1) + M_{O_{43}}(\Phi_2). \tag{21.1}$$

Let us consider in more detail the HMI structural diagram, which provides the interaction of a human and BTWS exoskeleton, shown in Fig. 21.5. The HMI consists of a back module that includes the exoskeleton's and the operator's back absolute inclination angle sensors, a force interaction sensor, a hip joint equipped with a relative angle meter, an on-board computer, and LGC HJ. By analyzing the information received from the sensors in accordance with the developed algorithm, the on-board computer generates control voltages supplied to the LGC electric drive.

For the development and research of the BTWS, a kinematic and dynamic model of the system has been developed that allows us to work out in detail the process of interaction of a biological object, the electromechanical part of the exoskeletal system, and HMI. Mathematical modeling of BTWS allows one to evaluate the influence of HMI parameters on the human links position and to develop control voltages applied to the windings of exoskeleton electric drives (Fig. 21.4).

A measure of the mechanical interaction of the system elements is the force P_{54}, and the measurement of which is carried out by using a special sensor that allows us to analyze the level of an exoskeleton to a person exposed and vice versa.

The position in the sagittal plane of the HPS links is estimated by the absolute angles $\phi_4, \phi_5, \phi_3 = \text{const}$. The relative angle ϕ_{43} controls the position of the exoskeleton's back relative to the hip. The operator sets the movement nature of the exoskeleton's back using the back inclination angle ϕ_5.

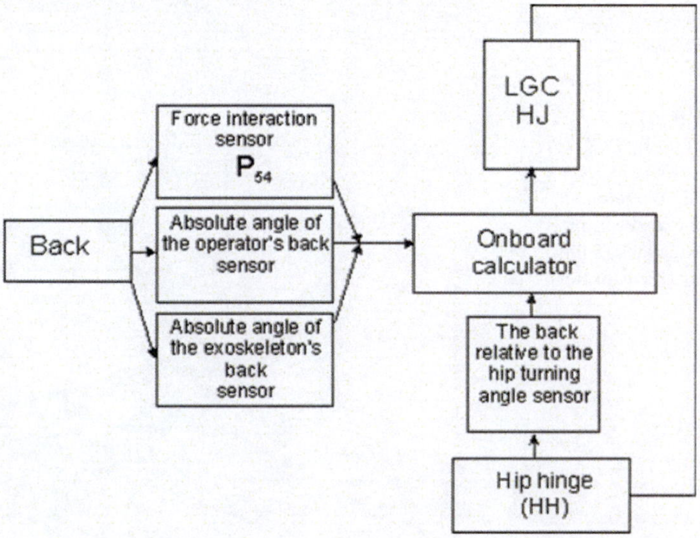

Fig. 21.4 HMI scheme providing the interaction of man and the BTWS exoskeleton

21.4 The BTWS V1 Kinematic Model

The considered system has two generalized coordinates ϕ_4 and ϕ_5. The position of the operator's back sets the angle ϕ_5, and the position of the exoskeleton's back sets the angle ϕ_4. It is necessary to connect the coordinates of points A, B, which determine the exoskeleton's kinematics and the generalized coordinate ϕ_4 between themselves. For this purpose, it is necessary to consider the circuit shown in Fig. 21.5.

To solve the problem of determining the relationship between the characterical system points corresponding coordinates, the vector-matrix method can be applied. We write the equality:

$$\overline{r}_B = \overline{r}_A + \overline{r}_{AB}. \tag{21.2}$$

From here, the distance between points A and B corresponding to the length L of the LGC can be found.

$$\overline{r}_{AB} = \overline{r}_B - \overline{r}_A \tag{21.3}$$

or, taking into account (21.2) (21.3), can be got:

$$\overline{r}_{AB} = \overline{r}_{4A} - \overline{r}_{3B}, \tag{21.4}$$

where

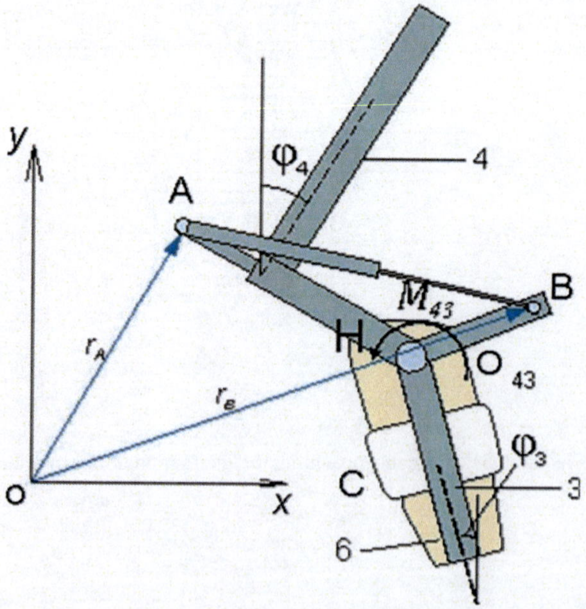

Fig. 21.5 Relative arrangement of the LGC for the operator in the lower extremities exoskeleton (1 the foot, 2 the lower limb, 3 the thigh, 4 the back, 5 the linear gravity compensator)

$$\bar{r}_{AB} = \begin{vmatrix} x_{AB} \\ y_{AB} \end{vmatrix}; \quad \bar{r}_{3B} = \begin{vmatrix} x_{O_{43B}} \\ y_{O_{43B}} \end{vmatrix}; \quad \bar{r}_{4A} = \begin{vmatrix} x_{O_{43A}} \\ y_{O_{43A}} \end{vmatrix}. \tag{21.5}$$

It follows that the displacement of the LGC depends only on the position of the vectors $\bar{r}_{3B}$ and $\bar{r}_{4A}$. The current LGC length can be determined by the formula:

$$\mathrm{L}^2 = x_{AB}^2 + y_{AB}^2. \tag{21.6}$$

Using the formula (21.2), we obtain:

$$\begin{aligned}
x_{AB} &= x_{O_{43B}} - x_{O_{43A}};\\
y_{AB} &= y_{O_{43B}} - y_{O_{43A}};\\
x_{O_{43B}} &= O_{43}B\cos\phi_3;\\
y_{O_{43A}} &= O_{43}B\sin\phi_3,\ x_{O_{43A}} = O_{43}A\cos\phi_4;\\
y_{O_{43A}} &= O_{43}A\sin\phi_4.
\end{aligned}$$

From here, the relationship between the LGC length and the rotation angle can be got:

$$L^2 = (O_{43}B\cos\phi_3 - O_{43}A\cos\phi_4)^2 + (O_{43}B\sin\phi_3 - O_{43}A\sin\phi_4)^2. \tag{21.7}$$

To define the minimum value of L, let us formulate the condition:

$$\begin{aligned}
\frac{\mathrm{d}L}{d\phi_4} = \frac{1}{2}&((O_{43}B\cos\phi_3 - O_{43}A\cos\phi_4)^2\\
&+ (O_{43}B\sin\phi_3 - O_{43}A\sin\phi_4)^2)^{-\frac{1}{2}}\\
&((O_{43}B\cos\phi_3 - O_{43}A\cos\phi_4)O_{43}A\sin\phi_4\\
&+ (O_{43}B\sin\phi_3 - O_{43}A\sin\phi_4)O_{43}A\cos\phi_4) = 0
\end{aligned} \tag{21.8}$$

Solving this equation, we obtain the angle ϕ_4 value that provides a minimum $L = L_0$.

LGC is characterized by the fact that the force created by it during its movement is determined by the formula: $\Phi = C\Delta L$ as the force created by the elastic element of the LGC.

where C—coefficient determined by the elastic element properties. The force moment created by the LGC relative to the HJ rotation axis is determined by the formula:

$$M_{O_{43}}(\Phi) = \Phi O_{43}H. \tag{21.9}$$

Here $O_{43}H$—force Φ arm relative to the center of HJ.

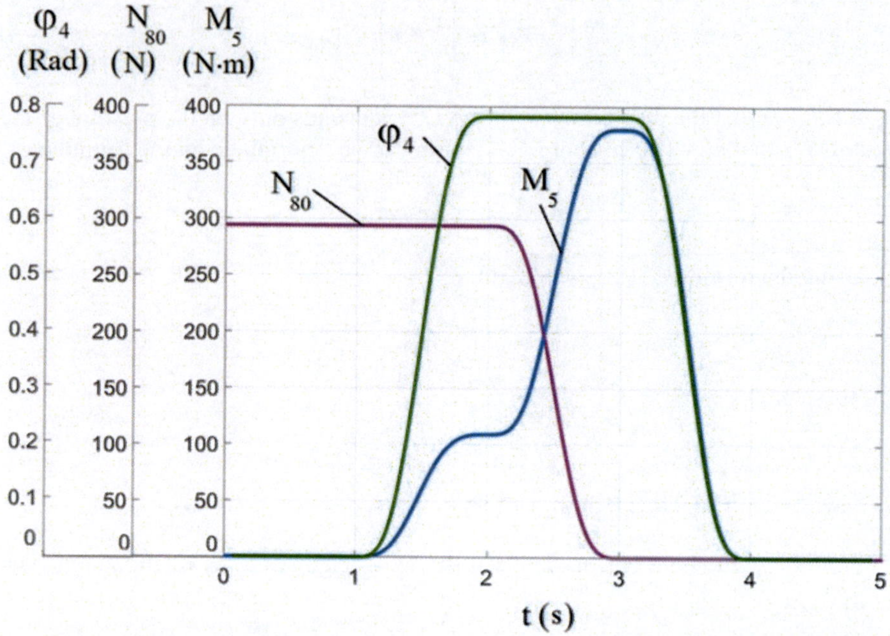

Fig. 21.6 Schedule of the parameters $\varphi 4$, N_{80}, and M_5 change in dependence of time over time

$$O_{43}H = \frac{2 \cdot \sqrt{(P \cdot (P - L_{AB}) \cdot (P - L_{O_{43}A}) \cdot (P - L_{O_{43}B}))}}{L_{AB}}. \tag{21.10}$$

$$P = \frac{1}{2} \cdot (L_{AB} + L_{O_{43}A} + L_{O_{43}B}). \tag{21.11}$$

In Fig. 21.6 the dependences of the parameters determining the human-machine exoskeletal system operation when lifting a load are presented.

21.5 Conclusion

The structural schemes of BTWS and HMI are proposed. A kinematical model of BTWS movement has been developed. The dependences of the linear gravity compensator (LGC) length change on the magnitude of the exoskeleton's back rotation angle for various geometric dimensions determining the position of the LGC are determined. An analysis of influence of the variety of the exoskeleton's linear gravity compensator (LGC) which hinges installation geometric dimensions by the amount of accumulated energy during the back tilt operation was made. A mathematical model of the force and its moment created by the combined LGC is proposed.

Acknowledgements The work of the group of authors was supported by RFBR, research project № 18-08-00773-A.

References

1. Spada, S. et al.: Analysis of exoskeleton introduction in industrial reality: main issues and EAWS risk assessment. In: International Conference on Applied Human Factors and Ergonomics, pp. 236–244 (2017)
2. Theurel, J., Desbrosses, K., Roux, T., Savescu, A.P: Physiological consequences of using an upper limb exoskeleton during manual handling tasks. Appl. Ergonom. **67**, 211–217 (2018)
3. Strausser, K.A., Kazerooni, H.: The development and testing of a human machine interface for a mobile medical exoskeleton. In: 2011 IEEE/RSJ International Conference on Intelligent Robots and Systems, pp. 4911–4916. IEEE (2011)
4. Yatsun, S.F., Antipov, V.M., Karlov, A.Ye.: Modeling of lifting with the help of industrial exoskeleton. Bull. Southwestern State Univ. **22**, **6**(81), 14–20 (2018)
5. Yatsun, S.F., Antipov, V.M., Karlov, A.Y.: Lifting a load in an exoskeleton with gravitational compensation. Proc. Southwest State Univ. **23**(2), 8–17 (2019)
6. Jatsun S.F., Al Maji K.K.M., Pavlovsky V.E.: Analysis of the effect of the exoskeleton geometrical dimensions on the nature of a linear compensator operation. In: Developments in eSystems Engineering 2019. Robotics, Sensors and Industry 4.0 (2019)
7. Jatsun, S., Savin, S., Lushnikov, B.б Yatsun, A.: January. System analysis of sagittal plane human motion wearing an exoskeleton using marker technology. ITM Web Conf. EDP Sci. **6**, 03006 (2016)
8. Talaty, M., Esquenazi, A., Briceno, J. E.: Differentiating ability in users of the ReWalk TM powered exoskeleton: An analysis of walking kinematics. In: 2013 IEEE 13th International Conference on Rehabilitation Robotics (ICORR), pp. 1–5. IEEE (2013)
9. Esquenazi, A., et al.: The ReWalk powered exoskeleton to restore ambulatory function to individuals with thoracic-level motor-complete spinal cord injury. Am. J. Phys. Med. Rehabil. **91**(11), 911–921 (2012)
10. Neuhaus, P.D. et al.: Design and evaluation of Mina: a robotic orthosis for paraplegics. In: 2011 IEEE International Conference on Rehabilitation Robotics, pp. 1–8. IEEE (2011)
11. Yan, T., et al.: Review of assistive strategies in powered lower-limb orthoses and exoskeletons. Robot. Autonom. Syst. **64**, 120–136 (2015)
12. Gavrilov, S.V., Do, T.Z.: Computer simulation of the dynamics of movement of a five-step walking robot. Autom. Electr. Power Ind. Electr. Eng. **1**, 72–76 (2016)
13. Sankai, Y. HAL: Hybrid assistive limb based on cybernics. Robot. Res. 25–34 (2010)
14. Farris, R.J., Quintero, H.A., Goldfarb, M.: Preliminary evaluation of a powered lower limb orthosis to aid walking in paraplegic individuals. IEEE Trans. Neural Syst. Rehabil. Eng. **19**(6), 652–659 (2011)
15. Lushnikov, B.V., Skryl'nikov, N.Ye.: Simulation model of the process of verticalization of the exoskeleton of the lower extremities in Simmechanics, MATLAB medium taking into account dry friction in kinematic pairs. Vibration technologies, mechatronics and controlled machines, pp. 261–271 (2016)
16. Pavlovskiy, V.Y. et al.: Biomechatronic neurorehabilitation complex—concept, construction, models and control. In: Preprints of the Institute of Applied Mathematics named after MV Keldysh RAS 0, pp. 111–119 (2014)
17. Borisov, A.V.: Mechanics of the spatial model of the exoskeleton and anthropomorphic robot. In: Questions of defense technology. Series 16: Counter Terrorism Techniques 3–4, 46–55 (2018)

18. Plagenhoef, S., Evans, F.G., Abdelnour, T.: Anatomical data for analyzing human motion. Res. Q. Exerc. Sport **54**(2), 169–178 (1983)
19. Ulrey, B.L., Fathallah, F.A.: Subject-specific, whole-body models of the stooped posture with a personal weight transfer device. J. Electromyogr. Kinesiol. **23**(1), 206–215 (2013)
20. Kolesnikova, G.P., Formal'skiy, A.M. About one way of modeling a person's gait. Eng. J. Sci. Innov. **1**, 1 (2014)

Chapter 22
Mathematical Modeling of Load Lifting Process with the Industrial Exoskeleton Usage

Sergey Jatsun, Andrei Malchikov, Andrey Yatsun, and Ekaterina Saveleva

Abstract The article is devoted to investigation of the opportunity of exoskeleton usage in industrial conditions, which can partially compensate the load during carrying out of the technological operations. Such devices can significantly reduce human fatiguability, improve operations quality, and reduce the injury risk and the professional diseases development. The article presents a mathematical model describing the lifting load process with the use of industrial exoskeleton. A mathematical apparatus, which makes it possible to obtain load lifting kinematic parameters and evaluate the exoskeleton joints forces, is presented in the work. The paper presents the numerical modeling results and analyzes them. The values of the device hinges torques were obtained. The knees and ankle joints turned out to be the most loaded during lifting process. The necessity of additional external devices usage to compensate these efforts is mentioned.

22.1 Introduction

One of the industrial current tasks is the labor productivity increasing, which can be achieved through the use of specialized equipment raising the worker's physical activity and reducing the employee's fatiguability. One of the most promising and nowadays relevant solutions is the exoskeletal costumes usage [1–3]. The exosuit should partially compensate the load during carrying out of the technological operations that will significantly reduce human fatiguability, improve operations quality, and reduce the injury risk and the professional diseases development.

Currently, scientists from different countries (Japan, USA, China, Russia, countries of Europe) are conducting theoretical and experimental studies in the sphere of creating assisting exoskeleton devices and evaluating the technological operations

S. Jatsun · A. Malchikov (✉) · A. Yatsun · E. Saveleva
Department of Mechanics, Mechatronics and Robotics, Southwest State University (SWSU), Kursk 305040, Russian Federation
e-mail: teormeh@inbox.ru

A. Ronzhin and V. Shishlakov (eds.), *Proceedings of 15th International Conference on Electromechanics and Robotics "Zavalishin's Readings"*, Smart Innovation, Systems and Technologies 187, https://doi.org/10.1007/978-981-15-5580-0_22

effectiveness with its usage [4–6]. In this place, commercial products of such companies as Panasonic (Japan), Daewoo (South Korea), and Lockheed Martin (USA) that unload the operator by using of controlled electric drives are known. At the same time, there are few works devoted to the theoretical justification of the exoskeleton's use during performing technological operations. The studied subject domain's short review is presented in [7–16]

Mathematical modeling of the industrial exoskeleton functioning during performing of various type technological operations allows us to analyze the human-exoskeleton system behavior and estimate the motion range and the forces magnitude in the construction which hinges cargo lifting with the usage of industrial exoskeleton; the mathematical apparatus and the results of the device modeling are presented.

22.2 Scheme Justification of Investigated Structure

In the general case, when developing the industrial exoskeleton structure and control system, the necessity of the joints torques determination in dependence of different cargo masses and various lifting algorithms can appear. If the exoskeleton contains passive mechanical compensators, the task of determination of the exoskeleton attachment points' efforts during the technological operations performing which determines the comfort of using an exoskeleton suit can be interesting.

Let us consider the exoskeleton structure to justify the structural scheme (Fig. 22.1).

Structurally, the exoskeleton repeats the human skeleton and consists of the following basic elements: 1—the exoskeleton power frame, on which the cargo securing slings are fixed, 2—the textile shoulder mount of the operator's body, 3—the flexible load securing element, 4—the belt mount of the operator's body, 5—the hip joint, 6—femoral cuff, 7—knee joint, 8—ankle joint, 9—the exoskeleton supporting element.

The device links are connected through hinges. The exoskeleton hip joint has three degrees of freedom similarly with the operator. The femoral link can rotate around two horizontal axes: "forward-backward" and "abduction-reduction," as well as around the vertical axis relative to the link connected to the back of the operator.

The exoskeleton tibia link is connected to the thigh by a single axis of rotation cylindrical hinge. The foot is replaced by a flexible emphasis perceiving the weight of the exoskeleton. The ankle joint of the exoskeleton has two degrees of freedom, providing the convenience of movement in the exoskeletal complex (EC).

According to the load lifting rules, the most effective movement for a person is that ensures the achievement of the goal with the least muscle tension and skeleton load. It is necessary to observe a number of rules for effective load lifting: keeping of the operator's back in the correct position (so that the spinal column physiological curves remaining), maximizing of the legs usage during lifting process, and keeping person's arms outstretched [17].

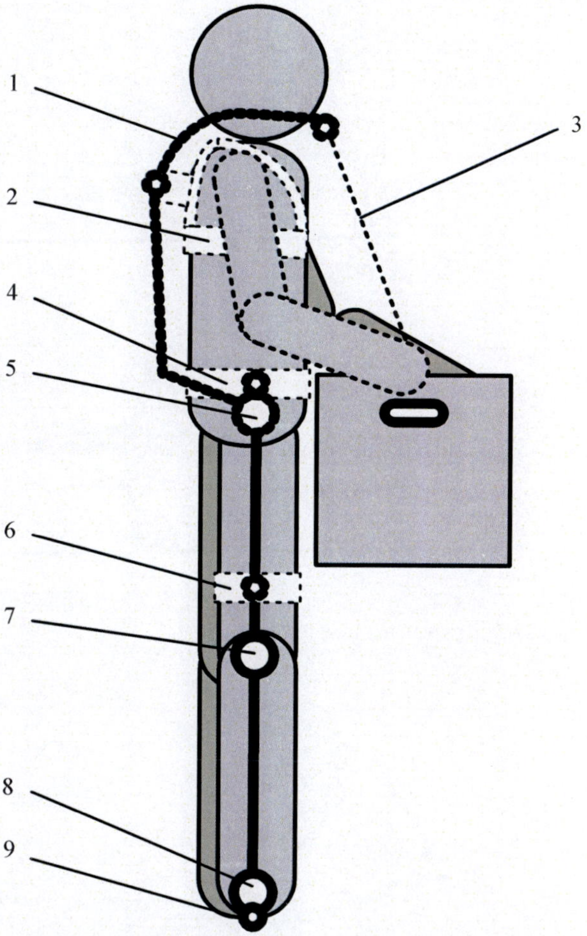

Fig. 22.1 Exoskeleton structural scheme

During the load lifting process, three groups of joints located in the sagittal plane are mainly involved: the hip, knee, and ankle joints [18, 19]. The hands also perceive the load, but to reduce it, flexible connections redistributing the load on the exoskeleton power frame can be used.

The load lifting process scheme with the usage of exoskeleton is shown in Fig. 22.2.

In this case, the exoskeleton can assist a person when lifting a load, in particular, compensating for a part of the hip and knee joints torques. Note that in this case, the exoskeleton can be considered as a three-link mechanism in which the force load is applied at the sling attachment point O_4 (Fig. 22.3).

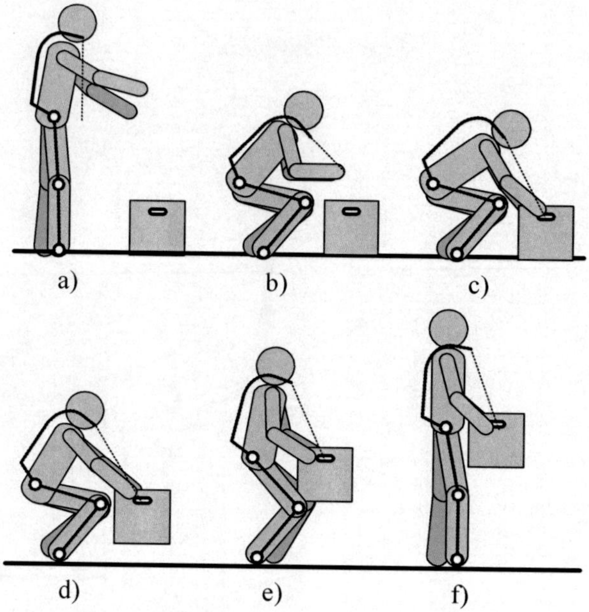

Fig. 22.2 Exoskeleton load lifting algorithm

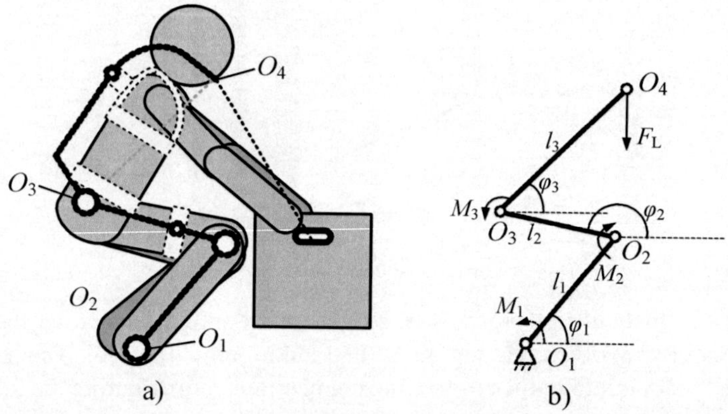

Fig. 22.3 Equivalent analytical model of an exoskeleton during the lifting load process

The load force $F_L = M_L g$ acts on the three-link system in such a way: at the point O_4(M_L—mass of the cargo). The links are affected by given moments, which can be created by either passive (springs, brakes, etc.) or active (electric, pneumatic, and other types of drives) elements. The position of the mechanism links is determined by absolute angles ϕ_i.

22.3 Kinematical Analysis of Load Lifting Process

The most important issue during exoskeleton modeling is the kinematical and force parameters of the device movement determination that allows to achieve the required load movement. Thus, in the framework of this study, the modeling task consists in determining the time dependences of the exoskeleton links rotation angles during the load movement along the given trajectory, as well as determining of the joints torques.

Let us consider analytical models shown in Fig. 22.4. A number of assumptions were accepted in the study of load lifting process:

- The load lifting process is carried out along a straight vertical line that means that during the lifting process, the load securing point O_4 changes only one coordinate y_{O4}. In this case, the law of the O_4 point coordinates change i, represented by the radius vector r_{O4} in time, can be represented in a polynomial form:

$$r_{O4} = \left(0, \sum_{i=5}^{5} k_i t^i\right)^{\mathrm{T}}. \tag{22.1}$$

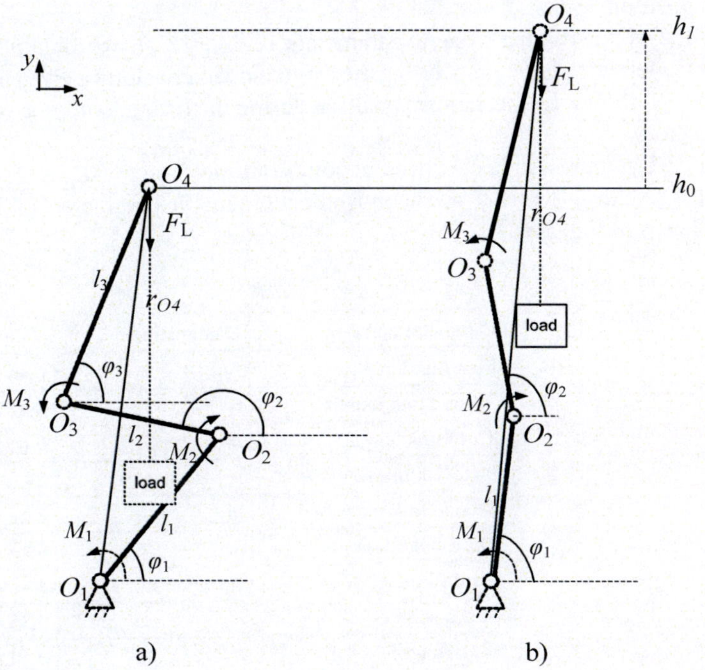

Fig. 22.4 Exoskeleton analytical model: **a** at the beginning of the load lifting process; **b** at the end of the load lifting process

- The cargo lifting process is realized with a fixed angle of the exoskeleton's back inclination $\phi_3 = \text{const} = 80°$. This limitation is due to the biomechanics of cargo lifting, because observing the vertical position of the back, the injury getting chance decreases.
- The cargo dimensions need to be neglected, taking into account that the cargo does not touch the operator's body and exoskeleton links during the lifting process, and the cargo center of mass projection is located within the operator's foot abutment surface. In reality, this condition can be observed by passing a small cargo between the legs, or with the use of specialized rigging devices. Also, such conditions can be obtained with a group load lifting.

Thus, the inverse kinematics task for load lifting can be represented by the equations:

$$\phi_1 = \frac{\pi}{2} - \frac{\arccos(y_{O4} - l_3 \sin(\phi_3))^2 + l_1^2 - l_2^2}{2l_1(y_{O4} - l_3 \sin(\phi_3))}; \tag{22.2}$$

$$\phi_2 = \frac{\pi}{2} + \frac{\arccos(y_{O4} - l_2 \sin(\phi_3))^2 + l_1^2 - l_2^2}{2l_2(y_{O4} - l_3 \sin(\phi_3))}, \tag{22.3}$$

where y_{O4} is the specified coordinate of the load securing point O_4, which determines the cargo position.

Given the values of the system parameters (Table 22.1), we obtain the links movement kinematic parameters during the cargo movement along a given trajectory.

Figure 22.5 shows the mechanism position during the lifting load process and the given trajectory.

In Fig. 22.5, numbers indicate different points of time: 1—$t = 0, 5$ s, 2—$t = 2$ s, 3—$t = 2, 5$ s, and 4—$t = 3, 5$ s. The movement kinematic parameters of the in this case are shown in Fig. 22.6.

Table 22.1 Modeling parameters

Parameter name	Designation	Value
First link length	l_1	0.44 m
Second link length	l_2	0.45 m
Third link length	l_3	0.65 m
First link mass	m_1	8 kg
Second link mass	m_2	8 kg
Third link mass	m_3	8 kg
Load force	F_L	300 N
Initial height	h_0	0.35 m
Final height	h_1	0.5 m

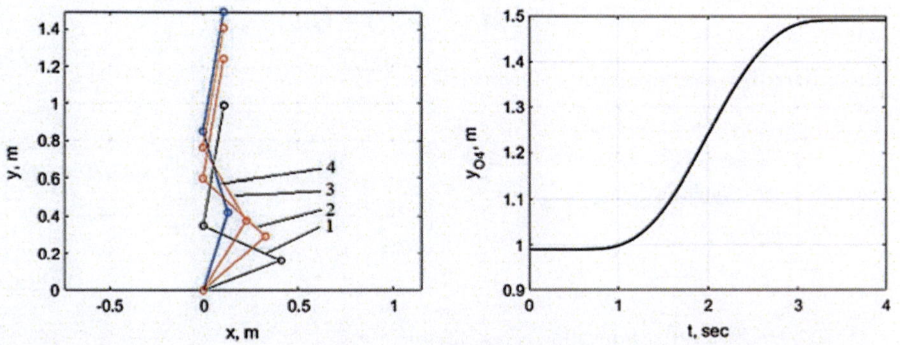

Fig. 22.5 Simulation results: **a** the mechanism positions points in different moments of time; **b** changing of coordinates y_{O4}

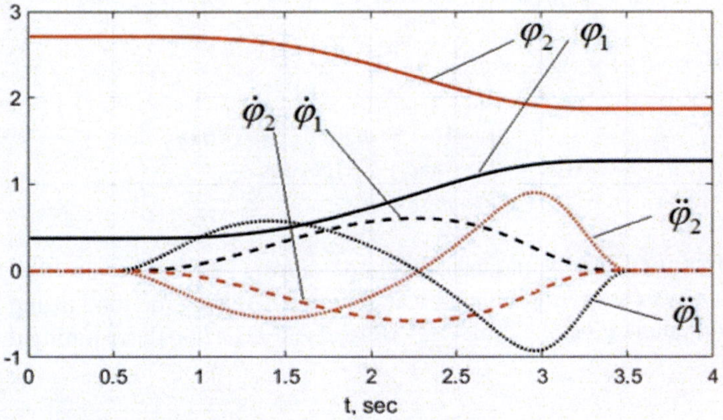

Fig. 22.6 Mechanism movement kinematical parameters: rotation angle (rad), speed (rad/sec), and link acceleration (rad/sec^2)

22.4 Dynamical Analysis of Load Lifting Process

The load lifting process dynamical analysis consists in determining the exoskeleton joints torques necessary to ensure the load lifting. Presenting the generalized coordinates vector $q = (\phi_1, \phi_2, \phi_3)^{\mathrm{T}}$, the vector of generalized moments applied to the system:

$$\mathbf{Q} = \begin{bmatrix} M_1 \\ M_2 \\ M_3 \end{bmatrix} \tag{22.4}$$

which is determined from the equation:

$$\mathbf{Q} = \mathbf{M}(q)\ddot{q} + \mathbf{V}(q, \dot{q}) + \mathbf{G}(q), \tag{22.5}$$

where $\mathbf{M}(q)\ddot{q}$—inertia matrix:

$$\mathbf{M}(q) = \begin{bmatrix} \frac{m_1 l_1^2}{3} + (m_2 + m_3) l_1^2 & (\frac{m_2}{2} + m_3) l_1 l_2 \cos(\phi_1 - \phi_2) & \frac{m_3 l_1 l_3}{2} \cos(\phi_1 - \phi_3) \\ \frac{(m_2 + 2m_3) l_1 l_2}{2} \cos(\phi_2 - \phi_1) & \frac{m_2 l_2^2}{3} + m_3 l_2^2 & \frac{m_3 l_2 l_3}{2} \cos(\phi_2 - \phi_3) \\ \frac{m_3 l_1 l_3}{2} \cos(\phi_3 - \phi_1) & \frac{m_3 l_2 l_3}{2} \cos(\phi_3 - \phi_2) & \frac{m_3 l_3^2}{3} \end{bmatrix}, \tag{22.6}$$

$\mathbf{V}(q, \dot{q})$—generalized inertia forces vector:

$$\mathbf{V}(q, \dot{q}) = \begin{bmatrix} \dot{\phi}_2^2 (\frac{m_2}{2} + m_3) l_1 l_2 \sin(\phi_1 - \phi_2) + \dot{\phi}_3^2 m_3 \frac{l_1 l_3}{2} \sin(\phi_1 - \phi_3) \\ \dot{\phi}_1^2 (m_2 + 2m_3) \frac{l_1 l_2}{2} \sin(\phi_2 - \phi_1) + \dot{\phi}_3^2 m_3 \frac{l_2 l_3}{2} \sin(\phi_2 - \phi_3) \\ \dot{\phi}_1^2 m_3 \frac{l_1 l_3}{2} \sin(\phi_3 - \phi_1) + \dot{\phi}_2^2 m_3 \frac{l_2 l_3}{2} \sin(\phi_3 - \phi_2) \end{bmatrix}. \tag{22.7}$$

$\mathbf{G}(q)$—gravity forces vector:

$$\mathbf{G}(q) = \begin{bmatrix} -(m_3 + m_2 + \frac{m_1}{2}) g l_1 \cos\phi_1 \\ -(m_3 + \frac{m_2}{2}) g l_2 \cos\phi_2 \\ -m_3 g \frac{l_3}{2} \cos\phi_3 \end{bmatrix}. \tag{22.8}$$

The described mathematical apparatus allows us to obtain the operator-exoskeleton system motion kinematical parameters during the load lifting process.

The parameters given in Table 22.1 were also used when determining the torques (Fig. 22.7).

22.5 Conclusion

In this work, the industrial exoskeleton analytical model that describes the load lifting secured by flexible connections on a power frame was developed. According to the analytical model and a number of accepted assumptions, kinematic motion parameters when lifting a 30 kg load up to a height of 1.5 m in 3 s were obtained. The obtained speed values do not exceed 0.5 rad/s that corresponds to the actual load lifting process.

The values of the device hinges torques were obtained. The knees and ankle joints turned out to be the most loaded during lifting process. As it was earlier mentioned, fixing the back inclination angle at the degree 80°, proposed in the work, allowed to reduce the required hinge torque to 40 Nm. The maximum knee and ankle joints torques were 170 and 200 Nm, respectively, which indicates the necessity of usage additional external devices to compensate these efforts.

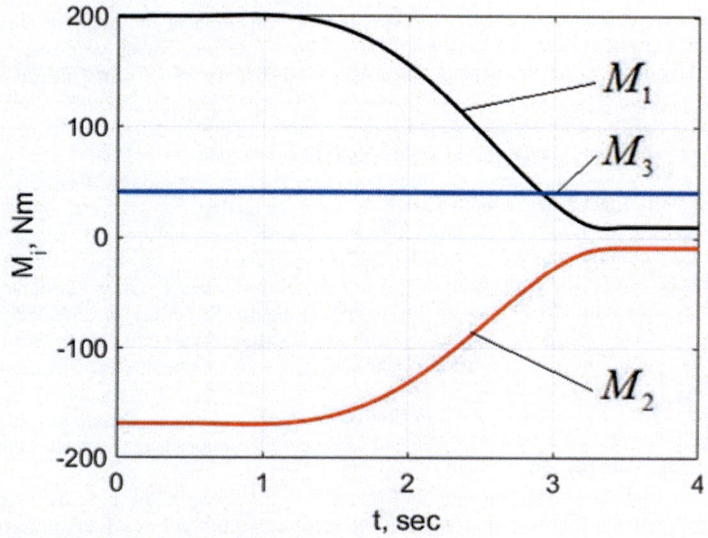

Fig. 22.7 Mechanism hinges torques

Acknowledgements The work of authors was supported by RFBR, research project no. 19-01-00540.

References

1. De Looze, M.P., Bosch, T., Krause, F., Stadler, K.S., O'Sullivan, L.W.: Exoskeletons for industrial application and their potential effects on physical work load. Ergonomics **59**(5), 671–681 (2016)
2. Van der Vorm, J., OSullivan, L., Nugent, R., de Looze, M.: Considerations for developing safety standards for industrial exoskeletons. Robo-Mate 1–13 (2015)
3. Yatsun, A., Jatsun, S. Investigation of human cargo handling in industrial exoskeleton. In: 2018 Global Smart Industry Conference (GloSIC), pp. 1–5. IEEE (2018)
4. Voilqué, A., Masood, J., Fauroux, J.C., Sabourin, L., Guezet, O.: Industrial exoskeleton technology: Classification, structural analysis, and structural complexity indicator. In: 2019 Wearable Robotics Association Conference (WearRAcon), pp. 13–20. IEEE (2019)
5. Bogue, R.: Robotic exoskeletons: a review of recent progress. Ind. Robot Int. J. (2015)
6. Jatsun, S., Malchikov, A., & Yatsun, A.: Investigation of movements of lower-limb assistive industrial device. In: International Conference on Interactive Collaborative Robotics, pp. 226–235 (2019)
7. Aliseychik, A.P. et al.: Mechanics and control of exoskeletons of the lower extremities for neurorehabitation of spinal patients. In: XI All-Russian Congress on Fundamental Problems of Theoretical and Applied Mechanics. Collection, pp. 132–134 (2015)
8. Aliseychik, A.P., et al.: Rehabilitation exoskeleton BioMech: models, management, design, experiments. Mechatron. Autom. Control **17**(10), 670–677 (2016)
9. Huysamen, K.: e6t al: Assessment of an active industrial exoskeleton to aid dynamic lifting and lowering manual handling tasks. Appl. Ergonom. **68**, 125–131 (2018)

10. Borisov, A.V., Rozenblat, G.M., Chigarev, A.V.: Application of the matrix method and the recurrent algorithm to the model of a plane-multipurpose mechanism with lines of variable length moving on a horizontal plane. Theor. Appl. Mechan. Int. Sci. Tech. Collection, 370–380 (2018)
11. Lavrovskiy, E.K.: About energetics of a human gait—operator carried out with the passive exosceleton apparatus. Proc. RAS. Solid Mech. **1**, 9–24 (2015)
12. Gavrilov, S.V., Do, T.Z.: Computer simulation of the dynamics of motion of a five-step walking robot. Autom. Electr. Power Ind. Electr. Eng. **1**, 72–76 (2016)
13. Ergashevam, B.I.: Lower limb exoskeletons: brief review scientific and technical. Jo. Inf. Technol. Mech. Opt. **17**(6), 1153–1158 (2017)
14. Carmichael, M.G., Liu, D.K.: Human biomechanical model based optimal design of assistive shoulder exoskeleton. In: Conference on Field and Service Robotics, pp. 245–258 (2015)
15. Kim, H.G., Lee, J.W., Jang, J., Park, S., Han, C.: Design of an exoskeleton with minimized energy consumption based on using elastic and dissipative elements. Int. J. Control Autom. Syst. **13**(2), 463–474 (2015)
16. Domrachev, T.B., Yashmetov, K.S., Loskutov, Yu.V.: Kinematics of human locomotion when standing up from a gray-haired vehicle and landing on a support. Eng. Personnel Future Russ. Innov. Econ. **1**, 40–42 (2017)
17. Jatsun, S., Malchikov, A., Yatsun, A.: Comparative Analysis of the Industrial Exoskeleton Control Systems. In: Proceedings of 14th International Conference on Electromechanics and Robotics “Zavalishin’s Readings”, pp. 63–74 (2020)
18. Yatsun, S.F., Antipov, V.M., Karlov, A.E.: Modeling of loading by industrial exoskelet. Proc. Southwest State Univ. **22**(6), 16–20 (2018)
19. Sylla, N., Bonnet, V., Colledani, F., Fraisse, P.: Ergonomic contribution of ABLE exoskeleton in automotive industry. Int. J. Ind. Ergon. **44**(4), 475–481 (2014)

Chapter 23
Deep Q-Learning Algorithm for Solving Inverse Kinematics of Four-Link Manipulator

Dmitriy Blinov, Anton Saveliev, and Aleksandra Shabanova

Abstract This paper presents deep Q-learning algorithm designed to solve inverse kinematics problem of four-link manipulator. This algorithm uses dynamic exploration coefficient instead of a constant value, which allows to avoid convergence of the neural network to a local optimum. In addition, a method for generating a Q-table has been developed to avoid the bottleneck effect when neural network construction. This in turn leads to reduction of training time and lower hardware requirements. To evaluate the effectiveness of the proposed algorithm, three environments were developed and for each of them specific neural networks model were used. Three different environments allow to evaluate the algorithm performance for solving inverse kinematics of varying complexity: with one initial and one target points, with several initial and one target points, and, conversely, with one initial and several target points. Obtained dependency graph of rewards on the number of training episodes shown successful training of agents in all environments. Successful training of the Q-learning algorithm in the third environment suggests that the algorithm can be used for solving the inverse kinematics for all points of the manipulator working space. The main advantage of the developed algorithm is the possibility of its application for solving inverse kinematics problems of varying complexity. In addition, this algorithm can be used to solve inverse kinematics of manipulator with a different number of links.

D. Blinov
St. Petersburg State University of Aerospace Instrumentation, 67, st. Bolshaya Morskaya, 190000 St. Petersburg, Russia
e-mail: d99b09@yandex.ru

A. Saveliev · A. Shabanova (✉)
St. Petersburg Institute for Informatics and Automation of the Russian Academy of Sciences, 39, 14th Line, 199178 St. Petersburg, Russia
e-mail: iialex.shabanovaii@gmail.com

A. Saveliev
e-mail: saveliev.ais@yandex.ru

A. Ronzhin and V. Shishlakov (eds.), *Proceedings of 15th International Conference on Electromechanics and Robotics "Zavalishin's Readings"*, Smart Innovation, Systems and Technologies 187, https://doi.org/10.1007/978-981-15-5580-0_23

23.1 Introduction

Solving inverse kinematics is an important prerequisite for ensuring a robot control [1]. The most popular ways to solve this problem are analytical and geometric methods. Currently, there are various methods for solving inverse kinematics. The reliability of a control system by a robotic means largely depends on the choice of these methods.

Today, the most popular methods are the analytical (inverse transformation method) and geometric proposed in [2, 3]. However, these methods have a significant drawback: When increasing the number of manipulator links, it is necessary to introduce restrictions. This leads to the loss of a large number of possible inverse kinematics solutions and a decrease in the overall control system performance.

Also, genetic algorithms can be used to solve the inverse kinematics problem of manipulator [4, 5]. Genetic algorithms relate to bio-inspired approaches of machine learning [6] and are used in various areas of robotics: When developing systems for end-effector control and constructing an optimal robot trajectory, where obtaining an analytical solution is difficult or impossible. This method is based on the evolution strategies: crossbreeding, mutation, inheritance. However, genetic algorithms are poorly scalable, and for each task, it is necessary to develop a separate genetic algorithm. In addition, this method tends to converge to a local optimum, which leads to a loss of accuracy of the inverse kinematics solution.

The problem of choosing the optimal position of a four-link manipulator is solved using genetic algorithm in [7]. In this case, the optimal position means the achievement of one of the following criteria: obtaining the minimum amount of attached manipulator parameters (link rotation angles), or obtaining the minimum total moment, or achieving the minimum or maximum height. The authors developed cross and mutation operators that are designed to solve the problem of finding the coordinates of the points of two circles intersection. The coordinates of the manipulator joints were used as individuals of the population. The scheme of the proposed genetic algorithm includes the formation of the initial population, the evaluation of individuals, selection, crossbreeding, mutation, the formation of a new population and checking the conditions for exit from the cycle. As a fitness function was used the sum of the joint rotation angles, the acceptance coordinate offset of the point of circle was used as a mutation parameter. The algorithm running time of 100 specimens and 100 iterations was 2.81 s.

Artificial neural networks (ANNs) do not have the disadvantages of genetic algorithm [8, 9]. In this case, ANN is trained on experimental, pre-collected data, which allows to expand the range of solutions in comparison with classical methods. The disadvantage of using neural networks to solve the inverse kinematics of manipulator is the difficulty of obtaining training data.

Most of the problems associated with the use of classical methods can be eliminated using machine learning methods [10, 11]. For example, tabular methods with reinforcement, the basis of these methods is the construction of a table of observation weights. This table is a set of possible states and the weights of observation to various actions. During the training process, the weights are updated, which allows to find the optimal policy. The application of these methods is impossible in environments with continuous states or with a huge number of discrete states. Thus, tabular methods are not effective for solving inverse kinematics.

One of the machine learning methods is deep Q-learning. The effectiveness of Q-learning in comparison with other machine learning methods has been proven in [12]. The authors trained agents to play the Atari 2600 game console. The result exceeded not only other machine learning methods, but also the results of people. Deep Q-learning has found application in robotics. For instance, in [13] an algorithm for continuous control of various robotic structures is proposed, and in [14] deep Q-learning was used to design the TurtleBot mobile control system to avoid obstacles without mapping. Also, deep Q-learning algorithms are widely used in the development of manipulator control systems, because it allows to get results that are not available when using traditional approaches. Using these algorithms enables data from various sensors without preprocessing [15–17].

Possibility of applying deep Q-learning to solve inverse kinematics problem is demonstrated in [18]. The objective of the study was to find the optimal position of the flexible surgical six-link manipulator with 6 degrees of freedom. Moreover, time for reaching target point by the end-effector was used as a criterion for the success of inverse kinematics solution. Test results showed that, using a set of 7 actions, the agent is able to reach 129 points out of 142 with an accuracy of 0.0056 m.

Thus, the most preferred method for solving robot control problem is reinforcement learning, because this method is free from the drawbacks of classical and neural networks. Despite the fact that there are works confirming the effectiveness of the application of deep Q-learning algorithms for solving various control problems, the use of these algorithms for solving inverse kinematics is an insufficiently well-researched problem. Therefore, the objective of this work is development deep Q-learning algorithm for solving inverse kinematics of manipulator in various environments.

This paper is organized as follows: The second section describes the general provisions of the Q-learning algorithm; the third section contains a description of the proposed algorithm and used methods; the fourth section shows the test results of the developed algorithm in various environments.

23.2 Reinforcement Learning Approach for Solving Inverse Kinematics of Manipulator

Inverse kinematic problem is to calculate the joint angles $\varphi_1 \ldots \varphi_4$ for a given end-effector position (x, y, z) and known kinematic diagram. The kinematic diagram of the investigated manipulator is shown in Fig. 23.1. The initial position of the end-effector is (x_0, y_0, z_0).

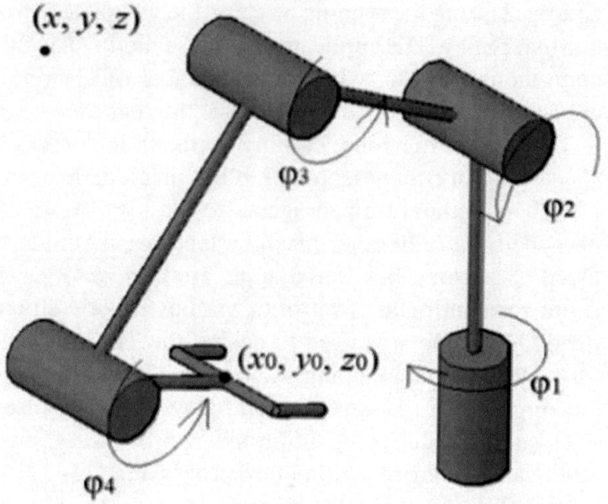

Fig. 23.1 Kinematic diagram of the investigated manipulator

Reinforcement learning is a machine learning approach, during which some entity (agent) interacts with some environment through feedback mechanism. In this case, the agent interacts with environment through actions, and environment in turn responds to these actions—observation and reward.

Q-learning is a method based on training an agent to further predict its behavior in a particular environment. Q-learning is one of the most effective types of reinforcement learning. Agent establishes a Q-function based on reward received from environment. This function determines agent's reward that agent can receive for its action. Training is to clarify the value of the Q-function at each step (action) of the agent to obtain the maximum reward.

In deep Q-learning, artificial neural network plays role of Q-function, trained on small data packets during agent operation. It is advisable to use neural networks instead of Q-tables is used in cases where it is impossible to sort through all possible environmental conditions.

To improve the efficiency of the Q-learning algorithm for solving inverse kinematics problem of a four-link manipulator, dynamic exploration coefficient was used and new method for Q-table generation was proposed. Such solutions help to avoid algorithm convergence to a local optimum and ensure smooth movement of links. Possible agent actions constitute an action space and represent a change in joint angles.

The goal of the agent is to maximize the reward function so that as a result of the training, the agent reaches the target point along the shortest path. The reward per step function is the difference in distance from the current to the target point, before and after the action. A step is one movement of end-effector. In order to bring the average value of rewards to 0.05, the expression for determining rewards was multiplied by the coefficient k_n. The reward that the agent receives in one step can be calculated by the formula:

$$\text{reward}_{\text{step}} = k_n(\text{dist}_{i-1} - \text{dist}_{i-1}), \tag{23.1}$$

where $\text{dist}_i = \sqrt{(x - x_i)^2 + (y - y_i)^2 + (z - z_i)^2}$, k_n is normalization coefficient, x_i, y_i, z_i are end-effector coordinates after the action.

The reward function for training episode is the sum of rewards for all training episodes minus the number of steps taken and plus reward for reaching the target point with a given accuracy:

$$\begin{aligned}\text{reward}_{\text{episode}} &= \sum \text{reward}_{\text{step}} - \text{step} \\ &+ \begin{cases} 0, \text{ if done} \\ \text{reward}_{\text{done}}, \text{ if } \sqrt{(x_{i-1} - x_i)^2 + (y_{i-1} - y_i)^2 + (z_{i-1} - z_i)^2} < 0.05. \end{cases}\end{aligned} \tag{23.2}$$

To evaluate the proposed Q-learning algorithm under various conditions, three experiments were performed, which were different in environmental details and in the structure of ANN models. For this, several environments were developed in which different variations of inverse kinematics problem of four-link manipulator were solved.

23.3 Deep Q-Learning Algorithm for Solving Inverse Kinematics of Manipulator

The developed algorithm is intended for agents training in various environments. The agent is a four-link manipulator. The following parameters were used as input:

1. discount factor—when it is reduced, the agent is less concerned about its future actions;
2. size of the training data packet—determines the number of steps (actions) during training;
3. minimum memory space to start learning—determines the minimum number of steps (actions) needed to start training;
4. data packet size for one training cycle is the number of steps (actions) that must be used for one training cycle;
5. maximum number of episodes is the number of episodes for which agent training is expected;
6. deep ANN model that will be used to predict agent actions;
7. environment;
8. values of the dynamic exploration coefficient;
9. parameters for saving ANN model.

The output of the algorithm is the saved trained ANN models, which allow the agent to reach the target point. Subsequently, trained models can be used to develop manipulator control systems. The developed algorithm is shown in Fig. 23.2.

At the beginning of each episode, the environment is reset to defaults and the observation vector is read. The agent then performs a random or predicted action based on the value of the dynamic exploration coefficient. Depending on the response of environment for each action, agent receives a reward and flag, which indicates whether episode is completed or not.

Each action is recorded in memory. Since the memory has a limited volume, when it is full, the old data is discarded, giving way to new. For training, the agent randomly selects some observation from data memory. Upon reaching a certain value of the total reward, the algorithm saves log files and trains ANN model. The final step of the algorithm is updating the dynamic exploration coefficient. The training cycle lasts until the specified maximum number of training episodes is exceeded.

The following methods were applied to improve Q-learning algorithm efficiency: dynamic exploration coefficient and reward distribution between links.

Agent performs an action in an environment. As a result of interaction with environment for each action agent receives reward and updated action. Each environment must provide for use of the following methods: method of resetting the environment to its initial observation and step method. The method of resetting the environment resets the environment and notifies the agent about it. The step method is designed to allow environment to respond to the actions. These methods allow the agent to receive reward and observation. The values of reward function depend on task that the agent needs to complete. Often, it is the reward function that determines agent training speed.

23.3.1 Dynamic Exploration Coefficient

Exploration coefficient is a necessary element of reinforcement learning. It forces agent to take risks, performing random actions with a certain probability. The exploration-vs-exploitation problem is one of the main difficulties encountered in the implementation of learning algorithms with reinforcement learning [19]. If the dynamic exploration coefficient is too small, ANN will be at its local optimum. If the intelligence coefficient is large, the agent will not be able to complete the task in given time or will not fit into set number of steps. Therefore, the exploration coefficient should vary depending on the number of training episodes. Therefore, it is necessary to use a dynamic exploration coefficient [20]. The dynamic exploration coefficient can be calculated by the formula:

$$\text{eps} = k^{\text{episodes}}, \tag{24.3}$$

where $k < 1$ is smoothing coefficient, *episodes* is number of completed episodes.

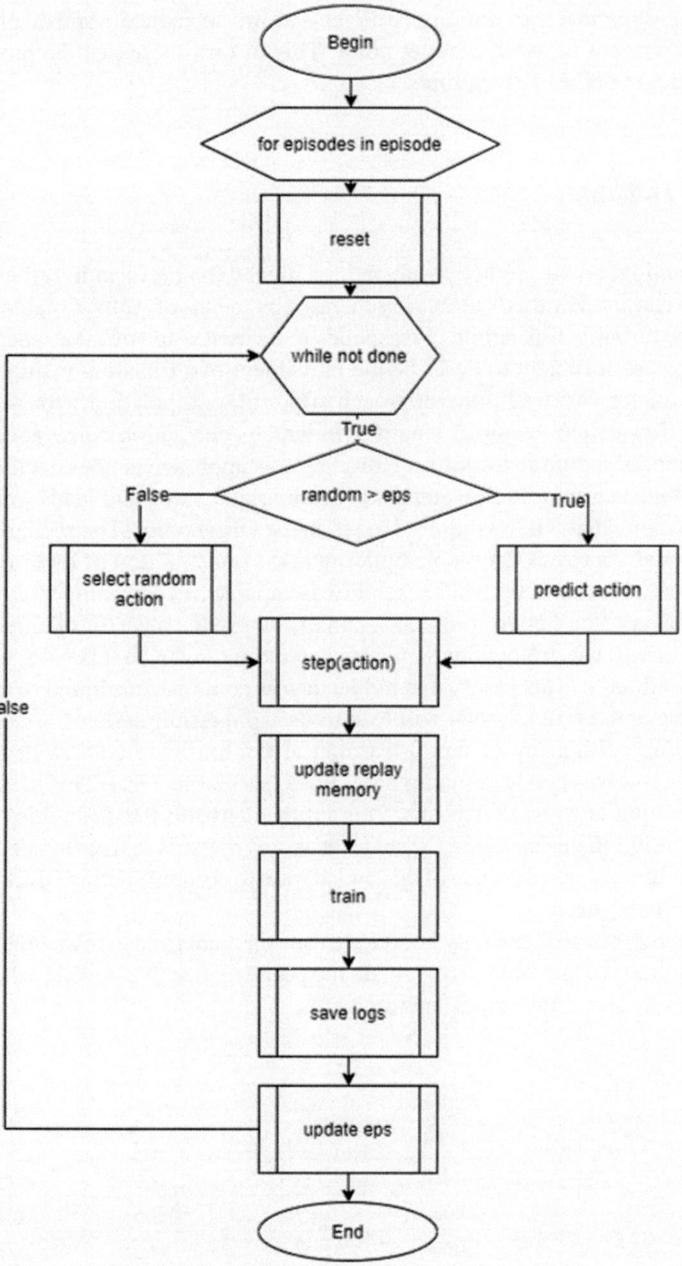

Fig. 23.2 Flowchart of deep Q-learning algorithm

Using a dynamic exploration coefficient allows to reduce number of steps that required for agent to reach a target point. This in turn increases the probability of reaching target point in given time.

23.3.2 *Q-Table*

During training, agent predicts what actions it needs to perform in order to achieve maximum reward. For this, Q-table is generated by means of ANN. Q-table is a vector, where each possible link action corresponds to its own value of the expected reward. Three steps are sufficient to describe the movement of a four-link manipulator.

In accordance with traditional approach to Q-table composition, a four-link manipulator requires action vector of length 8, in which each action corresponds to angle or movement of manipulator joints. However, this approach is not effective, because it does not ensure smooth movements of manipulator links and leads to increase in number of steps that will be required to reach the target point. The second version of the traditional approach to Q-table building is the construction of a vector of length 34, where all possible actions that each link is capable of combining with each other. This also incurs a number of problems, namely, to predict a vector of length $3^4 = 81$, it is necessary to use ANN with 81 neurons in output layer. In this case, to avoid the bottleneck effect, it is necessary that hidden layers contain a minimum of 81 neurons. An increase in the size of ANN will lead to its rapid retraining.

To improve efficiency of the Q-learning algorithm, method was developed for distributing rewards between manipulator links (see Table 23.1). This method allows to reduce output layer to 12 neurons. This approach avoids the difficulties associated with overfitting. In this case, the Q-table consists of rewards that each link can receive for all possible actions. Because of this, agent train to combine actions using relatively small vector length.

In this table, rewards can only be added to one element in each row, which allows to combine actions of any links. To evaluate the performance, the developed Q-learning algorithm was tested in various environments.

Table 23.1 Q-table distribution of rewards for actions between manipulator links

Action	a_1	a_2	a_3
Link 1	Q_{11}	Q_{12}	Q_{13}
Link 2	Q_{21}	Q_{22}	Q_{23}
Link 3	Q_{31}	Q_{32}	Q_{33}
Link 4	Q_{41}	Q_{42}	Q_{43}

Q_{ij} is reward of link of the manipulator i for action j

23.4 Results

To test the developed algorithm of deep Q-learning, three environments were developed that differ in the number of expected observations. In this study, for training deep neural networks, we used the Keras library and TensorFlow framework. In the first environment, the agent needs to build up trajectory between two points using the model of the PhantomX Pincher manipulator (see Fig. 23.3a). In this environment, discrete actions and continuous states were used. In this case, one initial point and one target point are considered. In this environment, ANN approximates the values of observation space; therefore, the accuracy of inverse kinematics solution decreases. External forces, such as gravity, act on manipulator links in addition to the forces, exerted by motors.

When testing Q-learning algorithm in the first environment, memory space was limited to 50,000 states. 32 data packets were used for training, and discount factor was 0.9. The developed algorithm trained ANN model, input and hidden layers of which contained 32 neurons, and the output layer—12 (see Table 23.2).

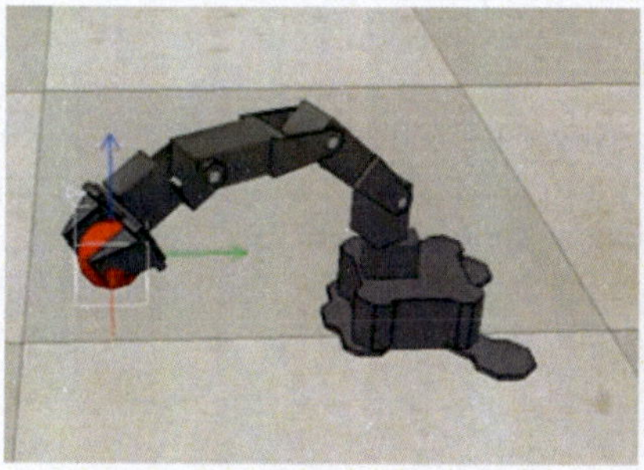

a)

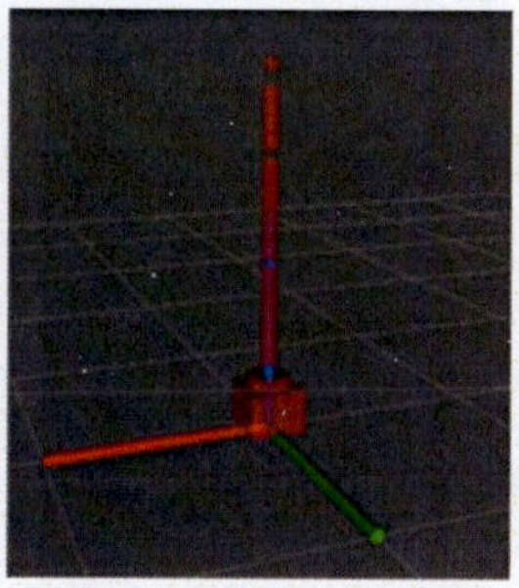

b)

Fig. 23.3 An example of the algorithm in: environment 1, V-REP (**a**); environments 2 and 3, ROS (**b**)

Table 23.2 Settings and test results of the deep Q-learning algorithm

Parameter	Environment 1	Environment 2	Environment 2
Observation vector length	7	7	9
Memory space for training	50 000	100 000	300 000
Training time	50 min	8 h	22 h
Number of training episodes	300	25,000	30,000

For two other environments, a simulation based on Robot Operating System (ROS) was developed (see Fig. 23.3b). In the second environment, control signals were sent directly to manipulator joints without taking into account external effects. This made it possible to increase the rate of environment reaction to actions and to increase inverse kinematics solution accuracy. Discrete actions and discrete states were used in this environment. This environment is designed to solve inverse kinematics problem in limited manipulator working range, which is represented by a plane. Moreover, several initial and one target points were considered. Put differently, the task was to move end-effector to the target point from arbitrary initial positions. Testing the proposed algorithm in such an environment consisted in simulating object grip. For testing the algorithm in this environment, memory space was limited to 100,000 observations (see Table 23.2). In the third environment, the number of expected observations is greater than the previous environment, and therefore, some parameters need to be changed. In this environment, discrete actions and discrete states were used. To solve inverse kinematics, agent needs information about target point position; therefore in this case, the length of the observation vector was 9. Testing the proposed algorithm in this environment consisted in simulating the object capture with one initial and several arbitrary target points.

Also, in this environment, in contrast to the previous one, the number of expected observations was increased; therefore, memory space was also increased to 300,000 observation samples and one hidden layer of 48 neurons was added to the neural network model (see Table 23.2). This allowed agent to remember more information and increased accurately predict actions with greater rewards. The discount factor in this and previous environments was 0.95, because target point reaching plays more significant role than in the first environment, and the agent needs to try to reach the target by all possible means.

The greater complexity of task (increased number of initial and target points) and the more variable observations are the more episodes are required for training. Increasing task complexity also leads to increase of training time. In addition, the more possible observation, the more connections must be established in ANN.

Agent training in the first environment took 50 min. Developed algorithm completed 800 episodes (60 steps each) and achieved inverse kinematics solution accuracy of 0.05 m. Moreover, number of possible actions of manipulator links at each step was $3^4 = 81$. Also, the developed algorithm was used to retrain ANN model while decreasing the step of changing angles. This allows to increase inverse kinematics solution accuracy. After retraining for 5 min with keeping smooth movement, the accuracy was 0.005 m.

Training time in the second environment was 8 h. During this time, the agent completed 40 thousand training episodes with an average duration of 40 steps each. Testing the algorithm in the third environment took 22 h. During this time, the agent completed 30 thousand episodes of training with an average duration of 55 steps each. Time increase is a consequence of increase in episode duration.

During testing, dependency graph of rewards on the number of episodes for three developed environments was obtained (see Fig. 23.4). The presented graphs allow to determine training quality of Q-algorithm.

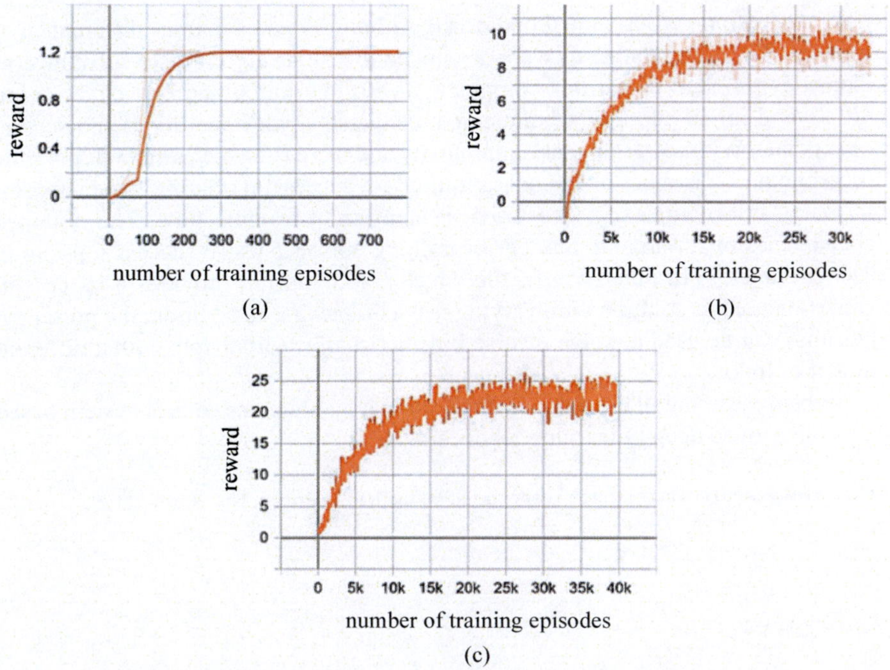

Fig. 23.4 Dependency graph of reward on number of training episodes: environment 1 (**a**), environment 2 (**b**), environment 3 (**c**)

The dependencies show that agent was trained enough to complete the task. To get the maximum reward in the first environment, agent needs only 300 training episodes (Fig. 23.4a), about 25,000 episodes in the second environment (Fig. 23.4b) and about 30,000 episodes in the third environment (Fig. 23.4c). The monotonously increasing dependencies for three environments and the achievement of a constant value of reward indicate success of training. In other words, the developed algorithm provides a successful inverse kinematics solution with varying task complexity.

23.5 Conclusion

This paper presents deep Q-learning algorithm for solving inverse kinematics of four-link manipulator. To overcome the bottleneck effect, a new method of forming Q-tables was proposed. The Q-table consists of rewards that each link can receive when performing all possible actions. Because of this, the agent trains to combine actions using a relatively small observation vector size. To avoid convergence of ANN model to local optimum, dynamic exploration coefficient was used instead of a fixed value. The value of the dynamic exploration coefficient depends on the smoothing coefficient and the number of training episodes.

For evaluation purpose, the proposed algorithm was tested in three different environments. These environments were developed to test the algorithm to solve inverse kinematics problem of various complexity: with one initial and one target point, with several initial and target points, and, conversely, with one initial and several target points. With increasing task complexity and increasing the number of possible observations, the number of episodes required for successful training also increases. Increasing task complexity also leads to increase of training time. The obtained dependencies of rewards on number of training episodes prove success training in all environments. In other words, the developed algorithm provides a successful inverse kinematics solution with varying task complexity. In addition, the presented algorithm can be used to solve inverse kinematics of manipulators with a different number of links.

Further research will be aimed at developing a manipulator control system based on the developed deep Q-learning algorithm.

Acknowledgements This research is supported by the RFBR Project No. 20-08-01109.

References

1. Negrete, M., Savage, J., Contreras-Toledo, L.A.: A motion-planning system for a domestic service robot. SPIIRAS Proc. **5**, 5–38 (2018)
2. Craig, J.J.: Introduction to robotics: mechanics and control. 3/E. Pears. Educ. (2009)
3. Yahya, S., Moghavvemi, M., Mohamed, H.A.: Geometrical approach of planar hyper-redundant manipulators: inverse kinematics, path planning and workspace. Simul. Model. Pract. Theory **19**(1), 406–422 (2011)
4. Momani, S., Abo-Hammour, Z.S., Alsmadi, O.M.: Solution of inverse kinematics problem using genetic algorithms. Appl. Math. Inf. Sci. **10**(1), 225 (2016)
5. Starke, S.: A hybrid genetic swarm algorithm for interactive inverse kinematics. Doctoral dissertation, Universität Hamburg, Fachbereich Informatik (2016)
6. Malyshev, A.A., Burgov, E.V.: Revisiting parameters of bioinspired behavior models in group foraging modeling. SPIIRAS Proc. **19**, 79–103 (2020)
7. Medvedev, S.N., Yakovlev, A.Yu., Korol'kov, O.G.: Development of the automated information system of the condition evaluation, diagnostics and selection of tactics for the treatment of patients with chronic diseases. Bull. Voronezh State Tech. Univ. **14**(6), 8–15 (2018)
8. Almusawi, A.R., Dülger, L.C., Kapucu, S.: A new artificial neural network approach in solving inverse kinematics of robotic arm (denso vp6242). Comput. Intell. Neurosci. (2016)
9. Duka, A.V.: Neural network based inverse kinematics solution for trajectory tracking of a robotic arm. Procedia Technol. **12**(1), 20–27 (2014)
10. Sutton, R.S., Barto, A.G.: Introduction to reinforcement learning. MIT Press (1998)
11. Dorigo, M., Bersini, H.: A comparison of Q-learning and classifier systems. From Animals Animats 3, 248–255 (1994)
12. Mnih, V. et al.: Playing atari with deep reinforcement learning. arXiv preprint arXiv:1312.5602 (2013)
13. Gu, S., Lillicrap, T., Sutskever, I., Levine, S.: Continuous deep q-learning with model-based acceleration. In: International Conference on Machine Learning, pp. 2829–2838 (2016)
14. Tai, L., Liu, M.: A robot exploration strategy based on q-learning network. In: 2016 IEEE International Conference on Real-time Computing and Robotics (RCAR), pp. 57–62. IEEE (2016)

15. Katyal, K., Wang, I., Burlina, P.: Leveraging deep reinforcement learning for reaching robotic tasks. In: Proceedings of the IEEE Conference on Computer Vision and Pattern Recognition Workshops, pp. 18–19 (2017)
16. Lopez, N.G. et al.: Gym-gazebo2, a toolkit for reinforcement learning using ROS 2 and Gazebo. arXiv preprint arXiv:1903.06278 (2019)
17. Peng, X.B., Andrychowicz, M., Zaremba, W., Abbeel, P.: Sim-to-real transfer of robotic control with dynamics randomization. In: 2018 IEEE international conference on robotics and automation (ICRA), pp. 1–8. IEEE (2018)
18. Ansari, Y. et al.: A multiagent reinforcement learning approach for inverse kinematics of high dimensional manipulators with precision positioning. In: 2016 6th IEEE International Conference on Biomedical Robotics and Biomechatronics (BioRob), pp. 457–463. IEEE (2016)
19. Bai, J., Luo, Z., Liu, L., Wang, Y.: Duality in deep reinforcement learning—theory. In: 2018 International Conference on Picture Processing, Robotics and Artificial Intelligence (PPRAI), pp. 59–66. IEEE (2018)
20. Tokic, M., Palm, G.: Value-difference based exploration: adaptive control between epsilon-greedy and softmax. In: Annual Conference on Artificial Intelligence, pp. 335–346. Springer, Berlin, Heidelberg (2011)

Chapter 24
Linearization-Based Forward Kinematic Algorithm for Tensegrity Structures with Compressible Struts

Sergei Savin, Oleg Balakhnov, and Alexander Maloletov

Abstract This paper presents a new local linearization method for elastic forces in tensegrity structures, which can be used to solve forward kinematic problems. Forward kinematic problems are often solved as a part of inverse kinematic algorithms and trajectory planning in robotics, and it is often desirable to be able to perform those algorithms online. The proposed method allows us to solve forward kinematics as a quadratic program, which makes it fast and reliable and allows us to take advantage of the existing convex programming software. The paper demonstrates the work of the proposed method using a three-link tensegrity structure.

24.1 Introduction

Tensegrity structures represent a new class of mechanical systems, consisting of cables (which experience tensile forces), struts (which experience compressive forces), and optionally, bars (which experience both tensile and compressive forces) [1]. The systematic study of those structures can be traced back to the works of Snelson, Fuller, and others [2–4]. The interest toward these structures comes from their mechanical properties: good weight-to-stiffness ratio, foldability, resistance to mechanical damage due to collisions [5–8]. Those properties made a subject of space robotics research, where their foldability, relative lightweight, and collision resistance was of value [9]; later tensegrity structures became interesting for collaborative robotics [10], and soft robotics in particular [11]. Particular instances of tensegrity robots include reservoir compliant tensegrity robot (ReCTeR) [5], duct climbing tetrahedral tensegrity (DuCTT) [12], Tetraspine [13], tensegrity quadruped MountainGoat [14], NASA's SUPERball [15, 16], and TR-3 [9]. Some of those robots had been built and tested, while others exist only as mathematical models and were tested in physics simulation environments.

S. Savin (✉) · O. Balakhnov · A. Maloletov
Innopolis University, 420500 Innopolis, Russia
e-mail: s.savin@innopolis.ru

A. Ronzhin and V. Shishlakov (eds.), *Proceedings of 15th International Conference on Electromechanics and Robotics "Zavalishin's Readings"*, Smart Innovation, Systems and Technologies 187, https://doi.org/10.1007/978-981-15-5580-0_24

As of now, there are a number of separate research directions within the field of tensegrity robotics. One is the use of machine learning (reinforcement learning, evolutionary algorithms, and other tools) for solving form-finding, gait generation, and other problems related to the tensegrity robots [9, 14]. Another direction is gait generation using central pattern generators (CPG) [14, 17], which can be combined with the first approach. There are also attempts to solve classic robotics problems as applied to tensegrity structures: path planning, forward and inverse kinematics, inverse dynamics, feedback control, and state estimation [16, 18, 19]. However, there is still a lack of general methods that can be reliably used to generate control laws for tensegrity robots. This paper is focused on solving one of the above-mentioned problems: solving forward kinematics for a tensegrity structure.

A definition of forward kinematics for a tensegrity robot will be given in the following sections, but in general it concerns finding the state of the robot which is statically stable for a given set of current robot parameters, including rest lengths of the cables and other elastic elements, as well as the lengths of the struts (rest lengths if the struts are deformable) [18]. We should observe that forward kinematic problem for a tensegrity robot in static can be solved by finding minimal energy static state. The potential energy of a tensegrity structure is a non-convex function, and thus the presented minimization problem is non-trivial. It can be formulated analytically, using local minimum conditions for a (constrained) scalar function. Resulting set of equations will produce force (force and torque) balance equations for the nodes (struts). This set of equations can be solved using nonlinear gradient decent methods, interior-point method for non-convex objective functions, as well as other approaches [18]. But, those methods do no guarantee the speed of convergence to the solution and if there is no solution, they could not provide a certificate that the solution does not exist.

The main contribution of this paper is a formulation of the forward kinematic problem for tensegrity structures are a convex problem, which can be solved in real time with guarantees of convergence and the ability to guarantee that the solution does not exist if it was not found. That is achieved by the new local linear approximation of the elastic forces equations. The paper presents the method and forward kinematic algorithm based on it and studies the properties of the approximation itself and the associated forward kinematic algorithm. The results are applied to three-link tensegrity prism.

The rest of the paper is organized as follows. Section 24.2 discusses mathematical model of tensegrity structures with compressible struts. Section 24.3 provides a detailed description of the proposed linearization and its geometric motivations; it also provides data on the approximation discrepancy errors. Section 24.4 studies the use of the proposed linearization for the three-link tensegrity structure, including the study of how the model error depends on the number of iterations in the proposed accuracy control method.

24.2 Mathematical Model of a Tensegrity Robot

A tensegrity robot consists of nodes $\mathbf{r}_i$ connected via struts, cables or not connected at all. The connectivity pattern can be given as a connectivity matrix C, whose elements $C_{i,j} \in \{0\ 1\}$ assume 1 if there is a connection between the nodes, and 0 if there is no connection.

Assuming that struts are deformable, elastic forces $\mathbf{f}_{i,j}$ acting between any two nodes (connected via a cable or a strut, or not connected at all) can be written as follows:

$$\mathbf{f}_{i,j} = -\mu_{i,j}(||\mathbf{r}_i - \mathbf{r}_j|| - \rho_{i,j})\frac{\mathbf{r}_i - \mathbf{r}_j}{||\mathbf{r}_i - \mathbf{r}_j||}, \tag{24.1}$$

where $\mu_{i,j}$ is an element of the connection stiffness matrix and $\rho_{i,j}$ is an element of the rest lengths matrix. Matrices with elements $\mu_{i,j}$ and $\rho_{i,j}$ have the same structure as the connectivity matrix C, and when $C_{i,j} = 0$, both $\mu_{i,j} = 0$ and $\rho_{i,j} = 0$.

Assuming that every node is a point mass, we can formulate its dynamics as follows:

$$m\ddot{\mathbf{r}}_i = \sum_{j=1}^{n} \mathbf{f}_{i,j}. \tag{24.2}$$

In the presence of gravity and other external forces $\mathbf{f}_{i,e}$ acting on the ith node, the dynamics assumes a more general form:

$$m\ddot{\mathbf{r}}_i = \mathbf{f}_{i,e} + \sum_{j=1}^{n} \mathbf{f}_{i,j}. \tag{24.3}$$

We can formulate forward kinematics of a tensegrity robot as a non-convex optimization problem, or as a feasibility problem. By "forward kinematics", here we mean the problem of finding the configuration a tensegrity structure will assume as a static equilibrium, given a set of cable rest lengths ρ. In this sense, the problem is about finding a relation between the set of cable rest lengths ρ and the set of nodes $\mathbf{r}$.

We can formulate the problem of finding a stable configuration of $\mathbf{r}$ given ρ as an optimization problem or a feasibility problem (an optimization with a constant cost function). The later can be given as:

$$\begin{array}{ll} \text{find} & \mathbf{r} \\ \text{subject to:} & \begin{cases} \mathbf{f}_{i,e} + \sum_{j=1}^{n} \mathbf{f}_{i,j}(\mathbf{r}) = 0 \\ \mathbf{r}_\alpha = \mathbf{r}_\alpha(0) \end{cases}, \end{array} \tag{24.4}$$

where $\mathbf{r}_\alpha$ are elements that should remain fixed in place (such as elements that are fixed to the supporting surface).

This problem cannot be transformed into a convex one, as the equality constraints include norms or a vector of decision variables.

24.3 Local Linearization of the Elastic Forces

As was shown previously, the problem of finding a statically balanced position of a tensegrity structure can be formulated as an optimization problem. The purpose of such formulations is the systematic use of algorithms and optimization solvers (software packages that solve optimization problems). One of the most effective sets of algorithms and their software implementations had been developed for the class of convex optimization problems [20–22]. Those problems can be solved globally (as they have no local minima) if solution exists, or if it does not exist, the algorithm can certify that the problem is infeasible.

In order to turn the non-convex problem (24.4) into a convex one, we can use local linear approximation of the elastic forces. In this section, we introduce a local linear model for the elastic force, motivated by the geometrical properties of the elastic force equations. The model uses a weighted sum of two local linear approximations, which are discussed in the remainder of this section.

24.3.1 Fixed-Center Approximation

Let us study the structure of the expression (24.1). It can be seen as including a product of two expressions: a scalar expression giving the force magnitude: $\mu_{i,j}(||\mathbf{r}_i - \mathbf{r}_j|| - \rho_{i,j})$ and a unit vector giving the force direction: $\mathbf{n}_{i,j} = -(\mathbf{r}_i - \mathbf{r}_j)/\,||\mathbf{r}_i - \mathbf{r}_j||$. We can decompose the original expression (24.1) into sum of a convex linear function and a non-convex function bilinear in $\mathbf{n}_{i,j}$ and $\rho_{i,j}$:

$$\mathbf{f}_{i,j} = -\mu_{i,j}(\mathbf{r}_i - \mathbf{r}_j) + \mu_{i,j}\rho_{i,j}\mathbf{n}_{i,j}. \tag{24.5}$$

This motivates a simple approximation procedure, where $\mathbf{n}_{i,j}$ is assumed constant. From here on such value of $\mathbf{n}_{i,j}$ is denoted as $\mathbf{n}^0_{i,j}$. Value of $\mathbf{n}^0_{i,j}$ depends on the values of $\mathbf{r}_i$ and $\mathbf{r}_j$ (denoted as $\mathbf{r}^0_i$ and $\mathbf{r}^0_j$) for which it was calculated: $\mathbf{n}^0_{i,j} = \mathbf{n}^0_{i,j}(\mathbf{r}^0_i, \mathbf{r}^0_j)$. The approximation is given below:

$$\mathbf{f}^{FC}_{i,j} = -\mu_{i,j}(\mathbf{r}_i - \mathbf{r}_j) + \mu_{i,j}\rho_{i,j}\mathbf{n}^0_{i,j}. \tag{24.6}$$

As all local approximations, it is only valid as long the changes in variables $\mathbf{r}_i$ and $\mathbf{r}_j$ remain relatively small, and therefore, the error associated with the discrepancy between $\mathbf{n}^0_{i,j}$ and $\mathbf{n}_{i,j}$ remains small. There are cases when even a small change in $\mathbf{r}_i$ and $\mathbf{r}_j$ will lead to a big discrepancy between $\mathbf{n}^0_{i,j}$ and $\mathbf{n}_{i,j}$. Those cases can be referred to as directionally ill-conditioned. An example of the ill-conditioned case is when $\rho_{i,j} \approx ||\mathbf{r}_i - \mathbf{r}_j||$. However, even though in this case the discrepancy between $\mathbf{n}^0_{i,j}$ and $\mathbf{n}_{i,j}$ can large, the error in the absolute value of the elastic force is small, as the value of the force approaches zero when $\rho_{i,j}$ approaches $||\mathbf{r}_i - \mathbf{r}_j||$.

Figure 24.1 illustrates the geometry of the fixed-center approximation for a planar case.

Figure 24.2 shows the error magnitude $\varepsilon_{FC} = ||\mathbf{f}^{FC}_{i,j}(\mathbf{r}_j) - \mathbf{f}_{i,j}(\mathbf{r}_j)||$ for the approximation (24.6), for fixed values of $\mathbf{r}_i$, $\mathbf{r}^0_i$ and $\mathbf{r}^0_j$.

In Fig. 24.2, red circle represents circle points where elastic force is equal to zero, and green dot represents the linearization point. We can observe radial structure of the approximation.

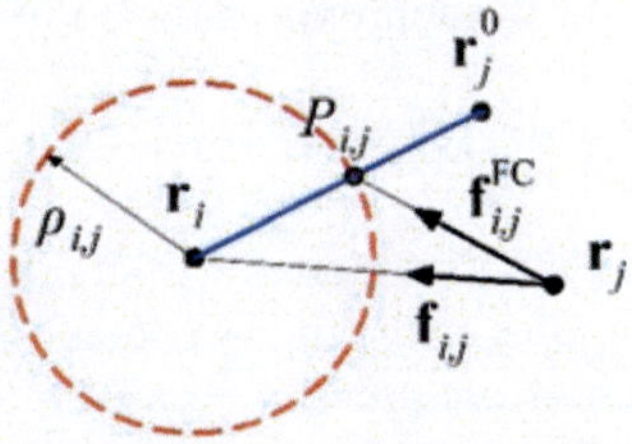

Fig. 24.1 Geometry of the fixed-center elastic force approximation for a planar case

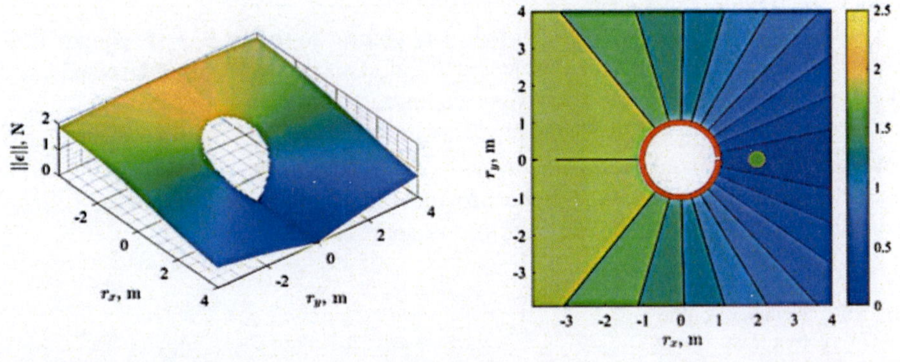

Fig. 24.2 Error magnitude for the approximation (24.6) as a surface and a contour graph for the compound approximation

24.3.2 Fixed-Direction Approximation

Another way of linearly approximating the elastic force is to fix the force direction while only changing its magnitude. We can calculate the magnitude of the force as a product of approximated deformation and the stiffness coefficient; here approximated deformation is calculated as a distance between the point $\mathbf{r}_j$ and a plane $\gamma_{i,j}$, generated by the normal vector $\mathbf{n}^0_{i,j}$ and point $P_{i,j}$, which is the closest to $\mathbf{r}_j$ point of intersection between the sphere $\sigma_{i,j}$: $\sigma_{i,j} = \{\mathbf{r} : ||\mathbf{r}^0_i - \mathbf{r}|| = \rho_{i,j}\}$ and a line $\lambda_{i,j}$ $\lambda_{i,j} = \{\mathbf{r}, \mathbf{r} = \mathbf{r}^0_i + t(\mathbf{r}^0_j - \mathbf{r}^0_i)\}$, where t is a parameter, $\mathbf{r}$ a variable representing point in space and $\mathbf{r}^0_i$ and $\mathbf{r}^0_j$ are the values of $\mathbf{r}_i$ and $\mathbf{r}_j$ for which the approximation is made. The sphere $\sigma_{i,j}$ represents the boundary between the tensile and compressive forces. For struts, this means that $\mathbf{r}_j$ should lie strictly inside the sphere, and for cables—strictly outside it (as long as approximation holds). Plane $\gamma_{i,j}$ is tangential to the sphere $\sigma_{i,j}$, vector $\mathbf{n}^0_{i,j}$ is normal to it.

The resulting expression for the approximation is given below:

$$\mathbf{f}^{FD}_{i,j} = -\mu_{i,j}(\mathbf{n}^0_{i,j} \cdot (\mathbf{r}_j - \mathbf{r}_i) - \rho_{i,j})\mathbf{n}^0_{i,j}. \tag{24.7}$$

Proposed method negates the directionally ill-conditioned cases when $\rho_{i,j} \approx ||\mathbf{r}_i - \mathbf{r}_j||$. As long as point $P_{i,j}$ lies on the line between $\mathbf{r}_i$ and $\mathbf{r}_j$, both fixed-direction and fixed-center approximations are precise. However, as the distance between $\mathbf{r}_j$ and plane $\gamma_{i,j}$ becomes smaller than the distance between $\mathbf{r}_j$ and point $P_{i,j}$, the approximation error increases.

Figure 24.3 illustrates the geometry of the fixed-direction approximation for the elastic force for a planar case.

In Fig. 24.3, an interval between points $\mathbf{r}_i$ and $\mathbf{r}_j$ on the line $\lambda_{i,j}$ is shown as a continuous blue line, sphere $\sigma_{i,j}$ projected to a circle is drawn with a red dashed line, plane $\gamma_{i,j}$ projected to line is shown as a continuous black line.

Figure 24.4 shows the error magnitude $\varepsilon_{\text{FD}} = ||\mathbf{f}^{\text{FD}}_{i,j}(\mathbf{r}_j) - \mathbf{f}_{i,j}(\mathbf{r}_j)||$ for the approximation (24.7), for fixed values of $\mathbf{r}_i$, $\mathbf{r}^0_i$ and $\mathbf{r}^0_j$. It shows that the approximation (24.7) has a directionally dependent error, but provides much better results close to the circle (sphere) $\sigma_{i,j}$, compared with the fixed-center approximation.

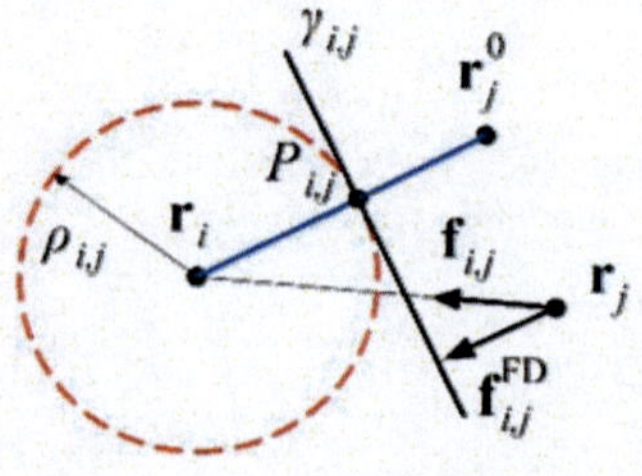

Fig. 24.3 Geometry of the fixed-direction elastic force approximation for a planar case

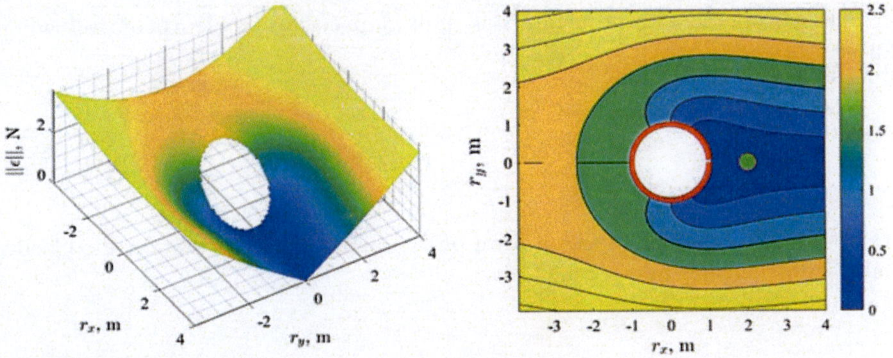

Fig. 24.4 Error magnitude for the fixed-direction approximation (24.7) as a surface and a contour graph for the compound approximation

24.3.3 Compound Linear Approximation

In order to produce a better approximation, a combination of the two proposed methods can be used. In a simple case, a weighted sum of (24.6) and (24.7) with equal weights can be used: $\mathbf{f}^{\text{CL}}_{i,j} = 0.5\mathbf{f}^{\text{FC}}_{i,j} + 0.5\mathbf{f}^{\text{FD}}_{i,j}$. Next expression shows the resulting compound linear approximation:

$$\mathbf{f}^{\text{CL}}_{i,j} = -\frac{1}{2}\mu_{i,j}(\mathbf{n}^0_{i,j} \cdot (\mathbf{r}_j - \mathbf{r}_i)\mathbf{n}^0_{i,j} - (\mathbf{r}_j - \mathbf{r}_i) + 2\rho_{i,j}\mathbf{n}^0_{i,j}). \tag{24.8}$$

Resulting approximation produces a larger area where the approximation error is small, as can be shown by analyzing error magnitude $\varepsilon_{\text{CL}} = ||\mathbf{f}^{\text{CL}}_{i,j}(\mathbf{r}_j) - \mathbf{f}_{i,j}(\mathbf{r}_j)||$ for the approximation (24.8), for fixed values of $\mathbf{r}_i$, $\mathbf{r}^0_i$ and $\mathbf{r}^0_j$ (see Fig. 24.5).

Figure 24.5 demonstrates the widest area of small approximation error near the approximation point, which is the desired property of the approximation

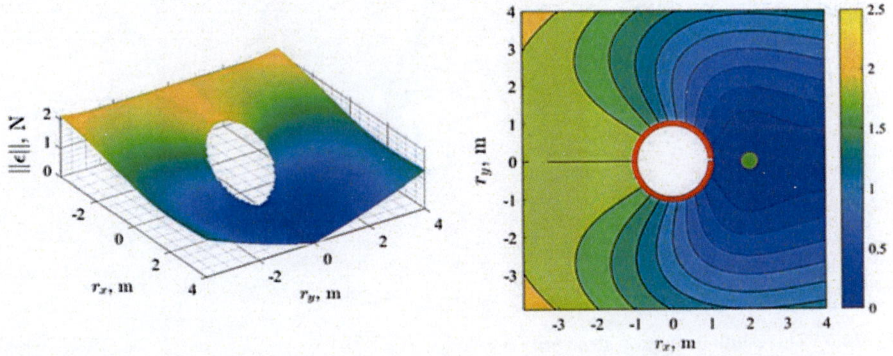

Fig. 24.5 Error magnitude as a surface and a contour graph for the compound approximation

Thus, the optimization problem (24.4) turns into a convex problem of the following form:

$$\begin{array}{ll} \text{find} & \mathbf{r} \\ \text{subject to:} & \sum_{j=1}^{n} (\mathbf{A}_{i,j}\mathbf{r}_{i,j} + \mathbf{b}_{i,j}) = 0 \end{array}, \tag{24.9}$$

where $\mathbf{A}_{i,j}\mathbf{r}_{i,j} + \mathbf{b}_{i,j}$ is a general form of the linear approximation for the elastic force acting between nodes $\mathbf{r}_i$ and $\mathbf{r}_j$.

24.4 Three-Link Tensegrity Structure

In order to demonstrate the function of the proposed linearization method, we study behavior of a three-link tensegrity structure. The ends of links (struts) are connected with elastic elements as shown in Fig. 24.6. That structure is often referred to as a tensegrity 3 prism [9].

The structure shown in Fig. 24.6 has six nodes, nine cables, and three struts. If both cables are struts are modeled as deformable, the total number of elastic connections for this structure is 12. Let us consider an experiment where three of those nodes are fixed, while others are in their stable positions, given the rest lengths of the cables. Since three nodes are fixed, from the topology of the structure follows that there are

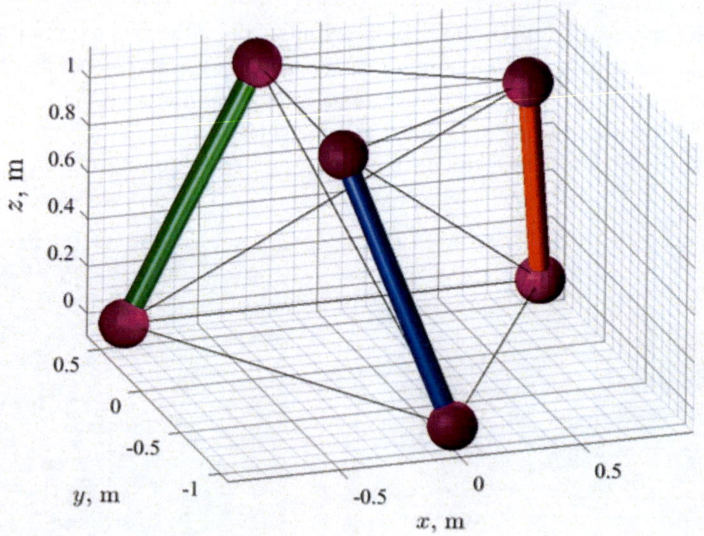

Fig. 24.6 Three-link tensegrity structure

three elastic connections (cables or struts) that only act between those points and remain inactive. Thus, there are only 9 active elastic connections.

24.4.1 Forward Kinematics with Local Linearization

The simplest way of solving inverse kinematic problem with a locally linearized elastic force model (24.8) is to find a solution for a feasibility problem (24.9). However, as the linearization introduces error to the elastic force model, the solution to the problem (24.9) might introduce error to the force balance equations; i.e., it might violate constraints in (24.4). Let us introduce a discrepancy error metric that enumerates the force balance constraint violation:

$$\varepsilon_i = \left\| \sum_{j=1}^{n} \mathbf{f}_{i,j}(\mathbf{r}) - \sum_{j=1}^{n} (\mathbf{A}_{i,j}\mathbf{r}_{i,j} + \mathbf{b}_{i,j}) \right\|, \tag{24.10}$$

where $\mathbf{f}_{i,j}(\mathbf{r})$ are calculated using precise formula (24.1).

Since the proposed linearization approaches the exact model at the linearization point, the solution of (24.9) can be used to update that linearization point, so that the problem (24.9) can be solved again with lower discrepancy error (24.10). This constitutes an iterative forward kinematic algorithm, based on the local linearization.

Figure 24.7 shows the convergence rate of the discrepancy error, as a function of the number of iterations of the algorithm. Without the loss of generality, we show the accuracy for only one of the three remaining free nodes of the structure, as the structure is symmetric.

Figure 24.7 demonstrates that the convergence of the algorithm is stable. While in absolute values the decrease in the error magnitude slows down as the number

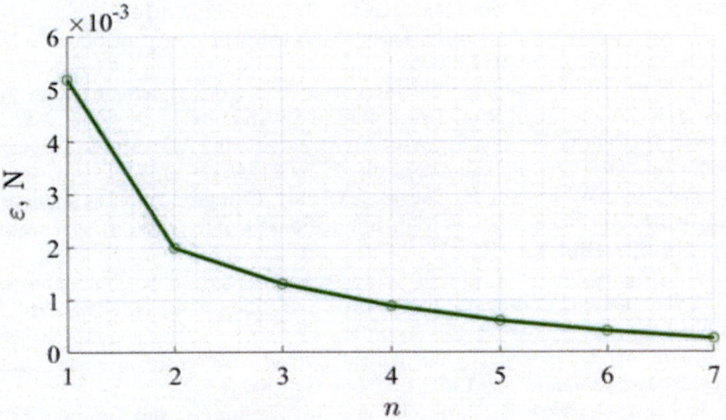

Fig. 24.7 Convergence of the discrepancy error with the iterative algorithm

of iteration increases, it can be shown that the relative improvement becomes better with consequent iterations.

Since each iteration of the algorithm requires solving a quadratic program of the same structure, the solution time and the stability of the algorithm can be accessed directly by studying the properties of the problem (24.9).

24.5 Conclusions

In this work, it was demonstrated that a local linearization of the elastic force expressions can be successfully used as a part of optimization-based forward kinematic algorithm. The geometry of the proposed compound linearization scheme is discussed at length in the paper.

An iterative scheme for the forward kinematic algorithm was proposed, and it was shown that it allows to control the accuracy of the obtained forward kinematic solution. The results were demonstrated using a three-link tensegrity structure with deformable struts and three fixed nodes.

Acknowledgements The research is supported by the grant of the Russian Science Foundation (project No:19-79-10246).

References

1. Guest, S.D.: The stiffness of tensegrity structures. IMA J. Appl. Math. **76**(1), 57–66 (2010)
2. Motro, R.: Tensegrity systems: the state of the art. Int. J. Space Struct. **7**(2), 75–83 (1992)
3. Sadao, S.: Fuller on tensegrity. Int. J. Space Struct. **11**(1–2), 37–42 (1996)
4. Snelson, K.: Snelson on the tensegrity invention. Int. J. Space Struct. **11**(1–2), 43–48 (1996)
5. Caluwaerts, K., et al.: Design and control of compliant tensegrity robots through simulation and hardware validation. J. R. Soc. Interface **11**(98), 20140520 (2014)
6. Paul, C., Valero-Cuevas, F.J., Lipson, H.: Design and control of tensegrity robots for locomotion. IEEE Trans. Rob. **22**(5), 944–957 (2006)
7. Smaili, A., Motro, R.: Foldable/unfoldable curved tensegrity systems by finite mechanism activation. J. Int. Assoc. Shell Spat. Struct. **48**(3), 153–160 (2007)
8. Bouderbala, M., Motro, R.: Folding tensegrity systems. In: IUTAM-IASS Symposium on Deployable Structures: Theory and Applications, pp. 27–36 (2000)
9. Paul, C., Roberts, J.W., Lipson, H. Cuevas, F.V.: July. Gait production in a tensegrity based robot. In: ICAR'05. Proceedings of 12th International Conference on Advanced Robotics, 2005, pp. 216–222. IEEE (2005)
10. Kim, K. et al.: Rapid prototyping design and control of tensegrity soft robot for locomotion. In: 2014 IEEE International Conference on Robotics and Biomimetics (ROBIO 2014), pp. 7–14. IEEE (2014)
11. Rieffel, J., Trimmer, B., Lipson, H.: Mechanism as mind-what tensegrities and caterpillars can teach us about soft robotics. In: ALIFE, pp. 506–512 (2008)
12. Friesen, J., Pogue, A., Bewley, T., de Oliveira, M., Skelton, R. and Sunspiral, V.: DuCTT: A tensegrity robot for exploring duct systems. In: 2014 IEEE International Conference on Robotics and Automation (ICRA), pp. 4222–4228. IEEE (2014)

13. Tietz, B.R, Carnahan, R.W, Bachmann, R.J, Quinn, R.D, SunSpiral V. Tetraspine: Robust terrain handling on a tensegrity robot using central pattern generators. In: 2013 IEEE/ASME International Conference on Advanced Intelligent Mechatronics pp. 261–267. IEEE
14. Hustig-Schultz, D., SunSpiral, V. Teodorescu, M.: Morphological design for controlled tensegrity quadruped locomotion. In: 2016 IEEE/RSJ International Conference on Intelligent Robots and Systems (IROS), pp. 4714–4719 (2016)
15. Bruce, J., Caluwaerts, K., Iscen, A., Sabelhaus, A.P., SunSpiral, V.: Design and evolution of a modular tensegrity robot platform. In: 2014 IEEE International Conference on Robotics and Automation (ICRA), pp. 3483–3489. IEEE (2014)
16. Caluwaerts, K., Bruce, J., Friesen, J.M., SunSpiral, V.: State estimation for tensegrity robots. In: 2016 IEEE International Conference on Robotics and Automation (ICRA), pp. 1860–1865. IEEE (2016)
17. Mirletz, B.T., Quinn, R.D., SunSpiral, V.: Cpgs for adaptive control of spine-like tensegrity structures. In: Proceedings of 2015 International Conference on Robotics and Automation (ICRA2015) Workshop on Central Pattern Generators for Locomotion Control: Pros, Cons and Alternatives (2015)
18. Savin, S., Balakhnov, O., Klimchik, A.: Energy-based local forward and inverse kinematics methods for tensegrity robots (2020). (in Publication)
19. Aldrich, J.B., Skelton, R.E., Kreutz-Delgado, K.: Control synthesis for a class of light and agile robotic tensegrity structures. In: Proceedings of the 2003 American Control Conference 6, pp. 5245–5251. IEEE (2003)
20. Diamond, S., Boyd, S.: CVXPY: A Python-embedded modeling language for convex optimization. J. Mach. Learn. Res. **17**(1), 2909–2913 (2016)
21. Boyd, S., Vandenberghe, L.: Convex optimization. Cambridge University Press (2004)
22. Grant, M., Boyd, S., Ye, Y.P: Disciplined convex programming. Glob. Opt. 155–210 (2006)

Chapter 25
Continuum Manipulator Motion Model Taking into Account Specifics of its Design

Dinar Bogdanov

Abstract The paper discusses the kinematic model continuum of the manipulator of the original design, as well as the issues of its application in the synthesis of the control system. The links of the manipulator are built on the basis of solid-state elements in contact with each other on the surface of a spherical shape and connected with the help of flexible cables. The operational characteristics of the manipulator (size and shape of the workspace, service and manipulation factors) are easily adjusted to technical requirements by changing the geometric parameters of the elements and their number in the links, as well as changing the number of links themselves. The complexity of controlling such a manipulator is determined by the need to obtain and process a significant amount of information about the state of the manipulator and its components. Due to the impossibility of obtaining explicit information about the position of the link elements relative to each other, it is proposed to compensate for this drawback by synthesizing an exact kinematic model. The authors propose such a link model, which takes into account rolling friction between link elements and the preliminary tension of the cables to ensure structural integrity in the entire range of motion. The results of the calculations revealed the peculiarities of the link movement in its entire range.

25.1 Introduction

Currently, along with the development of traditional manipulators based on rigid links, intended mainly for industrial use in a strictly determinate environment, researchers are developing the so-called continuum manipulators. The links of such manipulators necessarily have a flexible base (backbone) made of elastic materials or rigid elements moving relative to each other. Due to this, the links can change their shape (bending, possibly length) and adapt to the complex configuration of the environment or the shape of the object of manipulation. Many researchers use natural

---

D. Bogdanov (✉)
Mavlyutov Institute of Mechanics UFRC RAS, Prosp. Oktyabrya 71, 450054 Ufa, Russia
e-mail: dr_bog@mail.ru

A. Ronzhin and V. Shishlakov (eds.), *Proceedings of 15th International Conference on Electromechanics and Robotics "Zavalishin's Readings"*, Smart Innovation, Systems and Technologies 187, https://doi.org/10.1007/978-981-15-5580-0_25

objects such as animals (trunk, tentacles) and plants (vine, tendrils of plants) as a source of ideas.

Depending on the actuators used and the type of backbone used, the following basic principles for designing such manipulators were identified in [1]:

1. Tendon-Based Designs. The bending of the links of the manipulator is carried out by changing the lengths of the cables along with the links of the manipulator and fixed at their ends. The links can be a set of rigid elements [2, 3], interconnected in various ways, or be a flexible rod [4] or a compressible spring. Actuators can be moved outside the links to a separate base or located directly in the link [5].
2. Concentric Tube Designs. In this approach, the telescope principle is used: Other tubes are laid inside one tube, also nested in each other [6]. The tubes themselves are elastic and pre-bent. By moving the tubes relative to each other, a change in the length and configuration of the manipulator can be achieved. Another way to bend tubes is to use tendon-driven actuators. So, in the manipulator [7], which is a 2-link catheter for medical purposes, 4 independent drives moving the cables are used to move and bend each tube.
3. Locally Activated Backbone Designs. Here the actuators are located directly in the link, or the elements of the link themselves are actuators. So, in [8], each link is a set of 3 air channels, changing the air pressure in which you can bend or lengthen the link. If you make a link from an alloy with shape memory, you can change the shape of the link by heating or cooling it [9]. Using dielectric elastomeric actuators, the manipulator configuration can be controlled using voltage [10].

Due to the soft effect of such manipulators on the object and the adaptation of their configuration to the shape of the object, they can be attributed to software manipulators [11]. As can be seen from the review [11], the principles of constructing software manipulators are similar, with a greater emphasis on the safety of interaction with the object and the elastic properties of the materials used. Similar principles of design and manufacture can be used in the development of grips [5, 11]. Currently, the target areas of application for such manipulators are: medicine (minimally invasive surgery, dentistry), assistance in everyday human activities (collaborative robotics), work with fragile objects, examination of objects with complex internal cavities, etc.

This paper discusses a manipulator, the distinguishing features of which are:

(1) scalability in a wide range of geometrical sizes of links: depending on the required working space according to the kinematic model, the necessary design parameters of link elements and their number are selected;
(2) a wide range of materials used for the manufacture of link elements (metal, plastic, hard rubber);
(3) functional homogeneity of the design of the manipulator (lack of intermediate and/or docking nodes between links).

A key problem in the development of continuum manipulators is the complexity of controlling the shape of links in control [1, 11]. Also, relevant is the use of a fairly large number of sensors to measure basic parameters. A variety of designs leads to the need to develop special sensors for specific cases, as evidenced by the activity of research in this area. One of the approaches to solving this problem is to integrate specialized sensors directly into the design of the manipulator as its component [11], as well as the use of vision equipment [1]. A reasonable alternative to solving this problem in some tasks is the use of more precise kinematic and dynamic models of the developed manipulators in the synthesis of control systems. This approach is presented in this paper, where the author proposed a kinematic model of the manipulator taking into account the direction of action of the forces of gravity and friction between the elements in the links and also describes the mechanism of bending of the link in its entire range.

25.2 The Design of the Manipulator and Its Kinematics

The manipulator under consideration (see Fig. 25.1a) is a set of series-connected links, each of which consists of solid-state elements with a spherical surface in the form of circular disks of a given thickness (see Fig. 25.1b, c). Elements are characterized by the following parameters (see Fig. 25.2b): R is the radius of curvature of the surfaces of the elements by which they are in contact with each other; H is the height of the cylindrical part of the element; D is the diameter of the element; d is the diameter of the circle on which the holes for the cables lie; P is the weight of the element. Through the holes on the edges of the elements are spring-loaded flexible cables, the ends of which are fixed to the last elements of the links. Springs provide the necessary tension of the cables and are elements of the displacement and tension sensors of the cables, combined in one unit with the drives (based on stepper motors with gearboxes) (see Fig. 25.1d).

A consistent change in cable lengths forms the required link bend. The calculation of control actions is based on the kinematic model of the link. In the initial version of the kinematic model previously presented in [12], it was assumed that the rolling

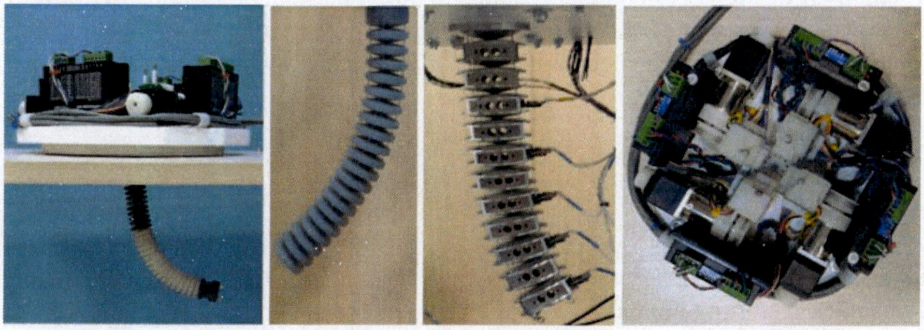

(a) (b) (c) (d)

Fig. 25.1 Manipulator design: **a** general view of the manipulator, **b**, **c** design of the manipulator link (plastic, metal), **d** drive unit with a displacement and tension sensors

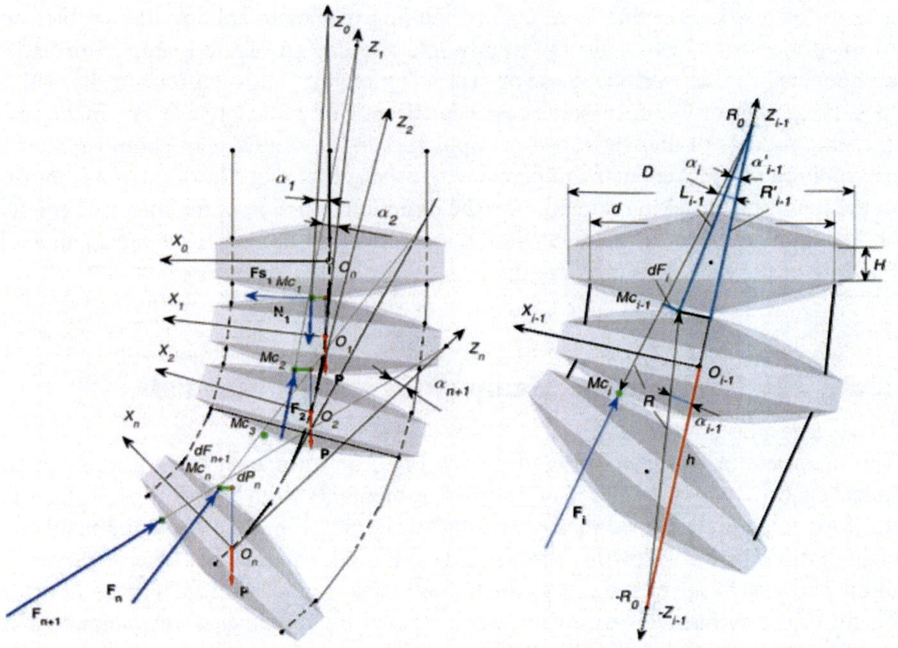

(a) (b)

Fig. 25.2 Scheme of the formation of the bend of the link: **a** acting forces, **b** geometry for calculating the value of the shoulder action of the force F_i

angles of α_{ji} elements are the same in each individual link:

$$\alpha_{j1} = \cdots = \alpha_{jM_j} = \alpha_{oj}, \quad (25.1)$$

where M_j is the number of elements in the jth link. The angle of bending of the links is defined as $\alpha_j = M_j \alpha_{oj}$. Taking into account the forces acting on the manipulator, condition (25.1) is not satisfied and the model of the manipulator has the following form:

$$\mathbf{T}_M = \prod_{j=1}^{N} \prod_{i=1}^{M_j} \mathbf{T}_{ji},$$

where i is the number of the element in the link; N-number of links in the manipulator; $\mathbf{T}_{ji}$ is the transformation matrix for the ith element of the jth link.

In turn, the matrix $\mathbf{T}_{ji}$ has the following form:

$$\mathbf{T}_{ji} = \begin{vmatrix} \mathbf{Rot}_{ji}(\mathbf{l}_j, \beta_{ji}) & \mathbf{p}_{ji}(\alpha_{ji}, \psi_j) \\ 000 & 1 \end{vmatrix}, \quad (25.2)$$

where $\mathbf{Rot}_{j,i}(\mathbf{l}_j, \beta_{ji})$ is the rotation matrix of the ith element in the jth link; $\mathbf{p}_{ji}(\alpha_{ji}, \psi_j)$ is the position vector of the center of rotation of the ith element in the jth link; α_{ji} is the rolling angle of element i relative to element $i-1$; ψ_j is the rotation angle of the jth link relative to the Z-axis of the coordinate system of the last element of the previous $j-1$ link [12]; β_{ji} is the rotation angle of the ith element in the jth link relative to the rotation vector at the contact point of the elements i and $i-1$.

The rotation vector $\mathbf{l}_j$ has the form: $\mathbf{l}_j = \left[\sin\psi_j, -\cos\psi_j, 0\right]^T$ [12]. Angle β_{ji} is defined as $\beta_{ji} = (1 + R_{i-1}/R_i)\alpha_{ji}$. In the manipulator under consideration $R_{ji} = R$ and $\beta_{ji} = 2\alpha_{ji}$. The components of the $\mathbf{Rot}_{j,i}(\mathbf{l}_j, \beta_{ji})$ matrix are calculated by the following formulas:

$$Rot_{ji}^{11} = \sin^2\psi_j(1-\cos\beta_{ji}) + \cos\beta_{ji},\ Rot_{ji}^{12} = -\sin\psi_j\cos\psi_j(1-\cos\beta_{ji}),$$

$$Rot_{ji}^{13} = -\cos\psi_j\sin\beta_{ji},\ Rot_{ji}^{21} = -\sin\psi_j\cos\psi_j\left(1-\cos\beta_{ji}\right),$$

$$Rot_{ji}^{22} = \cos^2\psi_j(1-\cos\beta_{ji}) + \cos\beta_{ji},\ Rot_{ji}^{23} = -\sin\psi_j\sin\beta_{ji},$$

$$Rot_{ji}^{31} = \cos\psi_j\sin\beta_{ji},\ Rot_{ji}^{32} = \sin\psi_j\sin\beta_{ji},\ Rot_{ji}^{33} = \cos\beta_{ji}.$$

The components of vector $\mathbf{p}_{ji}(\alpha_{ji}, \psi_j)$ are defined as:

$$p_{ji}^x = (-2R\sin\alpha_{ji} + h\sin 2\alpha_{ji})\cos\psi_j,$$

$$p_{ji}^y = (-2R\sin\alpha_{ji} + h\sin 2\alpha_{ji})\sin\psi_j,$$

$$p_{ji}^z = -2R\cos\alpha_{ji} + h(\cos 2\alpha_{ji} + 1).$$

The constant h is the distance from the center of curvature of the surface of the element to its center of coordinates (see Fig. 25.2b).

In accordance with this model, the position vector of the center of the 1st element of the 1st link is calculated as follows:

$$\mathbf{p}_{11} = \mathbf{T}_{11}\mathbf{p}_0^T,$$

where $\mathbf{p}_0^T = [0; 0; 0; 1]^T$ is the position vector of the center of the base fixed element.

The last element of each link is a reference for the next link, and the position vector of the center of the 1st element of the jth link is $\mathbf{p}_{j1} = \mathbf{T}_{ji}\mathbf{p}_{(j-1)M_{(j-1)}}^T$.

The position vector of the center of the ith element of the jth link is $\mathbf{p}_{ji} = \mathbf{T}_{ji}\mathbf{p}_{j(i-1)}^T$.

Accordingly, the position vector of the operating point of the manipulator is the position vector of the center of the last element of the last link: $\mathbf{p}_t = \mathbf{p}_{NM_N}$.

25.3 The Dynamics of the Manipulator Link's Bend Formation

Consider the formation of the bend of the 1st link, taking into account all the forces acting on it. The movable element 1 is affected by: the force of gravity P (weight of the element), normal reaction force N_1 arising at the contact point Mc_1 of elements 1 and 2, friction force of adhesion Fs_1 and force F_2 (see Fig. 25.2a). For the case of static equilibrium of the moving element 1 relative to the stationary one, for some rolling angle α_1, we write down the equations of equilibrium of moments and projections of forces on the selected axes (action vectors of forces N_1 and Fs_1):

$$N_1 + P\cos\alpha_1 - F_2\cos(\alpha_1 + \alpha_2) = 0, \tag{25.3}$$

$$-P\sin\alpha_1 - F_2\sin(\alpha_1 + \alpha_2) + Fs_1 = 0, \tag{25.4}$$

$$f_r N_1 \mp PdP_1 - F_2 dF_2 = 0. \tag{25.5}$$

Here f_r is the coefficient of rolling friction. The sign of the moment created by gravity P in Eq. (25.5) depends on its direction relative to the contact point of the elements. The sliding friction force at the contact point of the elements is defined as: $Fs_1 = N_1 f_s$. And f_s is the coefficient of sliding friction.

We express N_1 and F_2 from (25.3), (25.4) and substitute them in (25.5):

$$N_1 = F_2\cos(\alpha_1 + \alpha_2) - P\cos\alpha_1,$$

$$F_2 = (P(f_s\cos\alpha_1 + \sin\alpha_1))/(f_s\cos(\alpha_1 + \alpha_2) - \sin(\alpha_1 + \alpha_2)),$$

$$P_{B_1}(f_r\cos(\alpha_1 + \alpha_2) - dF_2)/(f_s\cos(\alpha_1 + \alpha_2) - \sin(\alpha_1 + \alpha_2)) + P_{A1} = 0,$$

where $P_{B_1} = P(f_s\cos\alpha_1 + \sin\alpha_1)$; $P_{A_1} = -P(f_r\cos\alpha_1 \pm dP_1)$.

The calculation of the shoulder of the action of gravity dP_1 acting on element 1 is made through the absolute coordinates of the contact point of elements Mc_1 and the center of the coordinate system O_1 of element 1, which, in turn, are determined through transformation matrices (25.2) for a given angle α_1. For the remaining elements, the calculation of dP_i is similar. The calculation of the dF_2 value is carried out geometrically (see Fig. 25.2b) and for all pairs of contacting link elements the same:

$$dF_i = L_{i-1}\sin(\alpha_i - \alpha'_{i-1}),\ L_{i-1} = \sqrt{(R'_{i-1})^2 + (R\sin\alpha_{i-1})^2},$$

$$\alpha'_{i-1} = \tan^{-1}(R\sin\alpha_{i-1}/R'_{i-1}),\ R'_{i-1} = 2\sqrt{(R^2 - (D/2)^2)} - R\cos\alpha_{i-1} - H,$$

where $i = \overline{2, \ldots, n}$, n is the number of elements in the link.

In the equation of the balance of moments (25.5), one unknown α_2 is the direction of action of the force F_2 from the side of element 2 at the contact point of elements 2 and 3 Mc_2. In this case, the solution is found numerically in the simulation. Then we find the values of forces N_1 and F_2. To ensure guaranteed contact of element 1 with the base element, we add additional force (pretension) N_{PT} to the calculated reaction force N_1 for the current angle α_1. The final value of the reaction force at the point of contact will be $N_1^* = N_1 + N_{PT}$. Next, we recalculate the values of F_2 and α_2 from the adjusted (25.6)–(25.9):

$$F_2 = (N_1^* + P\cos\alpha_1)/\cos(\alpha_1 + \alpha_2), \tag{25.6}$$

$$P_{A_1} = N_1^* f_r \mp PdP_1, \tag{25.7}$$

$$P_{B_1} = (N_1^* + P\cos(\alpha_1)), \tag{25.8}$$

$$P_{A_1} + P_{B_1} dF_2/\cos(\alpha_1 + \alpha_2) = 0. \tag{25.9}$$

The calculated force F_2 is equal in magnitude to the reaction force N_2 from the side of element 2 to the action of element 3 at the point of contact. Given this statement and Eqs. (25.3)–(25.5), the equations for calculating the forces for an arbitrary pair of link elements i and $i + 1$ can be written as follows:

$$F_i + P\cos(\alpha_{s_i}) - F_{i+1}\cos(\alpha_i + \alpha_{i+1}) = 0, \tag{25.10}$$

$$-P\sin(\alpha_{s_i}) - F_{i+1}\sin(\alpha_i + \alpha_{i+1}) + F_i f_s = 0, \tag{25.11}$$

$$F_i f_r \mp PdP_i - F_{i+1} dF_{i+1} = 0. \tag{25.12}$$

The formulas and equations for calculating α_{i+1} and F_{i+1} are as follows:

$$F_{i+1} = (F_i + P\cos(\alpha_{s_i}))/\cos(\alpha_i + \alpha_{i+1}),\ P_{A_i} = F_i f_r \mp PdP_i,$$

$$P_{B_i} = (F_i + P\cos(\alpha_{s_i})),\ P_{A_i} + P_{B_i} dF_{i+1}/\cos(\alpha_i + \alpha_{i+1}) = 0,$$

where $\alpha_{s_i} = \alpha_i + \alpha_{i-1} + \alpha_{s_{i-1}}$; $\alpha_{s_1} = \alpha_1$; $i = \overline{2, \ldots, n-1}$.

The rolling of the last element of the link is provided by a change in the tension of the cables (see Fig. 25.3).

A change in the magnitudes and ratios of the tension forces of the cables Tr_i leads to a change in the angles α and ψ, which uniquely determine the position of the elements relative to each other in space. The solution of the 3D problem can be

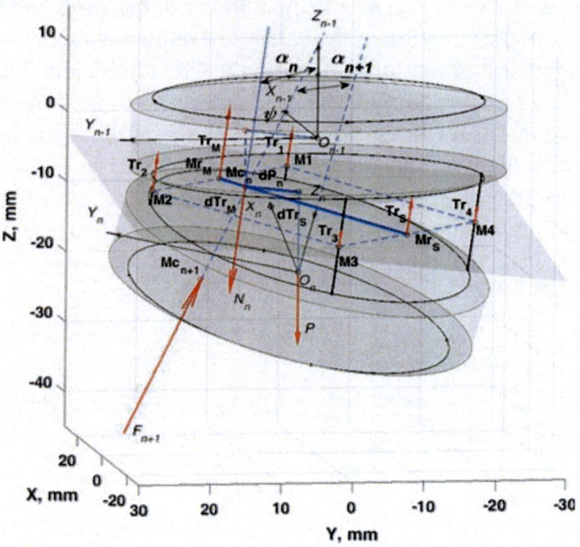

Fig. 25.3 Results of model calculations: visualization of the forces acting on the end element in the link when it is bent and rotated

reduced to the 2D version and consider the pairs of resultant tension forces Tr_M and Tr_S obtained as the vector sum of pairs of parallel cable tension forces, the choice of which depends on the value of angle ψ. For example, for angle $\psi =]0; 90^\circ[$, the driving force Tr_M will be the sum of the tension forces Tr_1, Tr_2 and the driven force Tr_S will be the sum of the tension forces Tr_3, Tr_4. Then equations of balance of forces and moments for the last element, equivalent to Eqs. (25.10) and (25.12), can be written:

$$F_n + P\cos(\alpha_{s_n}) - Tr_M - Tr_S = 0, \tag{25.13}$$

$$F_n f_r \mp P dP_n - Tr_M dTr_M + Tr dTr_S = 0. \tag{25.14}$$

We express Tr_M from (25.13) and substitute it in (25.14):

$$Tr_M = F_n + P\cos(\alpha_{s_n}) - Tr_S,$$

$$Tr_S = (F_n(dTr_M - f_r) + P(dTr_M \cos(\alpha_{s_n}) \mp dP_n))/(dTr_M + dTr_S).$$

The ratios of the tension forces of the cables and their resultants (see Fig. 25.3) define the tensile forces of each cable as follows:

$$Tr_1 = Tr_M|Mr_M M_2|/(|Mr_M M_1| + |Mr_M M_2|),$$
$$Tr_2 = Tr_M|Mr_M M_1|/(|Mr_M M_1| + |Mr_M M_2|),$$
$$Tr_3 = Tr_S|Mr_S M_4|/(|Mr_S M_3| + |Mr_S M_4|),$$

$$Tr_4 = Tr_S|Mr_SM_3|/(|Mr_SM_3| + |Mr_SM_4|).$$

The initial position of the link is the absence of its bend, i.e., when the axes O_iZ_i of the elements of the link of the manipulator are collinear, the total bending angle of the link is $\alpha = 0$. Link bending begins with the rolling of the lower element at a value of the resultant forces Tr_i exceeding the static friction force between this element and the upper element of the link with respect to it. After the element n begins to roll, the F_n force changes its value and direction (it is characterized by an angle α_n) with respect to the upper element $n-1$ at the contact point of the elements n and $n-1$. When the angle α_n exceeds the value of the onset angle $\alpha_{str_{n-1}}$ (25.15), the element $n-1$ will begin to roll. And so, sequentially, it happens with all the elements up to the highest moving element of the link. With this orientation of the base element, the motion onset angles for each element are defined as:

$$\alpha_{str_i} = \tan^{-1}(N_i f_r/(N_i + P)R'_{i-1}), \tag{25.15}$$

where $N_i = (i-1)P + N_{PT}$.

After shifting the uppermost element, the remaining elements of the link roll simultaneously, until the lowest element selects the full rolling range, which is characterized by the angle $\alpha_{max\psi}$. This value is determined by the design parameters of the link elements and its current rotation angle ψ [13]. After that, the lower element does not roll but moves along with the upper contacting element ("gluing" of the elements occurs). Such a movement with successive "gluing" of all elements occurs until the maximum rolling angles are selected by all elements. As a result, the maximum link bend will be obtained, which is equal to the sum of the maximum rolling angles of all elements.

25.4 Results

The obtained equations were used to develop a link motion simulation program. As a result of the program, data are generated in the form of rolling angle vectors of the link elements and the corresponding vectors of cable lengths and tension forces [12], and their number is determined by the specified simulation accuracy. As a goal, the calculated model data is used to synthesize the manipulator control system and analyze the formation of the link bend.

According to the results obtained during the simulation, it is possible to distinguish three phases in the movement of the link: the choice of the motion onset angles of the link elements (see Fig. 25.4a), the simultaneous rolling of all link elements (see Fig. 25.4b), and the gradual "bonding" of the elements to form the maximum bend of the link (see Fig. 25.4c).

The pretension of the N_{PT} link (other things being equal) has a significant effect on the values of the motion onset angles. The graphs of the dependence of the motion onset angle of the link elements on the pretension value obtained as a result

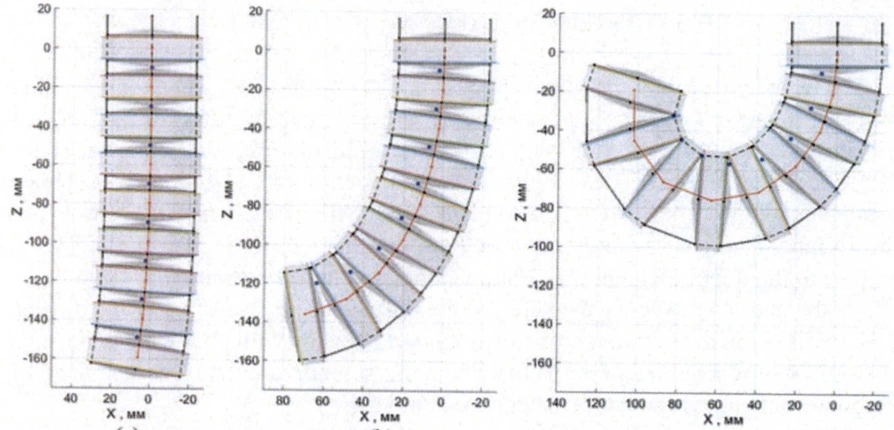

Fig. 25.4 Phase of link movement ($N_{PT} = 5N$): **a** phase of selection of the motion onset angles of elements; **b** simultaneous rolling of all elements; **c** "gluing" of link elements

of computer calculations are presented in Fig. 25.5a. In the simulation, the following link element parameters were used: $R = 78$ mm, $H = 11$ mm, $D = 52$ mm, $d = 46$ mm, $P = 1N$, $f_s = 1$, $f_r = 0.1$ mm, $n = 8$. Aluminum was used as the material of the link elements.

Modeling also showed the dependence of the change in the angle of bending of the link and vector position of the link tip on the values of the pretension (see Fig. 25.5b).

25.5 Conclusion

It should be noted that the above calculations are simplified and aim to identify the specifics of the mechanics of movement of the link of the developed manipulator. When managing the real design of the manipulator, it is necessary to take into account:

(1) the influence of external disturbances;
(2) impact on the last element of the first link from the side of the second link (a similar situation is typical for subsequent links);
(3) orientation of the last elements of links that are supporting elements for subsequent links.

The need for these additional studies is determined by the synthesis of an adequate control system, the algorithms of which are based on solving the inverse kinematic problem. The absence of analytical solutions to the inverse problem requires the use of computational methods, as well as implicit algorithms (e.g., neural networks). The development of refined kinematic models will provide a more adequate set of training samples for the neural network.

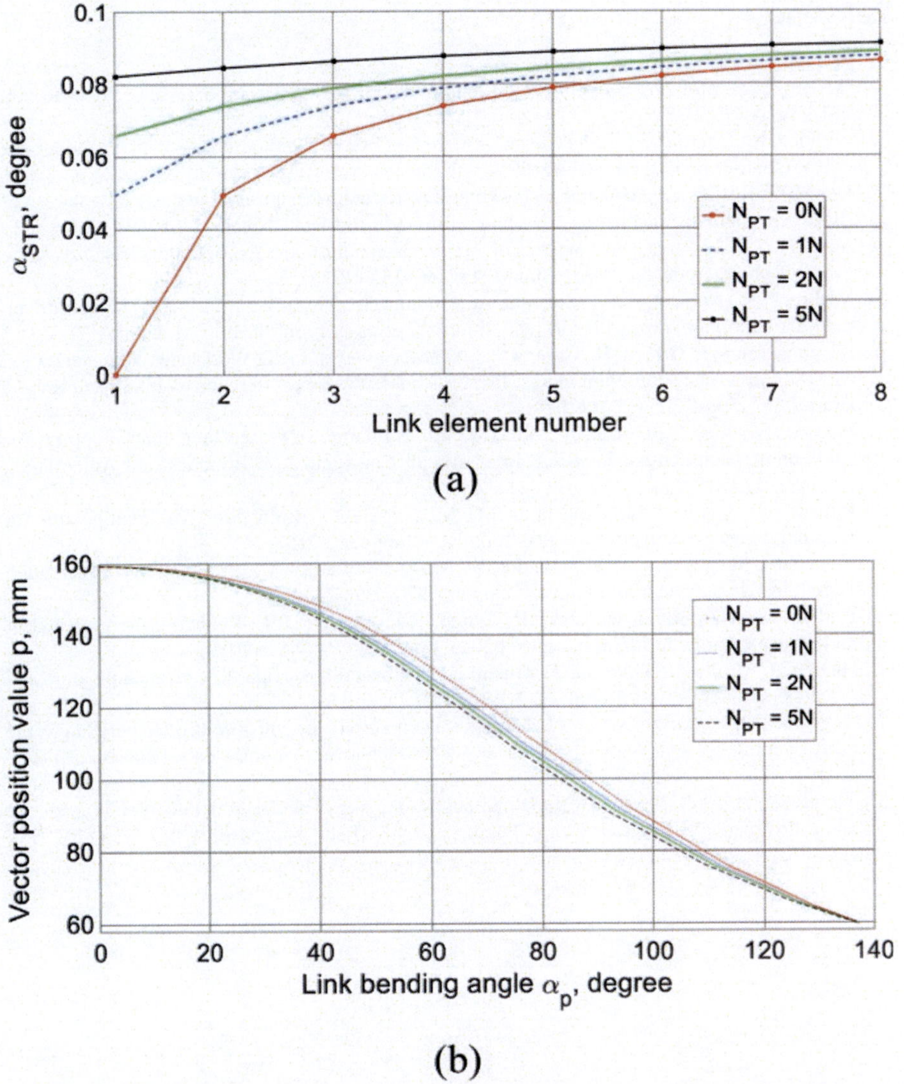

Fig. 25.5 Results of modeling the movement of the link: **a** dependence of the motion onset angles of the link elements on the value of the pretension, **b** dependence of the change in the angle of bending of the link and vector position of the link tip on the values of the pretension

This work was supported by the Program of the Presidium of the Russian Academy of Sciences and within the framework of state assignment No. 0246-2018-007.

References

1. Walker, I.: Continuous backbone "continuum" robot manipulators: a review. ISRN Robot. **2013**(1), 1–19 (2013)
2. Bamdad, M., Mardani, A.: Mechatronics modeling of a branching tendon—driven Robot. In: ICRoM 2014, pp. 522–527. IEEE (2014)
3. Dong, X., Raffles, M., Guzman, S., Axinte, D., Kell, J.: Design and analysis of a family of snake arm robots connected by compliant joints. Mech. Mach. Theory **77**, 73–91 (2014)
4. Yuan, H., Li, Z.: Workspace analysis of cable-driven continuum manipulators based on static model. Robot. Computer-Integr. Manuf. **49**, 240–252 (2017)
5. Mishra, A.K., Del Dottore, E., Sadeghi, A., Mondini, A., Mazzolai, B.: SIMBA: tendon-driven modular continuum arm with soft reconfigurable gripper. Front. Robot. AI **4**, 4 (2017)
6. Burgner-Kahrs, J., Gilbert, H., Granna, J., Swaney, P., Webster R.J.: Workspace characterization for concentric tube continuum robots. In: IEEE International Conference on Intelligent Robots and Systems, pp. 1269–1275 (2014)
7. Penning, R., Jung, J., Ferrier, N., Zinn, M.: An evaluation of closed-loop control options for continuum manipulators. In: IEEE International Conference on Robotics and Automation, pp. 5392–5397 (2012)
8. Ranzani, T., Gerboni, G., Cianchetti, M., Menciassi, A.: Bioinspired soft manipulator for minimally invasive surgery. Bioinspiration Biomimetics (2015)
9. Cianchetti, M., Licofonte, A., Follador, M., Rogai, F., Laschi, C.: Bioinspired soft actuation system using shape memory alloys. Actuators **3**(3), 226–244 (2015)
10. Duduta, M., Haiesmaili, E., Zhao, H., Wood, R.J., Clarke, D.R.: Realizing the potential of dielectric elastomer artificial muscles. PNAS **116**(7), 2476–2481 (2019)
11. Hughes, J., Culha, U., Giardina, F., Guenther, F., Rosendo, A., Iida, F.: Soft manipulators and grippers: a review. Front. Robot. AI **3**, 69 (2016)
12. Bogdanov, D.R., Darintsev, O.V.: Kinematics of a manipulator with a controlled bending on the basic of solid elements with a spherical surface. Mechatron. Autom. Control **16**(10), 671–677 (2015)
13. Bogdanov, D.R., Darintsev, O.V.: Dynamics (kinematics) of the manipulator link with the controlled bend. XII all-Russ Congr. Fundam. Probl. Theor Appl Mech **1**, 528–530 (2019)

Chapter 26
Modeling Wireless Information Exchange Between Sensors and Robotic Devices

Alexander Denisov and Oleg Sivchenko

Abstract Conceptual and set-theoretic models of wireless data transfer network for static and dynamic systems, particularly, for robotic devices, including radio modules, which use repeaters for communication with remote network nodes are described. These models are intended for agricultural applications. Radio modules, sensors, and robotic devices, connected using repeaters to the distributed wireless network, are described. The design problem of wireless data transfer network completed with static and dynamic radio modules is a complicated one. Certain specifics deal with how to maintain data transfer speed above the specified threshold by varying distance between radio modules and robotic device motions. Increased data transfer reliability, specific robot interaction patterns, and groundwork for IoT paradigm implementation are of particular importance here.

26.1 Introduction

Implementation of digital technologies and robotic devices in agriculture is aimed to solve problems, which directly concern control of multiple heterogeneous objects, their parameters, and data gathered by them (such as temperature, and humidity levels). Wireless sensory systems consist of distributed autonomous sensors which register changes in physical characteristics of the environment, parameters of receiver-transmitter radio modules, and base station (gateway). The systems, that use radio modules, can transmit data over vast distances. A particular advantage of such system is that the radio modules can be used not only for sensor data transmission, but as repeaters equally well; in the latter case, they redirect information from other modules to the base station.

A. Denisov · O. Sivchenko (✉)
St. Petersburg Institute for Informatics and Automation of the Russian Academy of Sciences, 39, 14th Line, 199178 St. Petersburg, Russia
e-mail: adrelian@gmail.com

A. Denisov
e-mail: sdenisov93@mail.ru

A. Ronzhin and V. Shishlakov (eds.), *Proceedings of 15th International Conference on Electromechanics and Robotics "Zavalishin's Readings"*, Smart Innovation, Systems and Technologies 187, https://doi.org/10.1007/978-981-15-5580-0_26

Wireless data transfer system setup and connection in context of wide area service are complicated by many factors, such as radio signal attenuation, noise, and insufficient data transmission speed, what can cause in some areas lack of connection with the remaining parts of the system. Multi-purpose intelligent agricultural facilities grow in number today, and they interact via wireless network. Hence, a problem arises, how to ensure sufficiently fast approach to system design and deployment, where the communication is established among heterogeneous distributed modules of different automation and mechanization grade.

The models enable development of a recommender system, which assists in establishment of a wireless network for sensory data transmission. This network, in turn, allows to boost planning and deployment of energy-efficient scalable network within the specified area for transmission of data, needed for completion of various agricultural tasks.

Sensor systems employed in agriculture have certain specifics, concerning their application requirements. One of the most promising modern solutions is the LoRa technology is being actively pursued in agricultural domain.

Usage of IoT systems based on different wireless data transmission technologies which ensure transformation of conventional cities into «smart» ones via wireless connectivity and IoT is considered in [1–9]. Novel architectures of two-way radio connections are also investigated here; sensor network architectures are given, and LoRa wireless network efficiency is assessed; an energy-efficient and well-scalable agricultural system is presented, built upon IoT technologies; service life expectancy of IoT devices is estimated, functioning in several wireless networks, which conform to the following standards: IEEE 802.15.4/e, Bluetooth Low Energy (BLE), IEEE 802.11 energy saving mode, IEEE 802.11ah, and in novel long-range transmission technologies, such as LoRa and SIGFOX. Models, algorithms, and mesh spatial design approaches for mesh networks which ensure comprehensive radio connectivity, propose an algorithm for effective classification characteristics system establishment based on the concept of truncated exhaustive search and the use of information on individual classification indicators, use global relief height models shuttle radar topographic mission (SRTM) to evaluate the general characteristics of forest stands are considered in [10–18]. Though, the papers do not consider possibilities of network composition of wireless data transmission modules, spaced so far from each other, that it becomes infeasible to maintain reliable two-way radio communication. Mathematical model describing design of wireless programmatic information interaction system based on establishment of radio transmission distance using Friis equation [19] (1) and distance equation [19] (2), telemetry network building recommendation system, and software for calculating the coordinates of relay locations was proposed in [20–23]:

$$\frac{P_R}{P_T} = \frac{G_T G_R \lambda^2}{(4\pi R)^2}, \tag{26.1}$$

$$R = \frac{3 \times 10^8}{4\pi F} \cdot 10 \frac{P_T + G_T + L_T + G_R + L_R + |V| dB - P_R}{20}, \tag{26.2}$$

where R is the target communication range in meters; F is the frequency in Hz; P_T is power of the radio transmitter in dBm; G_T is transmitter antenna gain in dB; L_T is loss in cable to the transmitter antenna in dB; G_R is receiver antenna gain in dB; L_R is loss in cable to the receiver antenna in dB; P_R is the sensitivity of the radio receiver in dBm (a certain data rate corresponds to its sensitivity value for the radio receiver, which is determined empirically or looked up in the manufacturer's datasheet for the radio module); $| V |$ dB is attenuation factor, taking into account additional losses due to the influence of the ground, vegetation, atmosphere, and other factors in dB; FSL is free space loss; system operating margin (SOM) is margin in radio communications energy.

This paper proposes two models of wireless data transmission—conceptual and set-theoretic ones—to design a wireless sensor network, allowing to attach additional radio modules, compose a mesh-network of arbitrarily placed LoRa modules, and establish connection with sensor systems and robotic devices in agriculture.

26.2 Sensor Complex Overview

For formal description of recommender system establishment, a conceptual model, which includes the following main entities: service area, sensor complex, object, robotic device, query, and archive, was proposed. The developed conceptual model given in Fig. 26.1 is the basis for establishment of transmitter distribution recommender system, ensuring of radio communication among sensor complex objects, robotic device localization.

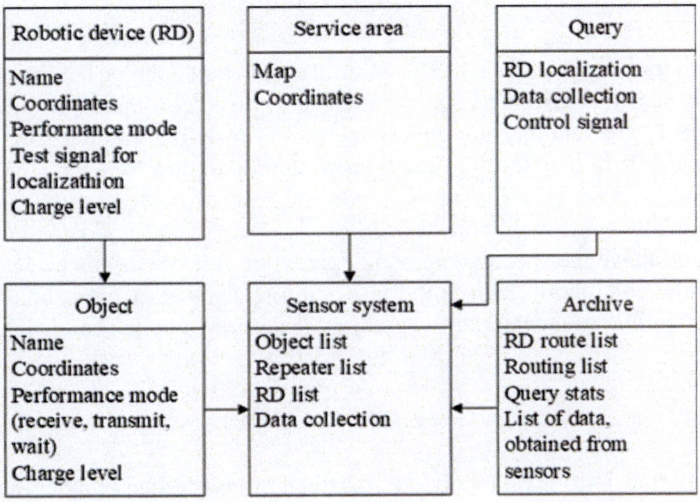

Fig. 26.1 Conceptual model of a sensor complex

During wireless sensor network modeling and design the following assumptions are made. Suppose that we have a flat area without obstacles, where data transmission with certain speed should be ensured. The given area has fixed points, where radio modules must be positioned, and these modules have to be connected to the gateway. Every robotic device also has its working range, which must be fully covered by radio signal for permanent connection with robot.

26.3 Set-Theoretic Model

Suppose, $D = \{d_1, d_2, \ldots, d_k, \ldots, d_K\}$—a set of radio modules, including repeaters, acting within a smart agricultural system, $R = \{r_1, r_2, \ldots, r_n, \ldots, r_N\}$—a set of mobile robotic devices, $S = \{s_1, s_2, \ldots, s_m, \ldots, s_M\}$—a set of objects in a data transmission sensor system, where K, N, M are quantities of radio modules, mobile robotic devices, sensor system objects, respectively, and A is description of the agricultural area in question. Then, formal statement of the problem «wireless data transmission network establishment» can be expressed as follows: knowing the radio module distribution D and robotic device motion paths R within area A, data transmission has to be established from the sensor system objects S and robotic devices R to some gateway B with B_{xy} coordinates with data transmission speed equal or above a specified limit L. A network router is used as a gateway in this setting, and it connects the network of LoRa radio modules with the sensor complex consumer (server).

The area A is described with a tuple of following terms:

$$A = \langle F, B_{xy}, Q, W \rangle, \tag{26.3}$$

where F is a topological map of an agricultural area, where specific landmarks denote locations of sensor radio modules, accompanied by their coordinates, B_{xy} are gateway coordinates, Q are data repeater coordinates, W is description of data transmission routing, schedule of data receive, and transmission workflow for each radio module and repeater.

The main challenge within this task consists in maintenance of data transmission rate equal or above the specified value in context of variable distances between the radio modules and robotic device motion. Therefore, delivery of a data batch tri from each radio module or repeater to the gateway is described by the following tuple of parameters:

$$tr_i = \langle Z_i, B_{xy}, X_i, C_i, V \rangle, \tag{26.4}$$

where Z_i are transmitting radio module coordinates, i is repeater number in the radio module chain, X_i is data transmission delay from the ith repeater, C_i is the count of the repeater modules from the ith module to the gateway, and V is the schedule of data reception and transmission workflow.

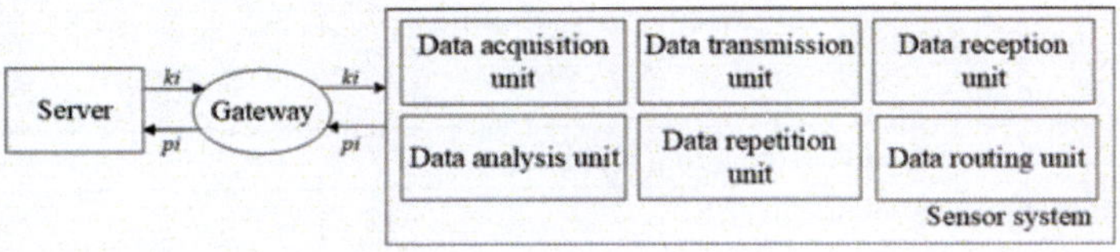

Fig. 26.2 Structural diagram of a sensor complex

A structural diagram of the sensor complex is depicted in Fig. 26.2.

As follows from the figure, the model includes a server, a gateway, and a sensor system, where k_i is a control signal and p_i is data obtained from the sensors. Hardware components of the complex can be divided into three main groups: (1) sensor devices, essential bundles of sensory elements, reading features, and parameters of the inspected object; (2) robotic devices, performing given functions in the agricultural area; (3) wireless data transmission network, consisting of a gateway, radio modules, repeaters, delivering data batches from sensor systems and robotic devices to the server, and control commands—from the server to the devices. As the hardware of the sensor complex includes many radio modules, as well mobile units (robotic devices), the sensor complex model can be described by the following tuple of parameters:

$$y_j = \langle Z_i, C_i, V, f_i, B_{xy}, RS_i, G_i, Q \rangle, \tag{26.5}$$

where Z_i are coordinates of the transmitting radio module, C_i is count of the repeater modules up to the gateway, V is schedule of data reception and transmission, f_i is current radio module performance mode, RS_i is robotic device motion velocity, G_i are additional parameters (charge level of batteries and other embedded hardware), necessary for radio module functioning, and Q are data of current positions and performance modes of all service systems within the facility.

In the proposed model, the current performance mode is selected from the following set:

$$f = \{f_a, f_b, f_c, f_d\}, \tag{26.6}$$

where f_a is sleep mode, f_b is transmitter mode, f_c is receiver mode, and f_d is repeater mode.

The interaction between the sensor system y_i and server u results in data representation E about sensor system object or robotic device (O_n):

$$E(u) = f(y_i, A, O_n). \tag{26.7}$$

For better understanding of the proposed solutions, we will aggregate all the introduced terms in the comprehensive list, as presented in Table 26.1.

Table 26.1 List of terms

Term	Description
D	Set of radio modules, including repeaters
R	Set of mobile robotic devices
S	Set of sensor system components (sensors)
A	Description of serviced agricultural area
Z_i	Coordinates of repeater radio module
X_i	Data transmission delay
C_i	Count of repeater modules to gateway
V	Schedule of data reception and transmission
f_i	Current radio module performance mode
O_n	Set of objects on the serviced area
F	Topological map of the agricultural area
G_i	Additional parameters (level of charge in batteries and other embedded hardware), necessary for the radio module function
Q	Data of current position and performance modes of all service system within the facility
RS_i	Robotic device motion velocity
B_{xy}	Gateway coordinates
f_a	Sleep mode
f_b	Transmitter mode
f_c	Receiver mode
f_d	Repeater mode
E	Data about object, obtained from sensors

26.4 Experiments

Based on the presented models, a software for automated calculation of coordinates of additional modules and for establishment of radio communication with all the modules, as well for transfer of data, obtained with sensors was developed. To test the repeater layout recommended by the developed software, consider an example. The test criterion in this case is the energy efficiency of the recommended layout, depending on the number of repeaters and gateway position. We have an area with known coordinates of gateway and radio modules of sensor systems, situated in arbitrary points and robot path. The respective example is shown in Fig. 26.3.

Red point on this figure indicates the gateway, blue points indicate the radio modules of sensor systems, and blue lines indicate the robotic device paths. Having run the software, we obtain coordinates of the additional modules (repeaters), and then, we get the following map of the area under consideration (see Fig. 26.4).

Fig. 26.3 Initial conditions

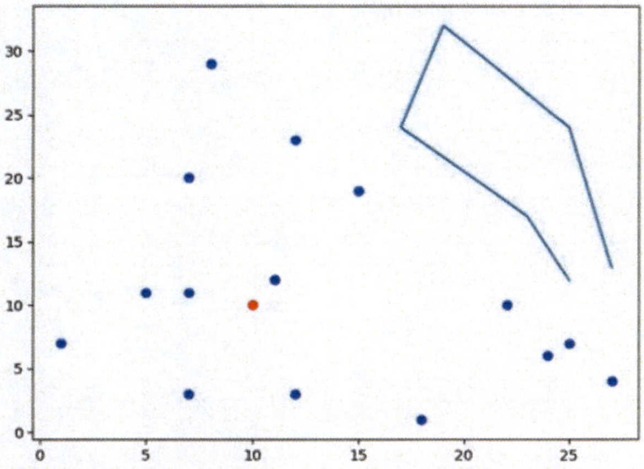

Fig. 26.4 Area map with coordinates of additional radio modules

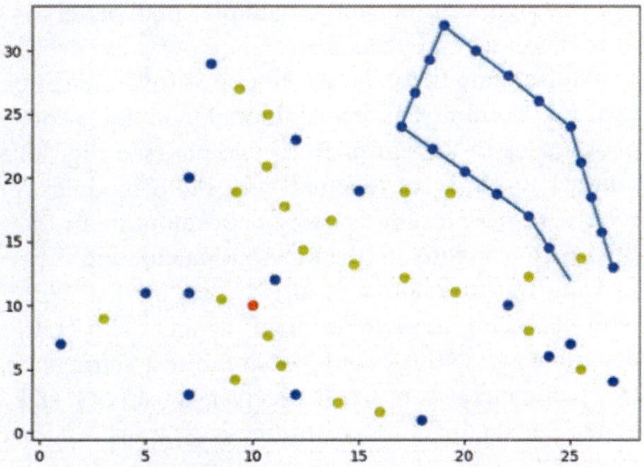

In Fig. 26.4, green points indicate the recommended locations of the additional modules for encompassing of the sensor system radio modules into the common network with the gateway.

Here, the software recommends to add 23 radio modules for establishment of the common network of all modules and 17 additional modules to maintain permanent connection with the robotic device. Hence, it amounts to 40 additional modules.

For testing purposes, Heltec ESP-32 modules were used, based on LoRa data transmission technology. According to the data, supplied in the technical documentation on this radio module, energy consumption value in sleep mode (800×10^{-6} W) is given, as the most of the time the sensor will remain not in active (transmitter) mode, but in the low energy consumption mode, it will act as a receiver or remain in sleep mode and transmit data at certain intervals. Hence, approximate energy consumption value for all the additional radio modules can be calculated. In the case described here, the energy consumption value will be equal to 32×10^{-3} W, what is

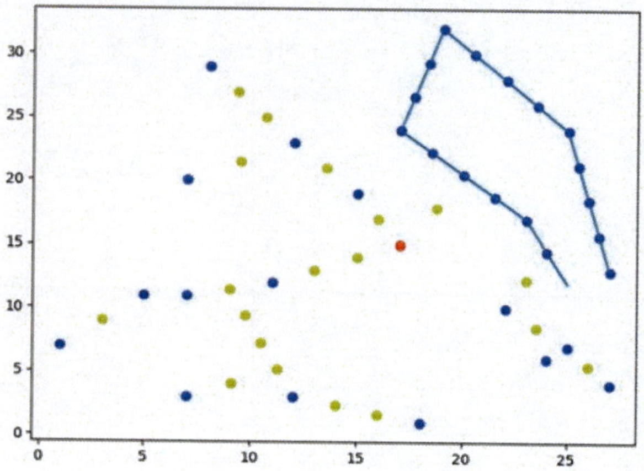

Fig. 26.5 Gateway position alteration

little enough. Consequently, a simple small battery is sufficient for the radio module to function several years long.

At the same time, the developed software enables to establish the best gateway position, to employ as few additional modules, as possible. If we change the gateway coordinates in the considered example (see Fig. 26.5), we can reduce the module number to 19, what together with radio modules, enabling connectivity with the robotic device, results in energy consumption of 28.8×10^{-3} W, what represents a little gain compared to the previous calculation.

Data transmission was also tested in real-world environment, in the system, built according to the developed models, on 5 Heltec ESP-32 LoRa modules. The experimental settings conform to the test settings described above. The following test parameters were used: receiver sensitivity and data transfer speed alteration depending on the distance, test message transmission delay on the way from the first module to the gateway, validity of routing tables, and data transmission/reception schedule. Two experiments with different module layouts were conducted. In the experimental settings, the modules were spaced at 500 m from each other. For the chosen modules at such distance, receiver sensitivity of −105 dBm should be maintained, what corresponds to data transmission rate of 40 kbit/s. In both cases, the same routing table was used for one by one data transmission from the fifth (terminal) radio module to the gateway (the first radio module). The routing table looks as follows (see Table 26.2).

The data reception/transmission schedule for each sensor was prepared. For the sake of experiment simplicity, data collection timeframe was taken to be equal to 1 min, i.e., module E transmitted a single data batch every minute. The schedule included alternate periods of inter-module data transmission 10 s each, separated by 2-s air silence periods. Hence, while the radio modules do not transmit messages, they function in receiver mode. Consequently, every minute module E transmits a test message to the module D within 10 s, and after that an air silence period of 2 s comes. Further module D transmits a test message to the module C within 10 s, and

Table 26.2 Routing table for module E

Step	Total data transmission cost	Next router
No. 1	0	E
No. 2	5	D
No. 3	10	C
No. 4	15	B
No. 5	20	A

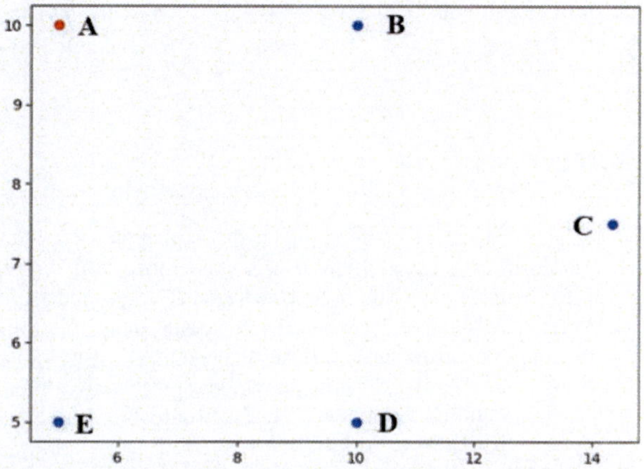

Fig. 26.6 Module layout in the second experiment

then, silence period follows, and so on. This schedule was intended to reduce network noise and avoid collisions. Experimental layout of modules is shown in Fig. 26.6.

It is noteworthy that module A could initially receive message from the module E because of close proximity between them, but such case was excluded programmatically, i.e., modules accepted messages only from certain senders and ignored batches from other "addresses."

At the start of the experiment, test messages successfully reached the module A from the module E, and receiver sensitivity was in the range of −100 to −108 dBm. But, after a few minutes, certain anomalies in the system arose, having to do with simultaneous message transmission by several modules. This occurred because of the limited hardware capacity of the selected modules. These modules are not intended for precision timing, and non-synchronized module activation also caused timing errors. To solve this problem, GPS units should be connected to each radio module for precise timing.

26.5 Conclusion

In this paper, conceptual and set-theoretic models of informational interaction of sensor complexes which differ in parameter registration approaches in context of interaction with heterogeneous sensor elements and mobile robotic devices were considered. These models ensure connection of the modules to the wireless network on a topological map of the serviced area, and LoRa-based interactions for applied problem solving in agricultural domain.

Further research will be concerned with the enhancement of proposed models in automation of their workflows and informational interactions of robotic devices with distributed sensor networks, noise immunity, and accounting for collisions (simultaneous data transfer from two or more repeaters to one radio module).

Acknowledgements This research is supported by the RFBR No. 18-58-76001 Era_a.

References

1. Magno, M., et al.: Combining LoRa and RTK to achieve a high precision self-sustaining geo-localization system. In: Proceedings of the 17th ACM/IEEE International Conference on Information Processing in Sensor Networks, pp. 160–161. IEEE, New York (2018)
2. Pütz, S., Wiemann, T., Sprickerhof, J., Hertzberg, J.: 3d navigation mesh generation for path planning in uneven terrain. IFAC-PapersOnLine **49**(15), 212–217 (2016)
3. Rabie, T., Suleiman, S.: A novel wireless mesh network for indoor robotic navigation. In: 2016 5th International Conference on Electronic Devices, Systems and Applications (ICEDSA), pp. 1–4. IEEE, New York (2016)
4. Nykorak, A., Hiromoto, R. E., Sachenko, A., Koval, V.: A wireless navigation system with no external positions. In: 2015 IEEE 8th International Conference on Intelligent Data Acquisition and Advanced Computing Systems: Technology and Applications (IDAACS), vol. 2, pp. 898–901. IEEE, New York (2015)
5. Konieczny, M., Pawłowicz, B., Potencki, J., Skoczylas, M.: Application of RFID technology in navigation of mobile robot. In: 2017 21st European Microelectronics and Packaging Conference (EMPC) & Exhibition, pp. 1–4. IEEE, New York (2017)
6. Malandra, F., Sansò, B.: A Markov-modulated end-to-end delay analysis of large-scale RF mesh networks with time-slotted ALOHA and FHSS for smart grid applications. IEEE Trans. Wireless Commun. **17**(11), 7116–7127 (2018)
7. Lavric, A., Popa, V.: Internet of things and LoRa™ low-power wide-area networks: a survey. In: 2017 International Symposium on Signals, Circuits and Systems (ISSCS), pp. 1–5. IEEE, New York (2017)
8. Garrido-Hidalgo, C. et al.: IoT Heterogeneous Mesh Network Deployment for Human-in-the-Loop Challenges Towards a Social and Sustainable Industry 4.0. IEEE Access (2018)
9. Leon, E., Alberoni, C., Wister, M., Hernández-Nolasco, J.: Flood early warning system by twitter using LoRa. In: Multidisciplinary Digital Publishing Institute Proceedings, vol. 2, no. 19, p. 1213 (2018)
10. Koucheryavy, A.E., et al.: Flying sensor networks. Elektrosvyaz **9**, 2–5 (2014)
11. Koucheryavy, A.E., Vladyko, A.G., Kirichek, R.V.: Theoretical and practical directions of research in the field of flying sensor networks. Elektrosvyaz **7**, 9 (2015)
12. Krasnov, P.A., Roslyakov, A.V.: Model of ZigBee wireless sensor network. Problems of engineering and telecommunications technology. In: Optical Technologies in Telecommunications, pp. 176–177 (2018)
13. Efremov, D.A., Roslyakov, A.V.: Network calculus modeling of BLE sensor network. In: Optical Technologies in Telecommunications, pp. 154–156 (2018)

14. Vyrelkin, A.D., Kucheryavyj, A.E., Prokop'ev, A.V.: Investigation of the possibility of using an unmanned aerial vehicle as a temporary head node for clusters of the ground sensor network. Inf. Technol. Telecommun. **1**, 27–34 (2015)
15. Zayats, A.M., Logachev, A.A.: The information system of monitoring forests and wildfires using the wireless sensor networks. Izvestia Sankt-Peterburgskoj Lesotehniceskoj Akademii **216**, 241–254 (2016)
16. Kirichek, R.V., Vladyko, A.G., Zaharov, M.V., Koucheryavy, A.E.: Model networks for internet of things and SDN. Inf. Technol. Telecommu. **3**, 17–26 (2015)
17. Chernikhovsky, D.M., Alekseev, A.S.: Methods for processing information of radar topographic survey for solving the tasks for determining heights and wood stocks for forest stands. SPIIRAS Proc. **18**(2), 416–443 (2019)
18. Astafurov, V.G., Skorokhodov, A.V.: Formation of a set of informative classification features for solving cloud classification problem using MODIS satellite data. SPIIRAS Proc. **4**(53), 118–139 (2017)
19. Shakhnovich, I.: The myth of the attenuation of free space: what GT Friis did not write. First Mile **2**, 40–45 (2014)
20. Denisov, A.V., et al.: Algorithms for radio beacon mesh network establishment for navigation of robotic systems in agriculture. Vestnik MSTU «Stankin» **3**(50), 57–65 (2019)
21. Denisov, A.V.: Development of the recommendation system for calculating the parameters of a wireless network of sensor complexes. MOIT **7**(4) (2019)
22. Denisov, A., Iakovlev, R., Lebedev, I.: Mathematical and algorithmic model for local navigation of mobile platform and UAV using radio beacons. In: International Conference on Interactive Collaborative Robotics, pp. 53–62 (2019)
23. Kalinin, A.I., Cherenkova, E.L.: Radio wave propagation and radio link operation. Communication (1971)

Chapter 27
Multi-robot Coalition Formation for Precision Agriculture Scenario Based on Gazebo Simulator

Nikolay Teslya, Alexander Smirnov, Artem Ionov, and Alexander Kudrov

Abstract Nowadays, the study of interaction models of intelligent agents is one of the main research directions in the field of collaborative robotics. It includes studies of coalition formation principles, tasks decomposition and distribution, winnings sharing, and implementation of proposed techniques and models. This work focuses on presenting environment and robots states in smart space during joint task solving as well as modelling and visualizing the interaction process using open software Gazebo and Robotic Operation System. The ontology of robot is presented that combines description of the robot equipment and physical characteristics. To provide variability in robot interaction, some of the concepts are evaluated with fuzzy sets. The architecture of modelling approach is presented based on the combination of smart-space concept for ontology-based information exchange between robots, Robotic Operation System for robot control, and Gazebo simulator for modelling and visualizing robot actions in 3D environment. The example of interaction is presented based on precision agriculture scenario.

27.1 Introduction

Nowadays, mobile robots are becoming more and more demanded in many areas of our life. They are actively used to solve various tasks such as scouting, assistance

---

N. Teslya (✉) · A. Smirnov
SPIIRAS, 39, 14th Line, 199178 St. Petersburg, Russia
e-mail: teslya@iias.spb.su

A. Smirnov
e-mail: smir@iias.spb.su

A. Ionov · A. Kudrov
ITMO University, 49, Kronverksky pr., 197101 St. Petersburg, Russia
e-mail: ionovartem98@gmail.com

A. Kudrov
e-mail: sasha.qudrov@gmail.com

A. Ronzhin and V. Shishlakov (eds.), *Proceedings of 15th International Conference on Electromechanics and Robotics "Zavalishin's Readings"*, Smart Innovation, Systems and Technologies 187, https://doi.org/10.1007/978-981-15-5580-0_27

in technological accidents and disasters, control and surveillance in the enterprise, concepts of the smart city and smart farming. Using robots is useful to perform time-consuming tasks in adverse conditions for humans.

In order to solve complex and difficult tasks can be used multi-robotic systems, consisting of many robots, intelligent agents, connected to intelligent space [1, 2]. Collaborative robotics is more effective than single robots: extended functionality achieved by the diversity of robots and their functions, higher productivity in tasks by delegating different tasks to different robots and extended workspace due to allocation of robots in the area. It also decreases a cost of each robot since there is no more need to equip one robot with all required equipment.

One of the most demonstrative cases of collaborative robotics is a precision agriculture. To date, many highly specialized robots have been developed. They effectively solve a limited set of tasks within the framework of the conditions for which they were developed [3–7]. For example, there are robots that can conduct soil quality exploration, automated seeders, cultivators, tractors, etc. [3, 8–10]. However, to solve a complex problem that goes beyond the conditions of each robot, it is required to ensure their joint work. So, to ensure the full cycle of growing crops, it is necessary to successively solve the problems of field exploration, selection of crops grown, sowing, watering and harvesting.

Testing robot coalition scenarios in real equipment is highly limited due to the high cost of real robots, requirements of great areas to test as well as low speed of overall scenario due to plants growth. To increase the testing speed, various modelling and simulation methods are used. They allow creating model that reflects main specific of the problem and proposed solution as well as increasing testing speed.

This paper proposes an approach for modelling and visualizing multi-robot coalition formation based on integration of three technologies: smart spaces, Gazebo simulator and Robotics Operation System. The coalition formation principle is based on the previous research [11]. The research involves fuzzy cooperative games to calculate overall benefit of coalition and form coalition from the set of available robots in the way to maximize overall benefit and benefit of each coalition participant. This work continues previous research by presenting the example of robot's characteristics description with fuzzy sets and modelling the interaction in three-dimensional virtual environment.

The rest of the paper is structured as follows. Section 27.2 briefly describes related works in intelligent agriculture systems, cooperation in multi-agent systems and robots interaction modelling and visualization techniques. Section 27.3 provides the approach description with ontology for robot, fuzzy sets examples for some of the robot characteristics and modelling platform architecture. Section 27.4 provides a precision agriculture scenario and its implementation. The paper is concluded with discussion of the proposed approach and directions for the future work.

27.2 Related Work

27.2.1 Cooperation in Multi-agent Systems

Intelligent communication and interaction of intelligence agents including physical work cells, robots, intelligent components and control units requires cooperation model development and research for more effective task management.

In the paper [12], a wide review was conducted over the existing multi-robot systems, as well as levels of automatization were described. The authors of the article had identified three levels of automation in multi-robotic systems: fully automated, semi-automated and not automated. As a result of the review, it was found that human still plays critical role in the management, decomposition and distribution of tasks.

In the paper [13], the method of coalition formation was designed to improve collaborative search and prosecution. The algorithm was developed, taking into account the constraints and delay of communication when searching for potential coalition members. For autonomous work of the coalition in search and prosecution tasks was developed state machine mechanism so the functions of decomposition and distribution of tasks lie on the human.

The other mechanism is described in the paper [14]. The hybrid mechanism is based on fuzzy logic and swarm intelligence for movement coordination of several UAVs. Using techniques of fuzzy logics distance between UAVs and to the target, safety of interaction is described. Based on fuzzy sets, a base of fuzzy rules was created in the form of a procedural knowledge base.

The information exchange between agents in multi-agent system is also an important component of the joint problem solving. It provides notification of agents about the current state of the problem solution for organizing coordinated actions, or monitoring the implementation of the plan [15]. The information exchange can be organized through a common centralized repository of information on a separate device or by the distribution of information between coalition members [16]. It also considers the combination of two approaches with the formation of so-called smart spaces—the creation of a common repository of information that provides links to resources that are coalition members, which makes information distributed among all participants [17].

27.2.2 Robot Interaction Modelling Methods

First, a robot middleware system has to be chosen. There are not that many systems which are ready to use; therefore, developing our own middleware also had been considered. The middleware comparison is presented in Table 27.1.

As a middleware system for robots, the ROS has been chosen as it is de facto industry standard. The ROS has detailed documentation, provides high flexibility in function implementing, integration with other systems and big community which

Table 27.1 Robot middleware comparison

Middleware	Advantages	Disadvantages
ROS [18]	• High flexibility • Integration with other systems • Big community • Well documented • Supports Python and C++ scripts	• Platform-dependent • Not easy to learn
Mobile robot programming toolkit [19]	• Cross-platform • Visualization tools included	• Supports only C++ • Has small community • Not fully documented • Hard to learn
Developing own middleware	• Highly configurable	• Time inefficient • Requires integration with visualization framework

speeds up the development speed. Moreover, it is used in the real robots as an operation system that makes it possible to transfer model between virtual and real environments with minimal changes.

When developing any system, including robot interaction system, it is very important to plan the ways of testing this system in advance. In case of this paper, there are basically two ways: to using a robot simulation system or implementing the developed robot interaction model on real constructed robots. Simulation environment for multi-robot coalition formation testing had been chosen due to several reasons:

- It is cheaper compared to building real robots;
- This way replacing robot implementations becomes much easier so it is possible to test it on different kinds of robots and scenarios;
- In a simulation environment, it is much easier to keep track of all kinds of target robot performance metrics and to analyse robot behaviour in details.

It should be noted that choosing to simulate robot interaction does not exclude the fact that it is just an additional step for testing convenience before implementing multi-robot coalition formation on real robots.

Let us observe most popular robot simulation systems.

- V-REP is a feature-rich, flexible and scalable simulation environment. However, it is the most resource demanding [20].
- ARGoS is a lightweight robot simulation platform. It is well suited for simulating swarm robotics but has fewer capabilities compared to V-REP [21].
- Gazebo is somewhere between V-REP and ARGoS. It has many functions but with a simpler interface and higher number of standard robot models and also provides high performance for simulating and visualizing large scenes [22, 23].

The comparison of simulation systems is presented in Table 27.2. Based on the

Table 27.2 Robots simulation systems comparison

Robot simulation system	Advantages	Disadvantages
V-REP	• Feature-rich • Flexible • Scalable	• Resource-demanding • Not well integrated with ROS
ARGoS	• Lightweight • Well suited for swarm robotics	• Less capabilities • Not well integrated with ROS
Gazebo	• Simple interface • Powerful tool • High performance • A lot of standard models • Well documented • Full ROS integration supported	–

comparison Gazebo is chosen for the implementation of multi-robot coalition formation for precision agriculture scenario as it is fully integrated with ROS, provides many standard models and is well documented.

27.3 An Approach to Multi-robot Coalition Formation Modelling and Visualization

Approach realization consists of four steps: creation of knowledge bases for robot and environment based on Smart-M3 system [24] and fuzzy sets algorithm; development of an algorithm for creating a coalition of intelligent agents; realization of agricultural scene in ROS/Gazebo environment; and integration of Smart-M3 and ROS/Gazebo.

For the first step has generalized ontology, based on the IEEE Standard Ontology for Robotics and Automation [25]. For the precision agriculture scenario, it was decided to add some descriptive entities that will be used in creation of fuzzy rules knowledge base. A "State" class that presents complete state of a robot has been added. "DeviceCompleteness" is a class that summarizes complete robot specification on a functional level: percentage and state of charge, potential movement speed, set of sensors. "GoalReachability" is a class that summarizes the reachability of the task by a specific robot with respect to the physical characteristics of the target. The ontology is presented in Fig. 27.1.

27.3.1 Architecture

The modelling and visualization stand consists of two big nodes: server with a running Smart-M3 platform and client, which acts as a container for many robots. Let us observe each of them in details (Fig. 27.2).

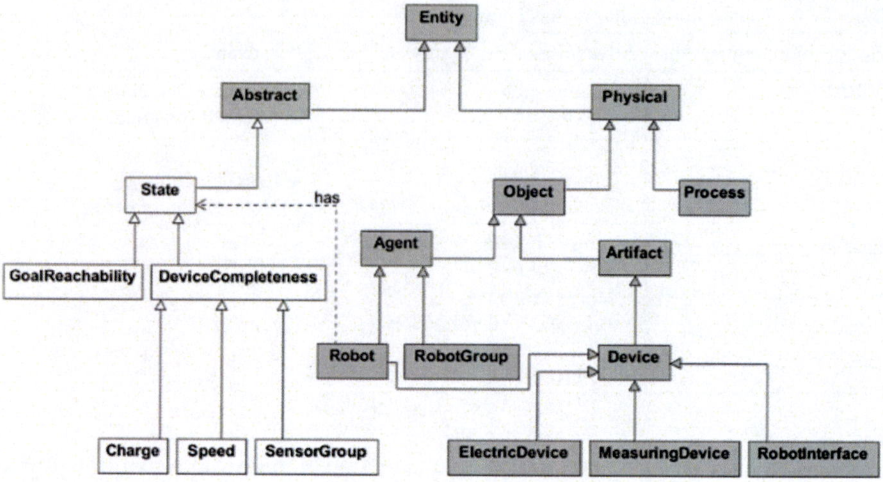

Fig. 27.1 Robot ontology for information sharing through Smart-M3

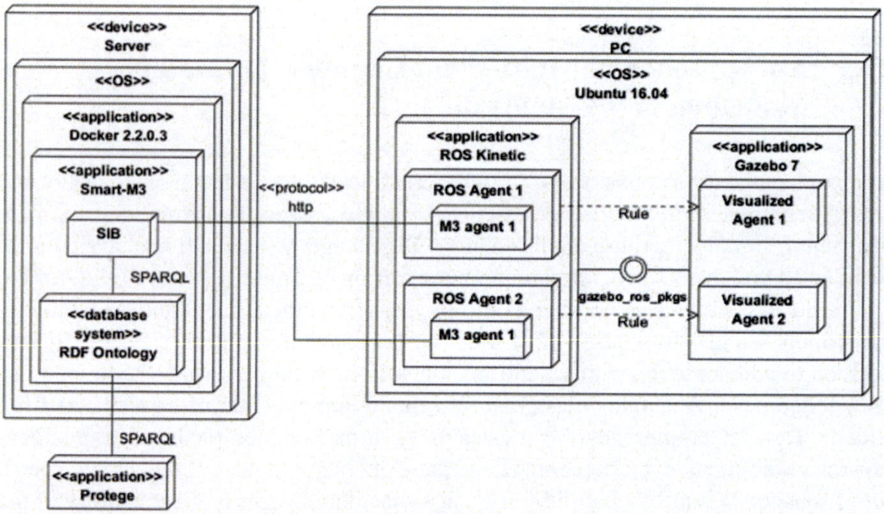

Fig. 27.2 Modelling and visualization architecture

Server runs the Smart-M3 platform in a Docker container. This solution makes smart-space infrastructure platform independent from the environment so it can run on any OS. All robots interact through the smart space using the ontology created via Protégé tool and uploaded to every robot [26].

Client contains ROS and Gazebo environments. They are integrated via *gazebo_ros_pkgs* package. For the precision agriculture scenario implementation, an open source models of harvester, scour and quadrotor was used. Some of them are already implemented as ROS models, for instance, Husky robot [27] as scout and quadrotor [28]. The model of harvester was created based on free 3D-model with harvester shape extended by control configuration in Gazebo and ROS environment. Each robot is associated with a set of ROS topics, which are used to send commands to a single robot or to gather information from robot's sensors. Each robot also has a SDF model which is used for simulation in Gazebo. Control programme for each robot is implemented as a knowledge processor (KP) that provides connection to smart space and ROS. It uses ontologies to store and exchange information.

Client and server interact over TCP/IP. They exchange information such as robots' locations, field parameters, tasks and other as RDF triples. Robot state is described in the form of fuzzy rules. Examples of parametric sets for charge and distance classes are presented below.

27.3.2 *Level of a Charge*

One of the rules that robots use during coalition formation is a level of battery charge. It allows to find robots that are able to complete a task and return back to charging station or continue doing other tasks. Percentage of robot's accumulator charge is received from power source sensors. Universal set is from 0% to 100%, respectively. Set was divided on three equal intervals. According to the intervals, three terms were identified: "Low", "Medium" and "High". The ranges and equations for calculating level of charge are presented in Fig. 27.3 and Table 27.3.

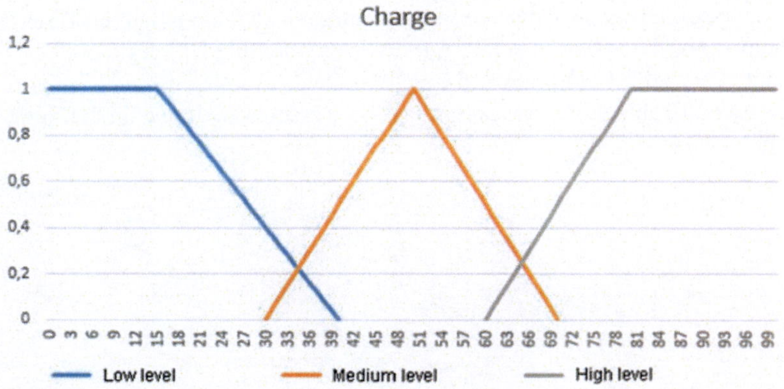

Fig. 27.3 Robot battery-level fuzzy set

Table 27.3 Robot battery-level fuzzy set equations

Term	Formula
Low level of a charge	$f(x) = \begin{Bmatrix} 1, & x \leq 15 \\ \frac{(40-x)}{(40-15)} & 40 > x > 15 \\ 0, & x \geq 15 \end{Bmatrix}$
Medium level of a charge	$f(x) = \begin{Bmatrix} 0, & x \leq 30 \\ \frac{(x-30)}{(50-30)}, & 50 \geq x > 30 \\ \frac{(70-x)}{(70-50)}, & 70 > x > 50 \\ 0, & x > 70 \end{Bmatrix}$
High level of a charge	$f(x) = \begin{Bmatrix} 0, & x \leq 60 \\ \frac{(x-60)}{(80-60)}, & 80 > x > 60 \\ 1, & x \geq 80 \end{Bmatrix}$

27.3.3 *Distance*

To prevent collisions between robots, the distance measurements are used. Since the visualization as well as real implementation will be controlled by human operator, the distance should be presented by language variable that clearly defines how close are robots to each other. On the other hand, coalition norms define that robots should keep some distance between each other, and it should not be a fixed value [12]. But fuzzy set theory allows to define distance that will be clear both for robots and human operator. Distance between robot and target is defined in universal set from 0 to N-metres. For the universal set, four terms were identified: "Very close", "Close", "Far" and "Very far". The ranges and equations for calculating distance are presented in Fig. 27.4 and Table 27.4. Values of N_1–N_7 are estimating based on the robot type.

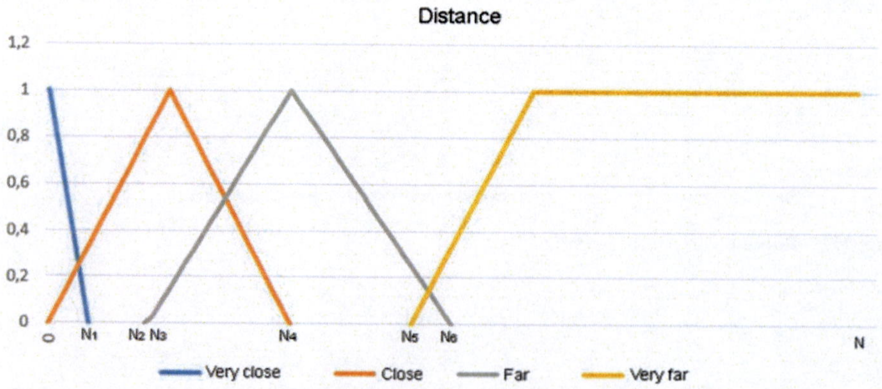

Fig. 27.4 Robot distance measurement fuzzy set

Table 27.4 Robot distance measurement fuzzy set equations

Term	Formula
Very close	$f(x) = \begin{cases} \frac{(N_1 - x)}{(N_1)}, & N_1 \geq x > 0 \\ 0, & x \geq N_1 \end{cases}$
Close	$f(x) = \begin{cases} \frac{x}{N_3}, & N_3 \geq x > 0 \\ \frac{(N_4 - x)}{(N_4 - N_3)} & N_4 > x > N_3 \\ 0, & x \geq N_4 \end{cases}$
Far	$f(x) = \begin{cases} 0, & x \leq N_2 \\ \frac{(x - N_2)}{(N_4 - N_2)} & N_4 \geq x > N_2 \\ \frac{(N_6 - x)}{(N_6 - N_4)} & N_6 > x > N_4 \\ 0, & x \geq N_6 \end{cases}$
Very far	$f(x) = \begin{cases} 0, & x \leq N_5 \\ \frac{(N_7 - x)}{(N_4 - N_5)}, & N_7 > x > N_5 \\ 1, & x \geq N_7 \end{cases}$

27.4 Scenario Implementation

To test the modelling and visualization approach for multi-robot coalition formation, the precision agriculture scenario has been used. Different parts of the field might need different crops for the efficiency purposes. This could be automated using intelligent agriculture robots cooperation. First, field needs to be explored to gather information about the soil structure, light and humidity conditions. In order to execute this task successfully, a special type of robot needs to be constructed; it should have all the necessary tools and sensors to be able to gain all the needed information. Second, when the field condition data is processed, the field should be sowed with different types of seeds. Eventually, all the crops should be harvested, which also needs a different type of robot. The scene is shown in Fig. 27.5.

All the scenario steps would benefit from implementing robot interaction through coalition. During exploration stage, robots could split the field into parts so that different robots would be attached to different zones which would speed up the exploration. During the sowing, robots should form coalitions and split the work between each other to maximize the efficiency of the field processing. During harvesting process, robots could also speed up this stage by forming coalitions. The scenario timeline is presented in Fig. 27.6.

For the scenario implementation, Husky robots were used as scout robot. It is equipped with laser scanner to navigate through the surrounding environment. All measured distances are transformed to the fuzzy sets according to equations in Table 27.4 and can be visualized as shown in Fig. 27.7. Line colours in Fig. 27.7 correspond to colours of fuzzy sets, presented in Fig. 27.4.

Fig. 27.5 Precision agriculture scene in Gazebo simulator

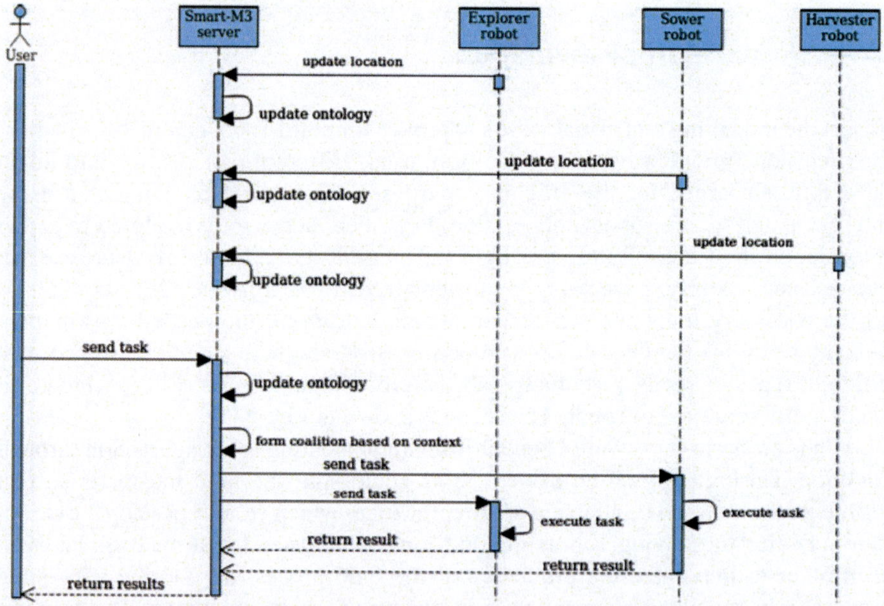

Fig. 27.6 Precision agriculture scenario timeline

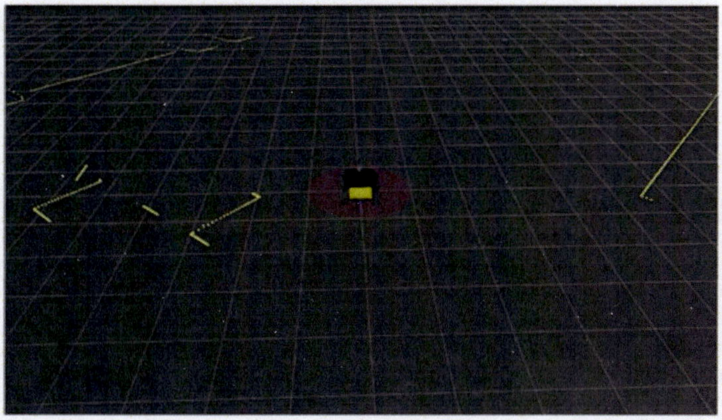

Fig. 27.7 Robot view of environment with fuzzy sets of distance

Quadrotor robots implement aerial scouting. Both Husky and quadrotor can use integrated navigational stack that allows to simplify the moving to the destination point. To do this, the robot have to post the location message "geometry_msgs/PoseStamped" to the according ROS topic "/move_base_simple/goal". Also own navigation algorithms can be developed to provide task-specific navigation [29].

27.5 Conclusion

The paper provides an approach for modelling and visualization of multi-robot coalition creation for agriculture scenario. The modelling is based on the open-source packages Gazebo and ROS that are de facto standards of virtual environments modelling in visualization in the field of robotics. The interaction between robots is implemented based on the smart space combined with the blockchain platform.

As it was shown by comparison integration of ROS and Gazebo is one of the best for modelling and visualization from the point of the performance and extensibility. The paper also shows how it can be integrated with smart-space platform to provide information exchange between robots. Information exchange is based on the ontology that provides interoperability between various robots.

The examples of fuzzy sets have showed that information exchange between robots can also use uncertain definitions to increase variability of possible coalitions. Also, it helps human operators to understand robots' behaviour.

The future work will be focused on scenario performance evaluation to prove that the proposed solution is appropriate for modelling and visualizing multi-robot cooperation and coalition formation.

Acknowledgements This research was funded by RFBR Grant No. 17-29-07073 for creating multi-robot coalition for precision agriculture and by Russian State Research No. 0073-2019-0005 in the field of multi-agent interaction and cooperation.

References

1. Bayram, H., Bozma, H.I.: Coalition formation games for dynamic multirobot tasks. Int. J. Rob. Res. **35**(5), 514–527 (2015). https://doi.org/10.1177/0278364915595707
2. Andreev, V.P., Pletenev, P.F.: Method of information interaction for distributed control systems of robots with modular architecture. SPIIRAS Proc. **2**(57), 134–160 (2018). https://doi.org/10.15622/sp.57.6
3. Bao, Y., Tang, L., Breitzman, M.W., Salas Fernandez, M.G., Schnable, P.S.: Field-based robotic phenotyping of sorghum plant architecture using stereo vision. J. Field Rob. **36**(2), 397–415 (2019). https://doi.org/10.1002/rob.21830
4. Pantazi, X.E, Moshou, D., Bochtis, D.: Artificial intelligence in agriculture. In: Intelligent Data Mining and Fusion Systems in Agriculture, pp. 17–101 (2020)
5. Ball, D., et al.: Vision-based obstacle detection and navigation for an agricultural robot. J. Field Rob. **33**(8), 1107–1130 (2016). https://doi.org/10.1002/rob.21644
6. Mohanraj, I., Ashokumar, K., Naren, J.: Field monitoring and automation using IOT in agriculture domain. In: Procedia Computer Science, pp. 931–939 (2016)
7. Jha, K., Doshi, A., Patel, P., Shah, M.: A comprehensive review on automation in agriculture using artificial intelligence. Artif. Intell. Agric. (2019). https://doi.org/10.1016/j.aiia.2019.05.004
8. Ren, G., Lin, T., Ying, Y., Chowdhary, G., Ting, K.C.: Agricultural robotics research applicable to poultry production: A review. Comput. Electron. Agric. **169**, 105216 (2020)
9. Bac, C.W., van Henten, E.J., Hemming, J., Edan, Y.: Harvesting robots for high-value crops: state-of-the-art review and challenges ahead. J. Field Rob. **31**(6), 888–911 (2014)
10. Bac, C.W., et al.: Performance evaluation of a harvesting robot for sweet pepper. J. Field Rob. **34**, 1123–1139 (2017). https://doi.org/10.1002/rob.21709
11. Smirnov, A., Sheremetov, L., Teslya, N. Fuzzy cooperative games usage in smart contracts for dynamic robot coalition formation: Approach and use case description. In: ICEIS 2019—Proceedings of the 21st International Conference on Enterprise Information Systems. SCITEPRESS—Science and Technology Publications, pp. 349–358 (2019)
12. Rizk, Y., Awad, M., Tunstel, E.W.: Cooperative heterogeneous multi-robot systems: a survey. ACM Comput. Surv. (CSUR) **52**(2), 1–31 (2019)
13. Liu, Z., Gao, X.G., Fu, X.W.: Coalition formation for multiple heterogeneous UAVs in unknown environment. In: Proceedings—5th International Conference on Instrumentation and Measurement, Computer, Communication, and Control, IMCCC 2015, pp. 1222–1227 (2015)
14. De Oliveira, L.F.F., De Lima, F.B., Oliveira, S.C., Bastos-Filho, C.J.A.: A fuzzy-swarm based approach for the coordination of unmanned aerial vehicles. J. Intell. Fuzzy Syst. **31**(3), 1513–1520 (2016). https://doi.org/10.3233/JIFS-151214
15. Verma, D., Desai, N., Preece, A., Taylor, I. A block chain based architecture for asset management in coalition operations. Ground/Air Multisensor Interoperability, Integration, and Networking for Persistent ISR VIII. International Society for Optics and Photonics 10190, 101900Y (2017)
16. Shabanov, V., Ivanov, D.: Organization of information exchange in coalitions of intelligent mobile robots. In: International Conference on Industrial Engineering, Applications and Manufacturing ICIEAM 2019, pp. 1–5 (2019). https://doi.org/10.1109/ICIEAM.2019.8743043
17. Ferrer, E.C.: The blockchain: a new framework for robotic swarm systems. Adv. Intell. Syst. Comput. **881**, 1037–1058 (2019). https://doi.org/10.1007/978-3-030-02683-7_77

18. Stanford Artificial Intelligence Laboratory et.al. (2018) ROS.org | Powering the world's robots. https://www.ros.org/. Accessed 5 Jan 2020
19. Claraco, J.L.B.: Development of Scientific Applications with the Mobile Robot Programming Toolkit. MRPT Ref book Mach Percept (2010)
20. Rohmer, E., Singh, S.P.N., Freese, M. V-REP: A versatile and scalable robot simulation framework. In: IEEE International Conference on Intelligent Robots and Systems, pp. 1321–1326 (2013)
21. Pinciroli, C., et al.: ARGoS: a modular, parallel, multi-engine simulator for multi-robot systems. Swarm Intell. **6**, 271–295 (2012). https://doi.org/10.1007/s11721-012-0072-5
22. Koenig, N., Howard, A.: Design and use paradigms for Gazebo, an open-source multi-robot simulator. In: 2004 IEEE/RSJ International Conference on Intelligent Robots and Systems (IROS), pp. 2149–2154 (2004)
23. Gazebo. http://gazebosim.org/. Accessed 28 Feb 2020
24. Korzun, D., Kashevnik, A., Balandin, S.: Novel Design and the Applications of Smart-M3 Platform in the Internet of Things. IGI Global (2018)
25. IEEE Robotics and Automation Society (2015) IEEE Standard Ontologies for Robotics and Automation
26. Smirnov, A., Kashevnik, A., Petrov, M., Parfenov, V.: Context-based coalition creation in human-robot systems: approach and case study. In: Interactive Collaborative Robotics. ICR 2017. Lecture Notes in Computer Science, pp. 229–238 (2017)
27. Gariepy, R., Mukherjee, P., Bovbel, P., Ash, D. GitHub—husky/husky: Common packages for the Clearpath Husky. https://github.com/husky/husky. Accessed 6 Jan 2020
28. Team Hector: https://www.teamhector.de/. Accessed 5 Jan 2020
29. Lavrenov, R.O., Magid, E.A., Matsuno, F., Svinin, M.M., Sutakorn, J.: Development and implementation of spline-based path planning algorithm in ROS/gazebo environment. SPIIRAS Proc. **18**(1), 57–84 (2019). 10.15622/sp.18.1.57-84

Chapter 28
Comparative Analysis of Monocular SLAM Algorithms Using TUM and EuRoC Benchmarks

Eldar Mingachev, Roman Lavrenov, Evgeni Magid, and Mikhail Svinin

Abstract Stable and robust path planning and movement in ground mobile robots require a combination of accuracy and low latency in their state estimation. However, state estimation algorithms must provide these qualities under the computational and power constraints of embedded hardware. Simultaneous localization and mapping (SLAM) algorithms are the best choices for state estimation in these scenarios, in addition to their ability to operate without external localization from motion capture or global positioning systems. Moreover, a single-camera setup is the most common solution for robotic platforms, which reduces our domain of interest to the specific SLAM algorithms type—monocular SLAM. Yet, it is still not clear from the existing literature, which monocular SLAM algorithms perform well under the accuracy, latency, and computational constraints of a ground mobile robot with onboard state estimation. This paper evaluates an array of the most recent publicly available monocular SLAM methods: ORB-SLAM2, DSO, and LDSO. The evaluation considers the pose estimation accuracy (alignment error, absolute trajectory error, and relative pose error) while processing the TUM Mono and EuRoC datasets on the specific hardware platform with a balanced amount of computational resources and power consumption. We present our complete results as a benchmark for the research community.

E. Mingachev · R. Lavrenov (✉) · E. Magid
Laboratory of Intelligent Robotic Systems (LIRS), Intelligent Robotics Department, Higher Institute for Information Technology and Intelligent Systems, Kazan Federal University, Kazan, Russian Federation
e-mail: lavrenov@it.kfu.ru
URL: https://kpfu.ru/erobotics

E. Mingachev
e-mail: ERMingachev@stud.kpfu.ru

E. Magid
e-mail: magid@it.kfu.ru

M. Svinin
College of Information Science and Engineering, Ritsumeikan University, Kyoto, Japan
e-mail: svinin@fc.ritsumei.ac.jp

A. Ronzhin and V. Shishlakov (eds.), *Proceedings of 15th International Conference on Electromechanics and Robotics "Zavalishin's Readings"*, Smart Innovation, Systems and Technologies 187, https://doi.org/10.1007/978-981-15-5580-0_28

28.1 Introduction

Dissanayake, Newman, Clark, Durrant-Whyte, and Csorba [1] defined the simultaneous localization and mapping (SLAM) as an ability for an autonomous vehicle to start in an unknown location in an unknown environment and then, using relative observations only, to incrementally build a map of this environment while simultaneously using this map to compute a bounded estimate of vehicle location.

In general, simultaneous localization and mapping (SLAM) methods estimate the egomotion, the 3D motion of a camera within an environment, using only images of single or multiple cameras. Underlying all formulations is a probabilistic model that takes noisy measurements Y as input and computes an estimator X for the unknown, hidden model parameters (3D world model and camera motion). Typically, a maximum likelihood approach is used, which finds the model parameters that maximize the probability of obtaining the actual measurements [2].

SLAM is currently applied to state and pose estimation problems in a variety of domains, including autonomous vehicles, virtual and augmented reality, and robotics. Delmerico and Scaramuzza also noticed [3] that the field has reached a level of maturity such that many commercial products now utilize proprietary SLAM algorithms, and there are several open-source software packages available that offer off-the-shelf pipelines that can be deployed on an end user's platform of choice. However, the most common is a single sensor platform setup [4, 5], which is cheaper and therefore more widespread than multiple sensor platforms. This implies the separate SLAM group—monocular methods.

As Engel, Schöps, and Cremers concluded [6], one of the major benefits of monocular SLAM—and simultaneously one of the biggest challenges—comes with the inherent scale ambiguity: The scale of the world cannot be observed and drifts over time, being one of the major error sources. The advantage is that this allows switching seamlessly between differently scaled environments, such as a desk environment indoors and large-scale outdoor environments. Scaled sensors on the other hand, such as depth or stereo cameras, have a limited range at which they can provide reliable measurements and hence do not provide this flexibility.

The present study offers some comparative results on the accuracy of the most recent and popular open-source monocular SLAM methods (ORB-SLAM2 [7, 8], DSO [2] and LDSO [9]), considering the power constraints of mobile ground robots.

28.2 Related Work

28.2.1 Slam

Indirect methods proceed in two steps. First, the raw sensor measurements are preprocessed to generate an intermediate representation, solving part of the overall problem, such as computing the image coordinates of corresponding points. Second, the computed intermediate values are interpreted as noisy measurements Y in a probabilistic model to estimate geometry and camera motion. Note that the first step is typically approached by extracting and matching a sparse set of keypoint—however, other options exist, like establishing correspondences in the form of dense, regularized optical flow. It can also include methods that extract and match parametric representations of other geometric primitives, such as line- or curve-segments [10, 11].

In our work, we estimated the accuracy of the recently proposed monocular visual SLAM method that tracks ORB features in real-time, ORB-SLAM [7], specifically, the latest revision—ORB-SLAM2 [8], which has the same core but an improved and optimized workflow. As an indirect method, it provides a good trade-off between accuracy and runtime.

Direct methods in contrast to indirect, which abstract images into a sparse set of feature points, skip the pre-computation step and directly use the actual sensor values, using the entire image information in order to minimize the photometric error. Therefore, as Krombach, Droeschel, and Behnke noticed [12], these methods are computationally very intense and thus much slower than indirect methods. They also often need to use GPUs to achieve real-time performance [13].

In our work, we estimate the accuracy of the recently proposed pure direct method and its latest revision with loop closure ability—DSO [2] and LDSO [9], respectively.

28.2.2 Benchmarks

Datasets

A Photometrically Calibrated Benchmark for Monocular Visual Odometry. Engel, Usenko, and Cremers [14, 15] presented a dataset for evaluating the tracking accuracy of monocular visual odometry and SLAM methods. It contains 50 real-world sequences comprising more than 100 min of video, recorded across dozens of different environments—ranging from narrow indoor corridors to wide outdoor scenes. All sequences contain mostly exploring camera motion, starting and ending in the same position. This allows evaluating the tracking accuracy via the accumulated drift from start to end, without requiring ground truth for the full sequence. In contrast to existing datasets, all sequences are photometrically calibrated which allows to reliably benchmark direct methods. The authors provided exposure times for each frame as reported by the sensor, the camera response function, and dense lens attenuation factors.

The EuRoC Micro-Aerial Vehicle Datasets. Burri et al. [16] proposed visual-inertial datasets collected onboard a *micro*-aerial vehicle. The datasets contain synchronized stereo images, IMU measurements, and accurate ground truth. The first batch of datasets facilitates the design and evaluation of visual-inertial localization algorithms on real flight data. It was collected in an industrial environment and contains millimeter accurate position ground truth from a laser tracking system. The second batch of datasets is aimed at precise 3D environment reconstruction and was recorded in a room equipped with a motion capture system. The datasets contain 6D-pose ground truth and a detailed 3D scan of the environment. Eleven datasets are provided in total, ranging from slow flights under good visual conditions to dynamic flights with motion blur and poor illumination, enabling researchers to thoroughly test and evaluate their algorithms. All datasets contain raw sensor measurements, spatiotemporally aligned sensor data and ground truth, extrinsic and intrinsic calibrations and datasets for custom calibrations.

The EuRoC MAV dataset includes 11 indoor sequences recorded with a Skybotix stereo VI sensor from a MAV. Accurate ground truth (approx. 1 mm) is recorded using a laser tracker or a motion capture system. Compared to the TUM benchmark, the sequences in EuRoC MAV are shorter and have less variety as they only contain recordings in one machine hall and one laboratory room. Furthermore, EuRoC MAV does not include full photometric data, which is important for benchmarking direct methods.

The KITTI Vision Benchmark Suite. Geiger and Lenz [17] *developed the benchmarks* for the tasks of stereo, optical flow, visual odometry/SLAM, and 3D object detection. The used recording platform was equipped with four high-resolution video cameras, a Velodyne laser scanner, and a state-of-the-art localization system. Proposed benchmarks comprise 389 stereo and optical flow image pairs, stereo visual odometry sequences of 39.2 km length, and more than 200 k 3D object annotations captured in cluttered scenarios (up to 15 cars and 30 pedestrians are visible per image). Results from state-of-the-art algorithms reveal that methods ranking high on established datasets such as Middlebury perform below average when being moved outside the laboratory to the real world. The authors' goal was to reduce this bias by providing challenging benchmarks with novel difficulties to the computer vision community.

The proposed dataset was recorded in an outdoor environment with two stereo cameras and a fast-moving car as a carrier. It provides low-frequency IMU information, which is, however, not time-synchronized with the camera images, and a GPS/INS-based ground truth with an accuracy below 10 cm, which is not so crucial for researches on computer vision algorithms in general, but critical for SLAM benchmark purposes.

28.3 Benchmark Comparisons

A Benchmark Comparison of Monocular Visual-Inertial Odometry Algorithms for Flying Robots. Delmerico and Scaramuzza [3] presented the paper, which evaluates an array of publicly available VIO pipelines (MSCKF, OKVIS, ROVIO, VINS-Mono, SVO + MSF, and SVO + GTSAM) on different hardware configurations, including several single-board computer systems that are typically found on flying robots. The evaluation considers the pose estimation accuracy, per-frame processing time, CPU, and memory load while processing the EuRoC datasets, which contain six degree-of-freedom (6DoF) trajectories typical of flying robots.

Table 28.1 Hardware platform specifications

CPU	Intel Core i7-7700HQ, 2800 MHz
RAM	16 GB, DDR4, 2400 MHz
Weight	2.56 kg
Battery	70 Wh Li-Ion
Power consumption	80 W (avg. load)

The conducted survey of the state estimation performance strictly focuses on VO methods usage in flying robotic platforms. The primary goal of the study is to benchmark VO methods performance on hardware and trajectories that are representative of state estimation for a flying robot with limited onboard computing power. In contrast, the current study mainly focuses on the trajectory estimation accuracy of SLAM methods for ground robots.

28.4 Experiments

28.4.1 Hardware

This study focuses on the use of SLAM methods with ground robots, which implies a restriction on energy consumption and the absence of significant constraints on the weight of the platform.

The hardware platform that we consider is the *HP Omen 15-ce057ur* laptop with the technical specifications briefly described in Table 28.1.

28.4.2 Datasets

Considering all the advantages and disadvantages of the approaches reviewed in 2.2, we decided to use the TUM and EuRoC dataset formats for benchmark purposes.

The TUM-Mono dataset [15] is a monocular dataset that consists of 50 indoor and outdoor sequences. It provides photometric camera calibration, but no full ground truth camera trajectories. The camera always returns to the starting point in all sequences, making this dataset very suitable for evaluating accumulated drifts of VO systems.

The EuRoC MAV dataset [16] consists of eleven visual-inertial sequences recorded with two monocular cameras onboard a micro-aerial vehicle while it was manually piloted around three different indoor environments. Within each environment, the sequences increase qualitatively in difficulty as the sequence number increases. For example, Machine Hall 01 is 'easy,' while Machine Hall 05 is a more challenging sequence in the same environment, introducing things like faster motions, poor illumination, etc.

28.4.3 Metrics

TUM. To evaluate the TUM benchmark results, we used the metrics proposed by Engel, Usenko, and Cremers [15] in the corresponding study.

For the evaluation of the metrics, we used the tracked positions $p_1 \dots p_n \in \mathbb{R}^3$ of frames 1 to n and the frame indices $S \subset [1, n]$, $E \subset [1, n]$ for the start and end segments for which aligned ground truth positions $\hat{p} \in \mathbb{R}^3$ are provided. It is important that both S and E contained sufficient poses in a non-degenerate configuration to well-constrain the alignment—hence the loopy motion patterns at the beginning and end of each dataset sequence.

First, we aligned the tracked trajectory with both the start and end segments independently and computed two relative transformations:

$$T_s^{gt} := \underset{T \in \mathrm{Sim}(3)}{\arg\min} \sum_{i \in S} \left(T p_i - \hat{p}_i\right)^2, \tag{28.1}$$

$$T_e^{gt} := \underset{T \in \mathrm{Sim}(3)}{\arg\min} \sum_{i \in E} \left(T p_i - \hat{p}_i\right)^2. \tag{28.2}$$

Then we computed the accumulated drift as:

$$T_{\mathrm{drift}} = T_e^{\mathrm{gt}} \left(T_s^{\mathrm{gt}}\right)^{-1} \in \mathrm{Sim}(3). \tag{28.3}$$

from which we explicitly computed the scale drift:

$$e_s := \mathrm{scale}(T_{\mathrm{drift}}), \tag{28.4}$$

the rotation drift:

$$e_r := \mathrm{rotation}(T_{\mathrm{drift}}), \tag{28.5}$$

and the translation drift:

$$e_t := \|\mathrm{translation}(T_{\mathrm{drift}})\|. \tag{28.6}$$

Furthermore, we calculated the combined error measure defined by the authors, the alignment error, which equally takes into account the error caused by scale, rotation, and translation drift over the full trajectory. It is the translational RMSE between the tracked trajectories when aligned to the start and end segments:

$$e_{\text{align}} := \sqrt{\frac{1}{n}\sum_{i=1}^{n}\left\| T_s^{\text{gt}} p_i - T_e^{\text{gt}} p_i \right\|_2^2}. \tag{28.7}$$

EuRoC. In contrast to the previous dataset format, the EuRoC includes full ground truth camera trajectories, which allows the use of the existing metrics and does not imply a novel error measurement approach. To evaluate the EuRoC benchmark results, we used the metrics proposed by Sturm, Engelhard, Endres, Burgard, and Cremers [18].

Absolute Trajectory Error. Due to the monocular SLAM nature, the estimated trajectory alignment and scale are subjective and each test result will not match directly with the other results and even the ground truth. This is a common mathematical problem for computer vision algorithms. Shinji Umeyama [19] proposed a closed-form solution of the similarity transformation parameter estimation, using the singular value decomposition.

The absolute trajectory error (ATE) measures global consistency by comparing the absolute distances between the estimated and the ground truth trajectory. As both trajectories can be specified in arbitrary coordinate frames, they first need to be aligned using the solution described previously, which finds the rigid-body transformation T that maps the estimated trajectory $P_{[1:n]}$ onto the ground truth trajectory $Q_{[1:n]}$. Given this transformation, we computed the absolute trajectory error at time step i as:

$$F_i := Q_i^{-1}\text{TP}_i. \tag{28.8}$$

Then we evaluated the root-mean-squared error over all time indices of the translational components:

$$\text{RMSE}(F_{1:n}) := \left(\frac{1}{n}\sum_{i=1}^{n}\|\text{translation}(F_i)\|^2\right)^{1/2}, \tag{28.9}$$

where translation(F_i) refers to the translational components of the absolute trajectory error F_i.

Relative Pose Error. The relative pose error (RPE) measures the local accuracy of the trajectory over a fixed time interval Δ. In other words, it measures the difference in the relative motion between pairs of poses, which is useful for estimating the drift. Sturm et al. [18] defined the relative pose error at time step i as:

$$E_i := \left(Q_i^{-1}Q_{i+\Delta}\right)^{-1}\left(P_i^{-1}P_{i+\Delta}\right), \tag{28.10}$$

where $P_{[1:n]}$ is the estimated trajectory and $Q_{[1:n]}$ is the ground truth trajectory.

From a sequence of n camera poses, we obtain individual relative pose errors along the sequence as $m = n - \Delta$. From these errors, as authors proposed, we computed the root-mean-squared error (RMSE) over all time indices of the translational component as:

$$\mathrm{RMSE}(E_{1:n}, \Delta) := \left(\frac{1}{m} \sum_{i=1}^{m} \|\mathrm{translation}(E_i)\|^2 \right)^{1/2}, \tag{28.11}$$

where translation(E_i) refers to the translational components of the relative pose error E_i.

To evaluate the global error of a trajectory, we averaged an error over all possible time intervals Δ:

$$\mathrm{RMSE}(E_{1:n}) := \frac{1}{n} \sum_{\Delta=1}^{n} \mathrm{RMSE}(E_{1:n}, \Delta). \tag{28.12}$$

Overall, the RPE metric provides an elegant way to combine rotational and translational errors into a single measure, while the ATE only considers the translational errors. As a result, the RPE is always slightly larger than the ATE (or equal if there is no rotational error). However, rotational errors typically also manifest themselves in wrong translations and are thus indirectly also captured by the ATE. From a practical perspective, the ATE has an intuitive visualization, which facilitates visual inspection. Nevertheless, as the authors noted, the two metrics are strongly correlated.

28.5 Evaluation

TUM. We compared ORB-SLAM2, DSO, and LDSO on the TUM-Mono dataset and evaluated the metrics described in 3.3 TUM section over all dataset sequences, each of which was run 10 times forward and backward to account for non-deterministic behavior.

Figure 28.1 shows RMSE when aligning the evaluated trajectory start and end segments with the provided ground truth trajectory segments (two first plots) along

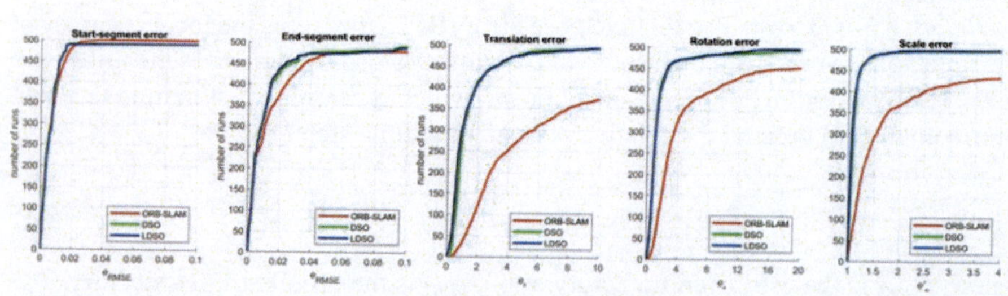

Fig. 28.1 Start and end segments e_{RMSE} and accumulated translational (e_t), rotational (e_r) and scale (e_s) drifts

with the cumulative error plots—accumulated translational, rotational, and scale drifts (the rest). The figure depicts the number of runs whose errors are below the corresponding x-values (thus, closer to left top is better). Note the difference in the order of magnitude—the RMSE within start and end segments is roughly 100 times smaller than the alignment RMSE.

Engel et al. [15] concluded that, due to ground truth origin and the similarity of the experimental results, almost all of the alignment errors originate from accumulated drift, and not from noise in the ground truth. As we can see, the results are very similar for all of the evaluated methods. This means that we can use these metrics for any benchmark with a ground truth of any accuracy as a reference.

Figure 28.2 shows the tracking accuracy when playing sequences only forwards compared to the results obtained when playing them only backward and the combined results (forwards and backward)—switching between predominantly forward motion and predominantly backward motion.

As we can see, the direct methods, DSO and LDSO, are largely unaffected by the motion bias, while the indirect, ORB-SLAM2, performs significantly better for backward motion.

Figure 28.3 shows the color-coded alignment error in all the sequences as defined in the 3.3 TUM segment.

Fig. 28.2 Dataset motion bias

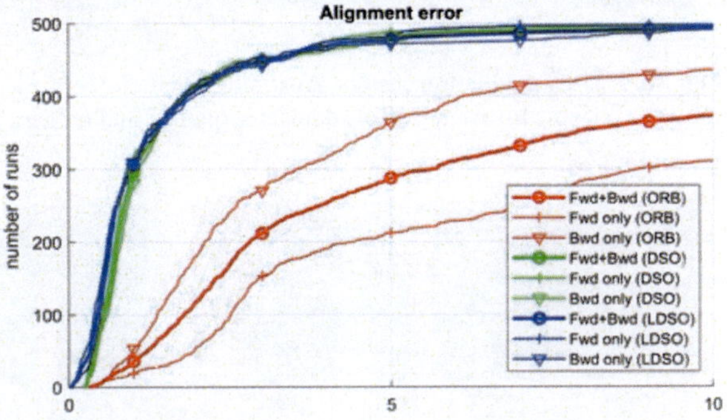

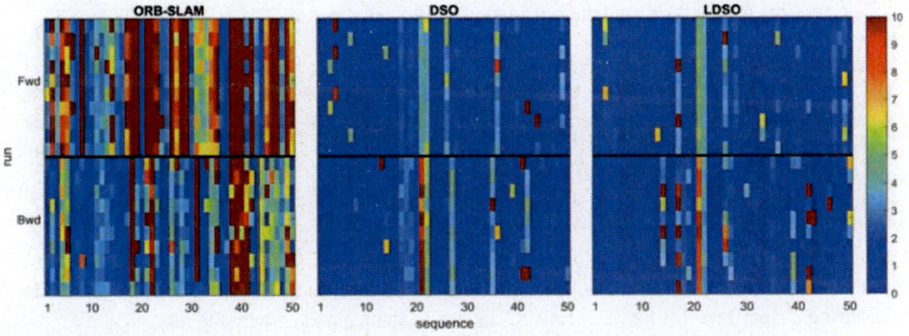

Fig. 28.3 Alignment error e_{align} for each TUM-Mono dataset sequence

Figures 28.1 and 28.3 also prove that comparing to DSO, the novel approach of corner features detection of LDSO does not reduce the VO accuracy of the original system and even slightly improves the overall accuracy due to the loop closure capabilities.

EuRoC. We compared ORB-SLAM2, DSO, and LDSO on the EuRoC MAV dataset and evaluated the metrics described in 3.3 EuRoC section over all dataset sequences for each of the two camera streams which were interpreted as separate dataset sequences with the same ground truth (see '0.0' and '0.1' notations in figures below). All of the sequences were run 10 times to account for non-deterministic behavior.

Figures 28.4, 28.5, and 28.6 show the evaluated absolute translation error e_{ate} and relative pose error e_{rpe} for the DSO, LDSO, and ORB-SLAM2 methods, respectively.

As we can see, all methods perform quite well on this dataset in terms of an absolute translation error. Still, ORB-SLAM2 comparing to direct methods has some advantage here—it withstands the most difficult sequences successfully—MH_{05},

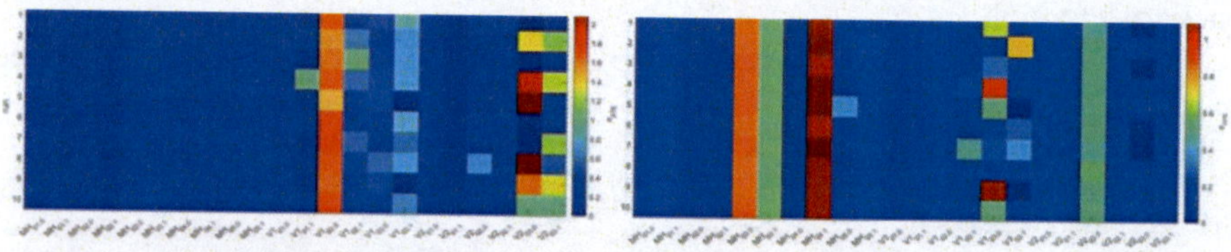

Fig. 28.4 DSO evaluation results. Absolute translation error e_{ate} (m) (left) and relative pose error e_{rpe} (m/s) (right) for each EuRoC dataset sequence and camera stream

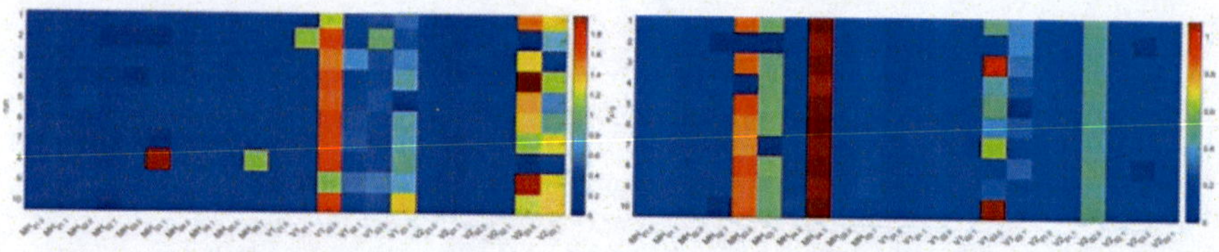

Fig. 28.5 LDSO evaluation results. Absolute translation error e_{ate} (m) (left) and relative pose error e_{rpe} (m/s) (right) for each EuRoC dataset sequence and camera stream

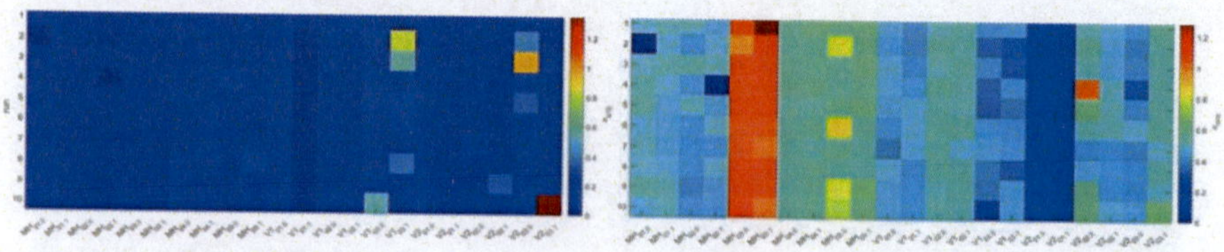

Fig. 28.6 ORB-SLAM2 evaluation results. Absolute translation error e_{ate} (m) (left) and relative pose error e_{rpe} (m/s) (right) for each EuRoC dataset sequence and camera stream

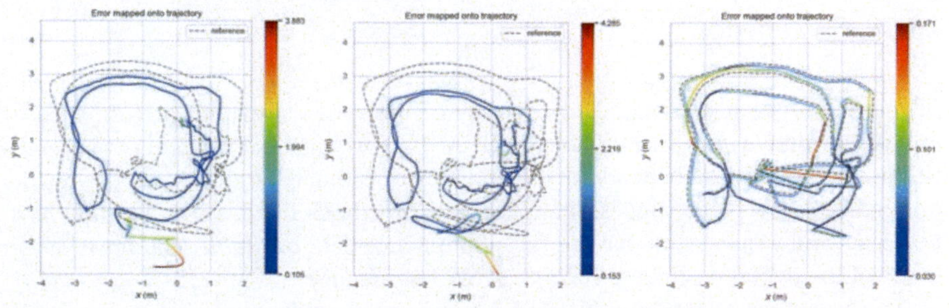

Fig. 28.7 Absolute translation error e_{ate} mapped onto estimated trajectory for each method: DSO (left), LDSO (middle) and ORB-SLAM2 (right) respectively

$V1_{03}$, and $V2_{03}$—the ones with the lack of light and much more rotational movements than translational.

In terms of relative pose error, purely direct methods generally perform much better than the indirect one, but it is still difficult for them to overcome the harsh environment. Both of the direct methods fail on sequence $V2_{03}$, but on most of the other sequences, comparing to DSO, LDSO significantly improves the camera tracking accuracy, again due to the LDSO loop closure capability.

Qualitative comparison. To show some qualitative results, examples of the reconstructed maps for the most difficult $V2_{03}$ sequence for each of the methods along with the ground truth are shown in Fig. 28.7.

The figure shows that the overall trajectory pattern is distinguishable, but, as we can see, DSO and LDSO path was distorted at the end of the sequence. This has led to issues in alignment and scale, which are also clearly visible.

The presented situation is a common case for all of the reviewed methods due to the specific movement patterns and the lack of light in some scenes. Yet, ORB-SLAM2 performed much better than the others did.

28.6 Further Work

The current study presents a fixed set of the most popular and recent open-source monocular SLAM methods. However, there is still a number of methods such as LSD-SLAM [6] and SVO [20], which are older, but well-proven, and DynaSLAM [21], which adds new capabilities to the ORB-SLAM2 [8] system. This benchmark comparison will be expanded by the mentioned methods in the future, which will be useful for understanding the full market picture of monocular SLAM methods.

28.7 Conclusion

In this work, we have conducted a survey of the state estimation performance of publicly available real-time monocular SLAM algorithms. In evaluating these algorithms, our goal was to benchmark the accuracy of estimated trajectories. The results presented in Sect. 28.3 suggest that pure direct methods like DSO and LDSO have better accuracy only while having full photometric data *available,* and however, in terms of the overall accuracy, ORB-SLAM2 has an edge.

The accuracy and robustness can be improved with additional computation resources, but it all comes to the capabilities of the robotic platform—the more computational resources are required, the more power units and weight the platform should carry. In our research, we selected the specific hardware platform with a balanced amount of computational resources and power consumption. We hope that the results and conclusions presented in this paper may help members of the research community in finding appropriate compromises for their robot systems.

Acknowledgements The reported study was funded by the RFBR according to the research project No. 19-58-70002 and research grant of Kazan Federal University.

References

1. Dissanayake, M.G., Newman, P., Clark, S., Durrant-Whyte, H.F., Csorba, M.: A solution to the simultaneous localization and map building (SLAM) problem. IEEE Trans. Robot. Autom. **17**(3), 229–241 (2001)
2. Engel, J., Koltun, V., Cremers, D.: Direct sparse odometry. IEEE Trans. Pattern Anal. Mach. Intell. **40**(3), 611–625 (2017)
3. Delmerico, J., Scaramuzza, D.: A benchmark comparison of monocular visual-inertial odometry algorithms for flying robots. In: IEEE 2018 International Conference on Robotics and Automation, pp. 2502–2509 (2018)
4. Buyval, A., Afanasyev, I., Magid, E.: Comparative analysis of ROS-based monocular SLAM methods for indoor navigation. In: 2016 Ninth International Conference on Machine Vision, ICMV, pp. 103411K (2017)
5. Bokovoy, A., Yakovlev, K.: Enhancing semi-dense monocular vSLAM used for multi-rotor UAV navigation in indoor environment by fusing IMU data. In: The 2018 International Conference on Artificial Life and Robotics, pp. 391–394 (2018)
6. Engel, J., Schöps, T., Cremers, D.: LSD-SLAM: Large-scale direct monocular SLAM. In: European Conference on Computer Vision, pp. 834–849 (2014)
7. Mur-Artal, R., Montiel, J.M.M., Tardos, J.D.: ORB-SLAM: a versatile and accurate monocular SLAM system. IEEE Trans. Rob. **31**(5), 1147–1163 (2015)
8. Mur-Artal, R., Tardós, J.D.: Orb-slam2: An open-source slam system for monocular, stereo, and RGB-D cameras. IEEE Trans. Rob. **33**(5), 1255–1262 (2017)
9. Gao, X., Wang, R., Demmel, N., Cremers, D.: LDSO: Direct sparse odometry with loop closure. In: International Conference on Intelligent Robots and Systems, pp. 2198–2204 (2018)
10. Zakiev, A., Lavrenov, R., Magid, E., Svinin, M., Matsuno, F.: Partially unknown environment exploration algorithm for a mobile robot. J. Adv. Res. Dyn. Control Syst. **11**(8), 1743–1753 (2019)

11. Alishev, N., Lavrenov, R., Hsia, K. H., Su, K. L., Magid, E.: Network failure detection and autonomous return algorithms for a crawler mobile robot navigation. In: 11th International Conference on Developments in eSystems Engineering, pp. 169–174 (2018)
12. Krombach, N., Droeschel, D., Behnke, S.: Combining feature-based and direct methods for semi-dense real-time stereo visual odometry. In: International Conference on Intelligent Autonomous Systems, pp. 855–868 (2016)
13. Ramil, S., Lavrenov, R., Tsoy, T., Svinin, M., Magid, E.: Real-time video server implementation for a mobile robot. In: 2018 11th International Conference on Developments in eSystems Engineering, pp. 180–185 (2018)
14. Schubert, D., Goll, T., Demmel, N., Usenko, V., Stückler, J., Cremers, D.: The TUM VI benchmark for evaluating visual-inertial odometry. In: 2018 IEEE/RSJ International Conference on Intelligent Robots and Systems, pp. 1680–1687 (2018)
15. Engel, J., Usenko, V., Cremers, D.: A photometrically calibrated benchmark for monocular visual odometry. arXiv preprint arXiv:1607.02555 (2016)
16. Burri, M., Nikolic, J., Gohl, P., Schneider, T., Rehder, J., Omari, S., Siegwart, R.: The EuRoC micro aerial vehicle datasets. Int. J. Robot. Res. **35**(10), 1157–1163 (2016)
17. Geiger, A., Lenz, P., Urtasun, R.: Are we ready for autonomous driving? The kitti vision benchmark suite. In: 2012 IEEE Conference on Computer Vision and Pattern Recognition, pp. 3354–3361 (2012)
18. Sturm, J., Engelhard, N., Endres, F., Burgard, W., Cremers, D.: A benchmark for the evaluation of RGB-D SLAM systems. In: 2012 IEEE/RSJ International Conference on Intelligent Robots and Systems, pp. 573–580(2012)
19. Umeyama, S.: Least-squares estimation of transformation parameters between two point patterns. IEEE Trans. Pattern Anal. Mach. Intell. **4**, 376–380 (1991)
20. Forster, C., Pizzoli, M., Scaramuzza, D.: SVO: Fast semi-direct monocular visual odometry. In: IEEE International Conference on Robotics and Automation, pp. 15–22 (2014)
21. Bescos, B., Fácil, J.M., Civera, J., Neira, J.: DynaSLAM: tracking, mapping, and inpainting in dynamic scenes. IEEE Robot. Autom. Lett. **3**(4), 4076–4083 (2018)

Chapter 29
Laser Rangefinder and Monocular Camera Data Fusion for Human-Following Algorithm by PMB-2 Mobile Robot in Simulated Gazebo Environment

Elvira Chebotareva, Kuo-Hsien Hsia, Konstantin Yakovlev, and Evgeni Magid

Abstract The paper presents a human-following algorithm for an autonomous mobile robot, which is equipped with a 2D laser rangefinder (LRF) and a monocular camera. As a rule, quality of a human tracking by a LRF is reduced in cluttered environments. We used a monocular camera to increase a human-tracking reliability. In contradiction with popular human-tracking algorithms that apply only a 2D LRF, our algorithm does not impose any restrictions on a type of human's clothes, and our approach does not require a human head and an upper body to be located within a monocular camera field of view. Several human trackers and variations of our algorithm were compared in the Gazebo virtual experiments within a free corridor and an office room environment. The virtual experiments demonstrated that our method successfully improved a human-tracking quality being employed with the human-following virtual PMB-2 robot.

E. Chebotareva (✉) · E. Magid
Laboratory of Intelligent Robotic Systems (LIRS), Intelligent Robotics Department, Higher Institute for Information Technology and Intelligent Systems, Kazan Federal University, Kazan, Russian Federation
e-mail: Elvira.Chebotareva@kpfu.ru
URL: https://kpfu.ru/erobotics

E. Magid
e-mail: magid@it.kfu.ru
URL: https://kpfu.ru/erobotics

K.-H. Hsia
Department of Electrical Engineering, National Yunlin University of Science and Technology, Douliu, Taiwan
e-mail: khhsia@yuntech.edu.tw

K. Yakovlev
Moscow Institute of Physics and Technology, Moscow, Russian Federation
e-mail: yakovlev@isa.ru

A. Ronzhin and V. Shishlakov (eds.), *Proceedings of 15th International Conference on Electromechanics and Robotics "Zavalishin's Readings"*, Smart Innovation, Systems and Technologies 187, https://doi.org/10.1007/978-981-15-5580-0_29

29.1 Introduction

The ability to follow a human is necessary for many types of assistant and companion robots. Such robots are in demand in various areas of human life. A typical example is a cargo robot that follows a human and helps to carry heavy objects, and a so-called "robotic suitcases" are already being mass-sold [8]. The concept of a robotic autonomous suitcase implies that the suitcase follows its owner, adapts to a walking pace, and avoids obstacles [18]. In RoboCup@Home competitions within "Carry My Luggage" test, a robot helps an operator to carry some luggage [24].

The ability to follow a person is important for search and rescue robots during emergencies, where a robot could follow a firefighter, which chooses a safest and easiest trajectory. Another socially significant example of human-following robots is robotic wheelchairs [26]. A capability of following a person in wheelchair does not need to be pushed, which allows an accompanying person to walk near the wheelchair, maintaining communication and visual contact with a passenger. Moreover, the ability to follow a human is critically important for social robots that interact with people through emotional communication [6].

Large number of static and dynamic objects in public spaces complicates autonomous navigation of mobile robots [21]. Presence of pedestrians makes it difficult to choose a path and can lead to a target loss. A navigation algorithm of a mobile cargo robot should simultaneously solve three problems: human detection, human tracking, and human following. A trajectory of a robot is determined by a trajectory of a target. In addition, the robot must navigate an environment, avoid collisions with obstacles and, if necessary, return to a starting point of a route [1].

2D laser rangefinders (LRF) are often used for SLAM in mobile robotics, but LRFs could also be used for human tracking. Since this tracking method is not always reliable, LRFs can be used in conjunction with visual sensors. It is economically attractive to use inexpensive monocular cameras as onboard visual sensors of a robot, which is already equipped with a LRF.

This paper presents our ongoing research on human-following problem for autonomous mobile robot PMB-2. The human-following algorithm is based on a joint use of monocular camera and LRF data. We assume that in a general case only a lower part of a human body is located within the camera's field of view. It is important to emphasize that we avoid an approach, which assumes a strictly defined type of a target human dress, allowing to clearly distinguish two legs of a human. We also propose a universal approach of quantitative characteristics for evaluating human-following algorithms that may be useful for comparing different algorithms.

A preliminary evaluation of a human-following algorithm performance could be obtained in a simulation testing. A realistic simulator allows to reduce costs of early stages of testing and effectively prepare for real-world experiments. In this paper, for preliminary testing, we employ the 3D Gazebo simulator [17].

29.2 Related Work

A significant number of works have been devoted to human-following autonomous mobile robots [11]. Many models of autonomous mobile robots are equipped with LRFs, which demonstrates popularity of people tracking methods using LRFs. 2D LRF data are conveniently presented as a two-dimensional image. Since LRFs are usually placed on a small height, many methods of human tracking are based on the geometric features of searching for human legs on LRF data.

Several parameters, such as a distance between (segmented) images of two legs, leg circumference, leg thickness can be used to track human legs. Authors in [14, 27] presented methods of human detection and tracking based on a geometric approach. This approach implies that in an image from a LRF, human legs are represented in a form of two arcs or curves. Another group of methods is based on machine learning. In this case, clustering of objects separating an image of legs from a background is performed [2, 9, 20].

In addition to a LRF, various types of cameras are used for human tracking. In [7], a stereo vision-based CNN tracker for human-following robot is proposed. The work [16] is devoted to a human-following algorithm that is based on using LRF and panoramic camera data. In [19], a system of human tracking and following for robotic platforms with RGB-D camera is presented. Monocular cameras are also used as onboard sensors of human-following robots. Authors in [4] proposed a hybridizing image information based method that combines color and shape data. In [15], a target person was identified by using a combination of convolutional channel features and online boosting. Yet, onboard monocular cameras have a number of disadvantages, including limited visibility, ambiguity of scale, and detection errors. To increase the accuracy and reliability of tracking, as well as to solve other problems that arise when using service robots in cramped conditions, it is recommended to use a monocular camera in a combination with a LRF.

29.3 Problems of Human-Following Algorithms Implementation

We consider the following formulation of a human-following task. At a start, a human is standing in front of a robot. At a random moment, the human begins to move along an arbitrary route. The robot must follow the human at a safe distance. The robot must also avoid collisions with any obstacles. At any time, if necessary, the robot should be able to return to the starting point. At the initial time, the robot does not have information about an environment. We also assume that there are no other objects between the robot and the human at the starting time and the human is a closest visible object for the robot.

In this paper, we consider a human-following task for the PMB-2 robot by PAL Robotics [23]. The robot PMB-2 is designed for an indoors transportation of goods (up to 50 kg). The robot PMB-2 is equipped with a Hokuyo URG-04LX-UG1 LRG. For our research, we equipped PMB-2 with a monocular camera. The camera is located at a height of 40 cm from the floor. The PMB-2 robot has a cylindrical shape with 54 cm diameter and 30 cm height. Relatively small dimensions allow the robot to move in rooms of various types, including office rooms (Fig. 29.1a). PMB-2 is navigated by ROS. The robot velocity is controlled by the *cmd_vel* ROS package. The navigation is carried out by using *move_base* ROS package [22].

We faced a fact that in practice, methods for detecting a human using 2D LRF are not always reliable. As an example, consider a tracker in [20]. This tracker shows good results in free spaces, but it might give false positives in cluttered spaces (Fig. 29.1). In addition, our clothing might have a significant impact on a person's visibility for the tracker. Figure 29.2 presents the data of the PMB-2 LRF as a 2D image. In Fig. 29.2a, a person in trousers is in the LRF field of view. It is clearly seen that the legs are represented as two arcs. In Fig. 29.2b, a person in a long skirt is in the LRF field of view as well. Yet, in this case, trackers based on the representation of human legs with two curves or one curve with two maximum points do not work.

Thus, in practice, there is a need to use additional sensors to track a human. The most affordable visual sensors in this case are inexpensive monocular cameras. Technical specifications of mobile robots hardware often impose restrictions on computer vision methods use. Besides, a configuration of cargo mobile robots imposes additional restrictions on a mounting height of a camera. This makes it difficult to obtain a human's full-height image. The head and upper body are important for a human detection and identification. If a camera is located at a low height, then to capture the human's head, the camera should be positioned at an angle. But in this case, the human disappears from camera's field of view when moving away from the robot. Placing the camera straight solves this problem, but in this case, at close distances, the head and upper body are not visible.

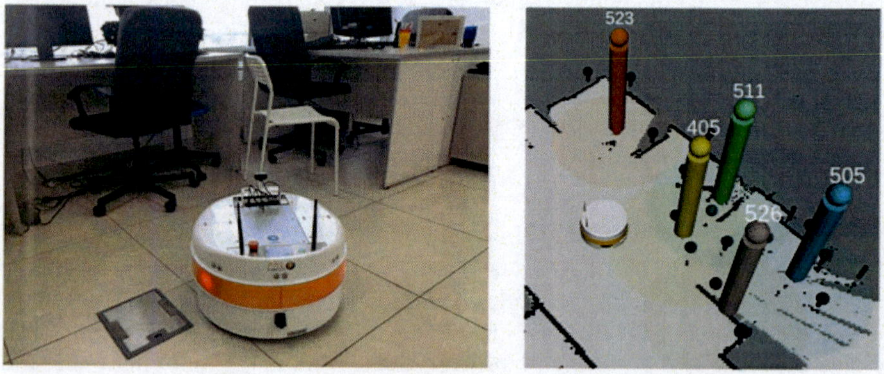

(a) (b)

Fig. 29.1 **a** PMB-2 robot in an office environment. **b** False positives of a leg tracker in the office environment of room 1403, Laboratory of Intelligent Robotic Systems, Kazan Federal University (KFU), 35 Kremlevskaya street, Kazan. Detected people are marked by cylinders with numerical ID labels

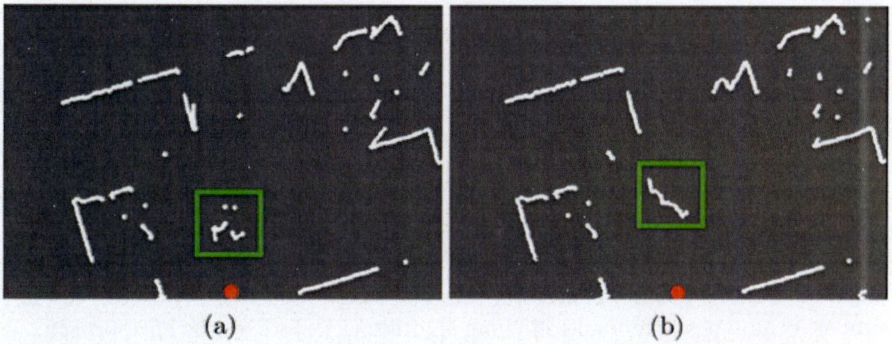

Fig. 29.2 **a** PMB-2 robot LRF data implemented as a two-dimensional image. Inside the green square are legs of a person in trousers. The red point denotes the robot. **b** Inside the green square is a person in a long skirt

Despite a large number of works that are devoted to human-following algorithms, the algorithms evaluation and comparison are often complicated by the absence of information about qualitative and quantitative parameters. In most of the works devoted to human-following algorithms, experiments are conducted in open spaces, empty rooms, or corridors. It is natural to assume that an environment, as well as a presence of a large number of other people, can influence human-following algorithms testing results. However, in practice, checking algorithm stable work in various environments faces significant difficulties.

If tests are carried out in complex environments with people, there is a risk of a collision with people and another objects. In real environments, it is not always possible to ensure that test conditions are unchanged. Time and resources cost on the early stages of testing could be reduced with a help of simulators.

The aim of our research is to develop an approach that allows to use LRF and monocular camera data for human-following by the PMB-2 mobile robot, taking into account the problems that were described above.

29.4 Proposed Solution and Its Evaluation in Gazebo Simulator

This section describes an approach that allows to track a human's movement with a mobile robot using a combination of a LRF and a monocular camera. Next, the results of human-following algorithm simulation testing, which demonstrate our algorithm performance, are presented.

29.4.1 *Evaluation of Human-Following Algorithms in Gazebo*

We suggest using the Gazebo simulator at early stages of human-following algorithms development, including debugging and testing. Many popular robot models are already featured in the Gazebo with publicly available simulation packages, including the PMB-2 robot [25]. In addition, Gazebo allows creating new models of robots, populating environments with static objects (walls, furniture, etc.) and models of walking people. Therefore, Gazebo could be successfully used to test and compare prototypes of human-following algorithms. Before testing an approach in a real-world environment, we suggest constructing a model of the environment using Gazebo and evaluating a human-following algorithm within the simulated environment.

For preliminary testing, we have built 3D models of real environment of the second study building of Kazan Federal University. For experiments, we have chosen a 2.4 m width corridor (Fig. 29.3a) and 8 × 8 m room (Fig. 29.3b). We had planned two types of a route for a human. In the first case, a human moved along the corridor (Fig. 29.4a). In the second case, a human started his route inside the room, left the room, and kept walking through the corridor (Fig. 29.4b). These routes contained pedestrian-specific locomotion patterns—a rectilinear movement, turning around a corner, and passing through a doorway. Each route had a length of 20 m.

To evaluate a human-following algorithm, we suggest paying a particular attention a human's velocity, a length of a successfully completed part of a route, a number of target losses, a number of human tracker false positives, a number of performed tests.

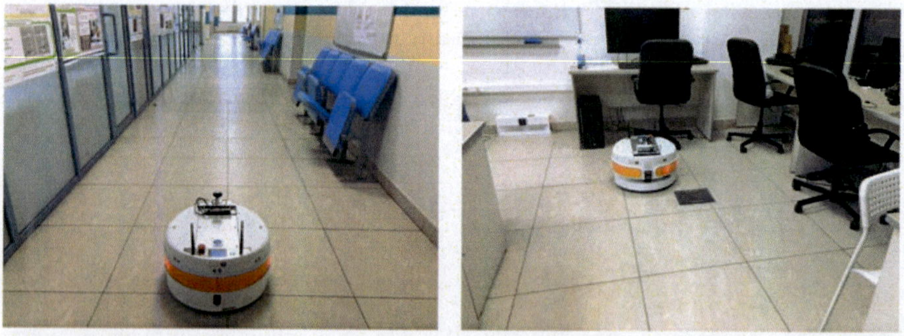

(a) (b)

Fig. 29.3 Experiments with the real robot PMB-2 inside the second study building of KFU, 35 Kremlevskaya street, Kazan. **a** In the corridor of the 14-th floor. **b** In the office room 1403, Laboratory of Intelligent Robotic Systems

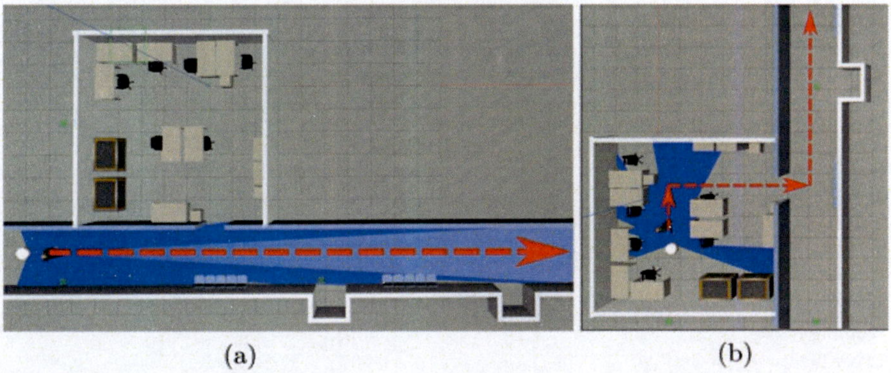

Fig. 29.4 **a** Gazebo model of Fig. 29.3a environment. **b** Gazebo model of Fig. 29.3b environment. The long corridor corresponds to the real environment of Fig. 29.3a. The red arrow shows the human's route

29.4.2 Human Detection and Tracking

It is convenient to present LRF data in a form of a two-dimensional image (Fig. 29.2). As it was shown above, we cannot use geometric features of human legs for people in long clothes (e.g., a skirt or a coat).

We assume that at the algorithm initialization time, a target human is the closest object to the robot, which allows to highlight in the LRF image a region of interest (ROI) that contains a nearest curve in front of the robot. Then, this curve can be tracked using any visual tracker. We present LRF data as a 2D image of 500 × 500 pixels. We connect the pairs of neighboring points by straight segments. Therefore, at the initialization time, the human is presented as a single curved line. We select the ROI so that this curve is located in its center. Knowing a scale and a position of the ROI relatively to the robot, we can calculate coordinates of a target point of the robot.

Two parameters affect the choice of a particular tracker—its accuracy and speed. To track a human in LRF data, we have compared five trackers that are released in the OpenCV library—KCF [10], TLD [13], Median Flow [12], MOSSE [5], and MIL [3]. The five trackers were tested in the Gazebo simulation of the environment shown in Fig. 29.4a. A total length of a path was 20 m with a human speed of 0.2 m/s. We conducted ten experiments for each tracker. KCF tracker immediately lost the object. TLD tracker produced false positives to other objects. Median Flow showed lost the object and produced false positives. The best results in this test were shown by MIL and MOSSE trackers. Therefore, KCF, TLD, and Median Flow trackers were not considered for further evaluation.

Next, we conducted virtual experiments on the same route with a human speed of 0.5 m/s. MOSSE tracker lost the target in ten cases out of ten, while MIL tracker successfully passed all tests not only at a speed of 0.5 m/s, but also at 1.08 m/s. Table 29.1 shows the quantitative parameters of virtual experiments results for MOSSE and

Table 29.1 Tracker testing results

Tracker	MOSSE	MIL	MIL	MIL
Route	Corridor	Corridor	Corridor	Room
Average human speed (m/c)	0.5	0.5	1.08	0.5
Average distance travelled (m)	6.1	20	20	8.8
Number of losses	10	0	0	9
Tracker false positives	0	0	0	9
Number of experiments	10	10	10	10
Success rate (%)	0	100	100	10

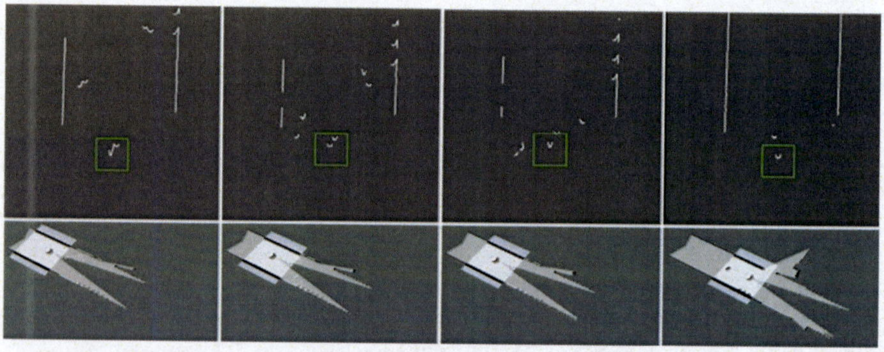

Fig. 29.5 Tracking a human within LRF data using MIL tracker [3] (top) and the environment map constructed during the simulation in Gazebo (bottom). The tracked human legs are detected within the green square

MIL trackers. When a virtual human in the simulation walked at a speed of 1.08 m/s, we added other pedestrians to the simulation scene. Figure 29.5 presents LRF data with the target human being tracked with MIL tracker. Figure 29.5 presents the environment map, which was built by the robot while it was moving. Figure 29.6a demonstrates the simulation in Gazebo.

In addition, we conducted virtual experiments for the route shown in Fig. 29.4b. With the human speed of 0.5 m/s in nine out of ten cases, the robot with MIL tracker lost a human at corner points of the route or in a doorway. Only once the robot was able to successfully overcome both corner points, exit the doorway, and follow the human to the end of the route. The robot lost the human due to sharp turns of the trajectory. False positives were also observed when the person passed near furniture or walls.

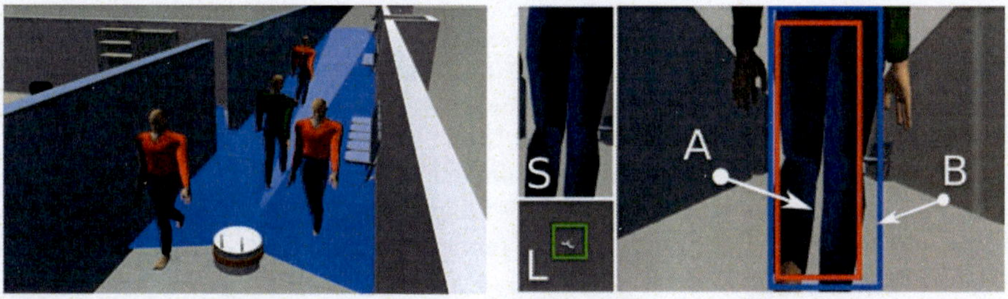

Fig. 29.6 **a** Simulation scene. **b** Examples of areas of ROI in LRF data based image *L*, ROI in camera image *A*, initial view of the ROI in the camera image *S*, and the area in the camera image that the tracker defined as the human *B*

The disadvantage of MIL tracker is a lack of information about an object loss. Additionally, we cannot reliably determine a false positive of the tracker. Therefore, we suggest using a monocular camera to control losses and false positives.

29.4.3 Joint Use of LRF and a Monocular Camera in a Human-Following Algorithm

In the absence of occlusions, MIL tracker provides quite accurate tracking. To control the reliability of MIL tracker, we suggest using a monocular camera. If at initialization a human simultaneously appears in the camera and LRD's fields of view, then we can fairly accurately determine the location of the person in the camera image. In this case, there is no need in human detectors. Having captured the ROI in a from a camera image, we can track the human using MIL tracker in a video stream, just like we do it on LRF data.

To detect a position of a human in initial frames of a video stream, we propose the following algorithm. In LRF data, we find a curve, which is the closest one to the robot. Using the LRF data, we trace the ROI in the camera image (so-called LRF data-based camera image area *L*) to keep this curve in the center. In the camera image, we highlight the ROI (area *A*) that corresponds to the LRF based ROI. We store the initial view of the ROI in the camera image (area *S*), and further, this image will be used as a sample.

We associate the LRF ROI with a part of the camera image. This area will be a kind of projection of the human tracker on the image from the LRF data. Thus, at each step, we have four areas: *L* is the area on the image that is based on the LRF and which is defined by the tracker as the area of the human's location; *A* is the projection of this area onto the camera image; *B* is the area in the camera image that the tracker defined as the image of the human; *S* is the sample image of the human. Figure 29.6b demonstrates examples of *A*, *B*, *S*, and *L* areas. Our task is to timely detect errors of the tracker that determines a position of a human in an image from the LRF. We also have a MIL tracker that tracks a human in the camera image. With the help of two MIL trackers, we update the areas B and L. Knowing the L area, we find the A area.

At each step of the algorithm, we calculate $P(A = S)$ and $P(A = B)$, where $P(A = S)$ is the probability that area A is a human image, $P(A = B)$ is the probability that area A coincides with area B. In the simplest case, the correlation of the images' histograms can be taken as $P(A = S)$ and $P(A = B)$. If $P(A = S) \geq \theta_1$, where θ_1 is a certain threshold, then we assume that the region L contains an image of the target human. If $P(A = B) \geq \theta_2$, where θ_2 is a certain threshold, then we assume that the area A coincides with area B. The human-following algorithm is represented in Algorithm 1. According to results of the analysis of camera and LRF data, the robot moves toward the human or begins a search of the human. Coordinates of the human are calculated based on the coordinates of the area L. The route to the human is built using a local planner.

Algorithm 1 Human-following algorithm

A = image of human
$B = A$
while not order to return **do**
  update tracker for A
  update tracker for B
  if $P(A = S) \geq \theta_1$ **and** $P(A = B) \geq \theta_2$ **then**
    calculate the human's position
    move to human
  else
    while $P(A = S) < \theta_1$ **do**
      turn
      A = image of the nearest object (or human)
    end while
    $B = A$
  end if
end while

29.4.4 Simulation Results

We conducted two sets of ten virtual experiments demonstrating the algorithm performance for data that was obtained using a LRF and a monocular camera. The simulations were carried out for two routes (Fig. 29.4). Table 29.2 presents the simulation results for each route. For comparison, Table 29.2 shows the results of simulations when only the LRF was used for a human tracking. As we expected, in all cases, the robot successfully completed not only the route in the corridor, but also the route in the room and the corridor. Unlike previous experiments, we did not observe false positives and target losses.

Table 29.2 Simulation results

Used sensors Route	LRF		Camera and LRF	
	Corridor	Room	Corridor	Room and corridor
Average human speed (m/s)	0.5	0.33	0.5	0.33
Average distance travelled (m)	20	13	20	20
Number of losses	0	5	0	0
Tracker false positives	0	5	0	0
Number of experiments	10	10	10	10

29.5 Conclusion and Future Work

In this paper, we presented a human-following algorithm based on simultaneous usage of laser rangefinder (LRF) and monocular camera data. We assumed that the camera was installed at a target human's legs level, which prevents constant tracking of such standard features as a face, an upper body or a height of a human. At the same time the algorithm does not place any restrictions on a human clothes type and is suitable for long clothes that cover legs area. Several human trackers and variations of our algorithm were compared in the Gazebo virtual experiments within a free corridor and an office room environment. The virtual experiments demonstrated that our method successfully improved a human tracking quality being employed with the human-following virtual PMB-2 robot.

As a part of our future work we will test the suggested human-following algorithm with a real PMB-2 robot in real environments that correspond to the Gazebo virtual environments, which were used in this paper.

Acknowledgements This research was funded by the Russian Foundation for Basic Research (RFBR), project ID 19-58-70002.

References

1. Alishev, N., Lavrenov, R., Hsia, K.H., Su, K.L., Magid, E.: Network failure detection and autonomous return algorithms for a crawler mobile robot navigation. In: 11th International Conference on Developments in eSystems Engineering (DeSE), pp. 169–174. IEEE, New York (2018)
2. Arras, K.O., Lau, B., Grzonka, S., Luber, M., Mozos, O.M., Meyer-Delius, D., Burgard, W.: Range-based people detection and tracking for socially enabled service robots. In: Towards Service Robots for Everyday Environments, pp. 235–280. Springer, Berlin (2012)
3. Babenko, B., Yang, M.H., Belongie, S.: Visual tracking with online multiple instance learning. In: Conference on Computer Vision and Pattern Recognition, pp. 983–990. IEEE, New York (2009)

4. Bakar, M.N.A., Saad, A.R.M.: A monocular vision-based specific person detection system for mobile robot applications. Proc. Eng. **41**, 22–31 (2012)
5. Borgefors, G.: Distance transformations in digital images. Comput. Vis. Graph. Image Process. **34**(3), 344–371 (1986)
6. Chebotareva, E., Safin, R., Shafikov, A., Masaev, D., Shaposhnikov, A., Shayakhmetov, I., Gerasimov, Y., Talanov, M., Zilberman, N., Magid, E.: Emotional social robot "Emotico". In: 12th International Conference on the Developments in eSystems Engineering. DeSE 2019 (Kazan, Russia) (2019)
7. Chen, B.X., Sahdev, R., Tsotsos, J.K.: Integrating stereo vision with a CNN tracker for a person-following robot. In: International Conference on Computer Vision Systems, pp. 300–313. Springer, Berlin (2017)
8. Cowarobot: Rover Speed. https://cowarobotrover.com/. Accessed on 13 Jan 2020
9. Guerrero-Higueras, Á.M., Álvarez-Aparicio, C., Olivera, M.C.C., Rodríguez-Lera, F.J., Fernández-Llamas, C., Rico, F.M., Matellán, V.: Tracking people in a mobile robot from 2D LIDAR scans using full convolutional neural networks for security in cluttered environments. Front. Neurorobot. **12** (2018)
10. Henriques, J.F., Caseiro, R., Martins, P., Batista, J.: Exploiting the circulant structure of tracking-by-detection with kernels. In: European Conference on Computer Vision, pp. 702–715. Springer, Berlin (2012)
11. Islam, M.J., Hong, J., Sattar, J.: Person-following by autonomous robots: a categorical overview. Int. J. Robot. Res. **38**(14), 1581–1618 (2019)
12. Kalal, Z., Mikolajczyk, K., Matas, J.: Forward-backward error: automatic detection of tracking failures. In: 20th International Conference on Pattern Recognition, pp. 2756–2759. IEEE, New York (2010)
13. Kalal, Z., Mikolajczyk, K., Matas, J.: Tracking-learning-detection. IEEE Trans. Pattern Anal. Mach. Intell. **34**(7), 1409 (2011)
14. Kawarazaki, N., Kuwae, L.T., Yoshidome, T.: Development of human following mobile robot system using laser range scanner. Proc. Comput. Sci. **76**, 455–460 (2015)
15. Koide, K., Miura, J., Menegatti, E.: Monocular person tracking and identification with on-line deep feature selection for person following robots. Robot. Autonomous Syst. **124** (2020)
16. Kristou, M., Ohya, A., Yuta, S.: Target person identification and following based on omnidirectional camera and LRF data fusion. In: RO-MAN, pp. 419–424. IEEE, New York (2011)
17. Lavrenov, R., Zakiev, A.: Tool for 3D Gazebo map construction from arbitrary images and laser scans. In: 10th International Conference on Developments in eSystems Engineering (DeSE), pp. 256–261. IEEE, New York (2017)
18. Lavrenov, R.O., Magid, E.A., Matsuno, F., Svinin, M.M., Suthakorn, J.: Development and implementation of spline-based path planning algorithm in ROS/Gazebo environment. SPIIRAS Proc. **18**(1), 57–84 (2019)
19. Lee, B.J., Choi, J., Baek, C., Zhang, B.T.: Robust human following by deep Bayesian trajectory prediction for home service robots. In: International Conference on Robotics and Automation (ICRA), pp. 7189–7195. IEEE, New York (2018)
20. Leigh, A., Pineau, J., Olmedo, N., Zhang, H.: Person tracking and following with 2D laser scanners. In: International Conference on Robotics and Automation (ICRA), pp. 726–733. IEEE, New York (2015)
21. Magid, E., Lavrenov, R., Afanasyev, I.: Voronoi-based trajectory optimization for UGV path planning. In: International Conference on Mechanical, System and Control Engineering, pp. 383–387. IEEE, New York (2017)
22. Moskvin, I., Lavrenov, R.: Modeling tracks and controller for Servosila Engineer robot. In: Proceedings of 14th International Conference on Electromechanics and Robotics "Zavalishin's Readings", pp. 411–422. Springer, Berlin (2020)
23. Pages, J., Marchionni, L., Ferro, F.: Tiago: the modular robot that adapts to different research needs. In: International Workshop on Robot Modularity, IROS (2016)
24. RoboCup: Rules & regulations. http://www.robocupathome.org/rules. Accessed on 13 Jan 2020
25. ROS: Tiago base. http://wiki.ros.org/Robots/PMB-2. Accessed on 13 Jan 2020

26. Sato, Y., Arai, M., Suzuki, R., Kobayashi, Y., Kuno, Y., Yamazaki, K., Yamazaki, A.: A maneuverable robotic wheelchair able to move adaptively with a caregiver by considering the situation. In: RO-MAN, pp. 282–287. IEEE, New York (2013)
27. Sung, Y., Chung, W.: Hierarchical sample-based joint probabilistic data association filter for following human legs using a mobile robot in a cluttered environment. IEEE Trans. Hum.-Mach. Syst. **46**(3), 340–349 (2015)

Chapter 30
Evaluation of Visual SLAM Methods in USAR Applications Using ROS/Gazebo Simulation

Ramil Safin, Roman Lavrenov, and Edgar Alonso Martínez-García

Abstract The problem of determining the position of a robot and at the same time building the map of the environment is referred to as SLAM. A SLAM system generally outputs the estimated trajectory (a sequence of poses) and the map. In practice, it is hard to obtain ground-truth for the map; hence, only trajectory ground-truth is considered. There are various works that provide datasets to evaluate SLAM algorithms in different scenarios including sensor configurations, robots, and environments. Dataset collection in a real-world environment is a complicated task, which requires an elaborate sensor and robot configuration. Different SLAM systems demand various sensors resulting in the problem of finding an appropriate dataset for their evaluation. Thus, in this paper, a solution that is based on ROS/Gazebo simulations is proposed. Two indoor environments with flat and uneven terrain to evaluate laser range and visual SLAM systems are created. Changing the sensor configuration and the environment does not require an elaborate setup. The results of the evaluation for two popular SLAM methods—ORB-SLAM2 and RTAB-Map—are presented.

R. Safin · R. Lavrenov (✉)
Laboratory of Intelligent Robotic Systems (LIRS), Intelligent Robotics Department, Higher Institute for Information Technology and Intelligent Systems, Kazan Federal University, Kazan, Russian Federation
e-mail: lavrenov@it.kfu.ru
URL: https://kpfu.ru/erobotics

R. Safin
e-mail: safin.ramil@it.kfu.ru
URL: https://kpfu.ru/erobotics

E. A. Martínez-García
Laboratorio de Robotica, Institute of Engineering and Technology, Universidad Autonoma de Ciudad Juarez, Ciudad Juarez 32310, Mexico
e-mail: edmartin@uacj.mx

A. Ronzhin and V. Shishlakov (eds.), *Proceedings of 15th International Conference on Electromechanics and Robotics "Zavalishin's Readings"*, Smart Innovation, Systems and Technologies 187, https://doi.org/10.1007/978-981-15-5580-0_30

30.1 Introduction

Robotic systems are employed in a variety of applications, such as urban search and rescue (USAR), space exploration, military, medicine, and education [14]. In all of the applications, robots are required to perform particular tasks interacting with an environment. In some cases, environments are complex and non-deterministic which makes the intended tasks even more complex for a robot to accomplish [5].

There are applications where a robot is required to determine its position in the environment for navigation and space exploration. If the map of the environment is provided with certain accuracy, one can localize itself on the map and accomplish the intended task. However, in most cases, the map of the environment is partially available or not available which leads to the necessity of the map construction. Hence, the task becomes even more complicated, to simultaneously perform localization and build a map of the environment. This problem is referred to as simultaneous localization and mapping (SLAM).

One of the widely used sensors in SLAM is cameras. They provide extensive information about the environment, and they are not vulnerable to slippage as wheel encoders. SLAM using cameras is referred to as Visual SLAM (VSLAM). VSLAM finds its applications in autonomous driving [2], 3D-scene reconstruction, augmented reality (AR), etc.

One of the problems of SLAM methods evaluation is that there are various datasets with different sensors configurations. For example, stereo SLAM methods require a pair of monocular cameras [18]. Generally, most of the datasets provide laser range and stereo image data. However, in the case of multi-camera (two or more cameras) SLAM systems, those datasets cannot be employed due to the lack of necessary sensor data. Thus, in this paper, we propose to use ROS/Gazebo simulation for collecting datasets using different sensors configuration for the multisensor SLAM systems development and evaluation. We show the usability of the solution by evaluating Hector SLAM, ORB-SLAM2, and RTAB-Map methods on the collected datasets in the USAR scenario with flat and uneven terrains.

30.2 SLAM Overview

In SLAM, a robot traverses a particular environment constructing its map and estimating the state of the robot. The map is a representation of the landmarks, obstacles, and objects in the surroundings. It helps a robot to navigate through the environment being the reference for localization [15]. Maps can be represented in several ways: metric, topological, or hybrid. The state that is estimated can be defined as a 2D position and orientation or a 6D pose (Cartesian coordinates and orientation).

SLAM approaches can be classified into filtering and smoothing methods. Filtering approaches are based on Bayesian theory; here, the position of landmarks and robot poses are computed over time by incorporating new measurements into the probabilistic model, i.e., estimating the posterior state distribution. There are two major steps in this framework: prediction of the state based on the control commands and update of the state by incorporating new sensor measurements. Considering the probability distribution representation, filter-based methods can be classified into unimodal and multimodal ones. The former is represented by the Kalman filter family methods (e.g., EKF, UKF, SIKF) while the latter—particle filters (e.g., Rao-Blackwellized [16]). The main disadvantages of unimodal filtering methods are the quadratic growth of time complexity with the growth of the map size (number of features) and poor accuracy in case of nonlinear motion. As for the multimodal approaches, the accuracy depends on the number of particles which directly correlates with time complexity.

Smoothing methods take into account a sequence of states—full trajectory or some part of it—in order to estimate the new state using nonlinear least-squares optimization techniques (e.g., Gauss–Newton, Levenberg–Marquardt). Optimization approaches can be divided into bundle adjustment (BA) and graph classes. The key idea behind BA is to jointly optimize a 3D structure (map) and camera poses. This is accomplished by minimizing the re-projection error. Full BA over the whole trajectory is a computationally expensive task; hence, it might be used in case of loop closure as a part of the global optimization process. On the contrary, local BA over a sliding window of poses is able to mitigate the accumulated drift providing real-time performance even on mobile devices. Graph approaches are modeled using a graph representation of SLAM. The graph optimization is solved by minimizing a particular cost function using the same optimization techniques as for the BA.

Generally, the architecture of VSLAM contains two main modules: back-end and front-end (Fig. 30.1). The front-end processes sensor measurements and builds the model of the state. Back-end solves for the state trying to estimate the best fit for the SLAM solution. Here, we have BA and graph optimization.

Visual odometry (VO) is the main building block of VSLAM. It is the process of estimating the relative motion between consequent frames using cameras [17]. It is less affected by the drift issue compared to the wheel odometry.

The pipeline of VO consists of the following blocks: feature detection, motion estimation, data association, and local BA. VO methods can be separated into two

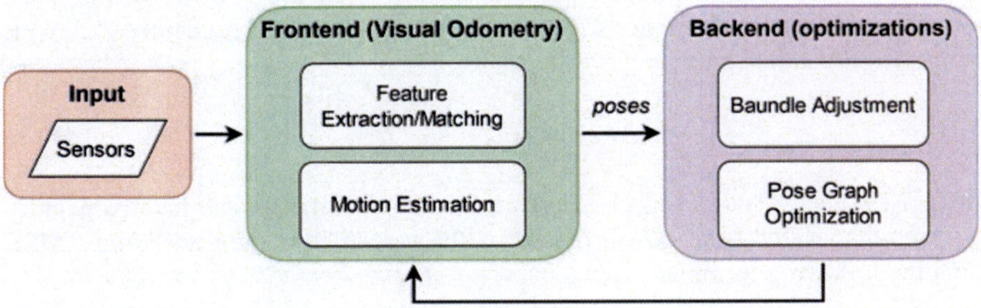

Fig. 30.1 Main components of VSLAM algorithm

groups: direct and indirect. Direct methods enable to build a dense map of the environment directly working with image pixels. On the other hand, indirect methods detect features in the images which increase the speed of computations. In feature-based VSLAM, the so-called feature detectors and descriptors are used to find salient points in the images and describe them into some representation (e.g., SIFT, ORB, SURF, etc.). This mechanism enables one to perform data association (feature matching) on them to recognize the same 3D points from different images. In the case of a direct approach, there is no feature detection routine in the pipeline.

Motion estimation can be divided into two categories: feature-based and optical flow methods. In the first case, the motion can be computed using one of the 3D-3D (Euclidean distance), 3D-2D (re-projection error) or 2D-2D techniques where the corresponding error is minimized to obtain the pose relative to the previous one. In the case of direct VO, generally, the photometric error is minimized. Local BA over the last keyframes can be performed to mitigate the drift errors.

VSLAM pipeline additionally may include global optimization and loop closure routines. Loop closure is the final step in SLAM algorithms. It optimizes the estimated SLAM solution incorporating additional constraints into the system. Here, feature matching techniques, a bag of words, etc., can be used for loop detection where features are matched across a set of keyframes.

30.3 VSLAM Evaluation

A SLAM system generally outputs the estimated camera trajectory (a sequence of poses) and the map. In practice, it is impossible to obtain the ground-truth map; hence, only trajectory ground-truth is obtained. Ideally, the sequences of SLAM output and ground-truth should be time-synchronized and evenly sampled with the same sequence length. The trajectory is a sequence of poses with respect to some reference frame which is obtained during the calibration. The reference frame is not required to be the same for the ground-truth and estimated trajectory.

The global consistency of SLAM can be evaluated by comparing the absolute distances between the estimated and ground-truth trajectories—absolute pose error (APE). The alignment is necessary in case if trajectories are in different reference frames [21]. Given the reference trajectory $GT_{1:n}$ and estimated trajectory $T_{1:n}$, APE at time step i can be computed as follows:

$$APE_i = GT_i^{-1} \cdot S \cdot T_i, \tag{30.1}$$

where S—is a least-squares rigid-body transformation that maps estimated trajectory $T_{1:n}$ onto the ground-truth $GT_{1:n}$. For the APE, we can compute translation RMSE using the following formula:

$$RMSE(APE_{1:n}) = \frac{1}{n}\left(\sum_{i=1}^{n} ||transl(APE_i)||^2\right)^{\frac{1}{2}}. \quad (30.2)$$

The relative pose error (RPE) measures the accuracy of the trajectory over a fixed period, which is drift. RPE considers both relative and translation errors while APE takes into account only translation error. Nevertheless, as the translation part is correlated to the rotation error, APE and RPE are correlated.

There are various works that provide with datasets to evaluate SLAM algorithms in different scenarios including sensor configurations and environments. One of the first datasets is provided in [4] with outdoor and indoor information of the ground-truth, GPS, IMU, LiDAR, and image data.

KITTI datasets [6] are obtained from a vehicle in an urban environment. The vehicle is equipped with a stereo camera and LiDAR. Translation and rotation error can be measured comparing the real data with ground-truth.

Amey K. provides a comparison of two SLAM methods—RGBD-SLAM and RTAB-Map [10]. Here, TUM RGBD dataset [20] is used for SLAM methods evaluation. The dataset provides with data from Microsoft Kinect RGBD sensor and time-synchronized ground-truth camera poses from a motion capture system. The evaluation metrics are absolute trajectory error (ATE or APE) and relative pose error (RPE). The analysis involves the RMSE computation for the two metrics above and processing time comparison.

In the other paper by Maksim et al. [9], SLAM systems in an indoor environment are compared for a mobile robot. Real-world experiments on their dataset were collected using LiDAR. They evaluate and compare LiDAR, monocular, and stereo algorithms: GMapping, Hector SLAM, ORB-SLAM, SVO, S-PTAM, etc. The robot was teleoperated through the environment to follow a closed-loop trajectory along a rectangular area. Mostly, the environment is consisted of walls. The rotation was accomplished in a radius of around 1m. Data from sensors was collected into ROS bag files. In the evaluation, the authors used ATE taking Hector SLAM algorithm as a ground-truth. The trajectory of the Hector SLAM was compared to the marked line on the floor.

In the work done by Arthur et al. [8], they evaluate different SLAM algorithm on TUM RGBD datasets. There are several metrics to evaluate SLAM systems: localization accuracy (RMSE), processing time (CPU usage), memory consumption, robustness (produce the same results for multiple runs), specific for visual SLAM: camera frame processing time (should be near FPS), map specific for occupancy grid maps: quality of the map (e.g., GMapping).

In the paper by Riccardo et al. [7], authors conduct a comparison of ROS compatible stereo SLAM methods on NVidia Jetson TX2 platform. The following metrics are evaluated: CPU and memory usage, trajectory estimation accuracy, and loop closure capability. The ground-truth is obtained utilizing LiDAR (reference). ZED camera is used as the main sensor. The rover is controlled remotely over a circular path while acquiring a stereo video stream and LiDAR. Ground-truth is provided by

LiDAR SLAM, Hector SLAM algorithm. Images are captured at 15 Hz. Both sensors are stored for further usage. ATE and RTE errors are computed for the computed trajectories.

In SlamBench2, the authors propose a framework for SLAM evaluation [1]. Currently, several algorithms are supported for evaluation (eight algorithms): ORB-SLAM2, OKVIS, etc. Its purpose is to unify the interface of benchmarking SLAM algorithms and provide reproducibility of results for different datasets. It allows to measure different metrics: computation speed (time per frame, FPS), ATE and RPE with online and offline alignment, power consumption (PAPI), memory usage. In their up-to-date paper SLAMBench3, they move toward scene understanding [3].

In the paper by David S. et al., TUM benchmark for evaluation of VO systems is presented [20]. It provides with camera images taken at 20 Hz synchronized with the IMU, HDR, and photometric calibration. For trajectory evaluation, they provide with a motion capture system at the start and the end of the sequences. It is focused RGBD odometry evaluation. There are lots of datasets available for VO evaluation: (a) KIITI odometry (stereo @10 Hz, IMU, software sync, GT: GPS); (b) Malaga Urban (stereo @20 Hz, IMU, software sync, GT: GPS); (c) EuRoC MAV (stereo @20, IMU, hardware sync, GT: motion capture); (d) PennCOSYVIO (stereo @20hz, fisheye @30 Hz, IMU, GT: Fiducial markers [22]). The main idea of TUM is to compute the drift using the start and the end positions of the trajectory.

An interesting solution is to move toward simulation environments where one can perform repetitive controlled SLAM methods evaluation. However, in the review literature, there is no solution which supports for multi-camera configurations. For instance, InteriorNet [13] provides with datasets for CV algorithms evaluation. There are multiple types of camera models and scenes which are supported. However, there is no possibility for the moment of writing this paper to incorporate three or more cameras into the simulation.

In a lot of works, Hector SLAM [11] is utilized to obtain ground-truth trajectories to evaluate against. However, generally in those works, only a flat terrain is considered, whereas in USAR applications, uneven terrains are frequently met. On the other hand, Hector SLAM provides with tools for six DoF that pose estimation mostly based on IMU data which is arguably not the most reliable source for ground-truth pose estimation. In more complicated environments, motion capture systems are employed, e.g., Vicon. However, those systems are expensive and require elaborate setup.

30.4 Proposed Solution

30.4.1 Environment

Simulators for robotic systems have been extensively used in research for repetitive evaluation of algorithms in different scenarios, including flat and uneven terrains,

Fig. 30.2 View of the Gazebo indoor environment (flat terrain)

various weather conditions, etc. Gazebo is one of the most popular simulators. It is widely used in conjunction with robot operating system (ROS) framework—a set of libraries and tools for the development of robotic systems. Gazebo provides with means for modeling various robotic systems equipped with different sensors, such as monocular and stereo cameras, IMU, and LiDAR [12].

In our solution, we create two environments for SLAM algorithms evaluation, i.e., an indoor environment of a building with flat and uneven terrain (Fig. 30.2). The building construction is 10 by 10 meters in size. The model of the environment was made in Blender 2.8.

Both environments have similar objects on the scene: windows, shelves, damaged wall, and truss. The flat environment's floor is a plane without any height changes throughout the scene. On the contrary, uneven environment's floor represents a surface with changes in height of up to 1.3 m relative to the ground. Illumination is even all over the environment without shadows.

30.4.2 Robot Model and Sensors

In this work, we use Husky mobile robot provided by the *husky_description* ROS package and modify it to include necessary sensors. The model of the robot is represented in the unified robot description format (URDF). The model consists of links connected by joints with their visual and collision meshes. The simplified transformation tree and robot model are presented in Fig. 30.3. The robot is equipped with the following sensors: (a) stereo view camera (resolution: 640 × 480, focal length: 320 pixels, RGB); (b) laser range finder (field of view 270°); (c) inertial measurement unit (IMU).

Gazebo sensors API enables to incorporate distortions into the sensory input. For example, attached cameras can have pixel intensity noise (e.g., Gaussian) as well as radial and tangential lens distortions. For stereo cameras, we employed a mutli-camera sensor API to provide synchronized global shutter images from both cameras [19].

30.4.3 Dataset Collection

We collected four datasets for each type of terrain for SLAM algorithms evaluation using ROS/Gazebo by remotely controlling the robot (in teleoperation mode) in a looped trajectory using *teleop_twist_keyboard* ROS package. The average linear velocity is around 0.15 m/s and angular—0.15 rad/s. Final trajectories do not have sharp turns to avoid drastic changes in the observed environment.

In our datasets collection, we employ a system with the following properties: (a) Intel Core i7-8750H @2.2 GHz CPU; (b) 16 GB LDDR4 RAM; (c) 1TB M.2 PCI-E SSD; (d) Ubuntu 16.04 LTS OS (with ROS Kinetic installed); The ground-truth 6D pose trajectory is obtained using the P3D Gazebo plugin. Poses are provided for the *base_footprint* frame w.r.t the *world* fixed frame (Fig. 30.3). Datasets were recorded into ROS bag files, they contain the following topics:

(a) */clock*—simulation timestamp;
(b) */cmd_vel*—control commands;
(c) */odometry/filtered*—IMU and wheel odometry EKF fusion;
(d) */ground_truth*—6D pose of the robot w.r.t world;
(e) */tf_static*—transformations between robot links;
(f) */scan*—LiDAR scans data;
(g) */camera/left/image_raw* and */camera/left/camera_info*—left stereo camera topics with image data and meta information;

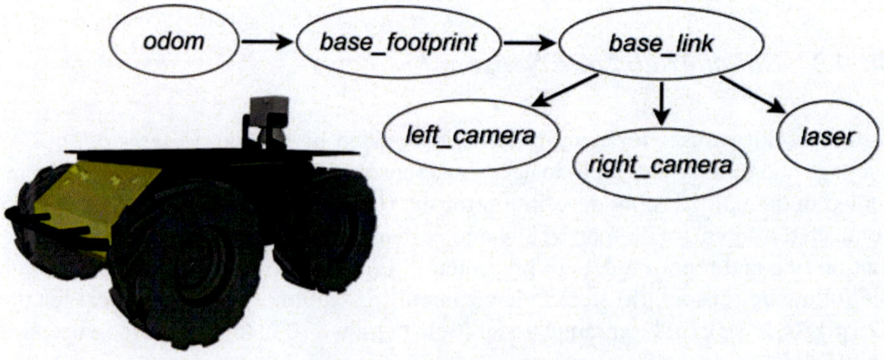

Fig. 30.3 Husky robot URDF model and its transformations tree (left—robot model visualization, right—robot model links hierarchy)

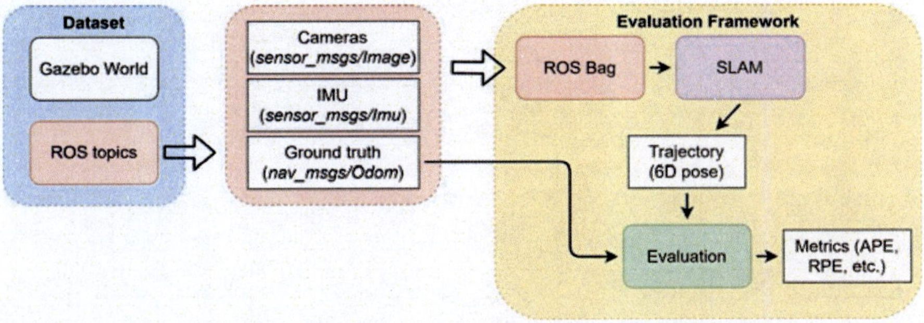

Fig. 30.4 SLAM evaluation components. Datasets contain the simulated environment and sensor data

(h) */camera/right/image_raw* and */camera/right/camera_info*—right stereo camera topics with image data and meta information;

We propose to record datasets and distribute them with the simulated environment (Fig. 30.4). The dataset contains the Gazebo world with all the meshes and textures included along with recorded ROS topics with sensor data. Datasets are stored in a ROS bag file which can be further issued to different SLAM algorithms for consequent evaluation.

30.4.4 Experiments

In our experiments, we evaluated the Hector SLAM, stereo RTAB-Map, and stereo ORB-SLAM2 algorithms on our recorded datasets of flat and uneven terrain environments. In the case of the flat terrain, Hector SLAM estimated trajectory error

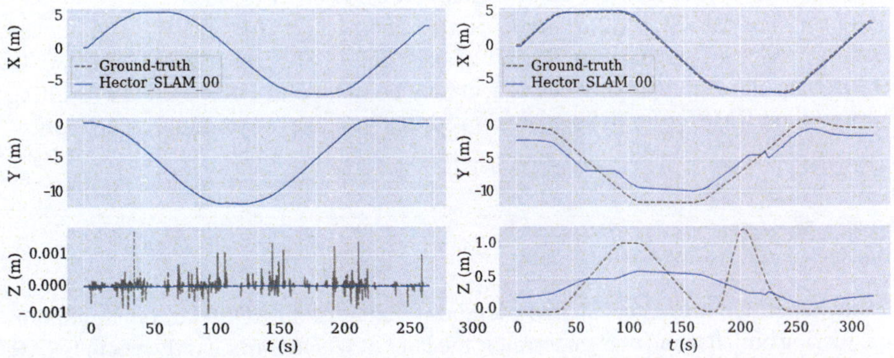

Fig. 30.5 Hector SLAM absolute pose error (APE) over time for flat and uneven terrains (sequence No. 0)

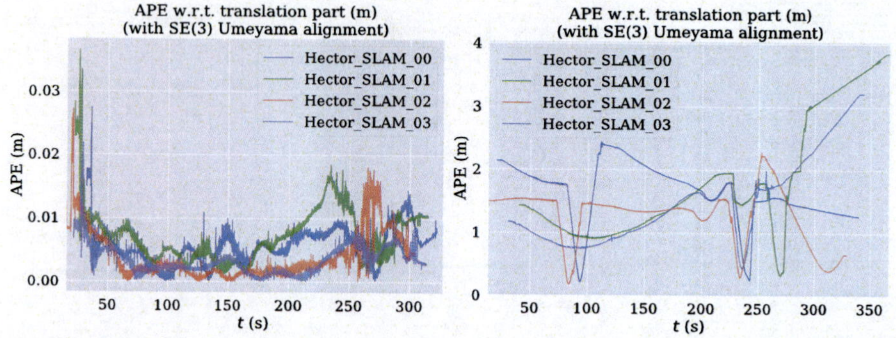

Fig. 30.6 Hector SLAM absolute pose error (APE) over time for flat (left) and uneven (right) terrains (all sequences)

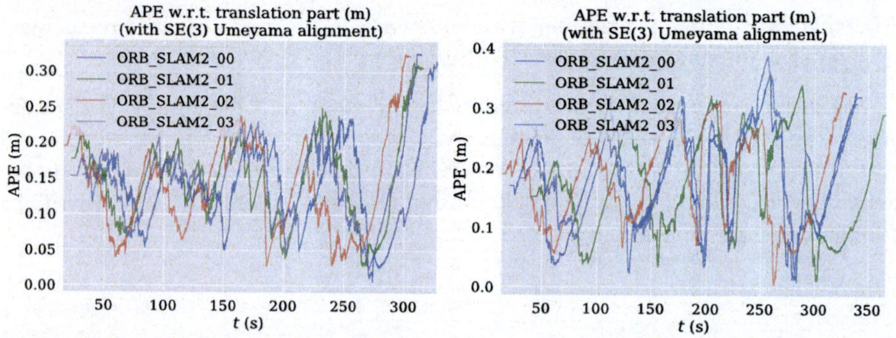

Fig. 30.7 ORB-SLAM2 absolute pose error (APE) over time for flat (left) and uneven (right) terrains (all sequences)

w.r.t the ground-truth is relatively small (Fig. 30.5). On the other hand in case of the uneven terrain, Hector SLAM diverges from the ground-truth trajectory in *Y* and *Z* axes. Hector SLAM APEs for all sequences are shown in Fig. 30.6.

As for the ORB-SLAM2, it demonstrates good results for both flat and uneven terrains resulting in translation RMSE of 0.15 and 0.19 m, respectively (Fig. 30.7). RTAB-Map showed relatively better accuracy compared to ORB-SLAM2 resulting in translation RMSE of 0.019 and 0.03 m for flat and uneven terrain, respectively.

30.5 Results

In this paper, we presented SLAM methods evaluation in ROS/Gazebo simulation considering both flat and uneven terrains, the later of which is frequently met in USAR applications. The evaluation process consisted of world creation, datasets collection, and computation of pose error metrics. The key advantages of simulations are the ease of configuration changes (using a different set of sensors and robot models) and repetitiveness of simulations.

We evaluated two popular Visual SLAM methods, ORB-SLAM2 and RTAB-Map, in both flat and uneven terrains using collected datasets, and Hector SLAM. The results showed that Hector SLAM in its simple form cannot be used in uneven terrains. However, in the case of flat terrains, due to its accuracy, it can be considered as a ground-truth for other SLAM methods evaluation. On the contrary, stereo ORB-SLAM2 and RTAB-Map estimated trajectories are fair in both cases, though in the case of flat terrains Hector SLAM pose error is less in an order of magnitude. In our future work, we plan to investigate other Visual SLAM algorithms for USAR applications.

Acknowledgements This research was funded by the Russian Foundation for Basic Research (RFBR), project ID 18-58-45017. This work was partially supported by the research grant of Kazan Federal University.

References

1. Bodin, B., Wagstaff, H., Saecdi, S., Nardi, L., Vespa, E., Mawer, J., Nisbet, A., Luján, M., Furber, S., Davison, A.J., et al.: Slambench2: multi-objective head-to-head benchmarking for visual slam. In: 2018 IEEE International Conference on Robotics and Automation (ICRA), pp. 1–8. IEEE, New York (2018)
2. Bresson, G., Alsayed, Z., Yu, L., Glaser, S.: Simultaneous localization and mapping: a survey of current trends in autonomous driving. IEEE Trans. Intell. Veh. **2**(3), 194–220 (2017)
3. Bujanca, M., Gafton, P., Saeedi, S., Nisbet, A., Bodin, B., O'Boyle Michael, F., Davison, A.J., Riley, G., Lennox, B., Luján, M., et al.: Slambench 3.0: systematic automated reproducible evaluation of slam systems for robot vision challenges and scene understanding. In: 2019 International Conference on Robotics and Automation (ICRA), pp. 6351–6358. IEEE, New York (2019)
4. Ceriani, S., Fontana, G., Giusti, A., Marzorati, D., Matteucci, M., Migliore, D., Rizzi, D., Sorrenti, D.G., Taddei, P.: Rawseeds ground truth collection systems for indoor self-localization and mapping. Autonom. Rob. **27**(4), 353 (2009)
5. Denisov, E., Sagitov, A., Lavrenov, R., Su, K.L., Svinin, M., Magid, E.: Dcegen: dense clutter environment generation tool for autonomous 3d exploration and coverage algorithms testing. In: International Conference on Interactive Collaborative Robotics, pp. 216–225. Springer, Berlin (2019)
6. Geiger, A., Lenz, P., Urtasun, R.: Are we ready for autonomous driving? The Kitti vision benchmark suite. In: 2012 IEEE Conference on Computer Vision and Pattern Recognition, pp. 3354–3361. IEEE, Berlin (2012)
7. Giubilato, R., Chiodini, S., Pertile, M., Debei, S.: An experimental comparison of ros-compatible stereo visual slam methods for planetary rovers. In: 2018 5th IEEE International Workshop on Metrology for AeroSpace (MetroAeroSpace), pp. 386–391. IEEE, New York (2018)
8. Huletski, A., Kartashov, D., Krinkin, K.: Evaluation of the modern visual slam methods. In: 2015 Artificial Intelligence and Natural Language and Information Extraction, Social Media and Web Search FRUCT Conference (AINL-ISMW FRUCT), pp. 19–25. IEEE, New York (2015)

9. Ibragimov, I.Z., Afanasyev, I.M.: Comparison of ros-based visual slam methods in homogeneous indoor environment. In: 2017 14th Workshop on Positioning, Navigation and Communications (WPNC), pp. 1–6. IEEE, New York (2017)
10. Kasar, A.: Benchmarking and comparing popular visual slam algorithms. arXiv preprint arXiv:1811.09895 (2018)
11. Kohlbrecher, S., Meyer, J., Graber, T., Petersen, K., Klingauf, U., von Stryk, O.: Hector open source modules for autonomous mapping and navigation with rescue robots. In: Robot Soccer World Cup, pp. 624–631. Springer, Berlin (2013)
12. Lavrenov, R., Zakiev, A., Magid, E.: Automatic mapping and filtering tool: From a sensor-based occupancy grid to a 3d gazebo octomap. In: 2017 International Conference on Mechanical, System and Control Engineering (ICMSC), pp. 190–195. IEEE, New York (2017)
13. Li, W., Saeedi, S., McCormac, J., Clark, R., Tzoumanikas, D., Ye, Q., Huang, Y., Tang, R., Leutenegger, S.: Interiornet: mega-scale multi-sensor photo-realistic indoor scenes dataset. arXiv preprint arXiv:1809.00716 (2018)
14. Magid, E., Pashkin, A., Simakov, N., Abbyasov, B., Suthakorn, J., Svinin, M., Matsuno, F.: Artificial intelligence based framework for robotic search and rescue operations conducted jointly by international teams. In: Proceedings of 14th International Conference on Electromechanics and Robotics "Zavalishin's Readings", pp. 15–26. Springer, Berlin (2020)
15. Magid, E., Tsubouchi, T.: Static balance for rescue robot navigation: discretizing rotational motion within random step environment. In: International Conference on Simulation, Modeling, and Programming for Autonomous Robots, pp. 423–435. Springer, Berlin (2010)
16. Montemerlo, M., Thrun, S.: FastSLAM: A Scalable Method for the Simultaneous Localization and Mapping Problem in Robotics, vol. 27. Springer, Berlin (2007)
17. Nistér, D., Naroditsky, O., Bergen, J.: Visual odometry. In: Proceedings of the 2004 IEEE Computer Society Conference on Computer Vision and Pattern Recognition, 2004. CVPR 2004, vol. 1, pp. I–I. IEEE, New York (2004)
18. Safin, R., Lavrenov, R., Saha, S.K., Magid, E.: Experiments on mobile robot stereo vision system calibration under hardware imperfection. In: MATEC Web of Conferences, vol. 161, p. 03020. EDP Sciences (2018)
19. Safin, R., Lavrenov, R., Tsoy, T., Svinin, M., Magid, E.: Real-time video server implementation for a mobile robot. In: 2018 11th International Conference on Developments in eSystems Engineering (DeSE), pp. 180–185. IEEE, New York (2018)
20. Schubert, D., Goll, T., Demmel, N., Usenko, V., Stückler, J., Cremers, D.: The TUM VI benchmark for evaluating visual-inertial odometry. In: 2018 IEEE/RSJ International Conference on Intelligent Robots and Systems (IROS), pp. 1680–1687. IEEE, New York (2018)
21. Sturm, J., Engelhard, N., Endres, F., Burgard, W., Cremers, D.: A benchmark for the evaluation of RGB-D slam systems. In: 2012 IEEE/RSJ International Conference on Intelligent Robots and Systems, pp. 573–580. IEEE, New York (2012)
22. Zakiev, A., Shabalina, K., Tsoy, T., Magid, E.: Pilot virtual experiments on ARUco and ARTag systems comparison for fiducial marker rotation resistance. In: Proceedings of 14th International Conference on Electromechanics and Robotics "Zavalishin's Readings", pp. 455–464. Springer, Berlin (2020)

Chapter 31
Mathematical Model for Evaluating Fault Tolerance of On-Board Equipment of Mobile Robot

Eugene Larkin, Tatiana Akimenko, Alexey Bogomolov, and Konstantin Krestovnikov

Abstract The approach to simulation of a mobile robot on-board equipment fault tolerance based on the theory of semi-Markov processes is presented. At the first stage of proposed approach a simulation the lifecycles of equipment units is considered and for every unit ordinary semi-Markov model is determined. Formulae for unit lifetime calculation in the general case are obtained. At the second stage, the model of competition between units for the fault is considered. To simplify the simulation, the digital approach to the competition analysis is worked out. Digital model is obtained by means of sampling densities described units lifetimes. It is shown that digital approach to analysis of competition permit, rather simply, calculates a lifetime of fault-tolerant assembles and/or on-board equipment as a whole. So, it is shown that number of sampled in histogram, presented unit lifetime allows to control both accuracy and computational complexity of mobile robot reliability parameters calculation. Example of calculation of fault-tolerant actuator control system reliability is presented.

31.1 Introduction

Mobile robots, executed target tasks in an aggressive environment are currently widely used in industry, anti-terrorism operations, technology disaster consequences

E. Larkin · T. Akimenko (✉) · A. Bogomolov
Tula State University, 92, ave. Lenin, Tula 300012, Russia
e-mail: tantan72@mail.ru

E. Larkin
e-mail: elarkin@mail.ru

A. Bogomolov
e-mail: avbogomolov@gmail.com

K. Krestovnikov
St. Petersburg University of Aerospace Instrumentation, 67, St. Bolshaya Morskaya, 190000 St. Petersburg, Russia

A. Ronzhin and V. Shishlakov (eds.), *Proceedings of 15th International Conference on Electromechanics and Robotics "Zavalishin's Readings"*, Smart Innovation, Systems and Technologies 187, https://doi.org/10.1007/978-981-15-5580-0_31

dissolution, military sphere, etc. [1–3]. The impact of environment leads to the fact that robot's on-board equipment reliability indicator, namely mean time between failures, falls sharply, which reduces the robot's lifetime in general. To increase the lifetime, the planned redundancy is brought into the equipment. Since the reliability parameters [4, 5] of an individual block are limited, this problem can be solved only systematically, using redundant fault-tolerant structures [6, 7]. For proper planning the redundancy it is necessary preliminary simulate failure-recovery process of both only units, and equipment as a whole, the general approach to modeling the reliability of a system is based on the theory of Markov [8] or semi-Markov [9, 10], processes which allows to describe a single unit of equipment lifetime. When modeling redundant structures, a competition effect arises, and semi-Markov models are replaced by more rough Markov models, which reduces the accuracy of the modeling procedure. It is proposed to use discrete semi-Markov models instead of Markov models to describe competition in fault-tolerant assemblies in which accuracy can be controlled by changing the number of samples at distribution densities. Therefore, it is necessary to develop a model whose accuracy can be estimated and increased/decreased in accordance with the solvable reliability problem, which explains the necessity and relevance of this study.

31.2 The Approach to Simulation of Fault-Tolerant Systems

Mobile robot equipment, in which fault-tolerance principle is realized, may be considered as M units, operated in parallel [11]. Fault/recoveries in assembly units develop in parallel, so such abstraction as M-parallel semi-Markov process may be used to describe a reliability of the assembly as follows:

$$\mu = \{\mu_1, \ldots, \mu_m, \ldots, \mu_M\}, \tag{31.1}$$

where $\mu_m = \{A_m, \mathbf{h}_m(t)\}$, $1 \leq m \leq M$, is the ordinary semi-Markov process [9], which is characterized with set of structural states $A_m = \{a_{0(m)}, \ldots, a_{j(m)}, \ldots, a_{J(m)}\}$ and semi-Markov matrix

$$\mathbf{h}_m(t) = \left[h_{j(m),k(m)}(t)\right]; \tag{31.2}$$

t is the time; $a_{0(m)}$ simulates the start of mth unit exploitation when it surely able to work; $a_{J(m)}$ is the absorbing state, which simulates the fully destroyed unit; $a_{j(m)}$, $1(m) \leq j(m) < J(m)$ simulate other physical states (able to work, short-time failed, under recovering, etc.);

$$\mathbf{h}_m(t) = \mathbf{p}_m \otimes \mathbf{f}_m(t); \tag{31.3}$$

$\mathbf{p}_m = \left[p_{j(m),k(m)}\right]$ and $\mathbf{f}_m(t) = \left[f_{j(m),k(m)}(t)\right]$ are $[J(m)+1]\times[J(m)+1]$ stochastic matrix and matrix of pure time densities, correspondingly.

Semi-Markov matrix 31.2) has the next features: elements of the matrix $\mathbf{h}_m(t)$ zero column and $J(m)$th row are equal to zeros. Physical it means that unit cannot return to the beginning of exploiting and cannot return to exploiting from the state of complete destruction. Weighted time densities $h_{j(m),k(m)}(t)$ describe both time of sojourn in the state $a_{j(m)}$, and prior probabilities of switching into conjugative states. For elements of rows from $0(m)$th till $[J(m)-1]$th, the next expression is true:

$$\mathbf{h}_m(t) \qquad \sum_{k(m)=1(m)}^{J(m)} \int_0^{\infty} h_{j(m),k(m)}(t)\mathrm{d}t = 1, \quad 0(m) \le j(m) \le J(m). \tag{31.4}$$

Both probabilities of $\mathbf{p}_m$-matrix and parameters of $\mathbf{f}_m(t)$-matrix (expectation, dispersion, initial and central moments of higher orders) depend on the substance, of which the element is made, a quality of element manufacturing and assembling, an exploiting conditions, side effects, etc., and define parameters of reliability.

There are only three tasks, which one should to decide, when reliability parameters evaluation: (1) calculation of the time till failure (wandering from $a_{0(m)}$ till $a_{J(m)}$); (2) calculation of the time and probability of wandering from $a_{j(m)} \neq a_{0(m)} \neq a_{J(m)}$ till $a_{k(m)} \neq a_{0(m)} \neq a_{J(m)}$; (3) calculation of the time and probability of returning to $a_{j(m)} \neq a_{0(m)} \neq a_{J(m)}$. In common, time till failure may be defined as follows [11]:

$$\tilde{f}_{0(m),J(m)}(t) = L^{-1}\left[I_{0(m)}^{R} \cdot \sum_{w=1}^{\infty} \{L[\mathbf{h}_m(t)]\}^{w} \cdot I_{J(m)}^{C}\right], \tag{31.5}$$

where $I_{0(m)}^{R}$ is the $[J(m)+1]$-size row-vector, in which $0(m)$th element is equal to one, and other elements are equal to zeros; $I_{0(m)}^{C}$ is the $[J(m)+1]$-size column-vector, in which $J(m)$th element is equal to one, and other elements are equal to zeros; $L[\ldots]$ and $L^{-1}[\ldots]$ are direct and inverse Laplace transforms, correspondingly.

To solve the second task, one should transform $\mathbf{h}_m(t)$ as follows:

$$\mathbf{h}_m(t) \to \mathbf{h}'_m(t). \tag{31.6}$$

When transformation, the only restriction is imposed onto wandering trajectories, is that neither to state $a_{j(m)}$, nor to state $a_{k(m)}$ process should not fall twice. To form $\mathbf{h}'_m(t)$ with such properties in semi-Markov matrix $\mathbf{h}_m(t)$, all elements of $j(m)$th column and $k(m)$th row should be replaced by zeros. Elements $h_{i(m),l(m)}(t)$ should be recalculated:

$$h'_{i(m),l(m)}(t) = \frac{h_{i(m),l(m)}(t)}{\sum_{\substack{k(m)=0(m),\\ k(m)\neq j(m)}}^{J(m)} p_{i(m),k(m)}}; \quad 0(m) \le i(m) \le J(m);\, i(m) \neq k(m). \tag{31.7}$$

Stochastic summation of densities, formed on all possible wandering trajectories, gives next expression:

$$\tilde{h}'_{j(m),k(m)}(t) = \mathbf{I}^R_{j(m)} \cdot L^{-1}\left[\sum_{w=1}^{\infty} \{L[\mathbf{h}'_m(t)]\}^w\right] \cdot \mathbf{I}^C_{k(m)}, \qquad (31.8)$$

where $I^R_{j(m)}$ is the row-vector, in which $j(m)$th element is equal to one, and other elements are equal to zeros; $I^C_{k(m)}$ is the column-vector, in which $k(m)$th element is equal to one, and other elements are equal to zeros.

In the semi-Markov process $\mathbf{h}'_m(t)$, there are as minimum two absorbing states, namely $a_{k(m)}$ and $a_{J(m)}$, so group of events of reaching $a_{k(m)}$ from $a_{j(m)}$ is not full, and in common case, $\tilde{h}'_{j(m),k(m)}(t)$ is weighted, but not pure density. The state $a_{k(m)}$ from the state $a_{j(m)}$ may be reached with probability [12]:

$$\tilde{p}'_{j(m),k(m)} = \int_0^{\infty} \tilde{h}'_{j(m),k(m)}(t)\mathrm{d}t, \qquad (31.9)$$

and pure time density of wandering from the state $a_{j(m)}$ to the state $a_{k(m)}$ may be defined as follows:

$$\tilde{f}'_{j(m),k(m)}(t) = \frac{\tilde{h}'_{j(m),k(m)}(t)}{\tilde{p}'_{j(m),k(m)}}. \qquad (31.10)$$

When solving the third task, one should execute the following transformation:

$$\mathbf{h}_m(t) \to \mathbf{h}''_m(t). \qquad (31.11)$$

When doing 31.11): one row and one column should be added to the matrix; added, $[J(m)+1]$th, row should be fulfilled with zeros; $j(m)$th column at first should be carried to the $[J(m)+1]$th column, and then, it should be fulfilled with zeros. Stochastic summation of densities, formed on all possible wandering trajectories, gives next expression:

$$\tilde{h}''_{j(m),k(m)}(t) = \mathbf{I}^R_{j(m)} \cdot L^{-1}\left[\sum_{w=1}^{\infty} \{L[\mathbf{h}''_m(t)]\}^w\right] \cdot \mathbf{I}^C_{J(m)+1}, \qquad (31.12)$$

where $I^R_{j(m)}$ is the $[J(m)+2]$-size row-vector, in which $j(m)$th element is equal to one, and other elements are equal to zeros; $I^C_{J(m)+1}$ is the $[J(m)+2]$-size column-vector, in which $[J(m)+1]$th element is equal to one, and other elements are equal to zeros.

In the semi-Markov process $\mathbf{h}''_m(t)$, there are two absorbing states, namely $a_{J(m)}$ and $a_{J(m)+1}$, so group of events of reaching $a_{J(m)+1}$ from $a_{j(m)}$ is not full, and in common case, $\tilde{h}''_{j(m),J(m)+1}(t)$ is weighted, but not pure density. The state $a_{J(m)+1}$ from the state $a_{j(m)}$ may be reached with probability:

$$\tilde{p}''_{j(m),J(m)+1} = \int_0^\infty \tilde{h}''_{j(m),J(m)+1}(t)\mathrm{d}t, \tag{31.13}$$

and pure time density:

$$\tilde{p}''_{j(m),J(m)+1} = \int_0^\infty \tilde{h}''_{j(m),J(m)+1}(t)\mathrm{d}t. \tag{31.14}$$

Let us consider generalized density:

$$\phi_m(t) \in \left\{ \tilde{f}_{0(m),J(m)}(t), \tilde{f}'_{j(m),k(m)}(t), \tilde{f}''_{j(m),J(m)+1}(t) \right\}. \tag{31.15}$$

In the most general case, $\phi_m(t)$ is continual function, with the next common properties:

$$0 \le t_{\min} \le \arg[\phi_m(t)] \le t_{\max} < \infty. \tag{31.16}$$

Time density $\phi_m(t)$ may be represented as a histogram. For this purpose domain $[t_{\min}, t_{\max}]$ should be divided onto X intervals, $0 \le t < \tau_1, \ldots, \tau_{x-1} \le t < \tau_x, \ldots, \tau_{X-1} \le t < \infty$. For simplification of the model, it is advisable to do both borders τ_x between histogram digits and sampling points θ_x, representing digits, uniform for all $1 \le m \le M$. Histogram digits width Δ is the next:

$$\Delta = \frac{\tau_1 - \tau_{X-1}}{X - 2}, \tag{31.17}$$

where X is quantity of histogram digits; τ_1 and τ_{X-1} are right border of first digit and left border of Xth digit. Values of digits are equal to:

$$\pi_{m,x} = \int_{l(x)}^{r(x)} \phi_m(t)\mathrm{d}t, \tag{31.18}$$

where $l(x) \le t \le r(z)$ are left and right limits of integration interval:

$$l(x) = \begin{cases} \tau_1 + \Delta(x-2), & \text{when } 2 \le x \le X; \\ 0, & \text{when } x = 1; \end{cases} \tag{31.19}$$

$$r(x) = \begin{cases} \tau_1 + \Delta(x-1), & \text{when } 1 \le x \le X-1; \\ \infty, & \text{when } x = X. \end{cases} \tag{31.20}$$

Sampling points, representing digits, are as follows:

$$\theta_x = \tau_x - \frac{\Delta}{2}; \quad 1 \le x \le X \tag{31.21}$$

In sampled model, every digit of histogram is represented as weighted shifted degenerative distribution law, so function, described histogram, is as follows:

$$\tilde{\phi}_m(t) = \sum_{x=1}^{X} \pi_{m,x} \cdot \delta(t - \theta_x), \tag{31.22}$$

where $\delta(\ldots)$ is the Dirac δ-function; $\pi_{m,x}$ is the weight of Dirac function:

$$\sum_{x=1}^{X} \pi_{m,x} = 1. \tag{31.23}$$

Mobile robot redundant units, assembled into fault-tolerant structure during operation, compete between them for failure. Result of competition is the fail the mth unit the first, or not the first. Lifetime of M-units redundant structure is as follows [15, 17]:

$$\phi_v(t) = \frac{\mathrm{d}\left\{1 - \prod_{m=1}^{M} [1 - \Phi_m(t)]\right\}}{\mathrm{d}t}, \tag{21.24}$$

where $\Phi_m(t)$ is the distribution function;

$$\Phi_m(t) = \int_0^t \phi_m(\xi)\mathrm{d}\xi, \tag{31.25}$$

The weighted time density, the probability and the pure time density of winning the competition for failure by the mth unit are as follows:

$$\vartheta_{v,m}(t) = \phi_m(t) \prod_{\substack{l=1,\\ l \ne m}}^{M} [1 - \Phi_l(t)], \tag{31.26}$$

$$\pi_{v,m} = \int_0^\infty \vartheta_{v,m}(t)\mathrm{d}t, \tag{31.27}$$

$$\phi_{v,m}(t) = \frac{\vartheta_{v,m}(t)}{\pi_{v,m}}, \tag{31.28}$$

where $\vartheta_{v,m}(t)$ is the weighted time density; $\pi_{v,m}$ is the probability; $\phi_{v,m}(t)$ is the pure time density of winning by the mth unit. When $\phi_m(t)$ is transformed into its discrete analogue $\tilde{\phi}_m(t)$, as it is shown in (31.22), time distribution function is transformed to $\tilde{\Phi}_m(t)$:

$$\tilde{\Phi}_m(t) = \int_0^t \tilde{\phi}_m(\xi)\mathrm{d}\xi = \sum_{x=1}^{X} \pi_{m,x} \cdot \eta(t - \theta_x), \tag{31.29}$$

where $\eta(t - \theta_x)$ is the shifted Heaviside function.

Dependence 31.29) may be transformed into sequence of samples as follows:

$$\tilde{\Phi}_m(t) \to \bar{\Phi}_m(t) = \sum_{x=1}^{X} \left[\left(\sum_{y=1}^{x} \pi_{m,y} \right) \cdot \delta(t - \theta_x) \right], \tag{31.30}$$

where $\sum_{y=1}^{x} \pi_{m,y}$ is the nomination of the function $\bar{\Phi}_m(t)$ sample at the point θ_x.

Function $1 - \Phi_m(t)$ may be transformed into discrete form:

$$\overline{[1 - \Phi_m(t)]} = \sum_{x=1}^{X} \left[\left(\sum_{y=x+1}^{X} \pi_{m,y} \right) \cdot \delta(t - \theta_x) \right]. \tag{31.31}$$

Combinations of $\tilde{\phi}_m(t)$ and $\overline{[1 - \Phi_m(t)]}$ permit construct the discrete analog $\bar{\vartheta}_{v,m}(t)$ of function (31.26). It is necessary to admit that when time intervals are described with continual functions $\phi_m(t)$, $1 \leq m \leq M$, then there may be only one winner in the competition (31.25) due to the fact that probability of competition draw, even in the case of paired races, is too small in comparison with probabilities of winning by one of participants. Where time intervals are described with discrete function the probabilities, when draw effect emerging, is comparable with probabilities of one unit winnings due to the fact, that time interval $\tau_{x-1} \leq t < \tau_x$ may include number of events. To determine possible combinations $\tilde{\phi}_m(t)$ and $\overline{[1 - \Phi_m(t)]}$, it is necessary to consider data, which includes M binary digits:

$$n = \langle n(1), \ldots, n(m), \ldots, n(M) \rangle, \tag{31.32}$$

where inside triangle brackets there is the code, obtained by means of Cartesian exponentiation to M-th degree the set $\{0,\ 1\}$; $n(m) \in \{0,\ 1\}$ is binary digit; $0 \leq n < 2^M$.

All codes n may be gathered onto set N, which is divided onto subsets N_l:

$$N = \{N_0, \ldots, N_l, \ldots, N_M\}, \tag{31.33}$$

where N_l is the subset of codes, which include l "nulls" and $M - l$ "ones." In turn,

$$N_l = \left\{n_{1(M,l)}, \ldots, n_{c(M,l)}, \ldots, n_{C(M,l)}\right\}, \tag{31.34}$$

where $n_{c(M,l)}$ is $c(M, l)$th M-digits code, including l "nulls" and $M - l$ "ones"; $C(M, l)$ is common quantity of such codes, equal to lth of M binary coefficient; $c(M, l)$ is the index, which numerates codes in the set N_l;

$$C(M, l) = \frac{M!}{l! \cdot (M - l)!}, \tag{31.35}$$

$$n_{c(M,l)} = \langle n[1, c(M, l)], \ldots, n[m, c(M, l)], \ldots, n[M, c(M, l)]\rangle, \tag{31.36}$$

$$n[m, c(M, l)] \in \{0, 1\}. \tag{31.37}$$

Function of two parameters, namely time and mth code digit state $n[m, c(M, l)]$ should be introduced to describe distribution:

$$\psi\{t, n[m, c(M, l)]\} = \begin{cases} \tilde{\varphi}_m(t), & \text{when } n[m, c(M, l)] = 0; \\ \overline{[1 - \Phi_m(t)]}, & \text{when } n[m, c(M, l)] = 1. \end{cases} \tag{31.38}$$

A competition outcome, alike (31.26), when l units of M failure during time interval $\tau_{x-1} \le t < \tau_x$, may be expressed as:

$$\tilde{\vartheta}_{v,l/M}(t) = \sum_{m=1}^{M} \prod_{c(M,l)=1}^{C(M,l)} \psi\{t, n[m, c(M, l)]\}. \tag{31.39}$$

The probability and pure discrete time distribution and mean time of l/M units simultaneous failure are as follows:

$$\tilde{\pi}_{v,l/M} = \int_0^\infty \tilde{\vartheta}_{v,l/M}(t)\mathrm{d}t, \tag{31.40}$$

$$\tilde{\varphi}_{v,l/M}(t) = \frac{\tilde{\vartheta}_{v,l/M}(t)}{\tilde{\pi}_{v,l/M}(t)}, \tag{31.41}$$

$$\tilde{\tau}_{v,l/M} = \int_0^\infty t \cdot \tilde{\varphi}_{v,l/M}(t)\mathrm{d}t. \tag{31.42}$$

Let some robot fault-tolerant assemble is workable until all M units at least one unit stay “alive”. There are 2^{M-1} combinations of reaching unworkable state, f. e., 3-units assemble may fail as $1+1+1$, $1+2$, $2+1$, 3. Analysis of every combination gives different probabilities and pure time densities for evaluation of assemble “lifetime”. So, it is necessary to evaluate time till failure for every combination and then stochastically summarize them.

31.3 Example

As an example, let us consider mobile robot fault-tolerant actuator control system. To increase reliability, system includes reduplicated controller and sensor, connected to the circuit in such a way that both sensors measure the same value, for example, angular velocity, and measurement results of every sensor are transmitted to both controllers. Electronics of actuator control system are under impact of power line correlated noise, arising from on-board power equipment switching on/off. Noise may cause both short-term malfunctions and equipment hard failure. System as a whole fails only in the case when both sensors and/or both controllers fail. Structure of the 4-parallel semi-Markov process describes fault/recoveries in the system:

$$\mu = \{\mu_1, \mu_2, \mu_3, \mu_4\},$$

Where

$$\mu_m = \left\{ \{a_{1(m)},\ a_{2(m)},\ a_{3(m)}\},\ \begin{bmatrix} 0 & \delta(t) & 0 \\ 0 & h_{2(m),2(m)}(t) & h_{2(m),3(m)}(t) \\ 0 & 0 & 0 \end{bmatrix} \right\},\quad m \in \{1, 2, 3, 4\}. \tag{31.43}$$

In (31.43), $a_{1(m)}$ and $a_{3(m)}$ are the starting state (start of the system operation) and the absorbing state (equipment hard failure), correspondingly; $\delta(t)$ is the Dirac δ-function; $h_{2(m),i(m)}(t) = p_{2(m),i(m)} f_{2(m),i(m)}(t)$, $i \in \{2, 3\}$, is the weighted time density between noise impulses; $f_{2(m),2(m)}(t)$, $f_{2(m),3(m)}(t)$ are pure time densities; $p_{2(m),2(m)}$, $p_{2(m),3(m)}$ are probabilities of a short-term malfunctions and a hard failure, correspondingly:

$$p_{2(m),2(m)} = \begin{cases} 0,8, \text{ when } m \in \{1,\ 2\}; \\ 0,9, \text{ when } m \in \{3,\ 4\}; \end{cases}$$

$$f_{2(m),i(m)}(t) = \begin{cases} 2, \text{ when } |t-1| \le 0,25; \\ 0, \text{ when } |t-1| > 0,25, \end{cases} \quad i \in \{2,\ 3\},\ m \in \{1,\ 2,\ 3,\ 4\}.$$

Sampled time densities obtained from analysis of semi-Markov processes evolution accordingly (31.12) and using the sampling method described above are shown in

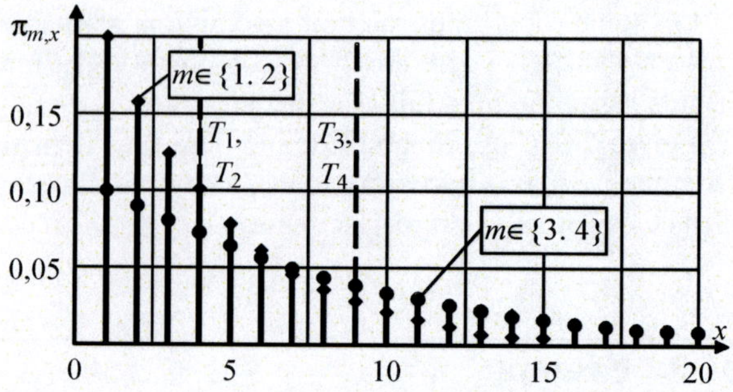

Fig. 31.1 Sampled time densities

Table 31.1 Scenarios of system failure

Scenario	Probability	MTBF (t.u.)
Sensor	0.306	1.3
Controller	0.694	1.3
Sensor/sensor	0.0553	3.02
Controller/controller	0.3695	3.52
Sensor/controller/sensor	0.0765	5.8
Sensor/controller/controller	0.1739	5.8
Controller/sensor/sensor	0.0994	6.3
Controller/sensor/controller	0.2254	6.3
MTBF summary (t.u.)		4.957

Fig. 31.1, where dashed lines indicate expectations of lifetime of the solar controller (T_1, T_2) and the solar sensor (T_3, T_4).

Failure evolution scenarios obtained from analysis of interaction in the fault-tolerant system are summarized in table. In Table 31.1, middle time (t.u. are time units) before failures (MTBF) and probabilities of every scenario calculated by the proposed method are given.

MTBF of the system without fault-tolerant arrangement is equal to 2.78 t.u., i.e., fault-tolerant structure increases durability of mobile robot actuator control system almost twice.

31.4 Conclusion

As a result, the task of designing fault-tolerant assemblies was proposed to be divided into two stages: (1) development of conventional semi-Markov models of individual units and conversion it to the discrete form; (2) analysis of a parallel discrete semi-Markov process to obtain the reliability parameters of a reliability parameters of equipment as a whole.

The proposed approach allows to create a model of a redundant system with any degree of accuracy, to develop a method for optimizing a fault-tolerant system based on the approach of a discrete model. Further research in this area may be aimed at modeling many practical redundant systems with complex interactions between components and complex "life cycle" algorithms.

Acknowledgements Research is supported by the RFBR Project No. №19-08-01215_A.

References

1. Gorobtsov, A., Andreev, A., Markov, A., Skorikov, A., Tarasov, P.: Features of solving the inverse dynamic method equations for the synthesis of stable walking robots controlled motion. SPIIRAS Proc. **18**(2), 85–122 (2019)
2. Landau, I.D., Zito, G.: Digital Control Systems, Design, Identification and Implementation. Springer, New York, 484 p. (2006)
3. Bychkov, I.V., Kenzin, MYu., Maksimkin, N.N.: A two-level evolutionary approach to the routing of a group of underwater robots under conditions of periodic rotation of the composition. SPIIRAS Proc. **18**(2), 267–301 (2019)
4. Sanchez-Silva, M., Klutke, G.-A.: Reliability and Life-Cycle Analysis of Deteriorating Systems. Springer International Publishing, Switzerland, 355p. (2016)
5. O'Conner, P., Kleyner, A.: Practical Reliability Engineering. Willey, London, 456p (2012)
6. Dubrova, E.: Fault-Tolerant Design, 185p. Springer Science + Business Media. New York (2013)
7. Zhang, Y., Jiang, J.: Bibliographical review on reconfigurable fault-tolerant control systems. Annu. Rev. Control **32**(2), 229–252 (2008)
8. Bielecki, T.R., Jakubowski, J., Niewęgłowski, M.: Conditional Markov chains: properties, construction and structured dependence. Stochastic Processes Their Appl. **127**(4), 1125–1170 (2017)
9. Janssen, J., Manca, R.: Applied Semi-Markov Processes, 310p. Springer, New York (2006)
10. Larkin, E.V., Malikov, A.A., Ivutin, A.N.: Petri-Markov model of fault-tolerant computer systems. In: 4th International Conference on Control, Decision and Information Technologies (CoDIT). IEEE, pp. 416–420 (2017)
11. Ivutin, A.N, Larkin, E.V.: Simulation of concurrent games. Bull. South Ural State Univ. Series: Math. Modell. Program. Comput. Softw. **8**(2), 43–54 (2015)
12. Larkin, E., Bogomolov, A., Gorbachev, D., Privalov, A.: About approach of the transactions flow to poisson one in robot control systems. Lecture Notes in Computer Science 10459 LNAI, pp. 113–122 (2017)

Chapter 32
Unmanned Transport Environment Threats

Maxim Kolomeets, Ksenia Zhernova, and Andrey Chechulin

Abstract Unmanned private and public transport may be susceptible to attacks through various interfaces including networks and physical sensors. With the spread of smart transport and the urban environment that can interact with vehicles, such threats will become increasingly relevant. The paper presents the overview of current cases of attacks on the connected unmanned transport environment that includes smart cars and smart city infrastructure. The paper includes the overview of implementation and classification of such environment components: smart vehicle components and smart city components that can interact with each other. Based on the implementation of components and what technologies they are used, paper overviews attack cases. The attack cases are based on the current reports of security incidents in the past and related research. The paper discusses the most urgent threats for such smart city environment based on the analysis of found attacks and classes of interfaces. The work highlights that today the most serious threat remains the problem of cyber-physical and network security.

32.1 Introduction

In the future, unmanned vehicles will become an integral part of smart cities. Many large companies are involved in unmanned vehicles development and implementation. At the same time, there is the development of both types of transport: private (cars, etc) and public (buses, trams, etc). For cities of particular interest is the possibility of integrating unmanned vehicles with smart city systems. In this case, it

M. Kolomeets (✉) · K. Zhernova · A. Chechulin
St. Petersburg Institute for Informatics and Automation of the Russian Academy of Sciences, 39, 14th Line, 199178 St. Petersburg, Russia
e-mail: kolomeec@comsec.spb.ru

K. Zhernova
e-mail: zhernova@comsec.spb.ru

A. Chechulin
e-mail: chechulin@comsec.spb.ru

A. Ronzhin and V. Shishlakov (eds.), *Proceedings of 15th International Conference on Electromechanics and Robotics "Zavalishin's Readings"*, Smart Innovation, Systems and Technologies 187, https://doi.org/10.1007/978-981-15-5580-0_32

becomes possible to exchange information between transport and the systems of smart city. It turns out the implementation of the concept of "connected cars," with the participation of urban infrastructure elements: traffic signs, weather stations, payment systems and others.

While unmanned vehicles and smart city systems have not become widespread, it is impossible to talk about existing integration cases. However, such integration carries security risks. During an attack on a separate unit of transport, other traffic participants and infrastructure do not experience consequences. However, in a connected transport environment, an attack on one device affects the entire environment.

In this paper, we consider the interfaces that make up an unmanned transport environment of a smart city and what threats they currently have. The novelty of the paper is the interaction model and classification of unmanned transport environment components. The contribution of the paper is the defined major threats and their connection with components of unmanned transport environment.

The paper consists of the following sections: Sect. 32.2 is *unmanned vehicle environment overview*—in this section, we present the interaction model and classes of components, highlight the components of unmanned transport and smart city that currently used and that can be attacked. Section 32.3 is *unmanned vehicle environment threats*—in this section, we examine existing cases of attacks on individual elements based on the reports and related research. Section 32.4 is *summarized threat*—in this section, we summarize threats of individual components, discuss the threats found and their relevance. The last section is *conclusion*—summarizing and future work plans.

32.2 Unmanned Vehicle Environment Overview

To understand the basic threats for unmanned vehicles environment we consider from which components it is consist and provide their classification.

The basic components of unmanned vehicle environment are *vehicles*, *infrastructure* of smart city and *surroundings*. The *vehicles* types are cars, busses and trolleybuses. The *infrastructure* includes information signs, toll points, monitoring and parking systems. The *surroundings* include weather, passengers, building, roadbed and other.

In our paper, we do not consider railway, water, air and underground transport, as they move through the allocated zones, and their impact on the rest of the infrastructure is not so significant. We consider only automobile and similar vehicles that share a single roadbed, as well as the infrastructure for it.

So, the environment interaction model includes interaction between vehicles, infrastructure and surroundings. There are five types of interfaces:

(1) Surroundings $\longleftrightarrow$ Vehicle. For example, measurement of distance to car by another car.

(2) Surroundings ⟷ Infrastructure. For example, measurement of temperature by weather station.
(3) Vehicle ⟷ Infrastructure. For example, transferring the information about bus location to bus stop by bus.
(4) Vehicle ⟷ Vehicle. Connected cars concept.
(5) Infrastructure ⟷ Infrastructure. Smart city concept.

In modern unmanned vehicles, the following components of interaction with passengers and with the environment are used (used in Surroundings ⟷ Vehicle and Vehicle ⟷ Infrastructure interfaces).

LiDAR. LiDAR is most commonly used to interact with other unmanned vehicles and with the environment. LiDAR [62, 63] is a laser-based technology that allows one to obtain three-dimensional environmental information. Based on this information, the artificial intelligence of pilotless vehicles makes a conclusion about the location of surrounding objects.

Radars. Radars are used for unmanned vehicles less frequently than LiDAR. This component emits radio waves and detects reflections of these waves from surrounding objects. [40, 59].

Ultrasonics. Ultrasonics use ultrasonic signals to determine the location of surrounding objects and the distance between objects and unmanned vehicles [51, 59]. The determination of distance is based on the calculation of the reflection time from an obstacle.

Vehicle Cameras. Also, for monitoring the environment, cameras with different viewing angles are used: front cameras, rear cameras, side front, side rear. [18, 59, 65].

Also, unmanned vehicles have an interface for interacting with passengers outside and inside the cabin. These interfaces include a visual, speech, tactile interface, as well as devices for the interaction of passengers with vehicles.

Displays. The visual interface often includes information displays on which the route, city map, weather conditions, etc., can be displayed. [35, 39, 40, 55] 3D panels can also be used that give the user the situation on the road [9–12, 61].

In addition to internal displays, the visual interface can also include external displays on cars [15]. The information that is displayed on them can be used by road users, pedestrians or autonomous systems.

The speech interface [6, 7] is used as a voice assistant in transport, which gives the user the necessary information, and also responds to voice commands.

Interaction with users [8, 33, 39, 52, 56] may include passenger counters, a terminal for fare payment and passenger information systems. Passengers are counted using an infrared sensor [52]. Payment for travel can be done using payment terminals that equipped with NFC technology [33, 39, 52], while the data of the passenger counter is compared with the number of passengers who paid for travel [52].

Smart city transport infrastructure also includes different smart components for traffic control and prioritization, taxation and fining, citizen notification and incident monitoring. Such components include next devices (used in Surroundings ⟷ Infrastructure and Vehicle ⟷ Infrastructure interfaces).

Information signs. Is one of the most spread devices for notification. It can be used in bus stops for notification of bus arrival time [57]. In highways for dynamical change of speed limit depending on weather conditions [23, 52]. In roads for notifications about incidents and difficulties on the road [26, 38]. In parking lots to show the number of available lots in the area [26]. The information signs usually include LED array with radio wave transmitter for receiving of information about the message that should be shown.

The toll points. The systems for taxation on the toll roads. For example, in Saint-Petersburg for car identification in the tall highway are used DSRC radio wave [5]: the driver should install the DSRC transponder on a car windshield to have the ability to pass the tall point without decreasing of car speed [52]. The cars without transponders pay more and need to stop at the cash toll point. Another toll points technology based on contactless car identification includes RFID and NFC [1, 27, 37].

Traffic Counters. Systems for counting of cars on the road. They can be based on laser, ultrasonic or radar technologies [52] to count the number of cars that pass some point on the road (example, Moscow transport system [26]). Another solution is to use the cameras and image recognition [48, 69].

Traffic lights. Traffic lights become smart when used information about traffic load and incidents for management of traffic flows [26]. The basic concept is to use traffic lights with counters to manage the flow based on the traffic congestion [69].

CCTV. Cameras are used with machine learning for identification of car incidents the road [26], car sign recognition for automatic fine and car counter. Usually are used simple IP–cameras.

Weather station. Weather station is used for collecting of data about weather conditions that can affect driving difficulty and as a result affect the speed limit [23] and other traffic management systems [26, 45, 58]. For example, slippery road due to rain, ice-crusted road, fog or snowfall.

Barriers and other limiters. They are used to limit the access to some areas, for example, courtyard, toll parking or toll road. In case of toll roads or parking, they used in connection with toll systems [27, 37]. In case of courtyard can be used the same contactless technologies, or SMS messaging [29]to get the access, resident can send the SMS to phone that is connected to barrier (guests to get the access should ask the resident to do it).

Location systems. Such systems are usually not a part of smart city and based on the satellites systems (GPS, GLONASS, etc.) [43]. But cities usually have systems that collect locations of public transport in one place, to share the information about public transport speed and arrival time [26].

Parking lots. Parking usually used the connection of different systems [26]. For payment and car identification can be used transponders, contactless cards or cameras with car sign recognition. For car counting and free lots detection are used radars or cameras. For driver's notification about number of available lots are used information signs.

At the radio level, *connected cars* communicate with each other both through dedicated short-range communication (DSRC) and via Wi-Fi, LTE, WiMAX (Vehicle $\longleftrightarrow$ Vehicle interaction) [22, 25]. The interaction of connected cars with the smart city (Vehicle ⟷ Infrastructure) is also provided by Wi-Fi, WiMAX, DSRC and LTE technologies [20]. In addition, Vehicle ⟷ Infrastructure interaction can occur through millimeter-wave (mmWave) microcellular networks [64]. The following protocols are used as a gateway between the physical layer and the cloud infrastructure for the interaction of connected cars with each other (Vehicle ⟷ Vehicle) MAC, LLC, WME, WSMP, TCP/ IP. The same protocols are used for the interaction of the machine with the infrastructure (Vehicle ⟷ Infrastructure) [13].

Smart city infrastructure used traditional computer networks. For uniting infrastructure elements are used Ethernet, WI-FI and mobile networks (3G, 4G and 5G).

The components are represented in Table 32.1, where the first column is the interface type, the second column is the component, and the third column is the implementation reference. In next section, we consider attacks on these components.

32.3 Unmanned Vehicle Environment Threats

As we know components of unmanned vehicle environment, we can explore which attacks can be performed.

Today vehicles components and infrastructure are based on SCADA or IoT systems. It is obvious that they have the same vulnerabilities than other SCADA and IoT systems. Also, as components are united into the network, network attacks also can be performed. So, SCADA\IoT and network attacks are common for all components, but there are also some specific attacks. In this section, we consider them based on the reports of the past security incidents and present them in Table 32.1.

LiDAR. Subject to blinding and different kinds of spoofing [44]. One can dazzle LiDAR using extraneous laser sources. The distortion of data obtained by LiDAR is possible when relaying the radiation of LiDAR. You can also create noise and interference using LiDAR radiation from surrounding unmanned vehicles. When attacking with spoofing, the same physical environment is used as in other attacks; however, the purpose of this attack is not to disable the sensor but to deceive the artificial intelligence of the machine [67].

Radars. Radars in unmanned vehicles are also susceptible to attacks, such as jamming, spoofing and interference [68]. Jamming is performed by saturating the receiver with a large number of radio signals. Also, the data received by the radar can be distorted by relaying the transmitted signals. Using various automotive radars, you can create interference of the desired level and disrupt the radar signal.

Ultrasonics. Attacks on sonars are aimed at drowning out the sonar signal or creating overlapping echo signals [67].

Car Cameras. Cameras are also subject to physical attacks, such as blinding. They are not adequately protected; therefore, they can be damaged or incapacitated by a sufficiently strong light source [67].

Displays. Attacks on the visual interface of unmanned vehicles can pursue a variety of purposes, for example, in order to show false information or to disable the

Table 32.1 Unmanned transport environment threats

Interface	Component	Implementation	Possible attack	Summarized threat
Surroundings ⟷ Infrastructure	Traffic counting	[26, 48, 52]	Cyber-physical	Cyber-physical and network
	Weather stations	[23, 26, 58, 45]	Cyber-physical	
	CCTV	[26, 58]	Cyber-physical and network [16]	
Surroundings ⟷ Vehicle	LiDAR	[62, 63]	Cyber-physical [44, 67]	Cyber-physical
	Radars	[40, 59]	Cyber-physical [68]	
	Ultrasonics	[51, 59]	Cyber-physical [67]	
	Vehicle cameras	[18, 59, 65]	Cyber-physical [67]	
	Displays	[9, 10, 11, 12, 15, 35, 39, 40. 55, 61]	Network	
	Speech interface	[6, 7]	Cyber-physical [54]	
	Counting passengers	[52]	Cyber-physical [30]	
	Payment devices	[8, 33, 39, 52, 56]	Cyber-physical and Network [19]	
Vehicle ⟷ Infrastructure	Toll points	[1, 5, 27, 52, 37]	Cyber-physical and network [24, 36, 60, 68]	Cyber-physical and network
	Barriers and limiters	[27, 29, 37]	Cyber-physical and network [46]	
	Location systems	[26, 43]	Cyber-physical [31, 32, 47]	
	Parking lots	[26]	Cyber-physical and network	

(continued)

Table 32.1 (continued)

Interface	Component	Implementation	Possible attack	Summarized threat
Vehicle ⟷ Vehicle	Connected cars	[1, 13, 19, 20, 22, 25, 61, 64]	Network [2, 4, 13, 17, 21, 28, 49]	Network
Infrastructure ⟷ Infrastructure	Computer networks	–	Network [66]	Network
	Information signs	[23, 26, 38, 52, 57]	Network [34, 53]	
	Traffic lights	[26, 69]	Network [41]	

display. For this, ordinary network attacks such as DDoS or MitM attacks can be used.

The speech interface can be attacked using speech recordings whose frequencies are moved to the ultrasonic range [54].

Interaction with users includes many different devices that are subject to a variety of attacks. For example, a device for counting passengers [52] is based on infrared radiation, and one can spoof it with using of special materials, for example, metallic foils [30]. Plastic cards or NFC payment terminals are subject to normal network attacks [19].

Information signs are a very popular target among attackers. The large collection of attack cases is presented in [34, 53]. The hackers change the messages on the signs with amusing or offensive messages. As the result, the drivers were distracted or confused and in one case, attack destabilized traffic.

Toll points are based on contactless technologies that are also susceptible to attacks. For toll points that are based on DSRC is possible to implement basic attacks that are reliable to car's DSRC systems [68]. Systems that are based on NFC technologies strongly depend on the encryption that is in the system. For example, it was possible to make a copy of the "St. Petersburg NFC Transport Card". As the result, one could get several copies of one paid card [24]. The software for cards cloning using Android phone was published on GitHub, so anyone could do it [36]. The possibility of card cloning was closed after the intervention of Saint-Petersburg transport committee. There are also possible general attacks that are not related to toll systems sensors [60].

As *traffic counting* is based on the CCTV and laser, ultrasonic or radar technologies, the attacks are similar to attack on cameras and smart cars sensors.

Smart *traffic lights* can be attacked indirectly by corruption of connected measurement systems (traffic counters, connected cars network, location systems) or directly by network attack. In [41], traffic lights where corrupted by simple wireless connection as vendor did not provide any password authentication or encryption on the traffic lights wireless communication.

Security of *CCTV* is well studied, and there are many reports on IP cameras corruption by network and light jamming. The review of CCTV security threats is presented in [16].

As *weather stations* are usually devices with multiple weather sensors, there is possible to do many physical attacks. For example, it is possible to change wind speed by hovering drone. The drones can also be used for affecting rain sensors (wetting the sensors), fog sensors (blocking the laser in case of laser measurements) and others.

Barriers and limiters work in connection with toll points systems and parking lots, so they also can be attacked indirectly through connected car identification systems. For barriers that use SMS or phone calls for authentication can be used SS7 attacks [46].

Location systems can be jammed [31, 32, 47] that can cause violations in the public transport schedule. At the same time, modern traffic systems use multiple location sensors that are based on GPS, cell towers triangulation and signals between stops and vehicles, so such scenario can be unlikely.

Parking lots are also a set of different technologies, such as signs, toll points, car counters, CCTV and barriers. So, attack on parking lot can be performed as attack on one of its sensors.

Attacks on the Vehicle ⟷ Vehicle interface are mostly network attacks, and they are well studied. For connected cars using DSRC, Wi-Fi, LTE, WiMAX, DoS attacks, location tracking and spoofing, as well as malware infection are possible [4, 13, 17, 21, 49]. In addition, network attacks such as Sybil attack (the car is displayed in several places at once) and Timing attack (receiving a message over the network, the car does not forward it immediately, but creates a delay) are possible [28].

As smart city networks work using traditional networks, for city infrastructure are possible the same network attacks (DDoS, MitM, etc.) [66].

The Table 32.1 summarizing found attacks for each component in column "Possible attack." We also classify them as *Cyber-Physical* attacks and *Network* attacks. In next section, we discuss the summarized threat that we define for each class of interface.

32.4 Summarized Threat

The most of attacks that we found relying on incidents and research are network and cyber-physical attacks. The difficulty is that cyber-physical attacks are complicated to detect because they do not leave traces in systems as cyberattacks, such as network and device logs. At the same time, some of them are very simple and do not require any professionalism if the goal is to jam the sensor. For example, for jamming of weather station can be used drones, for speech sensors can be used mobile phone, for LIDAR can be used lasers. At the same time, there are some more complicate attacks that not jamming sensors but changing their measurements.

In another hand, there are some attacks that are owing to lack of authentication mechanisms—as it was done for traffic lights in [41] and for payment cards in [24]. Such cases are common in IoT and are general for whole security area, but in case of unmanned vehicle and smart city infrastructure, they can cause more damage.

Based on the "possible attacks" column and "component" column, we can characterize each interface. We define the summarized threat based on the threats that we found for individual components.

(1) Surroundings ⟷ Infrastructure. In interaction of this type are used components with measuring instruments. So, attacks are directed on the corruption of measurements, and we define threat as *cyber-physical*. As such measuring instruments are connected (usually with mobile network, such as in IP cameras), we also define threat as *network*.
(2) Surroundings ⟷ Vehicle. In interaction of this type are also used components with measuring instruments for car navigation and interaction with passenger.

So, attacks are directed on the corruption of measurements, and we define threat as *cyber-physical*. Such measuring instruments are usually connected by CAN bus [14, 42], so we cannot say that major threats are based on *network* threats (but they possible [3, 50]).

(3) Vehicle ⟷ Infrastructure. In interaction of this type are used technologies for near communication. The most cases of such interaction are vehicle (toll points, barriers, parking) or infrastructure (location signal) authentication. Authentication has both threats: *cyber-physical* threat on the level of radio/laser/sound/other signal transmission and *network* threat on the level of access control data management.

(4) Vehicle ⟷ Vehicle. *Network* threats are major for connected cars while data transmission.

(5) Infrastructure ⟷ Infrastructure. *Network* threats are major for smart city infrastructure while data transmission.

So, nowadays, the most actual threat to smart city and unmanned vehicles is cyber-physical and network attacks. With the development of technology and the advent of commercial solutions, the trend can shift to networks and IoT security, but for today, there are not many cases. At the same time, cyber-physical attacks will not become less actual due the connectivity of components (when attack on one device leads the changing of measurements in a whole system) and relative ease of use.

32.5 Conclusion

This paper presents the part of ongoing research on security of interfaces that are used in unmanned transport environment. In paper, we present the classification of vehicles and infrastructure components that are used in unmanned transport environment, threats that are based on reports and summarized threat for each class. Such classification is helpful to specialists who wants to define what security actions need to be done while development and deployment of unmanned transport infrastructure.

The future work of this investigation is to define intruder model, dependency graph of components and present risk analysis technique based on them.

Acknowledgements This investigation was funded by RFBR Project № 19-29-06099.

References

1. Abari, O., Vasisht, D., Katabi, D., Chandrakasan, A.: Caraoke: an E-Toll transponder network for smart cities. In: Proceedings of the 2015 ACM Conference on Special Interest Group on Data Communication. pp. 297–310. Association for Computing Machinery, Inc (2015)
2. Appknox Company: Cyber attacks in connected cars—what Tesla did differently to win. https://www.appknox.com/blog/cyber-attacks-in-connected-cars. Last accessed 24 Jan 2020

3. Argus Company: The Challenge of Vehicle Hacking. https://argus-sec.com/car-hacking/. Last accessed 24Jan 2020
4. Berger, I., Rieke, R., Kolomeets, M., Chechulin, A., Kotenko, I. Comparative study of machine learning methods for in-vehicle intrusion detection. In: 4th Workshop on the Security of Industrial Control Systems and Cyber-Physical Systems, pp. 85–101 (2019)
5. Boldetsov, D.: DSRC-based fare systems. https://habr.com/ru/post/240047/. Last accessed 26 Jan 2020
6. Braun, M., Broy, N., Pfleging, B., Alt, F.: Visualizing natural language interaction for conversational in-vehicle information systems to minimize driver distraction. J. Multimodal User Interfaces **13**(2), 71–88 (2019)
7. Braun, M., Mainz, A., Chadowitz, R., Pfleging, B., Alt, F.: At your service: Designing voice assistant personalities to improve automotive user interfaces a real world driving study. In: Proceedings of the 2019 CHI Conference on Human Factors in Computing Systems, pp. 1–11 (2019)
8. Breitschaft, S.J., Clarke, S., Carbon, C.C.: A Theoretical framework of haptic processing in automotive user interfaces and its implications on design and engineering. Front. Psychol. **10** (2019)
9. Broy, N., Alt, F., Schneegass, S., Pfleging, B.: 3D displays in cars: Exploring the user performance for a stereoscopic instrument cluster. In: Proceedings of the 6th International Conference on Automotive User Interfaces and Interactive Vehicular Applications, pp. 1–9 (2014)
10. Broy, N., Guo, M., Schneegass, S., Pfleging, B., Alt, F.: Introducing novel technologies in the car. In: Proceedings of the 6th International Conference on Automotive User Interfaces and Interactive Vehicular Applications, pp. 179–186 (2015)
11. Broy, N., Schneegass, S., Guo, M., Alt, F., Schmidt, A.: Evaluating stereoscopic 3D for automotive user interfaces in a real-world driving study. In: Proceedings of the 33rd Annual ACM Conference Extended Abstracts on Human Factors in Computing Systems, pp. 1717–1722 (2015)
12. Broy, N., Zierer, B., Schneegass, S., Alt, F.: Exploring virtual depth for automotive instrument cluster concepts. In: CHI'14 Extended Abstracts on Human Factors in Computing Systems, pp. 1783–1788 (2014)
13. Butt, T.A., Iqbal, R., Shah, S.C., Umar, T.: Social internet of vehicles: architecture and enabling technologies. Comput. Electr. Eng. **69**, 68–84 (2018)
14. Chevalier, Y., Rieke, R., Fenzl, F., Chechulin, A., Kotenko, I.: ECU-Secure: characteristic functions for in-vehicle intrusion detection. In: International Symposium on Intelligent and Distributed Computing, pp. 495–504 (2019)
15. Colley, A., Häkkilä, J., Pfleging, B., Alt, F.: A design space for external displays on cars. In: Proceedings of the 9th International Conference on Automotive User Interfaces and Interactive Vehicular Applications Adjunct, pp. 146–151 (2017)
16. Costin, A.: Security of CCTV and video surveillance systems: Threats, vulnerabilities, attacks, and mitigations. In: Proceedings of the 6th International Workshop on Trustworthy Embedded Devices, pp. 45–54 (2016)
17. CPO Magazine: Connected cars: a new and dangerous vector for cyber attacks. https://www.cpomagazine.com/cyber-security/connected-cars-a-new-and-dangerous-vector-for-cyber-attacks/. Last accessed 24 Jan 2020
18. Cruise Company: Cruise company website. https://www.getcruise.com/technology/. Last accessed 24 Jan 2020
19. Deogirikar, J., Vidhate, A.: Security attacks in IoT: a survey. In: 2017 International Conference on I-SMAC (IoT in Social, Mobile, Analytics and Cloud) (I-SMAC), pp. 32–37. IEEE (2017)
20. Dey, K.C. et al.: Vehicle-to-vehicle (V2V) and vehicle-to-infrastructure (V2I) communication in a heterogeneous wireless network—performance evaluation. Transport. Res. Part C Emerg. Technol. **68**, 168–184 (2016)
21. Eiza, M.H., Ni, Q.: Driving with sharks: rethinking connected vehicles with vehicle cybersecurity. IEEE Veh. Technol. Mag. **12**(2), 45–51 (2017)

22. Flemisch, F., et al.: Cooperative control and active interfaces for vehicle assistance and automation. In: FISITA World Automotive Congress 2008, Congress Proceedings—Mobility concepts, man machine interface, process challenges, virtual reality, pp. 301–310 (2008)
23. Fontanka.ru Newspaper: How western speed diameter works. https://www.fontanka.ru/2019/05/24/175/. Last accessed 26 Jan 2020 (in Russian)
24. Fontanka.ru Newspaper: Transport card hacking. https://www.fontanka.ru/2017/03/23/133/. Last accessed 26 Jan 2020 (in Russian)
25. Forescout Company: Connected cars: the new automotive cybersecurity threat. https://www.forescout.com/company/blog/connected-cars-the-new-automotive-cybersecurity-threat/. Last accessed 24 Jan 2020
26. Government of Moscow: Presentation of the situation center of the Government of Moscow. https://docplayer.ru/31956526-Situacionnyy-centr-codd-pravitelstva-moskvy.html. Last accessed 24 Jan 2020 (in Russian)
27. Grewal, S., Segar, R.: Toll roads and contactless multiapplication smart cards. In: SATC 2000 (2000)
28. Haas, R.E., Möller, D.P.F., Bansal, P., Ghosh, R., Bhat, S.S.: Intrusion detection in connected cars. In: 2017 IEEE International Conference on Electro Information Technology (EIT), pp. 516–519 (2017)
29. Hanif, N., Hazrin, H.M., Badiozaman, M.H., Daud, H.: Smart parking reservation system using short message services (SMS). In: 2010 International Conference on Intelligent and Advanced Systems, pp. 1–5. IEEE (2010)
30. Harris, D.J.: Use of metallic foils as radiation barriers to reduce heat losses from buildings. Appl. Energy **52**(4), 331–339 (1995)
31. Hui, H., Na, W.: A study of GPS jamming and anti-jamming. In: 2009 2nd International Conference on Power Electronics and Intelligent Transportation System (PEITS), pp. 388–391. IEEE (2009)
32. Nuca, I., Nuca, I., Motroi, A., Eşanu, V.: Information technologies used in electric public vehicles. In: Sielmen, Oct 2015, pp. 307–312 (2015)
33. ITMO University: Smart vehicle test in Saint-Petersburg, https://news.itmo.ru/ru/science/it/news/5955/. Last accessed 24 Jan 2020 (in Russian)
34. Kelarestaghi, K., et al.: Intelligent transportation system security: hacked message signs. SAE Int. J. Transport. Cybersecur. Privacy **1**, 75–90 (2018)
35. Kolomeec, M., Chechulin, A., Pronoza, A., Kotenko, I.: Technique of data visualization: example of network topology display for security monitoring. J. Wireless Mob. Networks Ubiquitous Comput. Depend. Appl. **7**(1), 41–57 (2016)
36. Krikun, T.: Plantainreader app on GitHub. https://github.com/krikunts/plantainreader. Last accessed 26 Jan 2020
37. Lee, W.H., Tseng, S.S., Wang, C.H.: Design and implementation of electronic toll collection system based on vehicle positioning system techniques. Comput. Commun. **31**, 2925–2933 (2008)
38. Li, Q., Qiao, F., Wang, X., Yu, L.: Drivers' smart advisory system improves driving performance at STOP sign intersections. J. Traffic Transport. Eng. (English edition) **4**(3), 262–271 (2017)
39. MATRESHKA.ai: Presentation of the unmanned commercial vehicle modular system MATRESHKA. https://www.volgabus.ru/upload/Matreshka_Performance.pdf. Last accessed 24 Jan 2020 (in Russian)
40. Meinl, F., Stolz, M., Kunert, M., Blume, H.: An experimental high performance radar system for highly automated driving. In: 2017 IEEE MTT-S International Conference on Microwaves for Intelligent Mobility (ICMIM), pp. 71–74 (2017)
41. NBC Chicago News: New hacking threat could impact traffic systems. https://www.nbcchicago.com/news/local/new-hacking-threat-could-impact-traffic-systems/61892/. Last accessed 26 Jan 2020
42. Nie, S., Liu, L., Du, Y.: Free-fall: hacking tesla from wireless to can bus. Briefing, Black Hat USA, pp. 1–16 (2017)

43. Nkoro, A.B., Vershinin, Y.A.: Current and future trends in applications of intelligent transport systems on cars and infrastructure. In: 17th International IEEE Conference on Intelligent Transportation Systems (ITSC), pp. 514–519. IEEE (2014)
44. Petit, J., Stottelaar, B., Feiri, M., Kargl, F.: Remote attacks on automated vehicles sensors: experiments on camera and LiDAR. Black Hat Europe, pp. 1–13 (2015)
45. Pisano, P., Goodwin, L.C.: Surface transportation weather applications. federal highway administration in concert with mitretek systems. Presented at the 2002 Institute of Transportation Engineers Annual Meeting, pp. 1–11 (2002)
46. Positive Technologies: SS7 vulnerabilities and attack exposure. https://www.ptsecurity.com/upload/corporate/ww-en/analytics/SS7-Vulnerability-2018-eng.pdf. Last accessed 26 Jan 2020
47. Postnauka Magazine: Are hacker attacks dangerous for cars? https://postnauka.ru/faq/95774. Last accessed 24 Jan 2020
48. Purnama, I.K.E., Zaini, A., Putra, B.N., Hariadi, M.: Real time vehicle counter system for intelligent transportation system. In: International Conference on Instrumentation, Communication, Information Technology, and Biomedical Engineering 2009, pp. 1–4. IEEE (2009)
49. Ring, T.: Connected cars—the next target for hackers. Network Secur. **2015**, 11–16 (2015)
50. Schellekens, M.: Car hacking: navigating the regulatory landscape. Comput. Law Secur. Rev. **32**(2), 307–315 (2016)
51. Shahian, B.: Ultrasonic sensors in self-driving cars. https://medium.com/@BabakShah/ultrasonic-sensors-in-self-driving-cars-d28b63be676f. Last accessed 24 Jan 2020
52. SHTRIH-M Company: Technology "SHTRIH-M: Transport", https://www.shtrih-m.ru/solutions/avtomatizatsiya-transporta/tekhnologiya-shtrikh-m-transport-1/. Last accessed 24 Jan 2020 (in Russian)
53. Sitawarin, C., Bhagoji, A.N., Mosenia, A., Chiang, M., Mittal, P.: DARTS: deceiving autonomous cars with toxic signs. arXiv Prepr. arXiv1802.06430 (2018)
54. Song, L., Mittal, P.: Poster: inaudible voice commands. In: Proceedings of the 2017 ACM SIGSAC Conference on Computer and Communications Security, pp. 2583–2585 (2017)
55. Steinberger, F., Proppe, P., Schroeter, R., Alt, F.: CoastMaster: an ambient speedometer to gamify safe driving. In: Proceedings of the 8th International Conference on Automotive User Interfaces and Interactive Vehicular Applications, pp. 83–90 (2016)
56. Summerskill, S., Porter, J., Burnett, G.: Feeling your way home. Design Emotion 287–292 (2003)
57. Sungur, C., Babaoglu, I., Sungur, A.: Smart bus station-passenger information system. In: 2015 2nd International Conference on Information Science and Control Engineering, pp. 921–925, IEEE. (2015)
58. T-Traffic Company: T-Traffic company website. http://t-traffic.spb.ru/?page_id=865. Last accessed 26 Jan 2020 (in Russian)
59. Tesla Company: Autopilot. https://www.tesla.com/autopilot. Last accessed 24 Jan 2020
60. The Hacker News: Israeli road control system hacked, caused traffic jam on Haifa highway. https://thehackernews.com/2013/10/israeli-road-control-system-hacked.html. Last accessed 26 Jan 2020
61. Tönnis, M., Broy, V., Klinker, G.: A survey of challenges related to the design of 3D user interfaces for car Drivers. In: 3D User Interfaces (3DUI'06), p. 134. IEEE (2006)
62. Velodyne Lidar Company: How autonomous vehicles perceive and navigate their surroundings, https://velodynelidar.com/how-autonomous-vehicles-perceive-and-navigate-their-surroundings/. Last accessed 24 Jan 2020
63. Velodyne Lidar Company: How LiDAR technology enables autonomous cars to operate safely. https://velodynelidar.com/how-lidar-technology-enables-autonomous-cars-to-operate-safely/. Last Accessed 24 Jan 2020
64. Wang, Y., Venugopal, K., Molisch, A.F., Heath, R.W.: MmWave vehicle-to-infrastructure communication: analysis of urban microcellular networks. IEEE Trans. Veh. Technol. **67**(8), 7086–7100 (2018)
65. Waymo: Waymo company website. https://waymo.com/tech/. Last accessed 24 Jan 2020

66. Weimerskirch, A.: An overview of automotive cybersecurity. In: Proceedings of the 5th International Workshop on Trustworthy Embedded Devices, pp. 53–53 (2015)
67. Yan, C., Wenyuan, X., Jianhao, L.: Can you trust autonomous vehicles: contactless attacks against sensors of self-driving Vehicle. In: DEF CON 24 (2016)
68. Yeh, E., Choi, J., Prelcic, N., Bhat, C., Heath Jr., R.: Security in automotive radar and vehicular networks. Microwave J. **60**(5) (2016)
69. Zaid, A.A., Suhweil, Y., Yaman, M.: AI: smart controlling for traffic light time. In: 2017 IEEE Jordan Conference on Applied Electrical Engineering and Computing Technologies (AEECT), pp. 1–5. IEEE (2017)

Chapter 33
Cloud-Based Task Distribution System Infrastructure for Group of Mobile Robots

Airat Migranov

Abstract One of the possible approaches to the construction of control systems for groups of mobile robots using distributed cloud technologies is considered, for which a scheme of access to information resources and a mechanism for distributing resources of a cloud computing system with linear decomposition are developed: The solution to the problem is divided into a series of smaller, simpler, subtasks in a hierarchical tree based on the linear distribution method. The specifics of the workspace model are shown, the goals of the functioning of robots are formalized, and the optimal energy evolutionary algorithm for solving the problem of distributing tasks in the team is proposed taking into account the initial and current levels of battery power, the energy consumption of each robot and the energy needed to perform individual tasks. Parameters for evaluating the effectiveness of the obtained solutions are determined, and genetic algorithms are synthesized, for which a coding form of the solution in the form of a chromosome is proposed and specific fitness functions are compiled. An algorithm has been developed for calculating the fitness function, implemented taking into account the specifics of its work in the cloud. Experimental results were obtained when checking the operability of the algorithms on the available onboard computing means of mobile robots, and the effectiveness of using distributed computing resources of a group of robots was estimated when implementing cloud services on their basis.

33.1 Introduction

In recent years, a significant amount of theoretical and practical researches of leading robotics laboratories have been devoted to the problems of group control of mobile robots. In the course of the work, the effectiveness of the simultaneous use of mobile robots combined in groups was shown, which allowed to implement more complex

---

A. Migranov (✉)
Mavlyutov Institute of Mechanics, Ufa Centre of the Russian Academy of Sciences, 71, Ave. Oktyabrya, 450054 Ufa, Russia
e-mail: abm.imech.anrb@mail.ru

A. Ronzhin and V. Shishlakov (eds.), *Proceedings of 15th International Conference on Electromechanics and Robotics "Zavalishin's Readings"*, Smart Innovation, Systems and Technologies 187, https://doi.org/10.1007/978-981-15-5580-0_33

technological operations, as well as significantly improve time and energy indicators, increase the survivability of the team and the likelihood of successful completion of tasks.

With group management, several specific tasks arise related to the implemented strategy for the interaction of group robots, control problems in an unsteady and non-deterministic environment in conditions of resource shortage [1–3]. Therefore, the expansion of the capabilities of onboard robot systems, implemented through the use of distributed computing technologies, will provide more effective methods for solving such group management problems as:

1. implementation of "fast" intelligent algorithms, ideally—the creation of "mobile" artificial intelligence;
2. increasing the efficiency of group management, including the division and decomposition of tasks within the group.

Intellectualization of robot places high demands on the speed of onboard control systems; therefore, the use of distributed computing technologies is considered promising, where the onboard robot computers, which have software and hardware support for the processes of distributing information flows between group members, are considered to be computing nodes. The methodology for providing ubiquitous and convenient access to a common pool of computing resources is most often based on cloud computing technologies, since cloud structures can be used both as a basis for control and for creating the infrastructure of a distributed system for managing groups of mobile robots [4]. The implementation of cloud infrastructure in a team of robots is complicated by the problems of group management, depending on the selected paradigm of group interaction of robots:

1. distribution of goals and tasks between robots, taking into account the nature of the goals, the functionality of each robot, and the environment of their functioning;
2. planning of motion paths and allocation of subspaces for each robot, taking into account possible conflicts when working in a common workspace.

The article shows one of the ways to build a cloud infrastructure of a control system for a group of mobile robots used to solve the connected tasks of distributing goals and planning trajectories. Currently, research teams use different definitions of group interaction between robots [4]:

1. Multi-Robot System (MRS);
2. Multi-Agent System (MAS);
3. Swarm Robotics System (SRS) or Robotic Swarm (RS).

In this paper, the group of robots is the multi-robot system (MRS), which has the following main properties:

- Robots are autonomous.
- Centralized control is implemented.
- Scalable architecture built for both small and large number of robots.

The most common algorithms for centralized distribution of tasks are based on "pricing" approaches, in which the control center, according to the "pricing" algorithm, distributes *m j*-tasks in a team of *n* identical *i*-robots based on C_{ij} "cost" ("prices") performed by the robot tasks [5–7]. No less popular are the algorithms of centralized control of the movement of the slave robot, following the lead robot, used to move the system ("convoy") [8, 9].

Multi-agent control algorithms for robotic systems implemented on an auction-type negotiation model are also known, when information is exchanged between individual agents using optimality criteria: The lowest price of the set task is selected until all of them are distributed among the robots of the team in a certain sequence [5, 10]. Such an approach gained greater efficiency as a result of the modification, in which each robot was able to put up a task using an inquiry message and the ability to form groups with neighboring robots [11]. A similar in capabilities algorithm is implemented in a self-organizing control system based on the network model of the team of robots [12].

The special types of multi-agent control algorithms for robotic systems operating under uncertainty include dynamic programming [13], the use of advanced and virtual reality technologies [14], potential fields [15, 16], cognitive-adaptive method [17].

To control the movement of mobile robots operating in conditions of uncertainty, in recent years, solution methods based on such intelligent algorithms as neural networks [18], fuzzy logic [19], and genetic algorithms (GA) [20] have become widespread. It was also shown that with an increase in the number of robots in a group and the complexity of the task, the time required to find solutions exponentially increases, and the power requirements for onboard computer systems increase significantly. Therefore, the model of providing access to a common pool of computing resources based on cloud technologies can become one of the effective approaches to building distributed control systems for groups of intelligent robots.

The aim of the research was to develop a "cloud" strategy for managing a group of robots, to synthesize a methodology for distributing tasks in cloud applications, and also to develop a common infrastructure for a task distribution system for a group of mobile robots. Next, service models and a method for decomposing a task within a group of robots, an algorithm for distributing tasks between robots, taking into account the specific applications, will be shown.

33.2 Decomposition of Tasks for Distribution in Cloud

In the process of decomposition, the solution of one large problem is divided into the solution of a series of simpler subtasks of smaller volume, and one of the widely used methods of decomposition of tasks is a linear distribution [21], the result of which is often presented in the form of a hierarchical tree (see Fig. 33.1a).

The resulting individual subtasks will be distributed to find a solution in the form of separate cloud applications: The task is divided up until the computation of the subtask exceeds the requirements of the virtual machine (see Fig. 33.1b).

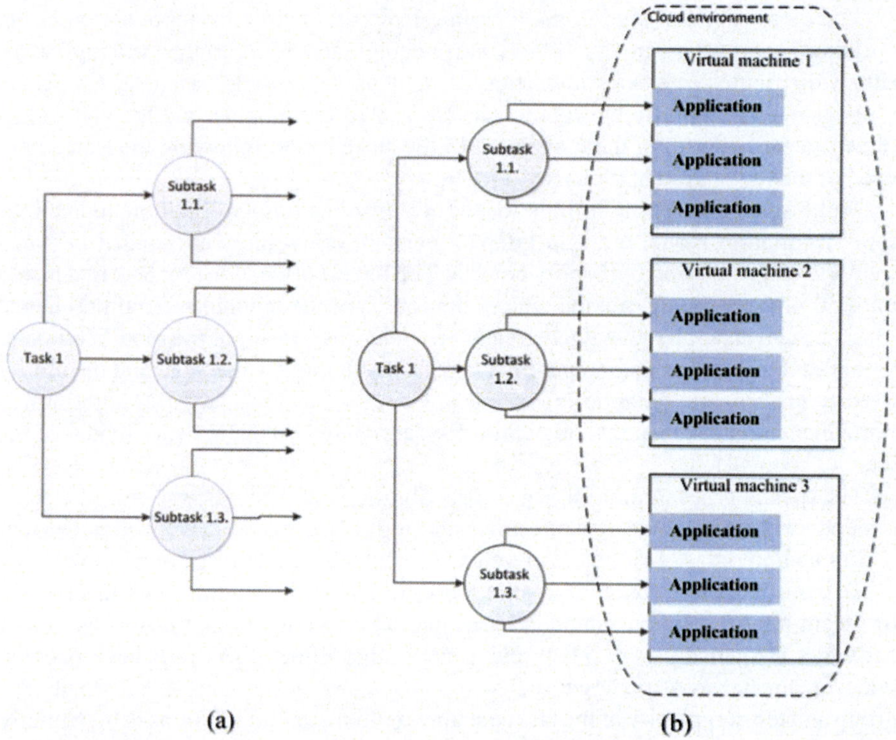

Fig. 33.1 Task distribution technology: linear decomposition (**a**); distribution in the cloud (**b**)

In the process, it is necessary to debug the operation of the control system in the form of the implementation of three service models (see Fig. 33.2):

1. Software as a service (SaaS);
2. Platform as a service (PaaS);
3. Infrastructure as a service (IaaS).

The choice of service model is determined by the level of access of agents to virtual applications, data processing, and storage facilities, as well as peripheral equipment that is part of the computing node, while the created cloud infrastructure must provide solutions to applied computing problems of different complexity levels. Given that the control system must operate through communication with the cloud, the development of a single conceptual approach, which is divided into the following main stages, is required:

- creating a compute pool;
- organization of work and network stack setup;
- storage device (reference data processing center);
- setting up a robot control system;
- virtualization;

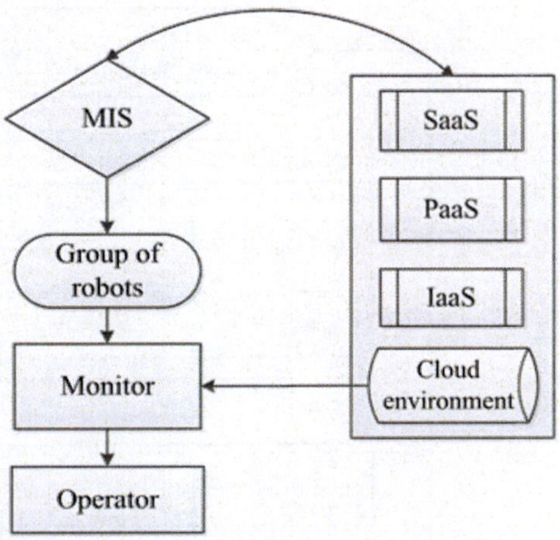

Fig. 33.2 Task distribution technology: linear decomposition (**a**); distribution in the cloud (**b**)

- automation;
- operator service (managing a group of robots through the cloud).

The autonomy of the robot implies its automatic and independent functioning, group management, coordination of its work with other robots from its group through a remote cloud service. The target set for the group is divided into many small subtasks, which are distributed based on relations of priority and optimal implementation from the point of view of energy. Suppose that on an $N \times M$-sized working field, there are n robots and m tasks, robot coordinates are $(x_i, y_i, i \in [1, n])$; task location coordinates are $(x_i^*, y_i^*, j \in [1, m])$. The purpose of the functioning of robots is to perform tasks detected in the working area, while in each cell of this zone only one robot and/or one task can be at a time. The tasks are divided into N_E stages (see Fig. 33.3).

The power consumption of the robot is conventionally divided into three modes: functioning, moving, and completing a task. Each mode is characterized by its power consumption (W_i^{FUN}, W_i^{MV}, W_i^{WR} for the ith robot, respectively).

The total energy spent on fulfilling the target task by all robots:

$$W_\Sigma = \sum_{m \in N_W} \left(W_m^{\mathrm{MV}} + W_m^{\mathrm{WR}} + W_m^{\mathrm{FUN}}\right), \tag{33.1}$$

where N_W is the number of robots involved in all tasks.

Each robot has an initial charge level (P_i^R), and its energy consumption for the task is determined by a set of onboard equipment (robot class) and task class. The task has its own level of energy consumption (W_j^{WR})—a certain average amount of energy spent by a robot with the necessary equipment to perform this class of task.

Given the above requirements and the format of the problem statement, it is necessary to develop an algorithm for finding the best option for distributing tasks

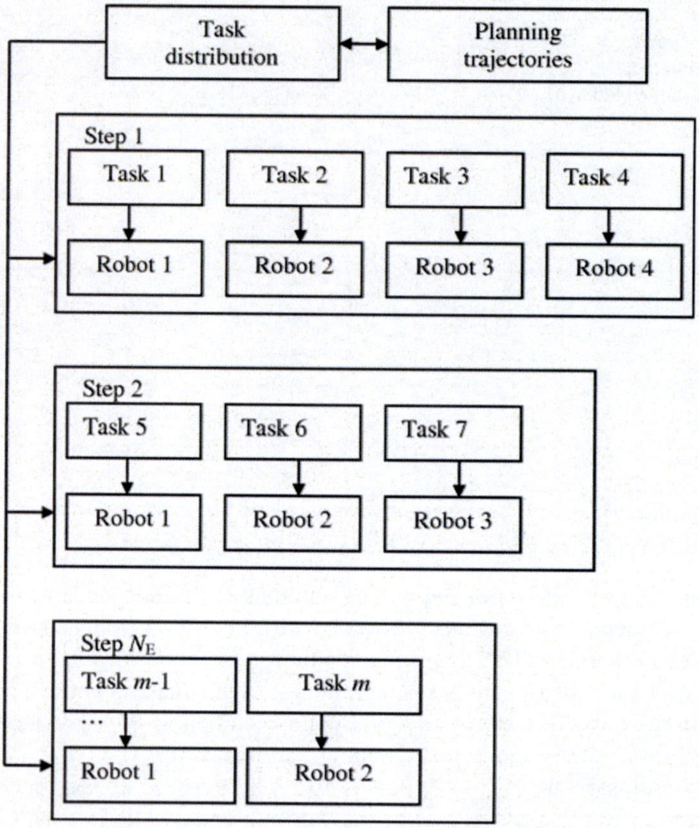

Fig. 33.3 Linear decomposition and distribution of tasks in a group of robots

between robots while minimizing the energy and time spent. For a solution, a heuristic method based on GA was used.

33.3 Genetic Task Distribution Algorithm

The solution to the distribution problem based on GA can be reduced to the following main stages [22]:

1. The formation of a model of the external environment (workspace), input coordinates of the location of tasks, robots, and classes established for them.
2. Coding of potential solutions (chromosome formation).
3. Initial population generation.
4. Determination of survival criteria for the population at each evolutionary stage and the formation of matching functions.

The purpose of the group of robots is to complete all tasks located in the working field. Tasks correspond to an identifier vector composed of tasks included in it (vector V_j), and each robot has a unique identifier for the F_i functions performed. Additional restrictions are introduced: The robot can complete the task if the task class is included in the vector of its functions (robot class); tasks are monosyllabic and have unique coordinates; only one robot can be assigned to a task. An indicator of the quality of the jth job by the ith robot is introduced—the coefficient q_{ij}, which determines the efficiency of the jth task by the ith robot under current conditions. This coefficient is calculated taking into account the time costs (the time taken by the robot to move to the location of the task) and the effectiveness of the execution.

The time spent on moving the robot is estimated by the path length calculated by the trajectory planning system or in the form of the Euclidean distance between the current and target point:

$$S_{ij} = \sqrt{\left(x_j^* - x_i\right)^2 + \left(y_j^* - y_i\right)^2}. \tag{33.2}$$

The efficiency coefficient for a robot with the required ability $F_i \cap Vj \neq 0$ is specified as the ratio of the total number of functions to the number used to solve the problem:

$$G_{ij} = \frac{\text{length}(F_i)}{\text{length}\left(F_i \cap V_j\right)}, \tag{33.3}$$

where *length* is the length of the vector.

We determine the weighted average value of the efficiency of the jth task by the ith robot using the weight functions w:

$$q_{ij} = w_1 \overline{S}_{ij} + w_2 \overline{G}_{ij}, \quad \sum_k w_k = 1, \quad k = \{1, 2\}, \tag{33.4}$$

where $\overline{S}_{ij}, \overline{G}_{ij}$ are normalized to unity values of S_{ij}, G_{ij}, respectively.

As a second quality indicator, the maximum time that one of the robots will spend to complete all tasks is used:

$$t_{\max} = \max_{m \in N_W} (t_m). \tag{33.5}$$

The third indicator is the number of outstanding tasks at this step N_{out}.

The number of tasks and robots in the general case may differ, so you need to keep track of how many tasks remain outstanding.

It is proposed to obtain a solution in the form of a positional vector $\mathbf{Z}$ containing task numbers in which each element Z_i defines a task for the ith robot to the next step. For example, for three tasks, the vector $\mathbf{Z} = \{1, 3, 2\}$ means that the first robot is given the command to proceed to the location of task 1 and execute it, the second

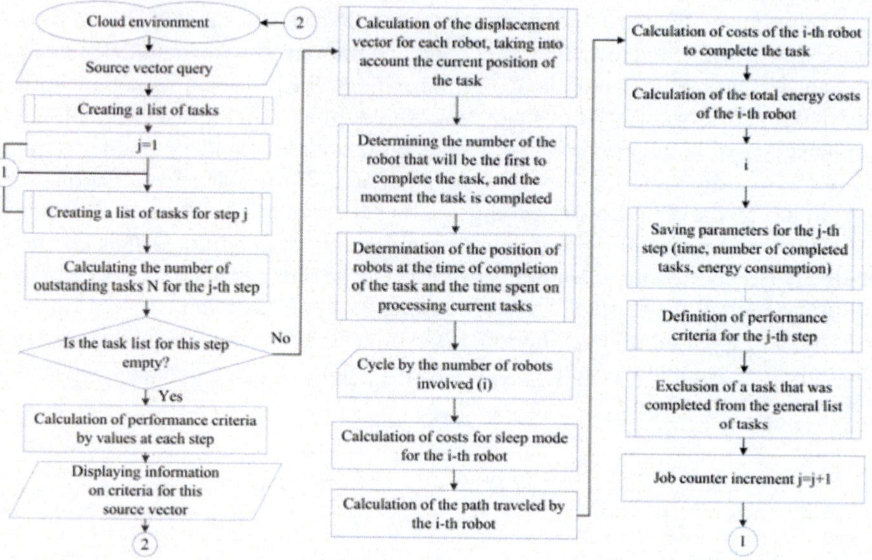

Fig. 33.4 Algorithm for calculating the fitness function, optimized for work in the cloud

to task 3, and the third to task 2. Note that the vector **Z** must consist of unique task numbers (tasks should not be repeated for several robots). Such an encoding implies the possibility of using an integer constraint in the implementation of GA. Then the vector **Z** will be a genotype, and each element of the vector will be a chromosome.

The fitness function should provide an assessment of each individual entering the input in encoded form, in terms of applicability to the conditions of the problem. Taking into account the peculiarities of decision coding and fitness assessment, a GA was selected with sorting by dominance and elite selection, presented in Fig. 33.4.

The criteria for stopping the algorithm are the immutability of the best individuals for a given number of generations (usually 20–50 generations), or the achievement of a critical size of the number of generations (e.g., 500). The exit criteria used can reduce the likelihood of falling into a local extremum.

The proposed algorithm allows us to search for the optimal strategy for given initial parameters of the problem. Let us consider one of the solution options: The group consists of three robots that need to complete 6 tasks located on a 10×10 field. The source data for this example are shown in Table 33.1.

The calculation results using the proposed GA are presented in Table 33.2.

Note that the minimum number of outstanding tasks is 3. This is determined by the difference in the number of robots and tasks (6–3). Figure 33.5 shows the distribution of robots and their trajectory for the fifth solution, which is a solution with the maximum number of completed tasks.

Thus, the developed algorithm for the step-by-step search of the optimal strategy for the behavior of a group of mobile robots makes it possible to simultaneously find the robot paths that are optimal along the length and the time distribution plan for

Table 33.1 Parameters of robots and tasks on the field

Robot	Coordinates	Class robot	Task	Coordinates	Class task
1	{6, 2}	3	1	{8, 3}	1
2	{4, 10}	1	2	{1, 6}	1
3	{1, 7}	2	3	{5, 4}	2
			4	{4, 9}	3
			5	{2, 10}	3
			6	{2, 7}	1

Table 33.2 Optimal solutions

Decision	Robot number			W_Σ	t_{max}	N_{out}
	1	2	3			
1	0	0	0	0	0	6
2	0	0	2	3,68	2	5
3	3	0	2	8,16	12,43	4
4	1	0	2	8,36	10,93	4
5	3	6	2	22,45	19,81	3

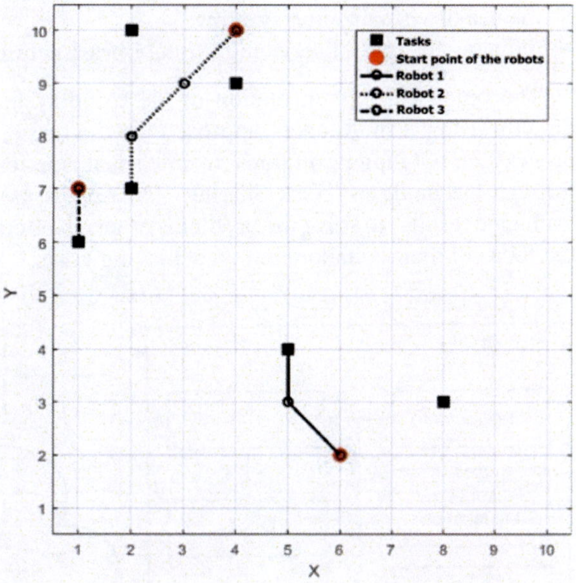

Fig. 33.5 Trajectories for solution No. 5

tasks. The proposed algorithm is based on the adaptability function, GA, an algorithm for finding optimal paths for a group of robots, and an algorithm for step-by-step solving a common problem.

The main goal of the created infrastructure for the control system of a group of robots is to work in real time and align the work of the task distribution modules according to the chosen concept through a cloud server. Consider the implementation of the specifics of the interaction of robots in a cloud environment.

33.4 Interaction of Robots in Cloud

Taking into account the features of the component management method "as a service," it seems appropriate to synthesize the architecture of the modules for managing a group of robots as an open system based on the model of an application server and the Internet (see Fig. 33.6).

During the work of the group management system, such advantages of cloud technologies as:

- reliability due to the location in specially equipped data centers;
- flexibility and unlimited computing resources (RAM, processor, storage) through the use of virtualization systems;
- reducing the cost of servicing virtual infrastructure.

The proposed implementation of the cloud system will allow you to set tasks directly using a multi-agent approach. In this case, it is possible to optimize the process of obtaining solutions from virtual applications by decomposing tasks through the modules "Task Sharing" (sharing tasks) and "Result Sharing" (using exchange results) to solve the problems of group interaction of a group of robots and the focus of their collaboration on achieving goals.

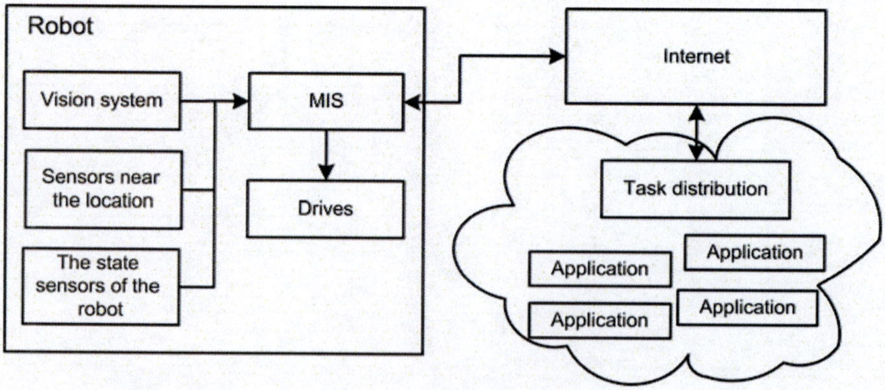

Fig. 33.6 Interaction of an individual robot with a cloud system

To carry out the experimental part with the models of the task distribution system, IaaS cloud service was used, on which the following hardware characteristics were configured: 2CPU—3 GHz CISCO B200, 4 GB RAM, 10 GB SSD, Windows Server 2008 R2 × 64. The previously developed software [5] was installed on a cloud server to conduct a series of model experiments with systems for planning and distributing tasks based on dynamic programming, GA, neural networks, and fuzzy logic.

The results showed the possibility of previously synthesized modules based on the cloud server infrastructure. Further research involves the inclusion of full-scale and full-scale experiments for fine-tuning the cloud infrastructure and developing applied algorithmic support for the task distribution system and group control of robots.

33.5 Conclusion

The paper considers one of the possible approaches to the implementation of the management of a group of mobile robots based on the use of distributed cloud systems. A distinctive feature of the proposed methods is the reflection of the specifics of the field of application and the tasks solved by the group of robots in the architectures of information management systems and methods of organizing information exchange. Based on the developed GA distribution of tasks, it was possible to generate optimal motion trajectories and strategies for the behavior of a group of robots with different initial data. In the future, it is planned to develop information exchange algorithms, create software modules and a hardware–software interface between the cloud and the real equipment of the prototype robotic complex.

Acknowledgements This research is supported by the Program of the Presidium of the Russian Academy of Sciences and within the framework of state assignment No. 0246-2018-007.

References

1. Tuncer, A., Yildirim, M.: Dynamic path planning of mobile robots with improved genetic algorithm. Comput Electr Eng **38**, 1564–1572 (2012)
2. Xiao, J., et al.: A hybrid membrane evolutionary algorithm for solving constrained optimization problems. Optik **125**(2), 897–902 (2014)
3. Darintsev, O.V., Migranov, A.B.: Synthesize the strucrure of cloud computing system to control of mobile robots group. Proc. Mavlyutov Inst. Mech. Ufa Centre Russ. Acad. Sci. **11**(1), 72–80 (2016)
4. Zakiev, A., Esoy, T., Magid, E.: Swarm robotics: remarks on terminology and classification. In: International Conference on Interactive Collaborative Robotics, pp. 291–300 (2018)
5. Olfati-Saber, R.: Flocking for multi-agent dynamic systems: algorithms and theory. Technical Report CIT (California Institute of Technology) CDS (Control and Dynamical Systems), pp. 401–420 (2006)

6. Schwager, M.: A gradient optimization approach to adaptive multi-robot control. Massachusetts Institute of Technology, Cambridge (2009)
7. McLurkin, J.: Analysis and implementation of distributed algorithms for multi-robot systems. Doctoral dissertation, Massachusetts Institute of Technology, Department of Electrical Engineering and Computer Science (2008)
8. Ahterov, A.V., Kiril'chenko, A.A., Pavlovsky, V.E., Rogozin, K.V.: Ways of controlling the distributed mobile system in the conditions of uncertainty. KIAM Preprint **67**, 1–33 (2012)
9. Zhu, H.: Control of the movement of a group of mobile robots in a system of the type "convoy". Dissertation for the Degree of Candidate of Technical Sciences, Moscow (2018)
10. Guy, S.J. et al.: Clearpath: highly parallel collision avoidance for multi-agent simulation. In: Proceedings of the 2009 ACM SIGGRAPH/Eurographics Symposium on Computer Animation, pp. 177–187 (2009)
11. Nazarova, A.V.: Methods and algorithms for multi-agent control of a robotic system. Bull. Bauman Moscow State Tech. Univ. **6**, 93–105 (2012)
12. Kalyaev, I., Kapustjan, S., Gajduk, R.: Self-organizing distributed control systems of intellectual robot groups constructed on the basis of network model. Large-Scale Syst. Control **30**(1), 605–639 (2010)
13. Kruglikov, S.V., Kruglikov, A.S.: An a priori planning of joint motions for USV as a problem of guaranteed control/estimation. Appl. Mech. Mater. **494**, 1110–1113 (2014)
14. Darintsev, O.V.: The use of advanced and virtual reality technologies in the implementation of algorithms for managing a team of robots. Piece Intell. **3**, 479–487 (2013)
15. Renzaglia, A., Martinelli, A.: Potential field based approach for coordinate exploration with a multi-robot team. In: 2010 IEEE Safety Security and Rescue Robotics, pp. 1–6. IEEE (2010)
16. Ivić, S., Crnković, B., Mezić, I.: Ergodicity-based cooperative multiagent area coverage via a potential field. IEEE Trans. Cybern. **47**(8), 1983–1993 (2016)
17. Renzaglia, A., Doitsidis, L., Martinelli, A., Kosmatopoulos, E.B.: Cognitive-based adaptive control for cooperative multi-robot coverage. In: 2010 IEEE/RSJ International Conference on Intelligent Robots and Systems, pp. 3314–3320. IEEE (2010)
18. Janglova, D.: Neural networks in mobile robot motion. Adv. Robot. **1**(1), 15–22 (2004)
19. Shah, H.N.M., et al.: Design and develop an autonomous UAV airship for indoor surveillance and monitoring applications. Int. J. Inform. Visual. (JOIV) **2**(1), 1–7 (2018)
20. Zhang, L., Min, H., Wei, H., Huang, H.: Global path planning for mobile robot based on A* algorithm and genetic algorithm. In: 2012 IEEE International Conference on Robotics and Biomimetics (ROBIO), pp. 1795–1799. IEEE (2012)
21. Darintsev, O.V., Migranov, A.B.: Distributed control system for group of mobile robots. Vestnik USATU **21**(2–76), 88–94 (2017)
22. Darintsev, O.V., Migranov, A.B.: Task distribution module for a team of robots based on genetic algorithms: synthesis methodology and testing. In: 2019 XXI International Conference Complex Systems: Control and Modeling Problems (CSCMP), pp. 296–300 (2019)

Chapter 34
Environment Classification Approach for Mobile Robots

Petr Neduchal and Miloš Železný

Abstract The type of environment plays an essential role in mobile robotics. Autonomous robots usually work in static and single type environments. Unfortunately, in the real world, there are many situations when the type of the environment changes. This paper deals with the design of the system for environment recognition working in real time on a mobile robot. Once the environment is recognized and classified, the parameters of the robot can be adapted. Consequently, the robot can handle operations in multiple environments. The paper contains information about previous-related work in the environment classification, description of the proposed system, and experiments. Experiments are focused on change detection and environment classification using visual and non-visual sensors. Moreover, two non-visual change detection approaches are proposed in the experiment section.

34.1 Introduction

An Autonomous mobile robot is dependent on accurate localization and mapping algorithms [1]. Such algorithms, called Simultaneous Localization and Mapping (SLAM), have to be able to create a good map for navigating a robot and for avoiding obstacles.

Many developed approaches and methods are capable of creating a map of the environment. They are based either on a LiDAR sensor—such as gMapping [2] , HectorSLAM [3], or Google Cartographer [4]—or vision-based such as ORB-SLAM [5], LSD-SLAM [6] and LDSO [7]. Moreover, some systems can fuse more sensors, such as RTAB-MAP [8]. Unfortunately, most of them are restricted by the assumption of the non-changing environment. Thus, they can have problems in situations such as the presence of dynamic objects or transition between two different

P. Neduchal (✉) · M. Železný
University of West Bohemia, 8, St. Univerzitní, 30614 Plzeň, Czech Republic
e-mail: neduchal@kky.zcu.cz

M. Železný
e-mail: zelezny@kky.zcu.cz

A. Ronzhin and V. Shishlakov (eds.), *Proceedings of 15th International Conference on Electromechanics and Robotics "Zavalishin's Readings"*, Smart Innovation, Systems and Technologies 187, https://doi.org/10.1007/978-981-15-5580-0_34

environments. There are some promising multi-environment mapping approaches, such as SmartSLAM [9] or Hierarchical SLAM based on Discrete Event Systems [10]. However, it seems that there is not any significant active research on this topic nowadays.

The most significant transition that should be addressed is the one between an indoor and an outdoor environment. Of course, there are a lot of other transitions that can occur, such as changing the light source or the transition between the hallway and room.

The typical situation that accompanies transition is a change in the form of the input data. In some cases, particular sensors could lose meaning for the mapping task. As an example, the LiDAR data can be useful in indoor environments where obstacles and walls are closer than the maximum range of the sensor. These assumptions cannot necessarily apply in outdoor environments. Another example is the textured environment under a good light source for the camera in contrast to textureless and dark environments.

This paper deals with the problem of transition between two different environments—indoor and outdoor in particular. The goal of this research is to propose an approach to detect and classify the environment. The behavior of the robot can be adapted based on this knowledge. The contributions of the paper are a design of the proposed classification system and two promising methods for non-visual environment change detection. Also, an experiment on visual environment classification is presented.

The paper is composed as follows. The second section is focused on existing related research. In the third section, the proposed system is described. In Sect. 4, the experiments are described, and the results are discussed. The conclusion of the paper is the central part of the last section. Moreover, future work is described there.

34.2 Related Work

This section is composed of several parts related to topics that can help to recognize the environment. It is in particular Semantic Localization and Mapping, change detection based on image processing and change detection based on other—usually non-visual—sensors.

34.2.1 Semantic Localization and Mapping

One way to deal with the transition between two environments could be based on the Semantic Simultaneous Localization and Mapping or Semantic Visual Odometry. In this case, all semantic data are obtained from the camera images. The goal of the semantic approaches is to obtain information about the scene and included objects to improve the properties of the localization and mapping methods. Paper focused on closing-loop semantic segmentation [11] is a good example. The authors also provide a monocular depth estimation of the environment. Many recent approaches are based on deep learning such a paper [12] about deep-learning-based semantic segmentation.

Semantic SLAM approaches are focused either on object-level SLAM, such as [13, 14], or the representation of the entire map. Despite the promising approaches in the field of semantic and map enhancement, none of them is capable of recognizing the environment and changing the behavior of the mobile robot.

34.2.2 Environment Change Detection Based on Image Processing

Every change detection can be divided into two categories. The first one is an offline category that processes all data. The other category is online processing, which operates with current sensor measurements. Many research papers focus on the first category, but its results can be transferred to the online domain. Moreover, several papers are focused on indoor/outdoor classification.

Older approaches were based on extracting image features such as color or texture. Then, they use feature vectors to train classifiers, such as support vector machines (SVM). An example of this approach is the paper [15] and papers of Serrano et al. [16]. Similarly, the authors of papers [17, 18] classify the scene using a bag-of-features model computed by k-means clustering of extracted image local features. Another option is to process edge straightness in the image [19] with promising results.

More recent research is focused on the combination of image preprocessing and a neural network as a classifier. This approach is used in the paper [20], where features based on color, entropy, and edge orientations are used as input of several neural networks. The research in paper [21] is based on the use of GIST features [22]. The more general approach can be found in the paper [23], where the scene is classified into multiple classes. This particular research uses Places dataset [24].

34.2.3 Environment Change Detection Based on One-Dimensional Signal

Approaches in this subsection are usually used in different applications than environment change detection—e.g., medical condition monitoring, speech detection, or human activity analysis [25]. Methods are based on the change point detection (CPD). It means that they are supposed to find abrupt changes in input signals. Change points are usually searched using edge detection, event detection, or anomaly detection.

Approaches are divided into supervised and unsupervised methods. Supervised methods are standard machine learning methods—i.e., decision trees, SVM, Bayesian nets, hidden Markov models, conditional random fields, or Gaussian mixture models. These methods usually need a big amount of data for training. Thus, there are suitable for cases with two environments with stable properties—i.e., stable temperature, humidity, or air pressure. In this case, the classifier is trained well on these properties.

On the other hand, the situation of indoor/outdoor classification is not suitable for supervised methods because the properties of the outdoor environment can change significantly during the time—e.g., daily or seasonally.

Unsupervised methods are more suitable for our task. They are based on the analysis of the statistical features of the signal. The most common method is CUmulative SUM (CUSUM) [26]. The methods usually use some dissimilarity measures such as Kullback–Leibler (KL) divergence—a method called KLIEP [27]—and Pearson divergence in Unconstrained Least-Squares Importance Fitting (uLSIF) and Relative uLSIF (RuLSIF) methods [28]. The change point is detected when the measure is above the threshold.

The different approach is adopted by papers [29] (Gaussian processes) and [30] (graph-based methods). In the case of our environment change detection, the most important is the computational cost of the method rather than its accuracy. It is always better to search for changes in 1D non-visual signal and process image only for time steps triggered by these methods, and then processing of all acquired images.

34.3 Environment Detection and Classification System

The goal of the system is to recognize the environment and to perform appropriate action based on this knowledge. The design of the proposed system is visible in Fig. 34.1. It is based on our previous paper [31].

The system is divided into three main parts. The first one is the environment change detection (ECD) module, the second one is image-based environment classification

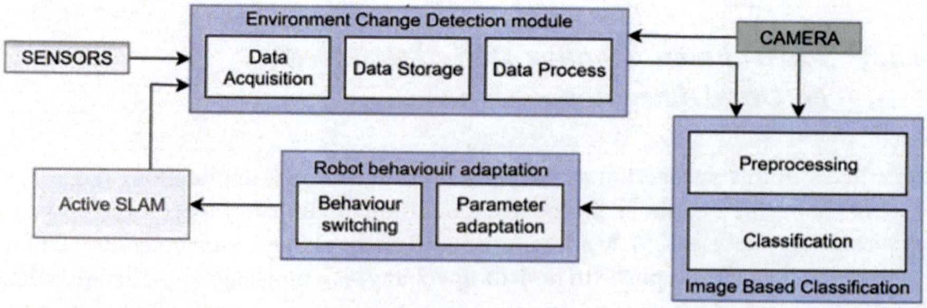

Fig. 34.1 The concept of the environment classification system

(IBEC) module, and the last one is robot behavior adaptation (RBA) module. The implementation of the system is developed for the robot operating system (ROS) [10, 32], and it is available as open-source on GitHub [33].

34.3.1 Environment Change Detection

The first part of the proposed system is supposed to communicate with sensors, store its data, and process data from both visual and non-visual sensors. The goal is to detect changes in the signal—i.e., high probability of transition between environments. The trigger should be generated when change occurs.

ECD module—as it is shown in Fig. 34.1 is composed of three parts—*data acquisition, data storage,* and *data process* sub-module. Measurements from sensors are sent to the data acquisition sub-module, where they can be processed in the time domain. Then, all data is sent to the data storage sub-module. The storage is based on the map created by localization and mapping software—i.e., data is stored in the map on the position where they were obtained. The last part is the data process sub-module, where stored data can be processed w.r.t. The map—e.g., examine whether they belong to the room or the hallway.

The data storage sub-module is only one in the system. On the other hand, multiple data acquisition and data process sub-modules can run in the system. The details about the ECD module can be found in our paper [33]. This module is based on the GridMap Library [34], and it is available as a standalone package on GitHub [35].

Each single data acquisition or data process sub-module can publish a trigger that is sent to the IBEC module for further processing or directly to the RBA—e.g., when the maximum brightness value falls under a threshold, then it is a change that is not needed to process in the IBEC module, but the robot should adapt its behavior.

The importance of this module is not in its accuracy—i.e., it is not necessary to create trigger only in the time step of real change. The reason for using it is to reduce the computational cost—i.e., IBEC module process only images in triggered time steps.

34.3.2 Image-Based Environment Classification

IBEC module is supposed to classify input image to the defined classes. They correspond with the environment types that are defined in the particular application. In general, the module has two parts. The first part is a preprocessing of the input image, and the second one is the classifier itself (multiple classifiers, respectively).

The preprocessing of the image is a necessary part of this module. In the first phase, it prepares input images in the manner of illumination properties. In the second phase, the input is transformed into the form that is suitable for the classifier/s—i.e., an image or a feature vector.

The second part is composed of one or multiple classifiers. They can be based on standard machine learning approaches such as nearest neighbors, SVM, or decision trees. The feature vector calculated from the input image is usually the input of this type of classifier. Another option is deep neural networks that can handle the whole image as an input. Neural network-based classifiers are tested in the experiment described in Sect. 34.4.

34.3.3 *Robot Behavior Adaptation Module*

The main goal of the RBA module is a switch between various behaviors based on the information received from previous parts of the system. In the best scenario, the only single parameter is changed. However, usually, it is necessary to switch between a set of variables or even whole parts of the robot system—e.g., switch between SLAM approaches.

The last scenario is caused by the fact that almost all existing SLAM systems can work in a non-changing environment. Thus, when the change occurs, they must be rerun with new parameters, which cause loss of the created map. The RBA is supposed to switch between systems and remember the state of the system before switching.

34.4 Experiments

In this section, the experiments focused on the particular parts of the proposed system are discussed. In the first part of a recorded dataset is described. The next part is focused on the non-visual detection module, especially on the system capability to reduce computational costs by generating triggers when the change occurs. Image-based classification experiment using neural networks is the main topic of the second experiment.

34.4.1 *Dataset*

Both the following experiments are processed using our recorded dataset. The dataset contains RGB image data from Intel Realsense camera, temperature measurements, humidity measurements, ceiling height measurements, and air pressure measurements, row LiDAR data, and wheels odometry data. The image data (see Fig. 34.1) is obtained with the 30 Hz rate, LiDAR with 10 Hz, and other sensors with a rate of 24 Hz. All sensors are equipped on the wild thumper robot, which is also shown in Fig. 34.2.

Fig. 34.2 Examples of an indoor—first and second image—environment, an outdoor—third image—environment and the robot used for recording on the last image

We recorded data during several runs focused on various situations—i.e., without transition, indoor/outdoor transition, and light change. We recorded more than 20 min in total. We plan to record more data on different day time and different parts of the year to create a bigger dataset suitable for benchmarking various approaches to mobile robot environment classification.

34.4.2 Triggers Generation Using Non-visual Sensor Data

In this particular experiment, we process data from the dataset described above. The goal is to analyze a signal by various approaches to reduce the computation cost caused by the IBEC module. The experiment is described in the example temperature signal shown in Fig. 34.3. During this recording, the robot starts indoor in the laboratory. It moves through building to the main entrance and outside of the building. Then it moves back to the laboratory. Red circles are the time steps when the transition between the two environments should be detected. Gray ones are places of minor changes in the environment—e.g., the transition between room and hallway.

As a baseline, we use the CUSUM approach. We also perform this task on two proposed approaches based on the variance of the signal (*VarSig*) and the ratio of the long-term and short-term differences of the signal (*DiRat*) method. As a value of the CUSUM algorithm, we used the subtraction of the upper control limit C^{+} and lower control limit C^{-} computed on the signal of defined length—i.e., signal

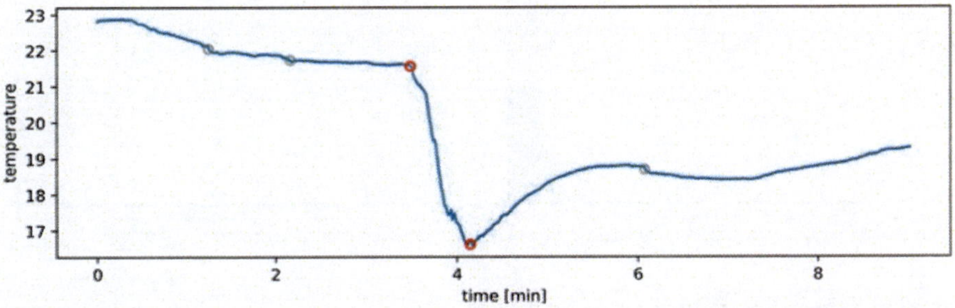

Fig. 34.3 Example of the temperature signal

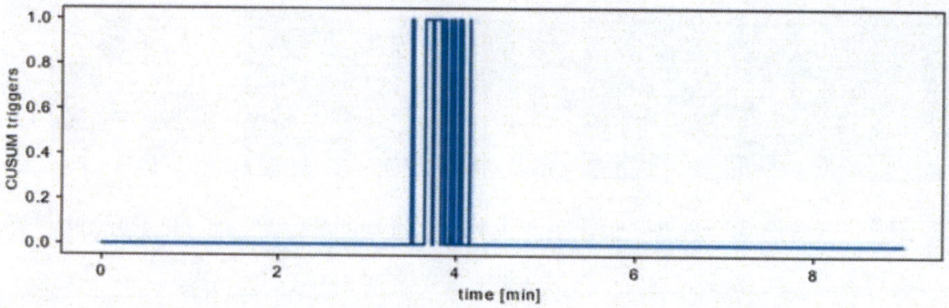

Fig. 34.4 Triggers generated using the CUSUM method

window. When the result value reaches a threshold, the trigger value is set to 1. The result is shown in Fig. 34.4. The disadvantage of this method is a necessity to set three parameters—a mean of the signal, a signal shift allowance value, and the threshold.

The first of our proposed methods *DiRat* based on the long-term and the short-term mean difference of the signal. It is computed using the following formula:

$$D(x, n_l, n_s) = \left(\frac{1}{n_l} \sum_{i=t-n_l}^{t} x_i + C \right) \cdot \left(\frac{1}{n_s} \sum_{j=t-n_s}^{t} x_j + C \right)^{-1}, \quad (1)$$

where x is an input signal, n_l, n_s are defined lengths of the long-term and short-term signal windows, t is a current timestamp, and C is a small constant to avoid division by zero. In our tests, n_s is constant, and n_l is a number of timestamps from the last change detection. Moreover, a maximum of D and D^{-1} is taken as a result. Comparing with a defined threshold is then used for the trigger generating. Triggers generated by *DiRat* approach is shown in Fig. 34.5. The method needs two parameters. The first one is *ns* parameter which also defines minimal length for n_l. The second parameter is the threshold.

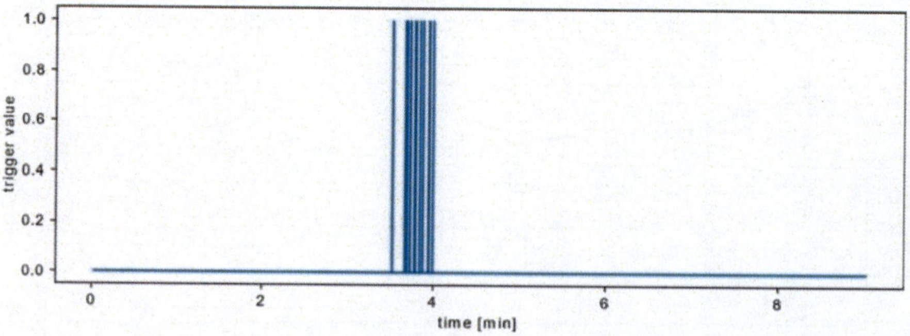

Fig. 34.5 Triggers generated by the *DiRat* method

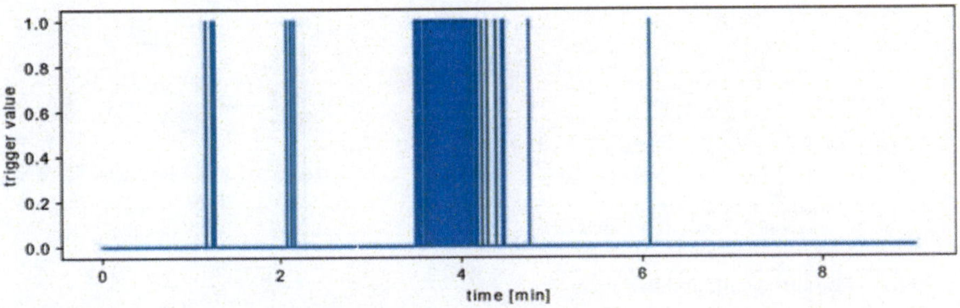

Fig. 34.6 Triggers generated by the *VarSig* method

The second proposed method is based on signal variance. It compares actual signal variance with the α multiple of the variance of the stable non-changing signal. The calculation is done as follows:

$$V(x, n, V) = \begin{cases} 1, \text{ if } \operatorname{var}(x(t - n : t)) > \alpha \cdot V \\ 0, \text{ otherwise} \end{cases}, \quad (2)$$

where n is a window length, var is a signal variance calculation, t is a current timestamp, v is a stable signal variance, and α is a defined constant.

The *VarSig* method needs the only parameter α, which defines the sensitivity of the method. The stable system variance can be computed in advance, so it is not necessary to recompute or reset it during the robot session. The triggers generated by the *VarSig* method with $\alpha = 1$ are shown in Fig. 34.6.

In conclusion, the most promising is the *VarSig* method. Based on the stable signal, it is possible to detect even minor changes, as it is also shown in Fig. 34.6.

34.4.3 Image-Based Classification Experiment

In the second experiment, we addressed the problem of image-based environment classification problems. The goal is to analyze the capability of a simple neural network (NN) architecture to classify the environment. In this particular experiment, we train NN to classify between the indoor and outdoor environments. We create training and testing data using recorded data. Then we add additional data from other image datasets from the internet. The whole dataset contains 18,980 images equally distributed to indoor and outdoor classes. Moreover, we use augmentation, such as horizontal flipping and offset of the image. The NN architecture used for system testing is shown in Fig. 34.7.

In the first phase, we train the network using raw data from the camera. Then we try to preprocess image using the following approaches: Fourier transform (FT), edge detection using Sobel operator, Roberts operator, and Canny operator. Raw data, FT, and Sobel have poor results around 50% on the testing set. On the other hand,

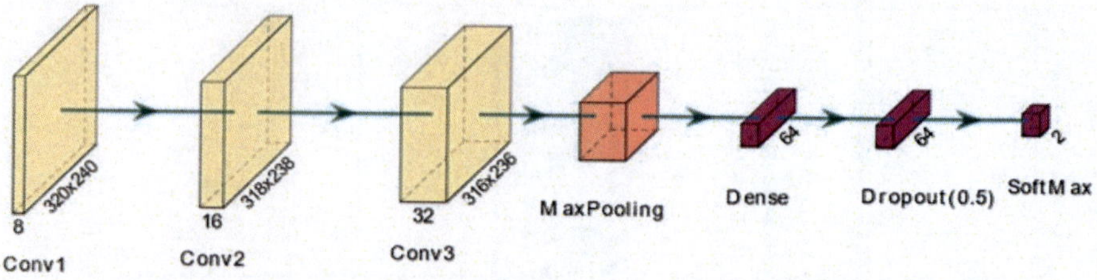

Fig. 34.7 Neural network architecture

Table 34.1 Accuracy of neural networks trained by preprocessed images

Input preprocessing	Training set accuracy (%)	Test accuracy (%)
Without preprocessing	50.43	51.34
Fourier Transform	88.10	50.45
Sobel	94.25	49.36
Roberts	95.93	86.39
Canny	93.34	92.45

training based on the data preprocessed by Roberts operator and Canny detection has significantly better accuracy. It is shown in Table 34.1

This experiment shows that it is possible to classify the environment based on the image data. Together with environment change detection from the previous section, it can enable the capability of the system to online environment classification on the mobile robot.

34.5 Conclusion

The paper proposed a system for environment detection and classification for mobile robots. The system recognizes the environment of the robot and switch between various behavior settings. The change of behavior can be based on the adaptation of system variables or even on the change of whole system parts such as the SLAM implementations.

In the experiments, we address two problems—the problem of change detection using non-visual sensor signal and the image-based classification. The first one is capable of generating triggers that signalize ongoing change in the environment. In this part, we propose two approaches that are capable of generating triggers called *DiRat* and *VarSig*. Moreover, the last one needs only one parameter α, which can be adaptively changed during the session. In the second experiment, we addressed the importance of image preprocessing in the problem of training a neural network to classify the environment into two classes (indoor/outdoor). The results show that a suitable preprocessing is to use the Canny edge detection method—accuracy 92.45% on the test set.

In future work, we want to run the proposed system on a real robot and implement it with SLAM approaches. Moreover, we are developing the whole system as a framework for multi-environment mobile robot tasks. Thus, we want to add new methods to transition detection to the system.

Acknowledgements This research is supported by the Ministry of Education of the Czech Republic, project No. LTARF18017. This research is supported by the Ministry of Education, Youth and Sports of the Czech Republic project No. LO1506. Computational resources were supplied by the project "e-Infrastruktura CZ" (e-INFRA LM2018140) provided within the program Projects of Large Research, Development and Innovations Infrastructures.

References

1. Thrun, S., Burgard, W., Fox, D.: Probabilistic Robotics. MIT Press, Cambridge (2005)
2. Grisetti, G., Stachniss, C., Burgard, W.: Improved techniques for grid mapping with rao-blackwellized particle filters. IEEE Trans. Rob. **23**(1), 34–46 (2007)
3. Kohlbrecher, S., et al.: Hector open source modules for autonomous map-ping and navigation with rescue robots. In: Robot Soccer World Cup, pp. 624–631. Springer, Berlin (2013)
4. Hess, W., Kohler, D., Rapp, H., Andor, D.: Real-time loop closure in 2D LIDAR SLAM. In: 2016 IEEE International Conference on Ro-botics and Automation (ICRA), pp. 1271–1278. IEEE (2016)
5. Mur-Artal, R., Tardos, J.D.: Orb-slam2: an open-source slam system for monocular, stereo, and rgb-d cameras. IEEE Trans. Rob. **33**(5), 1255–1262 (2017)
6. Engel, J., Koltun, V., Cremers, D.: LSD-SLAM: Large-scale direct monocular SLAM. In: European Conference on Computer Vision, pp. 834–849 (2014)
7. Gao, X., Wang, R., Demmel, N., Cremers, D.: LDSO: Direct sparse odometry with loop closure. In: 2018 IEEE/RSJ International Conference on Intelligent Robots and Systems (IROS), pp. 2198–2204 (2018)
8. Labbe, M., Michaud, F.: RTAB-Map as an open-source lidar and visual simultaneous localization and mapping library for large-scale and long-term online operation. J. Field Robot. **36**(2), 416–446 (2019)
9. Asmar, D.C., Zelek, J.S., Abdallah, S.M.: SmartSLAM: localization and mapping across multi-environments. In: IEEE International Conference on Systems, Man and Cybernetics, pp. 5240–5245 (2004)
10. Dai, X., Juan, Du.: Hierarchical simultaneous localization and mapping based on discrete event systems. In: International Conference on Measuring Technology and Mechatronics Automation, vol. 2, pp. 234–237. IEEE (2009)
11. Zhang, Z. et al.: Joint task-recursive learning for semantic segmentation and depth estimation. In: The European Conference on Computer Vision (ECCV), pp. 235–251 (2018)
12. Kaneko, M., et al.: Mask-slam: Robust feature-based monocular slam by masking using semantic segmentation. In: Proceedings of the IEEE Conference on Computer Vision and Pattern Recognition Workshops, pp. 258–266 (2018)
13. McCormac, J., et al.: Fusion ++: Volumetric Object-Level {SLAM}. In: International Conference on 3D Imaging, Modeling, Processing, Visualization and Transmission (3DIMPVT) (2018)
14. Nicholson, L., Milford, M., Sünderhauf, N.: QuadricSLAM: constrained dual quadrics from object detections as landmarks in semantic SLAM. IEEE Robot. Autom. Lett. **4**(1), 1–8 (2018)

15. Collier, J., Alejandro, R.S.: Environment classification for indoor/outdoor robotic mapping. In: Canadian Conference on Computer and Robot Vision, pp. 276–283. IEEE (2009)
16. Serrano, N., Savakis, A.E., Luo, J.: A computationally efficient approach to indoor/outdoor scene classification. In: Object Recognition Supported by User Interaction for Service Robots, vol. 4, pp. 146–149. IEEE (2002)
17. Zhou, L., Zhou, Z., Dewen, H.: Scene classification using multi-resolution low-level feature combination. Neurocomputing **122**, 284–297 (2013)
18. Zhou, L., Zhou, Z., Dewen, H.: Scene classification using a multi-resolution bag-of-features model. Pattern Recogn. **46**(1), 424–433 (2013)
19. Payne, A., Singh, S.: Indoor vs. outdoor scene classification in digital photographs. Pattern Recogn. **38**(10), 1533–1545 (2005)
20. Tao, L., Yeong-Hwa, K., Yeong-Taeg, K.: An efficient neural network based indoor-outdoor scene classification algorithm. In: Digest of Technical Papers International Conference on Consumer Electronics (ICCE), pp. 317–318 (2010)
21. Tahir, W., Majeed, A., Rehman, T.: Indoor/outdoor image classification using gist image features and neural network classifiers. In: 12th International Conference on High-capacity Optical Networks and Enabling/Emerging Technologies (HONET), pp. 1–5 (2015)
22. Li, Z., Itti, L.: Saliency and gist features for target detection in satellite images. IEEE Trans. Image Process. **20**(7), 2017–2029 (2011)
23. Zhou, B., et al.: Learning deep features for scene recognition using places database. In: Advances in Neural Information Processing Systems, pp. 487–495 (2014)
24. Zhou, B., et al.: Places: A 10 million image database for scene recognition. IEEE Trans. Pattern Anal. Mach. Intell. **40**(6), 1452–1464 (2017)
25. Aminikhanghahi, S., Cook, D.J.: A survey of methods for time series change point detection. Knowl. Inf. Syst. **51**(2), 339–367 (2017)
26. Page, E.S.: Continuous inspection schemes. Biometrika **41**(1/2), 100–115 (1954)
27. Kawahara, Y., Sugiyama, M.: Sequential change-point detection based on direct density-ratio estimation, statistical analysis and data mining. ASA Data Sci. J. **5**(2), 114–127 (2012)
28. Keogh, E., Lin, J.: Clustering of time-series subsequences is meaningless: implications for previous and future research. Knowl. Inf. Syst. **8**(2), 154–177 (2005)
29. Barry, D., Hartigan, J.A.: A Bayesian analysis for change point problems. J. Am. Statistical Assoc. **88**(421), 309–319 (1993)
30. Wei, L., Keogh, E.: Semi-supervised time series classification. In: Proceedings of the 12th ACM SIGKDD International Conference on Knowledge Discovery and Data Mining, pp. 748–753 (2006)
31. Neduchal, P., Bureš, L., Železný, M.: Environment detection system for localization and mapping purposes. IFAC-PapersOnLine **52**(27), 323–328 (2019)
32. Quigley, M., et al.: ROS: an open-source Robot Operating System. In: ICRA Workshop on Open Source Software, vol. 3, no. 3.2, 5p (2009)
33. GitHub: https://github.com/neduchal/ecs. Last accessed 13 Jan 2016
34. Fankhauser, P., Hutter M.: A Universal GridMap library: implementation and use case for rough terrain navigation. Robot Oper Syst (ROS) 99–120 (2016)
35. GitHub: https://github.com/neduchal/env_detection. Last accessed 13 Jan 2016

Chapter 35
Architecture and Algorithms of Geospatial Service for Navigation of Robotic Complexes

Dmitriy Levonevskiy, Evgenii Karasev, and Egor Aksamentov

Abstract This paper considers the problems of robotic device navigation for performing various agricultural tasks. The proposed approach for map building ensures automated establishment of 3D terrain models, based on snapshots, taken from unmanned aerial vehicles. The presented method of route planning employs such 3D models effectively enough. The techniques of edge weight calculation facilitate selection of the smoothest areas for route planning, therefore, reducing energy consumption. The suggested software architecture relies on a service bus for data transfer. This solution reduces the dependence of software modules, enables integration of heterogeneous data sources and allows controlling data flows and system processes more efficiently. Decent usability of the developed systems is achieved by means of cross-platform implementation of the front-end application.

35.1 Introduction

The concept of smart agriculture is actively developed now. Specific approaches are developed for establishment of hardware and software architectures for systems included in smart agriculture [1, 2]. This paper discusses an approach to implementation of a navigation service for robotic devices (RD) using area maps for solving agriculture-specific problems. The functionality of such service includes map processing, whereas the maps under consideration are stitched from images, taken by unmanned aerial vehicles (UAVs). The other aspects discussed here include management of zones, map layers and tasks for RDs, real-time visualization of agent data (actual battery charge, location, identification number of the device, distance between agents), etc.

Performing agricultural operations using RD assumes solution of the RD navigation problem; this problem, in turn, consists of several subproblems. Firstly, it is

D. Levonevskiy (✉) · E. Karasev · E. Aksamentov
St. Petersburg Institute for Informatics and Automation of the Russian Academy of Sciences, 39, 14th Line, 199178 St. Petersburg, Russia
e-mail: DLewonewski.8781@gmail.com

A. Ronzhin and V. Shishlakov (eds.), *Proceedings of 15th International Conference on Electromechanics and Robotics "Zavalishin's Readings"*, Smart Innovation, Systems and Technologies 187, https://doi.org/10.1007/978-981-15-5580-0_35

required to build a model of the area within which agricultural work is performed. This task includes construction of the area map and its tagging, together with the allocation of sites for various purposes, impassable areas, etc. Secondly, it is necessary to ensure RD path tracing. Thirdly, it is required to ensure communication between individual robots and the control center and provide the administrator with the means to manage tasks, monitor their implementation and monitor the status of the RD. Further follows an overview of the methods and tools, needed to solve these subtasks.

35.2 Related Works

When preparing a map for RD navigation, one of the first problems to be solved is the requirement to stitch individual map fragments, obtained with the UAV, into a whole landscape image. Almost any algorithms for stitching frames into a single orthomosaic map are based on the method of homography calculation between consecutive images. So, in the article [3], the authors proposed a method of map stitching, based on a new technology, allowing to choose the image type. This approach allows to compare each area with the best image of this area. Each image is projected onto a surface area separately from other frames. This algorithm provides high accuracy when constructing area maps, based on images, obtained in flight by an UAV at low altitude with large roll angles.

In [4], the authors used data obtained using UAVs to construct an indoor map of a building. The specificity of the presented algorithm consists in combining the data obtained from images with the data of inertial measurement units (IMU). The method, designed this way, proved to be well applicable in practice in conditions of low UAV flight altitude and by obstruction of some parts of the scene.

When stitching images, sometimes a parallax effect occurs—this means, visual position of the object changes, depending on the point, from which the shot was made. Due to the parallax effect, artifacts may appear on the resulting stitched image. To minimize the parallax effect, a novel map stitching approach was proposed in [5]. The method, developed in this paper, combines homography and curvature, retaining the map content without distortions. Then, in the aligned local areas, the potential seam is searched, along which the images are subsequently combined into a whole map. This method has proven itself well in practice in terms of using images without significant overlap.

Among the other issues, that need to be addressed, RD path tracing is particularly relevant [6, 7]. In [8], a navigation system for mobile robots on a surface in the form of a three-dimensional grid is presented. The system uses a cloud of 3D points to create a real-time navigation grid, consisting of triangles. Each section of the navigation grid is analyzed for obstacles, surface irregularities and passability. The weight of each edge between two grid points is calculated based on the distance between them, the difficulty in reaching these points, the local surface artifacts and altitude difference. To find the path on such a weighted graph, one can use Dijkstra's algorithm or $A*$. The presented approach to path planning was successfully implemented on a hardware platform VolksBot XT in urban environment.

A number of papers exist, related to the organization of robot control using mobile devices. So, in [9], the possibility of using mobile devices in controlling an industrial robot was investigated. According to the data of this paper, the use of mobile devices in controlling an industrial robot is a promising approach to solve this problem. The remote-control system presented in this work can be used both in the educational and practical domains. The remote control feature makes this device more mobile and flexible. In addition to the Internet connection, the results of testing the connection via Bluetooth 2.0 were also presented in the work. In this case, Bluetooth was revealed to be the worst solution due to the lower data transfer speed and limited performance range. The authors conclude that the remote control of the robot allows to extend robot usage and provides the robot with new functionality.

Currently, exist some studies, in which the Web platform is used as the main target for the implementation of applications for working with geographic data. In the paper [10], the authors presented a web-system model for displaying maps of various types (e.g., cadastral maps), with various locality-related data (socio-economic status, information about agriculture, etc.). The paper describes the implementation of the application to display the surroundings of the village of Rahim Pur and information about its agricultural conditions. The application allows you to display various objects on the map: houses, transformers, power lines, telephone poles, etc. Data were prepared and downloaded from the system using QGIS software and OpenLayer technology. The model, proposed by the authors of this paper, can be expanded for use in other areas to provide better rural development planning for the local government and assistance in disaster recovery. The results presented in the work were partially used in the development of the client application.

In the paper [11], the authors presented an oceanographic information system, based on web-based geo-visualization, built using dynamic particles and HTML5. The system enables simple users to intuitively and easily explore and recognize oceanographic data that have spatial and temporal variability. The authors managed to achieve improved visualization of the spatial and temporal variability of oceanographic data through the use of dynamic particles to represent vector-type information, such as the direction of the wind above the surface of the sea, ocean currents and waves. In addition, to overcome the limitations of low scalability and system availability, various factors were considered, such as service speed vulnerabilities, a non-standard Web environment on the Internet, HTML5 with open libraries such as WebGL, Canvas, D3 and Leaflet. The authors believe that improved visibility of information, speed of query execution and high scalability of the system will help users simplify interaction with the system and improve the perception of information.

In the paper [12], the authors considered the possibility of using VR technology as a medium for human–machine interaction between user and mobile robotic device. The work presents a VR system designed for remote control of a unicycle with an intuitive interface. The system relies on Internet and wireless networks to ensure human–machine interactions. The system allows the operator to remotely control the execution of tasks, relying on robotic means. It also allows the operator to directly control the robot through a 3D model, which facilitates the prediction of all types of operations. Consider the architecture of the proposed service in greater details.

35.3 Service Architecture for Robotic Platform Control

The service consists of a number of components running in Docker containers [13], each of which has an address on the local network. The components interact with each other over the network, while data transfer is carried out using a common bus, which uses WSO2 Enterprise Integrator, an integration platform that represents components in the form of data services and proxy services, for which program interfaces are exposed. The architecture of the whole navigation platform is represented in Fig. 35.1.

Using these components, it is possible to store, visualize, import and export information about land plots and their properties both through the API and through a Web application. One of the additional features of the service is the ability to process area maps taken from an unmanned aerial vehicle (UAV), specifically, cutting and stitching images into a single orthomosaic map. A digital camera mounted on a UAV

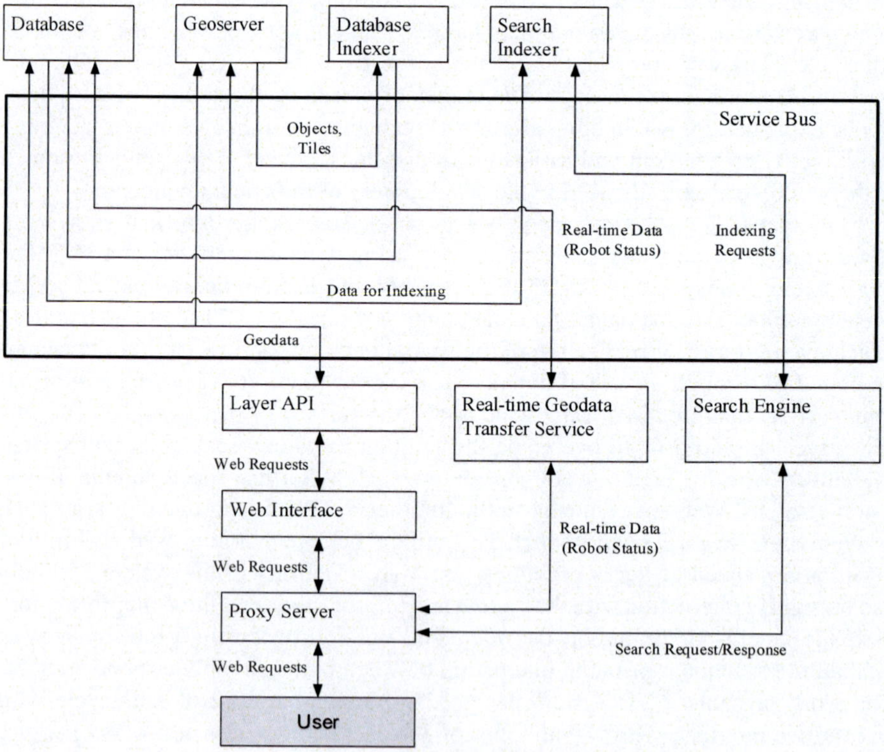

Fig. 35.1 Architecture of the whole navigation platform

collects photos of the area under consideration. After that, the resulting images are stitched into a single orthophotomap. Using the Mask R-CNN neural network, on the map are revealed areas, potentially dangerous for RD—i.e., water bodies, buildings, vegetation. This allows automatically determine the areas, impenetrable for the RD, without user intervention. The service allows automatically trace path from point A to point B, based on information about impenetrable areas, thereby ensuring the RD safety. The user can add or edit an existing zone at any time, using custom map layers (creating areas, setting their properties), leave notes on the map and build a data history on RD motion process and monitor map changes using the timeline. The service also enables the user to manage RD tasks (create and schedule tasks, display the progress of the current task), monitor and visualize the RD status in real time (actual charge, location, identification number, distance).

The interaction of the service with the user is performed using the Web interface, which, in turn, works with geo-data using the Layer API Server component. In addition, the Web interface establishes a WebSocket connection with the server when it is necessary to transmit real-time data on the status of the RV. To search over the present data, ElasticSearch engine is used. A specificity of this system consists in the support of geo-point and geo-shape data types for location data storage, which further enables not only text search queries, but also geo-referenced queries (e.g., search within a given area). The ES_Indexer component is used to index search data.

Components that work directly with data (such as search services) access the data services API provided by WSO2 Enterprise Integrator.

Two user services have been created within the data bus, described above. The geoserver proxy service broadcasts requests to the geoserver. It transfers requests without changes, but it has the ability to attach additional functionality (e.g., converting input and output data, logging, aggregating requests, assigning trigger functions, monitoring service availability). An appropriate data service has been implemented to work with data sources.

The geo-catalog is filled with data of various types and may contain bitmaps, vector geo-data, as well as metadata of an arbitrary type. Next, we consider the process of map establishment based on the images obtained from the UAV.

35.4 Data Preparation

RD control is performed in a specific area for which a map is being built. To build this map, it is necessary to stitch the overlapping images, received from the UAV, into a single image, and then label the impassable areas.

For stitching two consecutive images into one, you need to create your own set of feature points for each image. The obtained data of two images are compared and the points, having a minimal response function, are coupled into pairs. Each pair of points represents an instance of the same point located on different images. Based on papers [14, 15], the ORB algorithm was chosen for the detection of pairs of feature points.

Fig. 35.2 MASK R-CSS classification result

When stitching photos, distortion may appear—an effect that is characterized by a change in the shape of the object on the scene when changing the viewing angle. Due to the distortion effect on the stitched image, seams may arise on it. To minimize distortion, it is assumed that the overlap area of two images should be no less than 90%.

For automatic routing of the RDs, it is necessary to determine in advance all the areas, potentially dangerous for RD motion. In agricultural settings, examples of such obstacles are barns, trees and shrubs. For their detection, the neural network MASK R-CNN [13] was used, the specific feature of which is the automatic masking of the found objects (see Fig. 35.2). Information about the dangerous areas found is displayed to the user on the screen, which can add, delete or edit any of the zones.

To actually determine the location of the impenetrable zones, found on the map, it is necessary to associate the orthomosaic image with the GPS coordinates. During area imaging, in the center of each image, a corresponding GPS coordinate is attached with metadata (description of the place where the picture was taken). Using the GDAL library, a series of GPS coordinates is analyzed and the corresponding GPS coordinate is calculated for each pixel. Thus, GPS coordinates of the impassable zones are transmitted by the RD. The system should also build optimal paths for the RD motion, for example, when driving around hills, where the total time of RD ascent and descent exceeds the duration of obstacle avoidance. It is also necessary to consider the altitude difference and discover places, where a RD rollover is possible. These tasks can be solved by building a 3D area model. To build a 3D scene, it is necessary to get a series of pictures of the same area at different angles, upon which, using photogrammetry, a point cloud is established (see Fig. 35.3).

Based on the constructed surface models, it is possible to implement control and path planning methods for RD.

Fig. 35.3 Building of 3D model of an area

35.5 Front-End Application

To provide the user with a convenient interface for terrain map viewing, monitoring the RD and managing their tasks, a front-end application was developed. The specificity of the tasks associated with displaying various objects on the map and scenarios of interaction with them determines the key features of this application. One of the main tasks for software developers is to ensure that the application can work on devices without reliable network access (e.g., to use the application at various agricultural enterprises) and with limited performance (on mobile devices). To solve these problems, the software of the client part of the application was developed for reasons of reducing client–server interaction.

The essence of user interaction with the application interface is the implementation of such actions as obtaining information about existing objects in the area (mobile platforms, impassable areas, etc.), monitoring their status, assigning tasks to managed mobile objects. For most tasks, user interaction can be limited to working only with the data of objects related to the execution of the currently being solved task. The graphical application interface enables the user to interact with data without knowing the internal structure of the system or other technical details, not specific to the task.

To do this, the application implements a data layer management system. The data layer in the application is represented by a software module for visualizing the geographical data of objects on the map (borders, paths, etc.) based on metadata and implements the logic of interaction with object data through the application interface. Figure 35.4 shows a diagram of a component used in software.

The diagram shows the connections between the components of the module. The key component in the diagram is the "composite layer component," which represents the component of the layer with objects to be displayed on the map and child layers. The "component of a composite layer" receives the identifier of the layer as input. Using the "data load component," the "composite layer component" gets the layer configuration object. The component uses the resulting object to create a "component for displaying objects on the map," which is responsible for displaying objects on the map, and to form components of child layers. Also, in accordance with the parameters specified in the configuration object, "component of a composite layer," linking "component of displaying objects on a map" with components of services

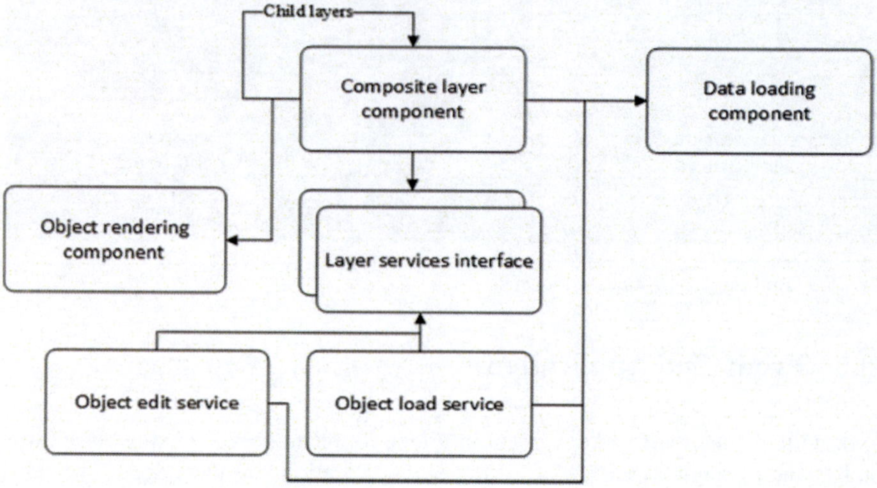

Fig. 35.4 Schematic layout of data layer module

that implement the "layer services interface." Currently, the application software supports services for editing objects and loading geo-data of objects.

The approach, enabling interactions of functional modules (particularly, services encapsulating the logic of object editions on the map and pulling out data of various objects from the application server, indicated on the diagram as the "service for editing layer objects" and "service for loading object data from the server," respectively), allows only limited connectivity of the module components. It is possible that in the future, along with the development of the application and the addition of new functionality, there will be a need to add new services.

Separation of data into individual layers, reflected in the architecture of the application, potentially provides for simplification of interoperability between the front-end and the application server, allows to boost data exchange and reduce the total number of requests. The server, having information about the user's work with the data of a particular layer in the present time, can allocate additional resources to ensure productive interaction with the relevant data.

To proceed, the layer module needs to get a layer configuration object. The layer configuration contains the name and identifier of the layer, information about the format of the objects displayed on the map for this layer, and information about the internal (child) layers. Based on the received configuration file, the layer module forms a tree of geographical objects, performing an entry for each of the child layers. The resulting data tree is used to visualize objects on the map in the application interface with sequential loading of data from top-level objects to lower objects. The metadata needed to configure the layer is provided by the application server.

Figure 35.5 demonstrates the graphical user interface of the developed application.

Fig. 35.5 Application output in the Web browser window

In Fig. 35.5, in the upper right corner, there is a toolbar for selecting a layer for work and tools for editing layer objects. In the viewport on the terrain map, an impenetrable area is depicted, relevant to the currently selected layer.

35.6 Conclusion

Within this paper, a number of issues of navigation of the RD were highlighted, which need to be addressed for successful application of the described approaches in the agriculture. The proposed approach to area map stitching provides for automated construction of a 3D terrain model based on images taken with UAVs.

The software architecture of the described complex is characterized by the use of a data bus, which allows to reduce the mutual dependence of components, makes it possible to work with various data sources and better control the data flows and the workflow of the entire system better. Convenience and functionality of system management are provided using a cross-platform front-end application.

Acknowledgements This work is supported by state research No. 0060-2019-0011.

References

1. Sinha, A., Shrivastava, G., Kumar, P.: Architecting user-centric internet of things for smart agriculture. In: Sustainable Computing: Informatics and Systems, pp. 88–102 (2019)
2. Bu, F., Wang, X.: A smart agriculture IoT system based on deep reinforcement learning. Future Gener. Comput. Syst. **99**, 500–507 (2019)
3. Li, Z., Isler, V.: Large scale image mosaic construction for agricultural applications. IEEE Robot. Autom. Lett. 295–302 (2016)
4. Ramaswamy, A., Gubbi, J., Raj, R., Purushothaman, B.: Frame stitching in indoor environment using drone captured images. In: 2018 25th IEEE International Conference on Image Processing (ICIP), pp. 91–95 (2018)
5. Zhang, F., Liu, F.: Parallax-tolerant image stitching. In: Proceedings of the IEEE Conference on Computer Vision and Pattern Recognition, pp. 3262–3269 (2014)
6. Negrete, M., Savage, J., Contreras-Toledo, L.A.: A motion-planning system for a domestic service robot. SPIIRAS Proc. **5**(60), 5–38 (2018)
7. Ovcharenko, K.L., Eremeev, IYu., Sazonov, K.V., Semenjuk, S.S., Abakumov, A.N.: Differential ranging method of locating satellite communication system earth stations using repeater on an unmanned aerial vehicle. SPIIRAS Proc. **18**(1), 176–201 (2019)
8. Pütz, S., Wiemann, T., Sprickerhof, J., Hertzberg, J.: 3d navigation mesh generation for path planning in uneven terrain. IFAC-PapersOnLine 212–217 (2016)
9. Zólkiewski, S., Galuszka, K.: Remote control of industry robots using mobile devices. WorldCIST **2**, 323–332 (2015)
10. Arora, P.K., Bhatia, R., Parkash, S., Sekhon, B.J.S.: Web based GPS and GIS model for rural areas. In: 2015 4th International Conference on Reliability, Infocom Technologies and Optimization (ICRITO) (Trends and Future Directions), pp. 1–5 (2015)
11. Kim, J.: Web-based geovisualization system of oceanographic information using dynamic particles and HTML5. J. Coast. Res. 1491–1495 (2018)
12. Benaoumeur, I., Zoubir, A.F., Reda, H.: Remote control of mobile robot using the virtual reality. Int. J. Electr. Comput. Eng. 2088–8708 (2015)
13. He, K., Gkioxari, G., Dollár, P., Girshick, R.: Mask R-CNN. In: Proceedings of the IEEE International Conference on Computer Vision, pp. 2961–2969 (2017)
14. Tareen, S.A.K., Saleem, Z.: A comparative analysis of sift, surf, kaze, akaze, orb, and brisk. In: 2018 International Conference On Computing, Mathematics and Engineering Technologies (iCoMET), pp. 1–10 (2018)
15. Rublee, E., Rabaud, V., Konolige, K., Bradski, G.: ORB: an efficient alternative to SIFT or SURF. In: 2011 International Conference on Computer Vision, pp. 2564–2571 (2011)

Part III
Electromechanics and Electric Power Engineering

Chapter 36
Quarter-Wave Symmetric Space Vector PWM with Low Values of Frequency Modulation Index in Control of Three-Phase Multilevel Voltage Source Inverter

Nikolay Lopatkin

Abstract The paper deals with the three-phase multilevel voltage source inverter (MLVSI) load current quality issue, which is important, in particular, for the MLVSI application to adjustable-speed AC drive. Here the problem of MLVSI control is solved by the use of the low-frequency space-vector PWM with the quarter-wave symmetric output voltages (QWS-SVPWM), based on the approach operating with the integer and fractional parts of the reference delta voltages relative values. The specified technique has low dynamic losses in power semiconductor switches. QWS-SVPWM is applied to control MLVSIs of any arbitrary numbers of the equal levels and any arbitrary topologies. Comparisons of dependences on a modulation depth for the THD and the first three orders' integrated voltage harmonics factors (IHF) of the simulated MLVSI output voltage waveforms are carried out between the three lowest appropriate values of frequency modulation index. To obtain the most appropriate voltage waveform for the best quality MLVSI load current, the amplitude modulation index ranges for preferable use of each of the three considered frequency modulation index values, under the terms of minimization of the appropriate orders' IHF values (depending on the particular kind of the load) are revealed.

36.1 Introduction

Multilevel voltage source inverters (MLVSI) perfectly fits into the modern trend of renewable energy application both in municipal engineering and in industry [1, 2] due to their specific features, to progress in implementation of high-rated in voltage and current values IGBT power switches and to a wide variety of their circuitry solutions and control techniques [3–7]. The above MLVSI features include the ability

N. Lopatkin (✉)
Shukshin Altai State University for Humanities and Pedagogy, 53, Korolenko Str, 659333 Biysk, Altai Region, Russia
e-mail: nikolay_lopatkin@mail.ru

A. Ronzhin and V. Shishlakov (eds.), *Proceedings of 15th International Conference on Electromechanics and Robotics "Zavalishin's Readings"*, Smart Innovation, Systems and Technologies 187, https://doi.org/10.1007/978-981-15-5580-0_36

to obtain high quality output AC power, the lowered requirements to the individual voltage ratings of power switches and the comparatively low value of the switches commutation frequency leading to the corresponding low level of the switching losses.

The modulation strategy is a key factor in the best utilization of the MLVSI full potential [7]. Due to affordability of microcontrollers and FPGAs, the method of the pulse-width modulation (PWM) with the sinusoidal reference and the carrier signals of different levels is now being superseded in engineering practice by the space vector (SV) PWM (SVPWM) of various techniques. The advantages of SVPWM in MLVSI are the best usage of the input direct voltage, better quality of the output voltage and the best output capability of the inverter, mostly, thanks to the possibility to use some redundant switching states for the MLVSI capacitors voltages' balancing [5–7].

MLVSIs of various power rates are now available, in particular, for applications in adjustable frequency induction motor drive systems. That use of MLVSI needs providing high quality currents consumed by the asynchronous motor in order to avoid torque ripples.

The purpose of this paper is to show an approach to achieving low enough higher harmonics content in MLVSI load current whereby frequency of switchings (commutations) is remaining low enough as well. It is accomplished here on the example of recently proposed SVPWM with the quarter-wave symmetric (QWS) output voltages (QWS-SVPWM) [8], which is applied to control of an MLVSI of some arbitrary number of levels and some arbitrary topology.

The cases of three various kinds of loads (three equivalent circuits of the load) are considered. For the three lowest values of frequency modulation index, which produce the conditions of the three-phase symmetry and quarter-wave symmetry of output voltages' waveforms (namely, the values multiple of 6: 12, 18 and 24), the THD and the first to third orders' integrated voltage harmonics factors' (IHF) dependences on a modulation depth are simulated in PSIM, where the seven-level cascaded H-bridge voltage source inverter idealized model with QWS-SVPWM control is developed. The amplitude modulation index ranges with preferable use of one of the considered frequency modulation index values, depending on the kind of the load, have been identified to provide the lowest possible THD values of the MLVSI load current.

36.2 Quarter-Wave Symmetric Space Vector PWM

QWS-SVPWM is the unique simple technique easy providing the QWS output voltages waveform for the three-phase multilevel inverter with an arbitrary number of the equal levels. It means that the generated instantaneous waveforms of both the phase-to-neutral voltages (wye voltages) and phase-to-phase voltages (delta voltages) of the MLVSI possess the half-wave symmetry (HWS) with the condition

$$u(\omega t + \pi) = -u(\omega t) \tag{36.1}$$

as well as one of the properties of an even signal or an odd signal, respectively:

$$u(-\omega t) = u(\omega t) \text{ or } u(\pi - \omega t) = -u(\omega t), \tag{36.2}$$

$$-u(\omega t) = u(-\omega t) \text{ or } u(\pi - \omega t) = u(\omega t), \tag{36.3}$$

where ω is the angular fundamental frequency of the output voltages.

The HWS condition (36.1) leads to the absence of the even-order harmonics, conditions (36.2) lead to the absence of the sine component in the remaining harmonics, and (36.3) lead to the absence of the cosine one. As a result, the QWS waveforms possess reduced low-order higher harmonics contents.

The background of QWS-SVPWM is as follows. The paper [9] explains why the vectors' duty cycles of SVPWM in its fast space-vector modulation algorithm [10] are equal to the fractional parts of the relative values of the MLVSI delta voltages. It is mathematically proved that the integer parts of these delta voltages' relative values are the coordinates of the three vectors which are the nearest vectors to the reference one, and the fractional parts are these nearest vectors' barycentric coordinates on these vectors' modulating triangle. A new symmetric and easy calculated modulating triangle's type identifier and a new five-segment switching sequence scheme, making it possible to implement HWS multilevel voltage waveform, are offered in [11]. At last, researches of [8, 12] improved the results of [11] and reached QWS-waveforms by applying the three-segment switching sequence scheme on the same new switching sequence algorithm.

The keystones of the QWS-SVPWM algorithm are described in the following subsections.

36.2.1 Vectors' Switching Sequence

The above vectors' switching three-segment sequence can be treated as the natural one in the sense that it is in full compliance with the reference voltage space vector (RVSV) moving process. The main traces of the executed voltage space vector (EVSV) in the context of the oblique-angled coordinates $\left(u_{\text{ab}}^*, u_{\text{bc}}^*\right)$ are shown in Fig. 36.1 for the first three rings of triangles.

Hereinafter for all the MLVSI phase (phase-to-neutral) voltages u_{x} and delta voltages u_{xy} (including the values with any additional subscript designations) their per-unit forms are introduced, and the respective relative values are marked with asterisks:

$$u_{\text{x}}^* = \frac{u_{\text{x}}}{U_{\text{d}}}, \quad u_{\text{xy}}^* = \frac{u_{\text{xy}}}{U_{\text{d}}}, \tag{36.4}$$

U_{d} is the direct voltage of the unit (base) level.

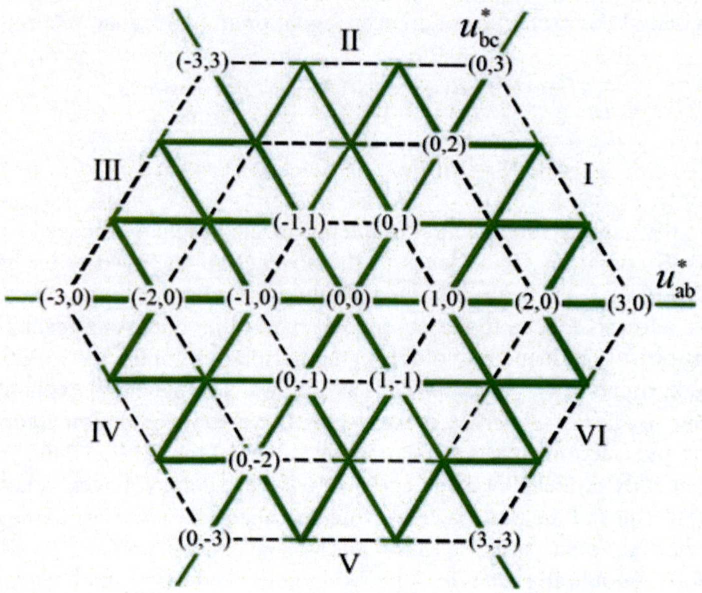

Fig. 36.1 Main traces of executed voltage space vector under proposed switching sequence

The solid lines correspond to the EVSV main transitions' traces and the dashed lines correspond to the EVSV auxiliary transitions, mostly transitions to the initial triangle vector after a next sample inside the same triangle [12].

The set (in ascending order of the sector number) of the sequences of the being executed vectors inside the triangles for the first ring of modulating triangles is {(1,0), (0,0), (0,1); (0,1), (0,0), (−1,1); (−1,1), (0,0), (−1,0); (−1,0), (0,0), (0,−1); (0,−1), (0,0), (1,−1); (1,−1), (0,0), (1,0)}. With the appearance of the down pointed triangle in the second ring of modulating triangles, the moving of the EVSV endpoint has performed along the zigzag line. So, the set of the vector sequences in sector I of the second ring is {(2,0), (1,0), (1,1); (1,0), (1,1), (0,1); (1,1), (0,1), (0,2)}. The same order is kept at every transition to the subsequent sector. Therefore, the set of the vector sequences in sector II of the second ring is {(0,2), (0,1), (−1,2); (0,1), (−1,2), (−1,1); (−1,2), (−1,1), (−2,2)}, and so on.

As can be seen, every subsequent ring adds two triangles (one triangle of each type) per sector, keeping both the EVSV trajectory of the moving inside the triangle of each type and the type of the initial (and the final) triangle in each sector. Thus, the increase in the numbers of the rings and inverter levels entails no changes in the pattern of EVSV moving, which has been used in the proposed SVPWM algorithm.

Whereas vectors' switching sequences, symmetrical on each clock cycle interval, are resulted in the only HWS waveforms, the offered unsymmetrical natural switching sequence spontaneously produces the QWS output voltages' waveforms.

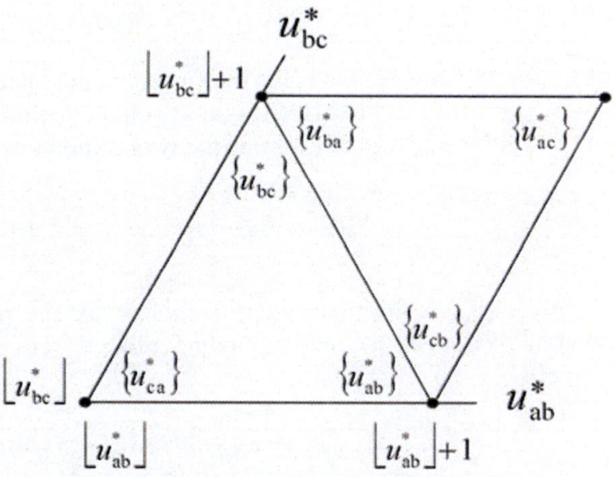

Fig. 36.2 Coordinates and duty cycles of modulating triangles' vectors

36.2.2 Voltage Space Vector PWM of Two Delta Voltages

The main used in QWS-SVPWM principle is the two delta voltage formation on the base of processing of the regular samples of their reference relative values, taking into account the conditions of the above natural switching sequence.

The coordinates and duty cycles values of modulating triangles' vectors are shown in Fig. 36.2 as the integer and fractional parts of the two delta voltages' relative values. Here $\lfloor x \rfloor$ means the rounding down x to the closest integer number, taking into account the sign (the "floor" function), so $\lfloor x \rfloor$ is the integer part of x; $\{x\}$ is the fractional part of x.

Hereinafter the desirable available for implementation and being formed voltages' waveforms are called being executed voltages.

The relative instantaneous value of the formed MLVSI output being executed delta voltage u_{xy} can be expressed as follows:

$$u^*_{EXExy}(t) = \sum_{k=1}^{m_f} u^*_{kEXExy}(t), \tag{36.5}$$

where $u^*_{kEXExy}(t)$ is the separate instantaneous function of the current time t, that is defined for its own clock cycle interval having the number k,

$$u^*_{kEXExy}(t) = \begin{cases} u^*_{kEXExy}(t_{kc}), \text{ if } (k-1) \cdot \mathrm{T_c} < t \le k \cdot \mathrm{T_c}, \\ 0, \text{ otherwise}, \quad k = 1, 2, \ldots, m_f, \end{cases} \tag{36.6}$$

t_{kc} is the current time from the start of the clock cycle with number k, $t_{kc} = t - (k - 1) \cdot T_c$, T_c is the clock period, $u^*_{kEXExy}(t_{kc})$ is some pulse function, m_f is the frequency modulation index,

$$m_f = f_c/f, \tag{36.7}$$

here f_c and f are, respectively, the clock and modulating (output) frequencies, $f_c = 1/T_c$, $f = 1/T$, T is the MLVSI output voltage period.

The pulse function in (36.6) is the two-component one:

$$u^*_{kEXExy}(t_{kc}) = \left\lfloor u^*_{ksREFxy} \right\rfloor + f_{kEXExy}(t_{kc}), \tag{36.8}$$

Here $u^*_{ksREFxy}$ is the preliminary sampled (for the midpoint of the k clock cycle interval) value of the corresponding relative reference delta voltage $u^*_{REFxy}(t)$, namely,

$$u^*_{ksREFab} = m_{aD}\sin((2\pi k - \pi)/m_f + 2\pi/3), \tag{36.9}$$

$$u^*_{ksREFbc} = m_{aD}\sin((2\pi k - \pi)/m_f), \tag{36.10}$$

where m_{aD} is the delta voltage amplitude modulation index,

$$m_{aD} = \sqrt{3}\, m_{aY} = \sqrt{3}\, U/U_d = \sqrt{3}\, U^* = U^*_{\Delta m}, \tag{36.11}$$

m_{aY} is the phase voltage (wye voltage) amplitude modulation index,
U and U^* are the value and relative value of the reference space vector magnitude,
$U^*_{\Delta m}$ is the amplitude relative value of the reference delta voltages.

The second component of (36.8), the function $f_{kEXExy}(t_{kc})$, possesses the only values 0 and 1. Its clock-cycle mean value f_{kEXExy_mean} correspond to the reference delta voltage as follows:

$$f_{kEXExy\_mean} = \left\{u^*_{ksREFxy}\right\} = u^*_{ksREFxy} - \left\lfloor u^*_{ksREFxy} \right\rfloor, \tag{36.12}$$

The $f_{kEXExy}(t_{kc})$ pulses' numbers and positions on the clock cycle k depend on the related voltage $u^*_{EXExy}(t)$ behavior, which is fitting for the accepted switching sequence of the executed vectors (see the previous subsection). Thus, for the most part, $f_{kEXExy}(t_{kc})$ formation is different for the different sectors and triangle types, and it uses the related identifiers [11]. The durations of these pulses of the value one and the instants of time of their rising and falling edges are defined via the fractional values of the relative values of the two delta voltages, see Tables 36.1 and 36.2.

Here the pulse positions take into account the changes in the two coordinates from their lowest to highest values on the clock cycle, and vice versa, in EVSV motion from the first vertex of the modulating triangle to the second and to the third one. As a matter of course, the plus or minus signs of the changes in coordinates of the executed vector EVSV correspond to the signs of the changes in coordinates of the

Table 36.1 Fractional components' functions of executed delta voltages, defined via fractional parts of two reference delta voltages (Sectors I, II)

RVSV's sector and argument's value range	Modulating triangle's type and EVSV's trajectory	Fractional components' functions of executed delta voltages on clock cycle k	
		$f_{kEXEab}(t_{kc})$	$f_{kEXEbc}(t_{kc})$
I 0°…60°		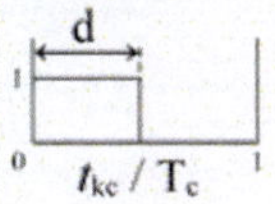 $d = \{u^{*}_{ksREFab}\}$	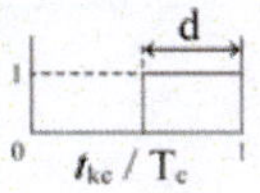 $d = \{u^{*}_{ksREFbc}\}$

II 60°…120°			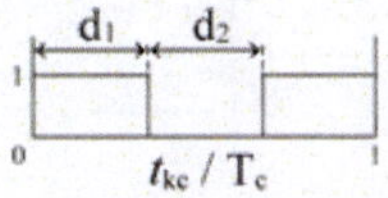 $d_1 = \{u^{*}_{ksREFab}\} + \{u^{*}_{ksREFbc}\} - 1$ $d_2 = 1 - \{u^{*}_{ksREFbc}\}$
			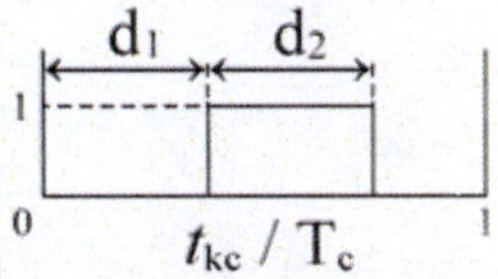 $d_1 = \{u^{*}_{ksREFab}\}$ $d_2 = \{u^{*}_{ksREFbc}\}$

reference vector RVSV motion inside the modulating triangle in whole, since we have chosen the appropriate vectors' switching sequence.

Fractional components' functions of executed delta voltages are described with the use of the fractional values of the two delta voltages' relative values, related to up-pointing modulating triangles only. This has been possible thanks to the equations, the first of which is valid for the non-integer values of coordinate u^{*}_{xy} inside the rhombus in Fig. 36.2:

$$\left\{u^{*}_{xy}\right\} + \left\{u^{*}_{yx}\right\} = 1, \quad \text{or} \quad \left\{u^{*}_{yx}\right\} = 1 - \left\{u^{*}_{xy}\right\}, \tag{36.13}$$

and the others are placed among the three-phase system's equations in Table 36.3, which are valid for all cases except when the reference space vector's endpoint hits exactly one of the vertices of the modulating triangle, i.e. when all the six coordinates (both the direct u^{*}_{xy} and inverse u^{*}_{yx}) are integer values.

To avoid the zero-duration pulses of $f_{kEXEab}(t_{kc})$ and $f_{kEXEbc}(t_{kc})$ in rare occasions when all the relative values of the delta voltages have the just integer values, the new above mentioned up-pointing triangle identifier can be applied:

Table 36.2 Fractional components' functions of executed delta voltages, defined via fractional parts of two reference delta voltages (Sectors III…VI)

<table>
<tr><th rowspan="2">RVSV's sector and argument's value range</th><th rowspan="2">Modulating triangle's type and EVSV's trajectory</th><th colspan="2">Fractional components' functions of executed delta voltages on clock cycle k</th></tr>
<tr><th>$f_{kEXEab}(t_{kc})$</th><th>$f_{kEXEbc}(t_{kc})$</th></tr>
<tr><td rowspan="2">III
120°…180°</td><td></td><td>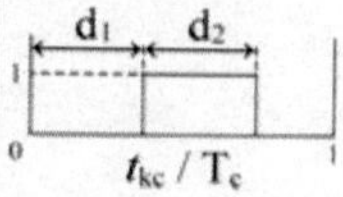

$d_1 = \{u^*_{ksREFbc}\}$
$d_2 = \{u^*_{ksREFab}\}$</td><td rowspan="4">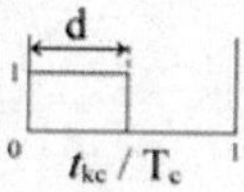

$d = \{u^*_{ksREFbc}\}$</td></tr>
<tr><td></td><td>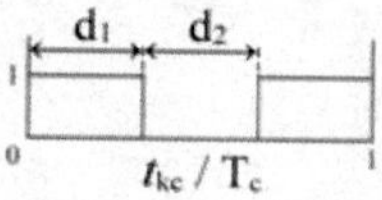

$d_1 = \{u^*_{ksREFab}\} + \{u^*_{ksREFbc}\} - 1$
$d_2 = 1 - \{u^*_{ksREFab}\}$</td></tr>
<tr><td rowspan="2">IV
180°…240°</td><td></td><td rowspan="4">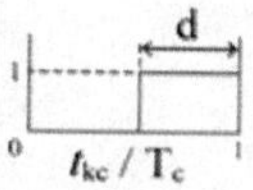

$d = \{u^*_{ksREFab}\}$</td></tr>
<tr><td></td></tr>
<tr><td rowspan="2">V
240°…300°</td><td></td><td>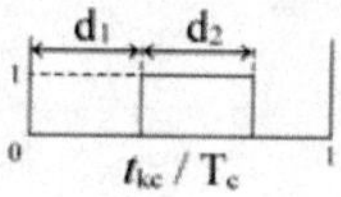

$d_1 = 1 - \{u^*_{ksREFab}\} - \{u^*_{ksREFbc}\}$
$d_2 = \{u^*_{ksREFbc}\}$</td></tr>
<tr><td></td><td>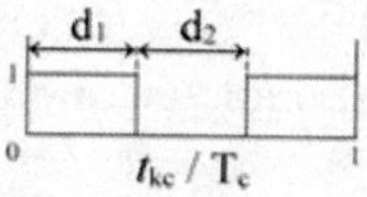

$d_1 = 1 - \{u^*_{ksREFab}\}$
$d_2 = 1 - \{u^*_{ksREFbc}\}$</td></tr>
<tr><td>VI
300°…360°</td><td></td><td>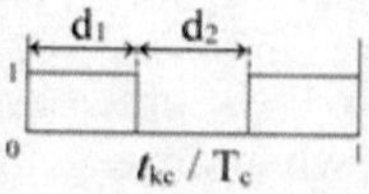

$d_1 = 1 - \{u^*_{ksREFbc}\}$
$d_2 = 1 - \{u^*_{ksREFab}\}$</td><td>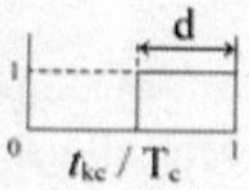

$d = \{u^*_{ksREFbc}\}$</td></tr>
</table>

(continued)

Table 36.2 (continued)

RVSV's sector and argument's value range	Modulating triangle's type and EVSV's trajectory	Fractional components' functions of executed delta voltages on clock cycle k	
		$f_{kEXEab}(t_{kc})$	$f_{kEXEbc}(t_{kc})$
		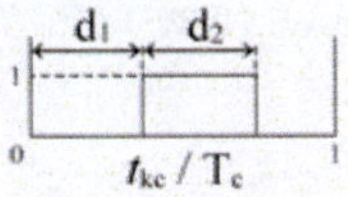 $d_1 = 1 - \{u^*_{ksREFab}\} - \{u^*_{ksREFbc}\}$ $d_2 = \{u^*_{ksREFab}\}$	

Table 36.3 Values of sums of fractional values and of integer values of direct delta voltages' and inverse delta voltages' relative values

Expression	Modulating triangle's type	
	Up-pointing, Δ	Down-pointing, ∇
$\{u^*_{ab}\} + \{u^*_{bc}\} + \{u^*_{ca}\}$	1	2
$\lfloor u^*_{ab} \rfloor + \lfloor u^*_{bc} \rfloor + \lfloor u^*_{ca} \rfloor$	−1	−2
$\{u^*_{ba}\} + \{u^*_{cb}\} + \{u^*_{ac}\}$	2	1
$\lfloor u^*_{ba} \rfloor + \lfloor u^*_{cb} \rfloor + \lfloor u^*_{ac} \rfloor$	−2	−1

$$F_\Delta = \lfloor u^*_{ab} \rfloor + \lfloor u^*_{bc} \rfloor + \lfloor u^*_{ca} \rfloor + 2 = \begin{cases} 0 \text{ if triangle is down - pointing,} \\ 1 \text{ if triangle is up - pointing,} \\ 2 \text{ if exact vector is pointed by reference vector.} \end{cases} \quad (36.14)$$

Thus, the k clock cycle interval's up-pointing triangle identifier is

$$F_{k\Delta} = \lfloor u^*_{ksREFab} \rfloor + \lfloor u^*_{ksREFbc} \rfloor + \lfloor -u^*_{ksREFab} - u^*_{ksREFbc} \rfloor + 2, \quad (36.15)$$

and $F_{k\Delta} = 2$ implies $f_{kEXEab}(t_{kc}) = f_{kEXEbc}(t_{kc}) = 0$.

Due to the taking place symmetry kinds, including the three-phase symmetry, the $f_{kEXEbc}(t_{kc})$ description for RVSV's argument's value range 0°…90° in Table 36.1 is enough for both $u^*_{EXEbc}(t)$ and $u^*_{EXEab}(t)$ implementing [8], for instance, with use of lookup tables. But in view of the simplicity of the operations in (36.8)-(36.10), as well as of the $f_{kEXEab}(t_{kc})$ and $f_{kEXEbc}(t_{kc})$ values finding, the consideration and direct use of both the Tables 36.1, 36.2 data for the two being executed voltages' calculating could be even more effective.

The directly executed phase-to-ground MLVSI voltages can be defined as follows:

$$u^*_{EXEag}(t) = k(t), \quad u^*_{EXEbg}(t) = k(t) - u^*_{EXEab}(t),$$
$$u^*_{EXEcg}(t) = k(t) - u^*_{EXEab}(t) - u^*_{EXEbc}(t) \quad (36.16)$$

here $k(t)$ can possess the only integer values, $k_{\min}(t) \le k(t) \le k_{\max}(t)$.

The phase-to-ground MLVSI voltage of the N-level cascaded H-bridge voltage source inverter is physically limited: $-(N-1) \le u^*_{\mathrm{EXExg}}(t) \le N-1$, where N is the number of non-negative various relative values generated by a phase leg, $N{=}N_c+1$, here N_c is the number of cascaded (series-connected) H-bridge cells in a phase leg. Thus, in this particular case, (36.17) produces the following limits of $u^*_{\mathrm{EXEag}}(t) = k(t)$:

$$\begin{aligned} k_{\min}(t) &= \max\left(1-N,\, u^*_{\mathrm{EXEab}}(t)+1-N,\, u^*_{\mathrm{EXEab}}(t)+u^*_{\mathrm{EXEbc}}(t)+1-N\right), \\ k_{\max}(t) &= \min\left(N-1,\, u^*_{\mathrm{EXEab}}(t)+N-1,\, u^*_{\mathrm{EXEab}}(t)+u^*_{\mathrm{EXEbc}}(t)+N-1\right) \end{aligned} \tag{36.17}$$

Available switching states' redundancies could be used for a more balanced utilization of input DC sources and a more balanced current-loading of switches.

Both the phase voltages and delta voltages, generated in accordance with the presented QWS-SVPWM algorithm with m_f multiple of 6, contain odd harmonics of the orders $6 \cdot h \pm 1$, where h possess natural numbers' values.

36.3 MLVSI Load Current's THD Assessment and Distinction of Amplitude Modulation Index Ranges

The simplified phase equivalent circuits of the MLVSI for the cases of three various kinds of its generalized load are shown in Fig. 36.3.

To assess the load current's higher harmonics component and THD the method, developed by Prof. G. S. Zinoviev (NSTU, Novosibirsk), namely the differential equations' algebraization (ADE) asymptotically approximate method [13] is used here.

The ADE first approximation estimates of the MLVSI phase load current's THD, which are designated here as the factor of current's harmonics K_{hia}, for circuits of Fig. 36.3a, b and c, are, respectively, as follows:

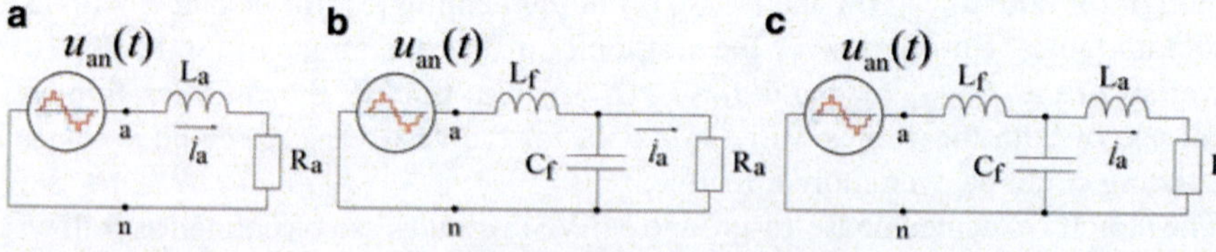

Fig. 36.3 MLVSI loads' simplified phase equivalent circuits: *RL* load circuit (**a**); *R* load circuit with MLVSI output *LC* filter (**b**); *RL* load circuit with MLVSI output *LC* filter (**c**)

$$K_{hia} \approx \overline{K}_{hu}^{(1)} \cdot \sqrt{1 + \left(\frac{R_a}{\omega L_a}\right)^2};\ K_{hia} \approx \overline{K}_{hu}^{(2)} \cdot \sqrt{1 + \left(\frac{1}{(R_a C_f)^2} - \frac{2}{L_f C_f}\right)\frac{1}{\omega^2} + \frac{1}{(L_f C_f)^2 \omega^4}};$$

$$K_{hia} \approx \overline{K}_{hu}^{(3)} \cdot \sqrt{1 + \frac{a_1^2 - 2a_2}{\omega^2} + \frac{a_2^2 - 2a_1 a_3}{\omega^4} + \frac{a_3^2}{\omega^6}},$$

$$a_1 = \frac{R_a}{L_a},\ a_2 = \frac{1}{C_f}\left(\frac{1}{L_f} + \frac{1}{L_a}\right),\ a_3 = \frac{R_a}{L_f C_f L_a},\ b_3 = \frac{1}{L_f C_f L_a}, \qquad (36.18)$$

where $\overline{K}_{hu}^{(n)}$ is the *n*-order weighted THD (WTHD) coefficient of the MLVSI output voltage $u_{an}(t)$, the so called n-order integrated voltage harmonics factor (IHF) [13],

$$\overline{K}_{hu}^{(n)} = WTHD^{(n)} = \frac{\overline{U}_{(hh)}^{(n)}}{\overline{U}_{(1)}^{(n)}} = \frac{\overline{U}_{(hh)}^{(n)} \cdot \omega^n}{U_{(1)}} = \sqrt{\sum_{k=2}^{\infty}\left(\frac{U_{(k)}}{k^n \cdot U_{(1)}}\right)^2}, \qquad (36.19)$$

here values $U_{(k)}, \overline{U}_{(1)}^{(n)}, \overline{U}_{(hh)}^{(n)}$ are the RMS values of the *k*-order harmonic component, and of the n times integrated values of the fundamental harmonic and the higher harmonics component.

As can be seen in (18), each estimate of load current's THD correlates to the voltage IHF of the order, equal to the order of the considered equivalent circuit of the generalized load. So, the appropriate order's IHF should be minimized for minimizing the MLVSI load current's THD.

The PSIM-simulated delta and phase voltages' waveforms of the seven-level cascaded H-bridge voltage source inverter idealized model with QWS-SVPWM control are shown in Fig. 36.4a for the particular control parameters, $m_f = 24$ and $m_{aY} = 3.2$.

The MLVSI load current waveform is estimated in the PSIM model of phase RL load circuit (Fig. 36.3a), where $R_a = 48.13$ *Ohm* and $L_a = 0.14$ H imitate an induction motor in a steady-state mode. The obtained dependences of the load current's THD K_{hia} and of the phase voltage's first-order IHF on the phase voltage

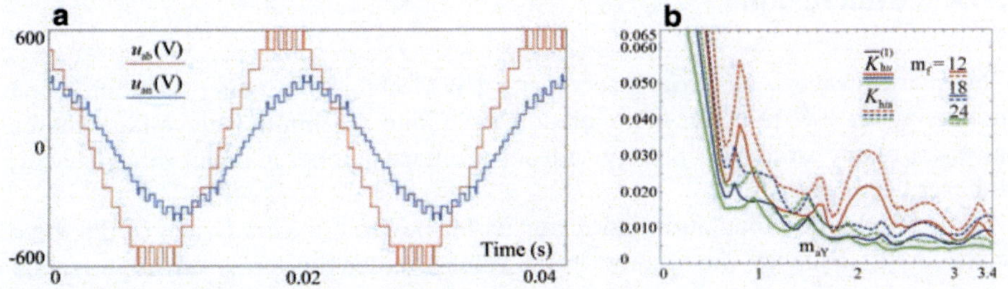

Fig. 36.4 PSIM simulation's results: example of MLVSI output delta and phase voltages' waveforms of QWS-SVPWM algorithm **a**; first-order IHF of MLVSI phase voltage, and load current's harmonics factor (THD) versus phase voltage amplitude modulation index **b**

Table 36.4 Phase voltage amplitude modulation index ranges of preferable use of considered values of frequency modulation index

Minimized integrated voltage harmonics factor (IHF)	Kind of generalized load	Frequency modulation index, m_f		
		12	18	24
Zero-order IHF, $\overline{K}_{hu}^{(0)} = K_{hu} = THD$	Pure *R* load	[0.05, 1.80]	(1.80, 2.45]	(2.45, 3.40]
First-order IHF, $\overline{K}_{hu}^{(1)}$	*RL* load	–	[0.90, 1.00), [1.35, 1.50)	[0.05, 0.90), [1.00, 1.35), [1.50, 3.40]
Second-order IHF, $\overline{K}_{hu}^{(2)}$	*R* load with *LC* filter	–	(0.75, 0.80], [0.90, 1.00), (1.30, 1.45), [1.75, 1.85), (2.00, 2.10)	[0.05, 0.75], (0.80, 0.90), [1.00, 1.30], [1.45, 1.75), [1.85, 2.00], [2.10, 3.40]
Third-order IHF, $\overline{K}_{hu}^{(3)}$	*RL* load with *LC* filter	[0.65, 1.25), (1.70, 1.85], (3.05, 3.15)	[0.05, 0.25), (0.55, 0.65), (1.30, 1.40], (2.00, 2.20]	[0.25, 0.55], [1.25, 1.30], (1.40, 1.70], (1.85, 2.00], (2.20, 3.05], [3.15, 3.40]

amplitude modulation index illustrate the complete alignment between the curves' changings of K'_{hia} and of $\overline{K}_{hu}^{(1)}$, see Fig. 36.4b, thereby validating (36.18) and the load current's THD-minimizing approach as a whole.

Based on PSIM-simulated results for the zero to third orders' IHFs, the m_{aY} ranges of the preferable use of each of values $m_f = 12$, 18 and 24 for the four load kinds (and the minimized IHFs) are summarized in Table 36.4.

36.4 Conclusion

The quarter-wave symmetric space vector PWM algorithm, being the simple and promising as low-frequency modulation technique for multilevel voltage source inverters of any arbitrary topology with any arbitrary number of equal voltage levels, is described in details.

The amplitude modulation index ranges for preferable use of each of the three considered frequency modulation index values, under the terms of minimization of the appropriate orders' integrated voltage harmonics factors values, have been revealed for the presented space vector PWM algorithm as a result of PSIM simulation. That makes it possible to reduce the load current's THD values, taking into account the particular kind of the generalized load.

The extremely low considered values of the frequency modulation index emphasize the possibility to improve the output power quality of multilevel inverters without increasing the clock frequency, the commutation frequency and related power losses.

References

1. Luo, F.L., Ye, H.: Advanced DC/AC Inverters: Applications in Renewable Energy. CRC Press (2013)
2. Ge, B., Peng, F.Z., Li, Y.: Multilevel converter/inverter topologies and applications. In: Abu-Rub, H., Malinowski, M., Al-Haddad, K. (eds.) Power Electronics for Renewable Energy Systems, Transportation and Industrial Applications, pp. 422–462 (2014)
3. Wu, B., Narimani M.: High-Power Converters and AC Drives. 2nd edn. Wiley-IEEE Press (2017)
4. Kouro, S., Malinowski, M., Gopakumar, K., Pou, J., Franquelo, L.G., Wu, B., Rodriguez, J., Pérez, M.A., Leon, J.I.: Recent advances and industrial applications of multilevel converters. IEEE Trans. Industr. Electron. **57**(8), 2553–2580 (2010)
5. Gonzalez, S.A., Verne, S.A., Valla, M.I.: Multilevel Converters For Industrial Applications. CRC Press (2013)
6. Holmes, D.G., Lipo, T.A.: Pulse Width Modulation For Power Converters. IEEE Press (2003)
7. Leon, J.I., Kouro, S., Rodriguez, J., Wu, B.: The essential role and the continuous evolution of modulation techniques for voltage-source inverters in the past, present, and future power electronics. IEEE Trans. Industr. Electron. **63**(5), 2688–2701 (2016)
8. Lopatkin, N.N.: Simple space vector PWM scheme with quarter-wave symmetric output voltage waveform for three-phase multilevel inverter. In: International Multi-Conference on Engineering, Computer and Information Sciences (SIBIRCON) Proceedings, pp. 433–438 (2017)
9. Lopatkin, N.N.: Some new representations of the multilevel inverter voltage space vector in the complex plane. In: International Siberian Conference on Control and Communications (SIBCON) Proceedings, pp. 1–11. IEEE, Russia, Omsk (2015)
10. Celanovic, N., Boroyevich, D.: A fast space-vector modulation algorithm for multilevel three-phase converters. IEEE Trans. Ind. Appl. **37**(2), 637–641 (2001)
11. Lopatkin, N.N.: Simple delta voltages space vector PWM algorithm for voltage source multilevel inverters. In: 2nd International Conference on Intelligent Energy and Power Systems (IEPS) Proceedings, pp. 149–154 (2016)
12. Lopatkin, N.N.: Voltage quality comparison of space vector PWM voltage source multilevel inverter under symmetric and nonsymmetric switching sequence variants: voltage waveforms, spectra and THD. In: 3rd International Conference on Industrial Engineering, Applications and Manufacturing (ICIEAM) Proceedings, pp. 1–8 (2017)
13. Zinoviev, G.S.: Power Electronics. Textbook For Undergraduate Students. 5th edn. Jurajt, Moscow (2015)

Chapter 37
Analysis of Resource Availability of Production Enterprise Based on Fuzzy Neural Network

Vladimir Bocharov, Alexander Danilov, Victor Burkovsky, Konstantin Gusev, and Pavel Gusev

Abstract Before launching a new product into production or when changing the production program of an aviation enterprise, you must clearly determine whether there are enough production resources to complete the new production program. The paper presents the use of modern intellectual technologies and methods. A workshop for the production of parts made of polymer composite materials is considered as the system under study. An up-to-date set of data describing the operation of the production unit was obtained. The analysis of the obtained data was carried out. The development of a fuzzy neural network is described, and the results of forecasting for a test data set are presented.

37.1 Introduction

Current trends in the development of aircraft engineering require operational decisions at all stages of the product life cycle. Tools for digitalizing production are designed to ensure that operational decisions can be made. Each stage of the life cycle has its own tools. For example, at the production stage of the life cycle, the digital counterpart of production is designed to provide support for decision-making. The digital production double allows you to fully evaluate the production capabilities and create a detailed work plan.

However, it should be noted that evaluating production capabilities and drawing up a work plan using a digital double require a lot of time. These time costs often exceed the time it takes to make a decision to change the production program. Also, there are often problems of reloading individual production resources and rescheduling all work becomes impractical. To solve such problems, you need to get a prompt answer to the question: Will the production cope with the new production program?

V. Bocharov · A. Danilov · V. Burkovsky · K. Gusev (✉) · P. Gusev
Voronezh State Technical University, 20 let Oktyabrya st., 84, 394006 Voronezh, Russian Federation
e-mail: gussev_konstantin@mail.ru

A. Ronzhin and V. Shishlakov (eds.), *Proceedings of 15th International Conference on Electromechanics and Robotics "Zavalishin's Readings"*, Smart Innovation, Systems and Technologies 187, https://doi.org/10.1007/978-981-15-5580-0_37

Modern intellectual methods of decision support can help you answer this question as soon as possible. These methods are based on machine learning or deep learning and use algorithms such as decision trees, logistic regression and gradient boosting. We should also note the neural network apparatus, which is used most often for image classification or video stream recognition. However, with the increase in the power of modern computers, the neural network apparatus can be used for a wider range of tasks—such as evaluating the feasibility of an aviation enterprise's production plan.

37.2 Literature Review

In Russian and foreign literature, forecasting issues are extensively and in detail considered by different authors [1–6]. At the same time, part of the work is devoted to forecasting production processes [7–10]. However, there are practically no works devoted to forecasting the implementation of the production plan.

In the foreign literature, there are works devoted to the analysis of production data [11–13], which can be the basis for further development of predictive models.

The neural network apparatus, as well as various hybrid fuzzy neural networks, which combine the advantages of both data analysis devices, have been widely used for the problem of analysis and forecasting. The use of such hybrid networks in production systems is mainly associated with fuzzy control of technological processes [14–17], and there is no mention of the use of such a device in the forecast of the production plan.

As a result of the analysis of the current state of the question about the possibility of predicting the implementation of the production plan, the goal of the work is to develop a fuzzy neural network that provides high efficiency of classification of the input data stream.

37.3 Methodology for Obtaining Initial Data

To analyze the effectiveness of the solutions proposed in the article, data from the workshop for the production of parts made of polymer composite materials (PCM) is used.

The workshop for the production of PCM parts consists of a procurement section, a section for laying out adhesive prepregs, sections for machining and applying paint and varnish, as well as sections for thermostating. All aircraft parts produced in the shop are made by autoclave molding. However, the manufactured parts have significant differences in temperature control modes, as well as methods of machining. Therefore, all parts produced in the shop are divided into 20 groups. In total, 20 groups contain 187 nomenclature items.

In the order of the production plan prediction algorithm to work with maximum accuracy, a sufficient set of source data is required for the required accuracy. The sufficiency of the source data set for each specific task is determined individually, but it can be stated unequivocally: the more data for various production programs is accumulated, the more accurate the forecast can be obtained.

However, even in those aviation enterprises where digitization technologies have already been implemented, there is still not enough data available for forecasting. Therefore, a simulation tool is used to obtain a set of source data. The simulation model allows you to accurately repeat all the technological stages of manufacturing parts of real production. In the created and verified production simulation model, it is possible to generate a set of data sufficient for the development of a predictive model [18].

To create a simulation model for the production of parts from PCM, real production data was used: the layout of production areas, the size of plots, the number of technological equipment, the number of workers, the initial data of parts, the initial data of tooling, etc., and the development of a simulation model for the production of parts from PCM is described in detail in [19, 20].

The created simulation model has a hierarchical structure. At the top level of the hierarchy, there are objects that simulate production sites. In this case, each simulation object is located in accordance with the location in the real production.

Each type of equipment is defined by its own object, which simulates either a technological process carried out on a single part, such as machining, or the process of processing multiple parts simultaneously, such as a temperature control operation. Simulation objects describe the behavior of the equipment and contain information about how to get the time for simulating the part processing, how to simulate loading the part on the equipment and how to move the part after the end of the simulation processing.

Management of material flows in the simulation model is carried out using software methods divided into two groups. The first group includes methods used on several sites. The second group consists of software methods that run within a single site and use information that is located within the site.

Figure 37.1 shows a fragment of a simulation model for the production of PCM parts.

As a result of the full cycle of work on the creation and verification of a simulation model of the workshop for the production of parts from polymer composite materials, a model is obtained that can adequately generate data used for the forecasting system.

Data generation for the forecasting system was performed using the following algorithm:

1. The generation of the initial data for the simulation.
2. Conducting a simulation experiment.
3. The collection of experimental results.
4. Export the collected data.

Generating an array of source data is one of the most important tasks. This is due to the fact that the source data in the simulation model must mimic real data.

The source data array is a monthly production plan. The monthly production plan is a list of 187 items, where each item contains the required number of units for

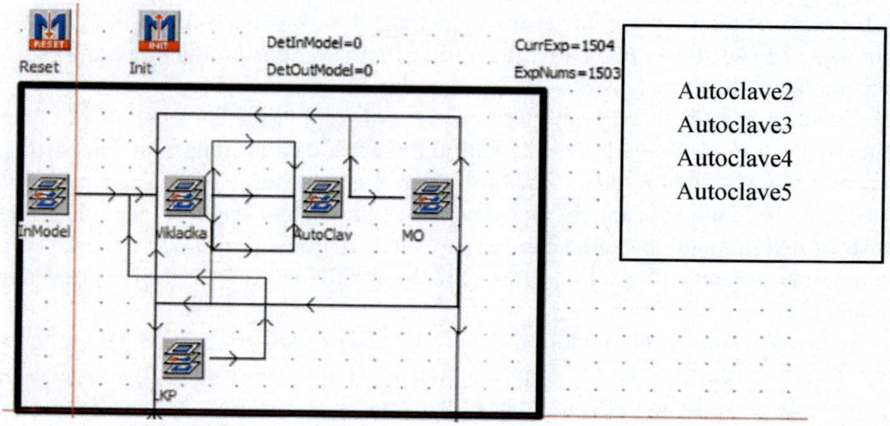

Fig. 37.1 Simulation model of the workshop

this part. In this case, the production of some parts may not be provided in the plan. The number of units of each specific part in the monthly plan is described by the normal distribution law, which provides a sufficient variety of the generated data array. All generated 1500 different input arrays, which corresponds to 1500 different production plans.

After generating an array of initial data, 1500 simulation experiments were performed. As a result of each experiment, the following data is recorded in the original table: the time of execution of the specified plan, the load of the main technological equipment and the number of parts that are processed in automated workplaces. The resulting information is exported to an excel file for further processing. Figure 37.2 shows a fragment of the table with the received data.

Each row in the table is a monthly production plan. The "Gr20" columns indicate the number of parts that need to be manufactured in a particular month. The "InLaser" column indicates the number of parts that pass the glue prepreg placement operation on automated workstations. The "target" column is the target variable and, if its value is 1, indicates the execution of the production plan.

In total, the resulting table contains 867 rows with data when the production plan is not completed and 633 rows when the production plan is completed.

The obtained data is analyzed in order to determine the relationship between the source data and the target variable—the implementation of the production plan. The results of the analysis can also be used to optimize the operation of the production unit.

First of all, you need to determine the correlation coefficients between the source data and the target value. This is necessary to check the generated data—if there is an initial value that correlates with the target value to one, then there is an error in the data generation. As a result of the calculation, it was found that the highest correlation coefficient for the first part of the 20 group is equal to 0.227. Also, the five parts with the highest correlation coefficient include two more parts from the 20

	FZ	GA	GB	GC	GD	GE	GF	GG
1	Gr20	Gr20	Gr20	Gr20	Gr20	Gr20	InLaser	target
2	2	0	7	2	4	5	506	0
3	0	1	6	4	6	4	526	0
4	5	6	0	1	0	2	568	0
5	3	0	6	5	5	6	461	0
6	7	8	5	7	3	1	532	0
7	6	2	4	5	8	0	566	0
8	7	11	11	0	2	5	505	0
9	2	2	3	5	4	4	535	0
10	6	2	6	2	0	4	459	1
11	2	4	2	0	5	0	548	0
12	2	5	0	0	0	6	479	0
13	3	0	2	0	1	1	472	0
14	1	8	1	2	1	6	508	0
15	4	5	6	10	2	5	509	0

Fig. 37.2 Table with the data set

group, which indicates a large impact of the number of parts in the 20 group on the implementation of the plan. Due to the fact that the maximum correlation coefficient does not exceed 0.227, the data is generated without errors.

The analysis of the influence of the quantity of each part from the nomenclature on the target indicator is carried out. Figure 37.3 shows a graph showing the average number of parts in the production plan. The red line indicates the number of parts when the production plan is not completed, and the blue line indicates when the production plan is completed.

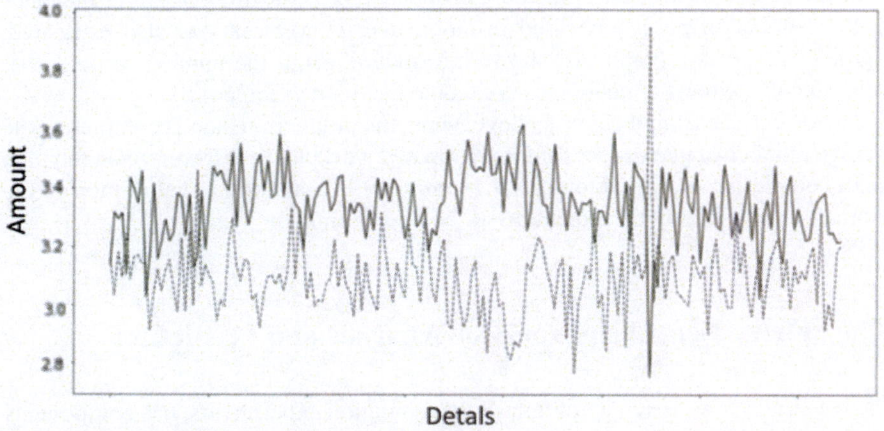

Fig. 37.3 Analysis of the impact of the number of parts

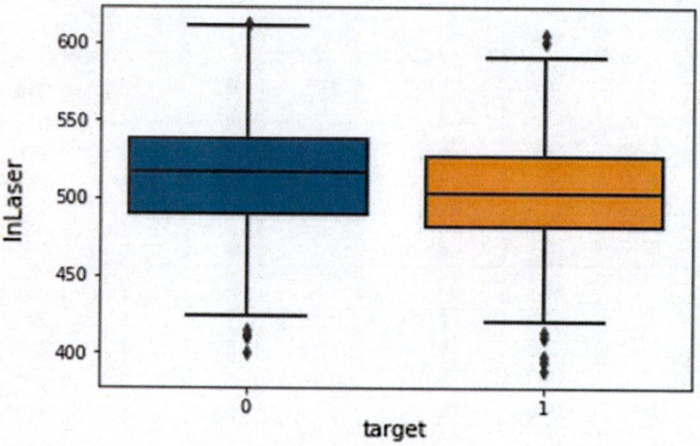

Fig. 37.4 Influence of the number of parts processed in automated workspaces

As you can see from the graph, the number of parts is critical to the execution of the production plan. However, there are parts that on the contrary have an average number of parts in the plan more when it is executed. Further analysis of the data showed that these are details from the 20 group. At the same time, the average number of the first 20 parts of the group is 1.2 parts more when the plan is executed than when it is not executed.

A study of the production process explained these indicators. With a small number of parts from the 20 group, the necessary temperature control modes in the autoclave are started much less often, which is due to the small number of parts in the cage. With an increase in the number of parts from the 20 group, the cages are assembled faster, which ensures that the autoclave is loaded more frequently with parts from the 20 group.

In the course of data analysis, the dependence of achieving the target indicator on the number of parts processed in automated workplaces was also estimated. Figure 37.4 shows a graph that shows information about the number of parts that were executed when the plan (1) was executed and not executed (0).

As you can see from the graph above, when the production plan is completed, the average number of parts processed in automated workplaces is lower when the plan is not completed. At the same time, the interquartile scale of the distribution of the number of parts during the execution of the plan is smaller.

37.4 Fuzzy Neural Network for Analysis and Prediction

The resulting series, made up of data from simulation experiments, has components that are predicted using technical analysis tools using the apparatus of fuzzy neural networks. To automate the forecasting process, we propose a model of a fuzzy system with an external knowledge base, where the Gauss function is used as a form of representation of fuzzy sets as follows:

$$\mu(a_i) = \exp\left[-\left(\frac{a_i - x_i^k}{s_i^k}\right)^2\right],$$

where n is the number of inputs of the fuzzy model, and the function parameters have a physical interpretation: x is the center, and s is the width of the Gaussian curve. The number and values of parameters of Gaussian functions are selected in such a way as to cover the entire amplitude range of the training sample. The first stage of operation of this model is to compile a knowledge base from the training sample in the form of pairs "input vector"–"predicted value". Data in the knowledge base is not stored in absolute values, but as functions of belonging to the corresponding Gaussian curves. At the second stage, using the generated knowledge base, a forecast is made by calculating the degree of belonging to the existing situation and the reference one. The fuzzy system makes it possible to calculate the result in two ways:

(1) direct finding of the situation in the knowledge base that has the highest degree of belonging to the input situation, where the result is the "predicted value" that coincides with the found situation ;
(2) calculating the result using the formula:

$$y = \frac{\sum y_i \cdot v_i}{\sum v_i}; \quad i = 1, \ldots, n.$$

The operation of a fuzzy neural network with fuzzy sets is required to perform mathematical operations performed by the following blocks:

(a) block for reducing state variables to fuzzy sets;
(b) decision-making block;
(c) the unit display of the output fuzzy set to the value of the forecast.

For each of the variables that characterize the state of the system, the degree of belonging curves was described, using which the prediction system presents analog parameters as fuzzy sets.

As a result, an algorithm for the operation of a fuzzy analysis and forecasting system is obtained:

1. Formation of typical parameters of a fuzzy forecasting system. A linguistic variable is introduced to describe the current situation for the release of parts with the following parameters:
   Name—corresponds to the name of a specific part;
   Term set—corresponds to the set of values of a particular part;
   Universal set—defined by the minimum and maximum number of parts in different shifts;
   There is no syntax rule;

A semantic rule is a Gauss function.

The parameters of linguistic variables described above are formed automatically on a sample of situations that are obtained as a result of simulation. The maximum and minimum of the parameter are searched for, and the Gauss functions are automatically spread over the definition area—the number of membership functions is determined empirically.

2. Fuzzification. Using the linguistic variable described above, parameters are converted to fuzzy values. You enter a two-dimensional array containing information on all the details and an array containing parameters after being translated into fuzzy logic.
3. At the next stage, the data of the new situation is fuzzified from the reference sample, that is, the parameters are translated into fuzzy logic using linguistic variables. An array containing information about the new situation is entered, and an array containing parameters is formed after being translated into fuzzy logic.
4. Next, the search is performed for the degrees of equality of the new situation with all available knowledge in the database. The degree of equality is the smallest of the degrees of inclusion of one fuzzy situation in another.
5. Defuzzification. The reverse conversion to exact values is performed by the method of equality of moments.

Based on the algorithm described above, a software module for analyzing and predicting the implementation of the production plan is implemented. The developed module allows you to use different data input formats, as well as conduct experiments.

The test sample was analyzed, from which the knowledge base is formed, and control checks were carried out on the test sample. Test results—93% of the forecast matches in the test sample.

37.5 Conclusions

The relevance of the use of modern intellectual technologies in domestic production enterprises is increasing every year. Intelligent technologies allow us to provide the production process with a qualitatively new level of decision support.

The proposed fuzzy neural network for classification of production programs can provide operational support for decision-making not only at the stage of changing the production program, but also for the operational management of smaller units.

References

1. Basova, L.A., Karyakina, O.E., Kochorova, L.V., Martinova, N.A., Kalinin, A.G.: Mathematical model for predicting the length of hospital stay after performing high-tech operations for arrhythmia correction. Kazanskii medicinskii J. **95**(1), 91–94 (2014)
2. Ivanov, N.N.: Mathematical reliability prediction perform a set of tasks with symmetric distribution of runtime. Open Educ. **2**(2), 52–55 (2011)

3. Strelnikov, D.D.: Technique of forecasting the performance time on transfer of bulk cargoes. Vestnik of Astrakhan State Technical University. Series: Marine Engineering and Technologies 3 (2018)
4. Sreekumar, S., Bhakar, R.: Solar power prediction models: classification based on time horizon, input, output and application. In: International Conference on Inventive Research in Computing Applications (ICIRCA), pp. 67–71 (2018)
5. Mercy, K.G., Rao, S.K.S.: A framework for rail surface defect prediction using machine learning algorithms. In: 2018 International Conference on Inventive Research in Computing Applications (ICIRCA), pp. 972–977 (2018)
6. Paliwal, S., Khatri, S.K., Sharma, M.: Sentiment analysis and prediction using neural networks. In: International Conference on Advanced Informatics for Computing Research, pp. 458–470 (2018)
7. Yadikin, E.A., Preys, V.V.: Theoretical bases of technical diagnostics automatic rotor and rotor-conveyor lines in mass productions. Izvestiya Tula State University 1, (2012)
8. Abramov, A.A., Shershakov, I.S.: Modeling the processes of production and sales of products based on the use of the method of autonomous adaptive control. Bull. Tambov Univ. Ser. Human. **110**(6), 72–77 (2012)
9. Zernov V.I., Paramonov, A.P.: Prediction of the effectiveness of modernization of space transport systems using an expert system. Elektronnyi zhurnal "Trudy MAI" **31**, (2008). Last access: http://trudymai.ru/published.php?ID=7495
10. Semakov, S.L., Semakov, I.S.: The simplest predictive model of the time series and its response to linear and parabolic input effects. Elektronnyi zhurnal "Trudy MAI" **100**(31), (2018). Last access: http://trudymai.ru/published.php?ID=93446
11. Straková, J., Pártlová, P., Dobrovič, J., Váchal, J.: Situational analysis and its role in the process of strategic business management. Polish J. Manage Stud **18** (2018)
12. Prasad, R.B., Arif, M.: Workspace and singularity analysis of five bar planar parallel manipulator. In: 2018 5th IEEE Uttar Pradesh Section International Conference on Electrical, Electronics and Computer Engineering (UPCON), pp. 1–6 (2018)
13. Wielki, J., Kozioł, P.: The analysis of opportunities to use the lean IT concept in modern enterprise. Polish J. Manage. Stud. **18** (2018)
14. Rutkovskaya, D., Rutkovskiy, L., Pilin'skiy, M.: Neural Networks, Genetic Algorithms and Fuzzy Systems. Goryachaya liniya – Telecom, 384p (2003)
15. Yarushkina, N.G.: Fundamentals of the Theory of Fuzzy and Hybrid Systems. Financy I statistica (2004)
16. Tran, D.T., Iosifidis, A., Kanniainen, J., Gabbouj, M.: Temporal attention-augmented bilinear network for financial time-series data analysis. IEEE Trans. Neural Networks Learn. Syst. **30**(5), 1407–1418 (2018)
17. Danilov, A., Povarov, V., Burkovsky, V., Podvalny, S., Gusev, K.: Intellectual decision-making system in the context of potentially dangerous nuclear power facilities. In: MATEC Web of Conferences 161, p. 02009 (2018)
18. Privalov, A.Y., Dorozhko, I.V., Zakharova, Y.A., Kopeyka, A.L.: Simulation model for assessing the availability factor complex technical systems taking into account process characteristics diagnosing. Elektronnyi zhurnal "Trudy MAI" **103** (2018). Last access: http://trudymai.ru/published.php?ID=101526. Accessed 20 April 2012
19. Chizhov, M.I., Skripchenko, Y.U.S., Gusev, P.Y.U.: Simulation modeling of the production of parts made of polymer composites. Comput Res Model **6**(2), 245–252 (2014)
20. Chizhov, M.I., Skripchenko, Y.S., Gusev, P.Y.: Genetic algorithms in the optimization of material flows of a mechanical milling processing site. In: Proceedings of Voronezh State Technical University, vol. 8(12-2), pp. 73–75 (2012)

Chapter 38
Synthesis of Nonlinear Impulse Systems

Vladislav Shishlakov, Elizaveta Vataeva, Nataliia Reshetnikova, Dmitriy Shishlakov, and Oksana Solenaya

Abstract The problem of synthesizing the parameters of the laws of control of continuous and pulsed automatic control systems with polynomial approximation of nonlinearities is considered. As a mathematical apparatus for solving the problem, the method is used that is the inverse of the direct variational analysis method—the generalized Galerkin's method, which allows you to completely algebraize the solution of the problem for the class of automatic control systems under study, the dynamics of which are described by nonlinear differential equations of arbitrary order. The generalized Galerkin method is extended to a new class of systems—impulse nonlinear systems. Recurrence expressions defining Galerkin integrals for this type of approximation are also given.

38.1 Introduction

The synthesis of the parameters of the laws of control of impulse systems containing elements and devices with nonlinear static and dynamic characteristics is a complex scientific and engineering task. This question is relevant in connection with the complication of electromechanical, electric power and robotic systems and complexes developed and implemented in various fields of science and technology [1–4].

Since we are talking about cyberphysical systems, which include nonlinear elements, the development of control laws is directly related to the method of approximating the characteristics. Since there are no universal approaches to the approximation problem, for each specific case, it is required to take into account the specific operating modes of the system [5–7].

The most widely used is the piecewise linear approximation, however, the accuracy of the result obtained is not always sufficient with such a mathematical model.

V. Shishlakov · E. Vataeva (✉) · N. Reshetnikova · D. Shishlakov · O. Solenaya
St. Petersburg State University of Aerospace Instrumentation, 67, st. Bolshaya Morskaya, 190000 St. Petersburg, Russia
e-mail: kaf31guap@gmail.com

A. Ronzhin and V. Shishlakov (eds.), *Proceedings of 15th International Conference on Electromechanics and Robotics "Zavalishin's Readings"*, Smart Innovation, Systems and Technologies 187, https://doi.org/10.1007/978-981-15-5580-0_38

Existing methods for the synthesis of nonlinear control systems are either limited in using a rather narrow class of systems, or are applicable only to systems described by low-order differential equations, restrictions can also be associated with the method of approximating a nonlinear characteristic [8–10].

To solve this problem, it is proposed to use the generalized Galerkin method as a mathematical apparatus, the application of which allows one to synthesize control laws of ACS of different classes (continuous ACS and systems with various types of signal modulation, the dynamics of which are described by linear equations as well as nonlinear arbitrarily high orders). In addition, the proposed approach can be applied to nonlinear systems containing an arbitrary number of elements with nonlinear characteristics. The method has shown its effectiveness in solving complex scientific and engineering technical problems, since it was used to develop laws for controlling large ground-based antenna installations, electric power and electromechanical systems and complexes, including with superconducting equipment, as well as the braking processes of the wheels of heavy aircraft [11].

To expand the capabilities of the generalized Galerkin method, we consider the features of its application for pulsed ACS with polynomial approximation of nonlinear characteristics (this problem was solved earlier for continuous systems).

38.2 Mathematical Description of the Synthesis Problem

Let us briefly consider the general scheme for solving the synthesis problem by the generalized Galerkin method (a detailed technique is described in [12–14]). For simplicity, we will consider the pulse element as ideal, and the output signal can be represented by the expression:

$$x^*(t) = \sum_{n=0}^{\infty} x(nT)\delta(t - nT), \tag{38.1}$$

where $x(nT) = \int_0^\infty x(t)\delta(t - nT)\mathrm{d}t$—is the value of the nth discrete value, δ—delayed impulse function existing at $t = nT$, T—interruption period, time interval between adjacent pulses.

The problem of synthesis of nonlinear impulse ACS is considered under the assumption that the structure of the synthesized ACS and the parameters of the control object are known. There are also restrictions on the values of the varied parameters based on the conditions of their technical feasibility:

$$c_k^- \le c_k \le c_k^+, \quad k = 1, 2, \ldots, m,$$

where c_k^+ and c_k^-—maximum and minimum permissible values of variable parameters, respectively.

Restrictions on the coarseness of the system for variable parameters are as follows:

$$\Delta = \frac{\delta c_k}{c_k} \leq \Delta^0,$$

where Δ^0—set value of the roughness of the system, δc_k—variations of parameters within which the stability of the system is ensured, c_k—the desired parameters of the control operator.

Let us write down a differential equation describing the motion of a system containing a modulator and a nonlinear element, taking into account the polynomials of the differentiation operator p with real constant coefficients of degrees h, u, v, respectively:

$$\sum_{i=0}^{n} a_i(c_k)p^i x(t) + \sum_{i=0}^{n^*} a_i^*(c_k)p^i x^*(t) + \sum_{i=0}^{u} b_i(c_k)p^i y(t) + \sum_{i=0}^{u^*} b_i^*(c_k)p^i y^*(t)$$
$$= \sum_{i=0}^{v} e_i(c_k)p^i f(t) + \sum_{i=0}^{v^*} e_i^*(c_k)p^i f^*(t),\ y(t) = F[x(t)],\ y^*(t) = F\left[x^*(t)\right], \tag{38.2}$$

where $y(t) = F[x(t)]$, $y^*(t) = F[x^*(t)]$—signals at the output of a nonlinear element with continuous $x(t)$ and impulse $x^*(t)$ input signals.

In the synthesis of nonlinear nth order automatic control systems, it is advisable to set the program motion by the expression:

$$x^0(t) = \left(x_y + H^* e^{-\alpha t} \cos(\beta t - \varphi_0)\right)1(t), \tag{38.3}$$

where x_y—is the desired process for $t = \infty$, H^* and φ_0 are determined by relations of the form:

$$H^* = \sqrt{\left(x_0 - x_y\right)^2 + \left[\frac{\alpha\left(x_0 - x_y\right) + \dot{x}_0}{\beta}\right]^2};$$
$$\varphi_0 = arctg\left[\frac{\alpha\left(x_0 - x_y\right) + \dot{x}_0}{\beta\left(x_0 - x_y\right)}\right],$$

Here, $x_0, \dot{x}_0$—the initial values of the studied coordinate, relative to which the equation of motion of the synthesized ACS and its derivative at a time is recorded $t \geq 0$, α—attenuation rate determined from the relation $\alpha = \frac{3 \div 4}{T_{\Pi\Pi}}$.

Therefore, the signal at the output of the ideal modulator, according to the expression (38.1):

$$x^{0*}(t) = \left[x_y + H^* e^{-\alpha t} \cos(\beta nt - \varphi_0)\right]\delta(t - nT). \tag{38.4}$$

After the desired programmed motion is set and its parameters are determined, a discrepancy is formed:

$$\begin{aligned}\psi(c_k, t) = {} & Q(c_k, D)x^0(t) + Q^*(c_k, D)x^{0*}(t) + R(c_k, D)F\big[x^0(t), D\{x^0(t)\}\big] \\ & + R^*(c_k, D)F^*\big[x^{0*}(t), D\{x^{0*}(t)\}\big] \\ & - S(c_k, D)f(t) - S^*(c_k, D)f^*(t),\end{aligned} \tag{38.5}$$

where D is the operator of generalized differentiation.

The required parameters c_k are determined from the conditions of orthogonality of the residual (38.5) to the coordinate functions:

$$\int_0^\infty \psi(c_k, t)\varphi_q(t)\mathrm{d}t = 0; \quad k, q = 1, 2, \ldots, m, \tag{38.6}$$

where $\varphi_q(t)$ is a system of m continuously differentiable linearly independent coordinate functions $\varphi_1(t), \varphi_2(t), \ldots, \varphi_q(t), \ldots, \varphi_m(t)$.

Substituting the desired software (38.3) into the ACS equation of motion (38.2) and solving Eq. (38.6), we obtain the following system of m linear algebraic equations:

$$\begin{aligned}& \sum_{i=0}^{n} a_i(c_k)A_{qi} + \sum_{i=0}^{n^*} a_i^*(c_k)A_{qi}^* + \sum_{i=0}^{u} b_i(c_k)B_{qi} \\ & + \sum_{i=0}^{u^*} b_i^*(c_k)B_{qi}^* - \sum_{i=0}^{v} e_i(c_k)C_{qi} - \sum_{i=0}^{v^*} e_i^*(c_k)C_{qi}^* = 0; \quad q = 1, 2, \ldots, m,\end{aligned} \tag{38.7}$$

where:

$$\begin{aligned}A_{qi} &= \int_0^\infty D^i\{x^0(t)\}\mathrm{e}^{-\rho_q t}\mathrm{d}t = A_q\rho_q^{i-1}; \quad i = 1, 2, \ldots, n \\ A_{qi}^* &= \int_0^\infty D^i\{x^{0*}(t)\}\mathrm{e}^{-\rho_q t}\mathrm{d}t = A_q^*\rho_q^i; \quad i = 1, 2, \ldots, n^* \\ B_{qi} &= \int_0^\infty D^i\{F[x^0(t)]\}\mathrm{e}^{-\rho_q t}\mathrm{d}t = B_q\rho_q^{i-1}; \quad i = 0, 1, \ldots, u; \\ B_{qi}^* &= \int_0^\infty D^i\{F[x^{0*}(t)]\}\mathrm{e}^{-\rho_q t}\mathrm{d}t = B_q^*\rho_q^i; \quad i = 0, 1, \ldots, u^*; \\ C_{qi} &= \int_0^\infty D^i\{f(t)\}\mathrm{e}^{-\rho_q t}\mathrm{d}t = C_q\rho_q^{i-1}; \quad i = 0, 1, \ldots, v, \\ C_{qi}^* &= \int_0^\infty D^i\{f^*(t)\}\mathrm{e}^{-\rho_q t}\mathrm{d}t = C_q^*\rho_q^i; \quad i = 0, 1, \ldots, v^*.\end{aligned}$$

Recurrent analytic expressions $A_{qi}, A_{qi}^*, C_{qi}, C_{qi}^*$ calculated in [14]. Calculations and generalizations of the expression B_{qi} for continuous systems with polynomial approximation are given in [15]. This article presents the calculation of the analytical expression B_{qi}^* for processes recorded relative to the ACS error signal:

Table 38.1 Analytical relations defining integrals B_q^* for impulse ACS

Exponent, g	B_q^* for a process of the form $x^0(t) = [H^* e^{-\alpha t} \cos(\beta t - \varphi_0)]\delta(t - nT)$
$g = 1$	$B_q^* = H\left[\frac{e^{2(\alpha+pq)T}\cos\varphi_0 - e^{(\alpha+pq)T}\cos(\beta T+\varphi_0)}{e^{2(\alpha+pq)T} - 2\cos\beta T e^{(\alpha+pq)T}+1}\right]$
$g = 2$	$B_q^* = \frac{1}{2}H^2\left[\frac{e^{(2\alpha+pq)T}}{e^{(2\alpha+pq)T}-1} + \frac{e^{2(2\alpha+pq)T}\cos 2\varphi_0 - e^{(2\alpha+pq)T}\cos 2(\beta T+\varphi_0)}{e^{2(2\alpha+pq)T} - 2\cos 2\beta T e^{(2\alpha+pq)T}+1}\right]$
$g = 3$	$B_q^* = \frac{1}{4}H^3\left[\begin{matrix} 3\frac{e^{2(3\alpha+pq)T}\cos\varphi_0 - e^{(3\alpha+pq)T}\cos(\beta T+\varphi_0)}{e^{2(3\alpha+pq)T} - 2\cos\beta T e^{(3\alpha+pq)T}+1} \\ +\frac{e^{2(3\alpha+pq)T}\cos 3\varphi_0 - e^{(3\alpha+pq)T}\cos 3(\beta T+\varphi_0)}{e^{2(3\alpha+pq)T} - 2\cos 3\beta T e^{(3\alpha+pq)T}+1}\end{matrix}\right]$
$g = 4$	$B_q^* = \frac{1}{8}H^4\left[\begin{matrix} \frac{e^{2(4\alpha+pq)T}\cos 4\varphi_0 - e^{(4\alpha+pq)T}\cos 4(\beta T+\varphi_0)}{e^{2(4\alpha+pq)T} - 2\cos 4\beta T e^{(4\alpha+pq)T}+1} \\ +4\frac{e^{2(4\alpha+pq)T}\cos 2\varphi_0 - e^{(4\alpha+pq)T}\cos 2(\beta T+\varphi_0)}{e^{2(4\alpha+pq)T} - 2\cos 2\beta T e^{(4\alpha+pq)T}+1} \\ +3\frac{e^{(4\alpha+pq)T}}{e^{(4\alpha+pq)T}-1}\end{matrix}\right]$
$g = 5$	$B_q^* = \frac{1}{16}H^5\left[\begin{matrix} 10\frac{e^{2(5\alpha+pq)T}\cos\varphi_0 - e^{(5\alpha+pq)T}\cos(\beta T+\varphi_0)}{e^{2(5\alpha+pq)T} - 2\cos\beta T e^{(5\alpha+pq)T}+1} \\ +5\frac{e^{2(5\alpha+pq)T}\cos 3\varphi_0 - e^{(5\alpha+pq)T}\cos 3(\beta T+\varphi_0)}{e^{2(5\alpha+pq)T} - 2\cos 3\beta T e^{(5\alpha+pq)T}+1} \\ +\frac{e^{2(5\alpha+pq)T}\cos 5\varphi_0 - e^{(5\alpha+pq)T}\cos 5(\beta T+\varphi_0)}{e^{2(5\alpha+pq)T} - 2\cos 5\beta T e^{(5\alpha+pq)T}+1}\end{matrix}\right]$

$$x^0(t) = [H^* e^{-\alpha t} \cos(\beta t - \varphi_0)]\delta(t - nT) \tag{38.8}$$

and relative to the ACS output:

$$x^0(t) = [x_y - H^* e^{-\alpha t} \cos(\beta t - \varphi_0)]\delta(t - nT). \tag{38.9}$$

In Tables 38.1 and 38.2 show the results of the calculation of recurrent analytical expressions defining the integrals B_q^*. Summarizing the expressions given in Table 38.1, for the process (38.8), we get:

- for an odd degree:

$$B_q^* = \frac{1}{2^{g-1}} H^g \sum_{k=0}^{\frac{g-1}{2}} C_k^g \frac{\cos\varphi_0 (g-2k) e^{2(g\alpha+pq)T} - e^{(g\alpha+pq)T}\cos[(g-2k)(\beta T+\varphi_0)]}{e^{2(g\alpha+pq)T} - e^{(g\alpha+pq)T} 2\cos[(g-2k)\beta T]};$$

- for an even degree:

$$B_q^* = H^g\left[\left(\sum_{n=0}^{\infty} e^{-(g\alpha+pq)T}\right)\frac{1}{2^{g-1}} \sum_{k=0}^{\frac{g-1}{2}} C_k^g \frac{\cos\varphi_0 (g-2k) e^{2(g\alpha+pq)T} - e^{(g\alpha+pq)T}\cos[(g-2k)(\beta T+\varphi_0)]}{e^{2(g\alpha+pq)T} - e^{(g\alpha+pq)T} 2\cos[(g-2k)\beta T]}\right].$$

Table 38.2 Analytical relations defining integrals B_q^* for impulse ACS

Exponent, g	B_q^* for a process of the form $x^0(t) = \left[x_y - H^* \mathrm{e}^{-\alpha t} \cos(\beta t - \varphi_0)\right]\delta(t - nT)$
$g = 1$	$B_q^* = x_y \frac{\mathrm{e}^{\rho q T}}{\mathrm{e}^{\rho q T} - 1} - H\left[\frac{\mathrm{e}^{2(\alpha+\rho q)T}\cos\varphi_0 - \mathrm{e}^{(\alpha+\rho q)T}\cos(\beta T + \varphi_0)}{\mathrm{e}^{2(\alpha+\rho q)T} - 2\cos\beta T \mathrm{e}^{(\alpha+\rho q)T} + 1}\right]$
$g = 2$	$B_q^* = x_y^2 \frac{\mathrm{e}^{\rho q T}}{\mathrm{e}^{\rho q T} - 1} - 2x_y H\left[\frac{\mathrm{e}^{2(\alpha+\rho q)T}\cos\varphi_0 - \mathrm{e}^{(\alpha+\rho q)T}\cos(\beta T + \varphi_0)}{\mathrm{e}^{2(\alpha+\rho q)T} - 2\cos\beta T \mathrm{e}^{(\alpha+\rho q)T} + 1}\right]$ $+ \frac{H^2}{2}\left[\frac{\mathrm{e}^{(2\alpha+\rho q)T}}{\mathrm{e}^{(2\alpha+\rho q)T} - 1} + \frac{\mathrm{e}^{2(2\alpha+\rho q)T}\cos 2\varphi_0 - \mathrm{e}^{(2\alpha+\rho q)T}\cos 2(\beta T + \varphi_0)}{\mathrm{e}^{2(2\alpha+\rho q)T} - 2\cos 2\beta T \mathrm{e}^{(2\alpha+\rho q)T} + 1}\right]$
$g = 3$	$B_q^* = x_y^3 \frac{\mathrm{e}^{\rho q T}}{\mathrm{e}^{\rho q T} - 1} - 3x_y^2 H\left[\frac{\mathrm{e}^{2(\alpha+\rho q)T}\cos\varphi_0 - \mathrm{e}^{(\alpha+\rho q)T}\cos(\beta T + \varphi_0)}{\mathrm{e}^{2(\alpha+\rho q)T} - 2\cos\beta T \mathrm{e}^{(\alpha+\rho q)T} + 1}\right]$ $+ \frac{3H^2 x_y}{2}\left[\frac{\mathrm{e}^{(2\alpha+\rho q)T}}{\mathrm{e}^{(2\alpha+\rho q)T} - 1} + \frac{\mathrm{e}^{2(2\alpha+\rho q)T}\cos 2\varphi_0 - \mathrm{e}^{(2\alpha+\rho q)T}\cos 2(\beta T + \varphi_0)}{\mathrm{e}^{2(2\alpha+\rho q)T} - 2\cos 2\beta T \mathrm{e}^{(2\alpha+\rho q)T} + 1}\right]$ $- \frac{1H^3}{4}\left[3\frac{\mathrm{e}^{2(3\alpha+\rho q)T}\cos\varphi_0 - \mathrm{e}^{(3\alpha+\rho q)T}\cos(\beta T + \varphi_0)}{\mathrm{e}^{2(3\alpha+\rho q)T} - 2\cos 2\beta T \mathrm{e}^{(3\alpha+\rho q)T} + 1} + \frac{\mathrm{e}^{2(3\alpha+\rho q)T}\cos 3\varphi_0 - \mathrm{e}^{(3\alpha+\rho q)T}\cos 3(\beta T + \varphi_0)}{\mathrm{e}^{2(3\alpha+\rho q)T} - 2\cos 3\beta T \mathrm{e}^{(3\alpha+\rho q)T} + 1}\right]$
$g = 4$	$B_q^* = x_y^4 \frac{\mathrm{e}^{\rho q T}}{\mathrm{e}^{\rho q T} - 1} - 4x_y^3 H\left[\frac{\mathrm{e}^{2(\alpha+\rho q)T}\cos\varphi_0 - \mathrm{e}^{(\alpha+\rho q)T}\cos(\beta T + \varphi_0)}{\mathrm{e}^{2(\alpha+\rho q)T} - 2\cos\beta T \mathrm{e}^{(\alpha+\rho q)T} + 1}\right]$ $+ \frac{6H^2 x_y^2}{2}\left[\frac{\mathrm{e}^{(2\alpha+\rho q)T}}{\mathrm{e}^{(2\alpha+\rho q)T} - 1} + \frac{\mathrm{e}^{2(2\alpha+\rho q)T}\cos 2\varphi_0 - \mathrm{e}^{(2\alpha+\rho q)T}\cos 2(\beta T + \varphi_0)}{\mathrm{e}^{2(2\alpha+\rho q)T} - 2\cos 2\beta T \mathrm{e}^{(2\alpha+\rho q)T} + 1}\right]$ $- \frac{4x_y H^3}{4}\left[3\frac{\mathrm{e}^{2(3\alpha+\rho q)T}\cos\varphi_0 - \mathrm{e}^{(3\alpha+\rho q)T}\cos(\beta T + \varphi_0)}{\mathrm{e}^{2(3\alpha+\rho q)T} - 2\cos\beta T \mathrm{e}^{(3\alpha+\rho q)T} + 1} + \frac{\mathrm{e}^{2(3\alpha+\rho q)T}\cos 3\varphi_0 - \mathrm{e}^{(3\alpha+\rho q)T}\cos 3(\beta T + \varphi_0)}{\mathrm{e}^{2(3\alpha+\rho q)T} - 2\cos 3\beta T \mathrm{e}^{(3\alpha+\rho q)T} + 1}\right]$ $+ \frac{1}{8}H^4\left[\frac{\mathrm{e}^{2(4\alpha+\rho q)T}\cos 4\varphi_0 - \mathrm{e}^{(4\alpha+\rho q)T}\cos 4(\beta T + \varphi_0)}{\mathrm{e}^{2(4\alpha+\rho q)T} - 2\cos 4\beta T \mathrm{e}^{(4\alpha+\rho q)T} + 1} + 4\frac{\mathrm{e}^{2(4\alpha+\rho q)T}\cos 2\varphi_0 - \mathrm{e}^{(4\alpha+\rho q)T}\cos 2(\beta T + \varphi_0)}{\mathrm{e}^{2(4\alpha+\rho q)T} - 2\cos 2\beta T \mathrm{e}^{(4\alpha+\rho q)T} + 1} + 3\frac{\mathrm{e}^{(4\alpha+\rho q)T}}{\mathrm{e}^{(4\alpha+\rho q)T} - 1}\right]$
$g = 5$	$B_q^* = x_y^5 \frac{\mathrm{e}^{\rho q T}}{\mathrm{e}^{\rho q T} - 1} - 5x_y^4 H\left[\frac{\mathrm{e}^{2(\alpha+\rho q)T}\cos\varphi_0 - \mathrm{e}^{(\alpha+\rho q)T}\cos(\beta T + \varphi_0)}{\mathrm{e}^{2(\alpha+\rho q)T} - 2\cos\beta T \mathrm{e}^{(\alpha+\rho q)T} + 1}\right]$ $+ 5x_y^3 H^2\left[\frac{\mathrm{e}^{(2\alpha+\rho q)T}}{\mathrm{e}^{(2\alpha+\rho q)T} - 1} + \frac{\mathrm{e}^{2(2\alpha+\rho q)T}\cos 2\varphi_0 - \mathrm{e}^{(2\alpha+\rho q)T}\cos 2(\beta T + \varphi_0)}{\mathrm{e}^{2(2\alpha+\rho q)T} - 2\cos 2\beta T \mathrm{e}^{(2\alpha+\rho q)T} + 1}\right]$ $- \frac{5x_y^2 H^3}{2}\left[3\frac{\mathrm{e}^{2(3\alpha+\rho q)T}\cos\varphi_0 - \mathrm{e}^{(3\alpha+\rho q)T}\cos(\beta T + \varphi_0)}{\mathrm{e}^{2(3\alpha+\rho q)T} - 2\cos\beta T \mathrm{e}^{(3\alpha+\rho q)T} + 1} + \frac{\mathrm{e}^{2(3\alpha+\rho q)T}\cos 3\varphi_0 - \mathrm{e}^{(3\alpha+\rho q)T}\cos 3(\beta T + \varphi_0)}{\mathrm{e}^{2(3\alpha+\rho q)T} - 2\cos 3\beta T \mathrm{e}^{(3\alpha+\rho q)T} + 1}\right]$ $+ \frac{5}{8}x_y H^4\left[\frac{\mathrm{e}^{2(4\alpha+\rho q)T}\cos 4\varphi_0 - \mathrm{e}^{(4\alpha+\rho q)T}\cos 4(\beta T + \varphi_0)}{\mathrm{e}^{2(4\alpha+\rho q)T} - 2\cos 4\beta T \mathrm{e}^{(4\alpha+\rho q)T} + 1} + 4\frac{\mathrm{e}^{2(4\alpha+\rho q)T}\cos 2\varphi_0 - \mathrm{e}^{(4\alpha+\rho q)T}\cos 2(\beta T + \varphi_0)}{\mathrm{e}^{2(4\alpha+\rho q)T} - 2\cos 2\beta T \mathrm{e}^{(4\alpha+\rho q)T} + 1} + 3\frac{\mathrm{e}^{(4\alpha+\rho q)T}}{\mathrm{e}^{(4\alpha+\rho q)T} - 1}\right]$ $- \frac{1}{16}H^5\left[\frac{\mathrm{e}^{2(5\alpha+\rho q)T}\cos 5\varphi_0 - \mathrm{e}^{(5\alpha+\rho q)T}\cos 5(\beta T + \varphi_0)}{\mathrm{e}^{2(5\alpha+\rho q)T} - 2\cos 5\beta T \mathrm{e}^{(5\alpha+\rho q)T} + 1} + 5\frac{\mathrm{e}^{2(5\alpha+\rho q)T}\cos 3\varphi_0 - \mathrm{e}^{(5\alpha+\rho q)T}\cos 3(\beta T + \varphi_0)}{\mathrm{e}^{2(5\alpha+\rho q)T} - 2\cos 3\beta T \mathrm{e}^{(5\alpha+\rho q)T} + 1} + 10\frac{\mathrm{e}^{2(5\alpha+\rho q)T}\cos\varphi_0 - \mathrm{e}^{(5\alpha+\rho q)T}\cos(\beta T + \varphi_0)}{\mathrm{e}^{2(5\alpha+\rho q)T} - 2\cos\beta T \mathrm{e}^{(5\alpha+\rho q)T} + 1}\right]$

(continued)

Table 38.2 (continued)

Exponent, g	B_q^* for a process of the form $x^0(t) = \left[x_y - H^* \mathrm{e}^{-\alpha t} \cos(\beta t - \varphi_0)\right]\delta(t - nT)$
$g = 6$	$B_q^* = x_y^6 \frac{\mathrm{e}^{\rho qT}}{\mathrm{e}^{\rho qT} - 1} - 6x_y^5 H \left[\frac{\mathrm{e}^{2(\alpha+\rho q)T} \cos\varphi_0 - \mathrm{e}^{(\alpha+\rho q)T} \cos(\beta T + \varphi_0)}{\mathrm{e}^{2(\alpha+\rho q)T} - 2\cos\beta T \mathrm{e}^{(\alpha+\rho q)T} + 1}\right]$ $+ \frac{15x_y^4 H^2}{2}\left[\frac{\mathrm{e}^{(2\alpha+\rho q)T}}{\mathrm{e}^{(2\alpha+\rho q)T} - 1} + \frac{\mathrm{e}^{2(2\alpha+\rho q)T} \cos 2\varphi_0 - \mathrm{e}^{(2\alpha+\rho q)T} \cos 2(\beta T + \varphi_0)}{\mathrm{e}^{2(2\alpha+\rho q)T} - 2\cos 2\beta T \mathrm{e}^{(2\alpha+\rho q)T} + 1}\right]$ $- \frac{20x_y^3 H^3}{4}\left[3\frac{\mathrm{e}^{2(3\alpha+\rho q)T} \cos\varphi_0 - \mathrm{e}^{(3\alpha+\rho q)T} \cos(\beta T + \varphi_0)}{\mathrm{e}^{2(3\alpha+\rho q)T} - 2\cos\beta T \mathrm{e}^{(3\alpha+\rho q)T} + 1} + \frac{\mathrm{e}^{2(3\alpha+\rho q)T} \cos 3\varphi_0 - \mathrm{e}^{(3\alpha+\rho q)T} \cos 3(\beta T + \varphi_0)}{\mathrm{e}^{2(3\alpha+\rho q)T} - 2\cos 3\beta T \mathrm{e}^{(3\alpha+\rho q)T} + 1}\right]$ $+ \frac{15}{8}x_y^2 H^4 \left[\frac{\mathrm{e}^{2(4\alpha+\rho q)T} \cos 4\varphi_0 - \mathrm{e}^{(4\alpha+\rho q)T} \cos 4(\beta T + \varphi_0)}{\mathrm{e}^{2(4\alpha+\rho q)T} - 2\cos 4\beta T \mathrm{e}^{(4\alpha+\rho q)T} + 1} + 4\frac{\mathrm{e}^{2(4\alpha+\rho q)T} \cos 2\varphi_0 - \mathrm{e}^{(4\alpha+\rho q)T} \cos 2(\beta T + \varphi_0)}{\mathrm{e}^{2(4\alpha+\rho q)T} - 2\cos 2\beta T \mathrm{e}^{(4\alpha+\rho q)T} + 1} + 3\frac{\mathrm{e}^{(4\alpha+\rho q)T}}{\mathrm{e}^{(4\alpha+\rho q)T} - 1}\right]$
	$- \frac{6}{16}x_y H^5 \left[\frac{\mathrm{e}^{2(5\alpha+\rho q)T} \cos 5\varphi_0 - \mathrm{e}^{(5\alpha+\rho q)T} \cos 5(\beta T + \varphi_0)}{\mathrm{e}^{2(5\alpha+\rho q)T} - 2\cos 5\beta T \mathrm{e}^{(5\alpha+\rho q)T} + 1} 5\frac{\mathrm{e}^{2(5\alpha+\rho q)T} \cos 3\varphi_0 - \mathrm{e}^{(5\alpha+\rho q)T} \cos 3(\beta T + \varphi_0)}{\mathrm{e}^{2(5\alpha+\rho q)T} - 2\cos 3\beta T \mathrm{e}^{(5\alpha+\rho q)T} + 1} + 10\frac{\mathrm{e}^{2(5\alpha+\rho q)T} \cos\varphi_0 - \mathrm{e}^{(5\alpha+\rho q)T} \cos(\beta T + \varphi_0)}{\mathrm{e}^{2(5\alpha+\rho q)T} - 2\cos\beta T \mathrm{e}^{(5\alpha+\rho q)T} + 1}\right]$ $+ \frac{H^6}{32}\left[10\frac{\mathrm{e}^{(6\alpha+\rho q)T}}{\mathrm{e}^{(6\alpha+\rho q)T} - 1} + 15\frac{\mathrm{e}^{2(6\alpha+\rho q)T} \cos 2\varphi_0 - \mathrm{e}^{(6\alpha+\rho q)T} \cos 2(\beta T + \varphi_0)}{\mathrm{e}^{2(6\alpha+\rho q)T} - 2\cos 2\beta T \mathrm{e}^{(6\alpha+\rho q)T} + 1} + 6\frac{\mathrm{e}^{2(6\alpha+\rho q)T} \cos 4\varphi_0 - \mathrm{e}^{(6\alpha+\rho q)T} \cos 4(\beta T + \varphi_0)}{\mathrm{e}^{2(6\alpha+\rho q)T} - 2\cos 4\beta T \mathrm{e}^{(6\alpha+\rho q)T} + 1} + \frac{\mathrm{e}^{2(6\alpha+\rho q)T} \cos 6\varphi_0 - \mathrm{e}^{(6\alpha+\rho q)T} \cos 6(\beta T + \varphi_0)}{\mathrm{e}^{2(6\alpha+\rho q)T} - 2\cos 6\beta T \mathrm{e}^{(6\alpha+\rho q)T} + 1}\right]$

Summarizing the expressions in Table 38.2 for the process (38.9), we obtain:

$$B_q^* = \sum_{k=0}^{g} x_y^{g-k} H^k (-1)^k C_g^k \left(\frac{\mathrm{e}^{2(k\alpha+\rho q)T} \cos k\varphi_0 - \mathrm{e}^{(k\alpha+\rho q)T} \cos k(\beta T + \varphi_0)}{2^{k-1}(\mathrm{e}^{2(k\alpha+\rho q)T} - 2\mathrm{e}^{(k\alpha+\rho q)T} \cos k(\beta T + 1)} + \frac{k\left(\mathrm{e}^{2(k\alpha+\rho q)T} \cos(k-2)\varphi_0 - \mathrm{e}^{(k\alpha+\rho q)T} \cos[(k-2)(\beta T + \varphi_0)]\right)}{2^{k-1}(\mathrm{e}^{2(k\alpha+\rho q)T} - 2\mathrm{e}^{(k\alpha+\rho q)T} \cos k(\beta T + 1)}\right).$$

38.3 Conclusion

In the course of solving the problem, the generalized Galerkin method was extended to the solution of the problem of synthesis of nonlinear pulse automatic control systems with polynomial approximation. Using the obtained recurrence relations for calculating the B_q^* integrals, one can significantly simplify the process of solving the parametric synthesis problem for nonlinear impulse systems of an arbitrarily high order and reduce all calculations to performing only simple mathematical operations.

Acknowledgements This research is supported by the RFBR Project No. №20-08-01056_A.

References

1. Pyrkin, A., Bobtsov, A., Kolyubin, S., Faronov, M.: Output controller for uncertain nonlinear systems with structural, parametric, and signal disturbances. In: IEEE Multi-conference on Systems and Control, pp. 1528–1533 (2012)
2. Bazylev, D.: Proceedings 2018 IEEE industrial cyber-physical systems (ICPS), IET, pp. 304–309 (2018)
3. Pyrkin, A., Isidori, A.: Adaptive output regulation of right-invertible MIMO LTI systems, with application to vessel motion control. Eur. J. Control **46**, 63–79 (2019)
4. Ankhimyuk, V.L., Opeiko, O.F., Mikheev, N.N.: Automatic control theory. Design PRO, p. 352 (2000)
5. Besekersky, B.A., Eliseev, A.A., Nebylov, A.V.: Radioautomatics: Textbook for Students. Vysh. shk, Moscow (1985)
6. Bogoslovskiy, S.V., Bogoslovskiy, V.S.: Non-stationary control theory: training manual. In: SUAI (2005)
7. Borgul, A., et al.: Mechatronic and robotic setups for modern control theory workshops. In: 9th IFAC Symposium on Advances in Control Education, pp. 348–353 (2012)
8. Shishlakov, V.F., Krivolapchuk, I.G., Vatayeva, E.Y.: Mechatronics, Automation and Robotics, Materials of the International Scientific and Practical Conference, vol. 1, pp. 130–132 (2017)
9. Shishlakov, V.F., Shishlakov, A.V., Timofeyev, S.S.: Synthesis of ACS for various types of approximation of nonlinear characteristics: theory and practice. In: SUAI (2017)
10. Pyrkin, A.A., Vedyakov, A.A., Ortega, R., Bobtsov, A.A.: Fradkov Theorem-based control of MIMO nonlinear Lurie systems. Autom. Remote Control **79**(6):1074–1085
11. Wang, J., Aranovskiy, S.V., Bobtsov, A.A., Pyrkin, A.A., Kolyubin, S.A.: Method to provide conditions for sustained excitation. Autom. Rem. Control **79**(2), 258–264 (2018)
12. Shishlakov, V., Vataeva, E., Reshetnikova, N., Shishlakov, D.: Synthesis of control laws of electromechanical systems under polynominal approximation of characteristics of nonlinear elements. In: MATEC Web Conference, vol. 161, p. 02006 (2018)
13. Nikitin A.V., Shishlakov V.F.: Parametric synthesis of nonlinear automatic control systems, St. Petersburg, 358p (2003)
14. Shishlakov, V.F.: Synthesis of non-linear ACS with various types of modulation, St. Petersburg, 267p (1999)

Chapter 39
Hidden Markov Model Based on Signals from Blocks of Semi-Markov System's Elements and Its Application for Dynamics Analysis Energy Systems

Yuriy Obzherin, Mikhail Nikitin, and Stanislav Sidorov

Abstract For modeling systems for various purposes, in particular, energy systems, semi-Markov processes are often used. During the functioning of the system for which the semi-Markov model is built, it is not always possible to get all the information contained in the status codes when changing its states, but you can only get the signal (information) in which block of system elements the state changed (failure, renewal, etc.). In this case, the states of the semi-Markov model can be considered hidden (unobservable). There are problems of analyzing the dynamics, predicting the states of the elements of the simulated system based on the received vector of signals from blocks of system elements. To solve these problems, the apparatus of the theory of hidden Markov models can be used. The paper considers the possibilities of applying this approach by the example of independent renewal processes superposition.

39.1 Introduction

Some of the main tasks of the energy industry theory and practice are the following: (1) creation of digital networks and digital substations; (2) development of information control systems for energy systems and transition to the intelligent control and engineering. The solution of these problems is possible based on the mathematical models construction relating to various aspects of the power systems structure and operation and the development on their basis of the algorithms and the decision support information systems while the energy systems design and operation,

Y. Obzherin (✉) · M. Nikitin · S. Sidorov
Higher Mathematics Department, Sevastopol State University, 299053 Sevastopol, Russia
e-mail: objsev@mail.ru

M. Nikitin
e-mail: m.nikitin.1979@gmail.com

S. Sidorov
e-mail: xaevec@mail.ru

A. Ronzhin and V. Shishlakov (eds.), *Proceedings of 15th International Conference on Electromechanics and Robotics "Zavalishin's Readings"*, Smart Innovation, Systems and Technologies 187, https://doi.org/10.1007/978-981-15-5580-0_39

as well as prediction of their states. Semi-Markov processes [1–6] are often used to model systems for various purposes [1–6], in particular, to construct models of energy systems [7–10].

In the process of functioning of the system for which the semi-Markov model is built, it is necessary to determine how much the constructed model is consistent with the data obtained during the operation of the system, analyze the dynamics of the system and predict its states. Hidden Markov models [11–16] and hidden semi-Markov models [17–20] make it possible to solve these problems.

An important part of the semi-Markov process is the embedded Markov chain (EMC), which is responsible for the transitions between the states of the system. EMC phase space of states coincides with semi-Markov process phase space of states. During the functioning of systems for which the semi-Markov model is built, it is not always possible, when changing its states, to fully obtain the information contained in the state encoding, but it is only possible to obtain some signal (information) associated with the states of the EMC. For example, in the phase state space for each element of the system, it is indicated whether it is in working condition or is being restored, and when the system is functioning, you can only receive a signal about the number of working elements of the system. In some cases, you can only receive a signal from a block of system elements in which a state change occurred, without indicating an element of this block in which a state change occurred. In these cases, the states of EMC and the semi-Markov model can be considered hidden (unobservable). There are problems of finding estimates of the characteristics of the semi-Markov model and predicting its states based on the obtained vector of signals appears. Previously, it was noted that these problems can be solved by hidden Markov and semi-Markov models.

Using hidden models allows you to expand the capabilities of the analysis of systems for which semi-Markov models are built, since there are opportunities: (1) refine the model parameters and the semi-Markov model itself in the process of functioning of the system (model training); (2) analyze the functioning of the system; (3) to predict the state of the system, based on the vector of signals obtained during the operation of the system.

In the papers [1, 2], the superposition of independent renewal processes was constructed. In this paper, using the stationary phase merging algorithm [1, 2], a merged semi-Markov model of this system is being constructed. Further, for the merged model, we consider a hidden Markov model built on the basis of signals received from blocks of the merged model elements. For the merged model, using the constructed hidden Markov model, the problems of analyzing of the model dynamics and predicting its states are being solved.

39.2 Construction of the Merged Semi-Markov Model

Let us consider the superposition of N independent renewal processes constructed in [1, 2]. The system S consists of N elements $E_1 - E_N$, the operation of each element is described by the renewal process [1, 2]. The elements i renewal time is a random variable (RV) α_i with distribution function (DF) $F_i(t)$ and distribution density $f_i(t)$, $i = \overline{1, N}$. RVs α_i are assumed being independent in aggregate and having finite expectations.

In the papers [1, 2], the semi-Markov model of the system S is constructed with the use of semi-Markov process $\xi(t)$ with discrete-continuous phase space of states of the form:

$$E = \left\{i\bar{x} : \bar{x} = (x_1, x_2, \ldots, x_k, \ldots, x_N), \quad x_k \geq 0, k = \overline{1, N}\right\}, \tag{39.1}$$

where i is the index of the renewed element, x_k is the time to kth element renewal, note that $x_i = 0$.

The following transitions of the EMC of the semi-Markov process $\xi(t)$ are possible [1, 2]:

- $i\bar{x} \to j\bar{y}, j \neq i$ under conditions $x_j = x = \wedge_{l \neq i} x_l$, $y_i = t$, $0 < t < \infty$, $y_l = x_l - x, l \neq i, j$ with transition probability density:

$$p_{i\bar{x}}^{j\bar{y}} = f_i(x + t); \tag{39.2}$$

- $i\bar{x} \to i\bar{y}$ under conditions $x_j = x = \wedge_{l \neq i} x_l, y_j = t, 0 < t < x_j, y_l = x_l - x + t$, $l \neq i, j$ with transition probability density:

$$p_{i\bar{x}}^{i\bar{y}} = f_i(x - t), \tag{39.3}$$

where $\wedge$ is the minimum sign.

In [1, 2], it is shown that the EMC stationary distribution of considered superposition has the following form:

$$\rho(i\bar{x}) = \rho_0 \prod_{\substack{k=1, \\ k \neq i}}^{N} \bar{F}_k(x_k), \quad i = \overline{1, N}, \ \bar{F}_k(x) = 1 - F_k(x), \tag{39.4}$$

the constant ρ_0 is being found through the normalization condition.

To simplify the model of system S, let us construct the merged semi-Markov model of the system using the stationary phase merging algorithm, proposed in [1, 2]. Let us split the initial model phase space of states into N classes:

$$E_i = \{i\bar{x} : \bar{x} = (x_1, x_2, \ldots, x_k, \ldots, x_N)\}, \quad i = \overline{1, N},$$

each of which is "glued" into one state i of the merged model. Phase space of states $\hat{E}$ of the merged model takes the form:

$$\hat{E} = \{\bar{1}, 1, \ 2, 3, \ldots, N\}. \tag{39.5}$$

Let us point the physical sense of the merged model:

$\bar{1}$—all elements are renewed, initial state;
i—the element with the number i is renewed, $i = \overline{1, N}$.

Let us determine the transition probabilities $\hat{p}_{kr}$ of the merged model that in accordance with [1, 2] are found by the formula:

$$\hat{p}_{kr} = \frac{\int_{E_k} \rho(de) P(e, E_r)}{\rho(E_k)}, \quad k, r = \overline{1, N}, \tag{39.6}$$

where $\rho(de)$ is the stationary distribution of EMC defined by the formula (39.4) and $P(e, E_r)$ are the transition probabilities of EMC.

Using formula (39.4), we obtain:

$$\rho(E_k) = \rho_0 \underbrace{\int_0^\infty \cdots \int_0^\infty}_{N-1} \prod_{\substack{i=1,\\ i \neq k}}^{N} \bar{F}_i(x_i) d\bar{x}^{(k)} = \rho_0 \prod_{\substack{i=1,\\ i \neq k}}^{N} E\alpha_i. \tag{39.7}$$

1. Let us find the transition probabilities $\hat{p}_{\bar{1}r}$, $r = \overline{1, N}$:

$$\hat{p}_{\bar{1}r} = P\left(\underset{k \neq r}{\wedge} \alpha_k > \alpha_r \right) = \int_0^\infty \prod_{\substack{k=1\\ k \neq r}}^{N} \overline{F_k}(t) f_r(t) \mathrm{d}t \tag{39.8}$$

2. Using formulas (39.2)–(39.4), (39.7), we determine the probabilities $\hat{p}_{kr}$ in the case $k = \overline{1, N}$, $r = \overline{1, N}$, $r \neq k$:

$$\hat{p}_{kr} = \frac{\int_0^\infty dx_r \int_{x_r}^\infty \cdots \int_{x_r}^\infty \int_0^\infty \prod_{\substack{i=1,\\ i \neq k}}^{N} \bar{F}_i(x_i) f_k(x_r + t) \mathrm{d}\bar{x}^{(k)} \mathrm{d}t}{\prod_{\substack{i=1,\\ i \neq k}}^{N} E\alpha_i} =$$

$$= \frac{\int_0^\infty \bar{F}_k(x_r) \bar{F}_r(x_r) \left(\prod_{\substack{i=1,\\ i \neq k,r}}^{N} \int_{x_r}^\infty \bar{F}_i(x_i) \mathrm{d}x_i \right) \mathrm{d}x_r}{\prod_{\substack{i=1,\\ i \neq k}}^{N} E\alpha_i}$$

$$= \frac{\int_0^\infty \bar{F}_k(t)\bar{F}_r(t)\left(\prod_{\substack{i=1,\\ i\neq k,r}}^N \int_{x_r}^\infty \bar{F}_i(x_i)\mathrm{d}x_i\right)\mathrm{d}t}{\prod_{\substack{i=1,\\ i\neq k}}^N E\alpha_i}. \quad (39.9)$$

In the case of $r = k$

$$\hat{p}_{kk} = 1 - \sum_{\substack{r=1,\\ r\neq k}}^{N} \hat{p}_{kr}, \quad k = \overline{1, N}. \quad (39.10)$$

Let us consider the case where all the RVs $\alpha_i, i = \overline{1, N}$ have the exponential distribution: $F_i(t) = 1 - \mathrm{e}^{-\lambda_i t}$, $\bar{F}_i(t) = \mathrm{e}^{-\lambda_i t}$. Using the formulas (39.8)–(39.10), we are obtaining that in this case:

$$\hat{p}_{\bar{1}r} = \frac{\lambda_r}{\sum_{i=1}^N \lambda_i}, \; r = \overline{1, N}, \quad \hat{p}_{kr} = \frac{\lambda_r}{\sum_{i=1}^N \lambda_i} = \frac{\frac{1}{E\alpha_r}}{\sum_{i=1}^N \frac{1}{E\alpha_r}}, \quad k, r = \overline{1, N}. \quad (39.11)$$

The found values of $\hat{p}_{kr}$ are used to construct the hidden Markov model.

39.3 Hidden Markov Model of a Merged Semi-Markov Model Based on Signals from Blocks of the Elements

Let $\{X_n, n = 1, 2, \ldots\}$ be the merged model EMC transition probabilities which are defined by the formulas (39.11). At the initial instant ($n = 1$), the merged model is being into the state $\bar{1}$ (further, it does not enter this state again). It is assumed that the states $\{1, 2, 3, \ldots, N\}$ of the merged model are split into several disjoint blocks (classes).

While the system S operation, one cannot observe the states of the merged model (hidden states) but can observe only the index of the system S block in which the renewal of an element took place.

To be definite let us suppose that $N = 6$ and the states are split into two blocks a and b:

$$a = \{1, 2, 3\}, \quad b = \{4, 5, 6\}.$$

Hence, the set of the signals takes the form:

$$J = \{f, a, b\},$$

Table 39. Emission function $R(s|x)$ connecting the merged model EMC states with the signals

State, x / Signal, s	$\bar{1}$	1	2	3	4	5	6
$s = f$	1	0	0	0	0	0	0
$s = a$	0	1	1	1	0	0	0
$s = b$	0	0	0	0	1	1	1

where f is the signal emitting on the first step, a is the signal emitting while the renewal of an element from the first block, b is the signal emitting while the renewal of an element from the second block.

Let us consider the relation between the merged model EMC states and the signals, i.e., the emission function $R(s|\boldsymbol{x})$ [13, 14]:

$$R(s|\boldsymbol{x}) = P(S_n = s|X_n = \boldsymbol{x}), \quad \boldsymbol{x} \in \hat{E}, s \in J, \sum_{s \in J} R(s|\boldsymbol{x}) = 1, \tag{39.12}$$

where S_n is the nth signal.

The emission function $R(s|\boldsymbol{x})$ connecting the merged model EMC states with the signals is represented in the Table 39.1.

39.4 Dynamics Analysis and Prediction of the States for the Merged Semi-Markov Model Based on Signals from Blocks of the Elements

Following the [13, 14], let us pass to dynamics analysis of the merged semi-Markov model based on the hidden Markov model constructed.

Let $\bar{S}^n = (S_1, S_2, \ldots, S_n)$ be a random vector for the first n signals. For the vector $\bar{s}_n = (s_1, s_2, \ldots, s_n)$ of signals obtained let $\bar{s}_k = (s_1, s_2, \ldots, s_k)$, $k \leq n$. It is required to estimate the merged (hidden) model EMC characteristics using the obtained vector of signals $\bar{s}_n$. It is assumed that at the initial instant, the model is within the state $\bar{1}$.

Let us introduce the functions $F_k(i)$ [13, 14]:

$$F_k(i) = P(\bar{S}^k = \bar{s}_k, X_k = i), \quad k = 1, 2, \ldots, n, \tag{39.13}$$

called the forward variables. For these functions, the next recurrent formula is valid [13, 14]:

$$F_k(i) = R(s_k|i) \sum_j F_{k-1}(j) P_j^i, \quad F_1(i) = R(s_1|i) p_i, \tag{39.14}$$

where P_j^i—are the merged model EMC transition probabilities defined by the formula (39.11), (p_i)—is the EMC initial state distribution.

Let us also consider the functions $B_k(i)$ [13, 14] called the backward variables

$$B_k(i) = P(S_{k=1} = s_{k=1}, \ldots, S_n = s_n | X_k = i), \quad k = \overline{1, n-1},$$

for which the recurrent formula [13, 14] takes place:

$$B_k(i) = \sum_j R(s_{k+1}|j) B_{k+1}(j) P_i^j, \quad B_{n-1}(i) = \sum_j P_i^j R(s_n|j).$$

Functions $F_k(i)$, $B_k(i)$ play an important role while using hidden Markov models for analysis of systems operation.

Let us pass to the merged semi-Markov model dynamics basing on the constructed hidden Markov model.

Let us consider the system S for which it is assumed that $M\alpha_1 = 15.0$ h, $M\alpha_2 = 2.0$ h, $M\alpha_3 = 14.0$ h, $M\alpha_4 = 19.0$ h, $M\alpha_5 = 23.0$ h, $M\alpha_6 = 12.0$ h before its operation begins.

Let us suppose that the system S operation was resulted in the next vector of signals:

$$\begin{aligned}\bar{s}_{30} = (&f, a, a, b, b, b, a, b, a, a, b, a, b, b, a, a,\\ &a, b, b, a, b, a, b, b, a, a, a, b, b, a), \quad n = 30.\end{aligned}$$

Let us consider the problems on characteristics evaluation for the hidden Markov model taking into account the parameters introduced.

1. Let us determine the probabilities of the hidden model states at the instant of the 30th signal emission. We apply the formula [13, 14]:

$$P\big(X_n = i | \bar{S}^n = \bar{s}_n\big) = \frac{F_n(i)}{\sum_j F_n(j)}. \quad (39.15)$$

   We obtain the result that at the 30th step, the merged model was within the state 1 with probability 0.1045, within the state 2 with probability 0.7836 and within the state 3 with probability 0.1119. For the states $\bar{1}$, 4, 5, 6, this probability is zero.

2. Let us find the probabilities with which the hidden model transits into the states at the next 31st step. To do this, let us use the formula [13, 14]:

$$P(X_{n+1} = j | \bar{s}_n) = \sum_i P(X_n = i | \bar{s}_n) P_i^j. \quad (39.16)$$

   We obtain the next transition probabilities for the hidden model at the 31st step: to the state 1—0.0815, to the state 2—0.6116, to the state 3—0.0874, to the state

4—0.0644, to the state 5—0.0532, to the state 6—0.1019. For the state $\bar{1}$, this probability is zero.

3. Let us determine the probabilities of signal emission within the next 31st step, to do this, we apply the formula [13, 14]:

$$P(S_{n+1} = s_{n+1}|\bar{s}_n) = \sum_i P(X_{n+1} = i|\bar{s}_n)R(s_{n+1}|i), \tag{39.17}$$

in this case, formula (39.16) is also used.
Thus, within the 31st step for the signal *f,* we obtain zero emission probability; for the signal *a,* this probability is equal to 0.7805; for the signal *b*—0.2195.

4. Let us find the probability of occurrence (emission) of the obtained vector of signals $\bar{s}_{30}$.
To do this, we can use the formulas [13, 14]:

$$P(\bar{S}^n = \bar{s}_n) = \sum_i F_n(i) = \sum_i R(s_1|i)B_1(i)p_i \tag{39.18}$$

and also the formula

$$P(\bar{S}^n = \bar{s}_n) = \sum_i F_k(i)B_k(i), \tag{19}$$

for any fixed *k*.
The probability of the obtained vector of signals occurrence for specified parameters of the system *S* is equal to 1.4639×10^{-11}.

5. Predicting of hidden model states under the obtained vector of signals.
It is required to find the most probable states of the hidden model within the transitions basing on the obtained vector of signals $\bar{s}_{30}$.
To solve this problem, the next formula is used [13, 14]:

$$P(X_k = i|\bar{S}^k = \bar{s}_k) = \frac{F_k(i)B_k(i)}{\sum_j F_k(j)B_k(j)}. \tag{39.20}$$

Thus, it is necessary to find i that maximizes the expression $F_k(i)B_k(i)$.

Table 39.2 represents the most probable states of the hidden Markov model within the specified transition and the probabilities of these states.

To find the most probable sequence of the states for the merged model, one can also use Viterbi algorithm [13, 14].

Table 39.2 Most probable states of the hidden model within the transition

Transition number	1	7	11	17	22	26	29
The most probable state	$\bar{1}$	2	6	2	2	2	6
The probability of the state	1	0.784	0.464	0.784	0.784	0.784	0.464

39.5 Conclusion

A large number of energy systems allow the construction of a semi-Markov model. When constructing the semi-Markov model, it is necessary to use a rather complicated phase space of states that reflect the physical states of the system and ensure the correctness of the construction of the model. An important component of the semi-Markov model is the embedded Markov chain, which is responsible for the transitions of the system from one state to another. The phase space of the embedded Markov chain coincides with the phase space of the semi-Markov model. During the functioning of the system for which the semi-Markov model is constructed, in some cases, when changing the states of the system, it is not possible to obtain all the information contained in the phase state codes of the semi-Markov model, but only some information (signal) related to the phase states can be obtained. For example, in the phase state space for each element of the system, it is indicated whether it is in working condition or is being restored, and when the system is functioning, you can only receive a signal about the number of working elements of the system. In some cases, you can only receive a signal from a block of system elements in which a state change occurred, without indicating an element of this block in which a state change occurred. In this case, the phase states of the embedded Markov chain (semi-Markov model) can be considered hidden and the problem arises of evaluating the characteristics of the embedded Markov chain and the semi-Markov model based on the obtained vector of signals. Hidden Markov and hidden semi-Markov models allow solving this problem.

The paper considers the possibilities of hidden Markov models application for characteristics evaluation, analysis of the dynamics and prediction of the states of energy systems basing on vector of signals received from blocks of elements of the system using the example of independent renewal processes stationary phase merged superposition, built in the works of V. S. Korolyuk, A. F. Turbin.

Note that when finding functions determined by recurrence relations and evaluating the characteristics of the Markov chain, whose states are hidden, it is possible to effectively use mathematical packages of analytical and numerical calculations.

In the future, it is planned to use the approach considered in the paper to analyze the operation of multicomponent energy systems taking into account various factors: renewal, control, maintenance, availability of a time reserve, etc.

Acknowledgements The reported study was funded by RFBR, project number 18-01-00392a.

References

1. Korolyuk, V.S., Turbin, A.F.: Markovian Restoration Processes in the Problems of System Reliability. Naukova Dumka, Kiev (1982)
2. Korolyuk, V.S., Korolyuk, V.V.: Stochastic Models of Systems. Springer Science + Business Media, Dordrecht (1999)

3. Korolyuk, V.S., Limnios, N.: Stochastic Systems in Merging Phase Space. World Scientific, Imperial College Press, London (2005)
4. Obzherin, YuE, Boyko, E.G.: Semi-Markov Models: Control of Restorable Systems with Latent Failures. Academic, London (2015)
5. Grabski, F.: Semi-Markov Processes: Applications in System Reliability and Maintenance. Elsevier Science, Amsterdam (2015)
6. Silvestrov, D., Silvestrov, S.: Nonlinearly Perturbed Semi-Markov Processes. Springer, Berlin (2017)
7. Kashtanov, V.A., Medvedev, A.I.: Reliability Theory of the Complex Systems (Theory and Practice). Fizmatlit, Moscow (2010)
8. Rudenko, Y.N., Ushakov, I.A.: Reliability of Power Systems, 2nd edn. Nauka, Novosibirsk (1989)
9. Obzherin, Y.E., Sidorov, S.M., Nikitin, M.M.: Analysis of reliability of systems with component-wise storages. In: E3S Web of Conferences 58, art. no. 02024 (2018)
10. Beichelt, F., Tittmann, P.: Reliability and Maintenance—Networks and Systems. Mathematical Models. CRC Press, Boca Raton (2012)
11. Cappe´, O., Moulines, E., Rydén, T.: Inference in Hidden Markov Models. Springer Science + Business Media, New York (2005)
12. Elliott, R., Aggoun, L., Moore, J.: Hidden Markov Models. Estimation and Control. Springer, New York (1995)
13. Ross, S.M.: Introduction to Probability Models, 9th ed. Academic, London (2006)
14. Kobayashi, H., Mark, B., Turin, W.: Probability, Random Processes, and Statistical Analysis: Applications to Communications, Signal Processing, Queueing Theory and Mathematical Finance. Cambridge University Press, Cambridge (2011)
15. Rabiner, L.R.: A tutorial on hidden markov models and selected applications in speech recognition. Proc. IEEE **77**(2), 257–286 (1989)
16. Obzherin, Y.E., Sidorov, S.M., Nikitin, M.M.: Hidden Markov model of information system with component-wise storage devices. In: Lecture Notes in Computer Science (including subseries Lecture Notes in Artificial Intelligence and Lecture Notes in Bioinformatics), 11965 LNCS, pp. 354–364 (2019)
17. Elliott, R., Limnios, N., Swishchuk, A.: Filtering hidden semi-Markov chains. Stat. Probab. Lett. **83**, 2007–2014 (2013)
18. Barbu, V.S., Limnios, N.: Semi-Markov Chains and Hidden Semi-Markov Models Toward Applications: Their Use in Reliability and DNA Analysis. Springer, New York (2008)
19. Van der Hoek, J., Elliott, R.: Introduction to Hidden Semi-Markov Models. Cambridge University Press, Cambridge (2018)
20. Yu, S.-Z.: Hidden Semi-Markov Models: Theory, Algorithms and Applications. Elsevier Science, Amsterdam (2015)

Chapter 40
Robot for Inspection and Maintenance of Overhead Power Lines

Sergej Solyonyj, Oksana Solenaya, Aleksandr Rysin, Vladimir Kuzmenko, and Evgeny Kvas

Abstract Problems, such as the destruction of insulators, the slope of the poles of power lines and a small layer of icing, will go unnoticed and can lead to serious consequences, such as a break in power lines and falling poles. If an accident occurs on the line, the repair team must drive along the entire path of the power line from the beginning to the point of the accident, if the terrain allows transportation. Provided that some power lines can be hundreds of kilometers long, troubleshooting can take several hours or even days. Using robotic systems will speed up the search and repair processes. This article discusses the types of damage that occur on overhead power lines and modern solutions to these problems.

40.1 Introduction

According to official data of the Ministry of Energy of Russia, in 2019, there were 13,206 accidents in electric networks with voltage of 110 kV and higher. Figure 40.1 shows a graph of the distribution of accidents by month of the year. Figure 40.2 shows the general accident schedule for 2019. Statistics also show that 53% of all accidents this year were caused by damage to power lines (PL). Due to damage to overhead line insulators (OL) and poor technical condition, 12% of the total number of accidents occurred. The accumulation of snow, ice and debris caused 3% of all accidents. Also 5% of accidents are caused by mechanical damage to poles and lightning rods of overhead lines. The main number of accidents (about 70%) occurs in the warm season from April to September.

The commissioning of robotic systems can reduce the number of accidents associated with such factors, as well as prevent prolonged shutdowns for consumers of reliability categories I and II.

S. Solyonyj (✉) · O. Solenaya · A. Rysin · V. Kuzmenko · E. Kvas
St. Petersburg State University of Aerospace Instrumentation, 67, St. Bolshaya Morskaya, 190000 St. Petersburg, Russia
e-mail: ssv555ssv@yandex.ru

A. Ronzhin and V. Shishlakov (eds.), *Proceedings of 15th International Conference on Electromechanics and Robotics "Zavalishin's Readings"*, Smart Innovation, Systems and Technologies 187, https://doi.org/10.1007/978-981-15-5580-0_40

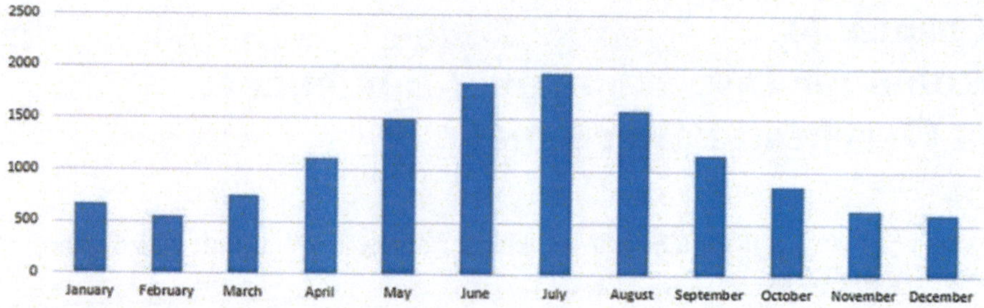

Fig. 40.1 Monthly distribution of accidents in electric networks of 110 kV and higher in 2019

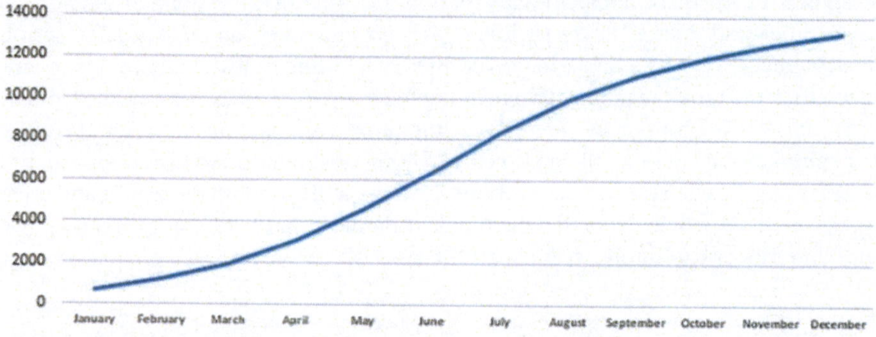

Fig. 40.2 Accident in electric networks of 110 kV and higher in 2019

At present, it is impossible to completely exclude the human factor, since human interaction with electrical equipment persists, but with the long-term development of such complexes, it will be possible to transfer them to a fully autonomous mode of operation. Information on the state of electrical equipment will be transmitted to a single center, where it will be processed for decision-making depending on the need for repair or complete replacement of equipment [1].

40.2 Problem Statement

Based on the foregoing, the aim of the study is to analyze the existing methods for determining damage to power lines and develop on its basis new possible diagnostic methods.

According to information on accidents on the Web site of the Ministry of Energy of the Russian Federation, 55% of all accidents on power lines are interfacial overlaps as a result of a wire splash due to wind gusts, trees falling from the forest and unauthorized persons. One of the modern methods for detecting cable slack is scanning the power transmission line using unmanned aerial vehicles for flying around the territory and shooting from the air [2–5].

The next, by the frequency of occurrence, cause of accidents on power lines (24%) is damage to overhead line insulators due to unsatisfactory technical condition. The destruction of insulators is difficult to diagnose without special equipment that allows you to fix small chips and microcracks. Today, special methods are used for spectral analysis using ultraviolet and infrared radiation [6–8].

The unsatisfactory technical condition of the power transmission line, under design wind loads and climatic conditions, makes up 6% of all accidents on power lines. The accumulation of wet snow, ice, gusty wind and lightning discharges leads to sagging wires and broken lines. The fight against icing wires is made by different methods:

1. Transmission of direct current electricity through the wire to heat live parts [9].
2. Passing a certain frequency current through the wire (selected individually for each section of the power transmission line) to excite high-frequency mechanical vibrations [10].
3. The use of mechanical vibration devices in special vehicles [11].
4. Chemical removal of ice by calcium carbide [12].

In addition to icing, extraneous debris such as cloth and plastic bags can also be wound around the wire. Special aircraft can be used to remove it. In the Chinese city of Sanyang, Hubei Province, a drone with a flamethrower was tested, which is designed to remove debris from overhead power lines. Light garbage from time to time hangs on power lines. As a rule, employees of energy companies are involved in garbage collection, which is removed manually from the cradle of a telescopic air platform. This method is simple, but it has tangible disadvantages—in each place of garbage accumulation, it is necessary to lay out stops and raise the cradle, which takes 5–10 min. One of the Chinese energy companies in Sanyang has tested a new way to clean power lines from debris using a flamethrower drone. The modified DJI S1000 + octocopter was adapted to destroy garbage, which can also lift up to 11 kg of cargo into the air.

Damage, corrosion and aging of insulator circuit breakers are 5% of all accidents on power lines. To detect them, they are also used either by inspection using unmanned aerial vehicles or by inspection by a person manually [13].

Mechanical damage to ground lines of overhead lines is 4% of the total number of accidents. The diagnosis of such malfunctions is carried out in the same way as the inspection of isolator switches.

Mechanical damage to the supports and traverse is a 3% share of all accidents. These damages can be divided into two categories: the initial stage of damage and visible damage. The initial stage is the inclination of the supports or traverse relative to the normal by an angle—invisible to the eye. Such deviations can only be recorded using installed equipment (gyroscopes) on each support.

Visible damage can already be seen with the naked eye and is diagnosed with the help of copters or inspection of the working team.

Based on the analysis, we can conclude such problems as:

- quick diagnosis of damage to insulators;
- universal system for cleaning the cable from ice.

To solve these problems, a robot moving along the wire was designed, equipped with cameras for conventional monitoring, a UV and IR camera for spectrum analysis of the condition of insulators, as well as contact cleaning of the wire from ice in the form of hard brushes.

Similar analogs exist in the world, such as Expliner, LineScout, Ilena and SkySweeper [14, 15].

They are able to move along overhead power lines, using spectral flaw detection to determine the condition of the wire and insulator. Laser sensors automatically detect the presence of corrosion on the wires, changes in wire diameter due to a violation of the internal structure of the wire, and eight high-resolution cameras allow the operator to visually detect mechanical damage on four lines simultaneously. The robot is able to independently bypass simple obstacles, due to the displacement of the center of gravity and the transfer of the front and rear wheels through the wire fastening system. In the case of a complex obstacle, such as a series of insulators, a manual transfer is required to reposition the robot.

40.3 Research Results

The robot is intended for its use without disconnecting the electrical equipment from the power source, so the materials of the robot must be non-conductive and have protection from electromagnetic radiation [16]. The materials used are composite materials, rubber of various compositions and plastics. Nylon and graphite plastic are suitable for these tasks in terms of compressive and tensile strength properties. However, graphite plastic is not suitable, since its dielectric properties remain in question. On high-voltage lines, it will become a conductor and create a short circuit between the robot and the wire. Nylon is not used, as it has a low coefficient of friction, which will not allow for movement along a wire that has sagging. Comparisons of various materials were carried out taking into account the loads on the wheel under the action of gravity, torque and crosswind. All the materials examined showed a good result in wheel deformation (Fig. 40.3) and safety margin. The main stresses are on the axis of rotation of the wheel and the contact surface with the wire. In the future, when using other materials, you can remove excess material from the outer edges of the wheels.

Further studies were conducted to determine the basic material for the robot body. Initially, the form was to be in the form of the Latin letter "G". This shape allows you to manually hang the robot on a cable, hanging it from the side of the wire, and had a streamlined shape that allows you to protect internal mechanisms and electronics from weather conditions with minimal junction points. However, tests showed on one edge of the robot excessive stresses and bends, leading to breakage under more

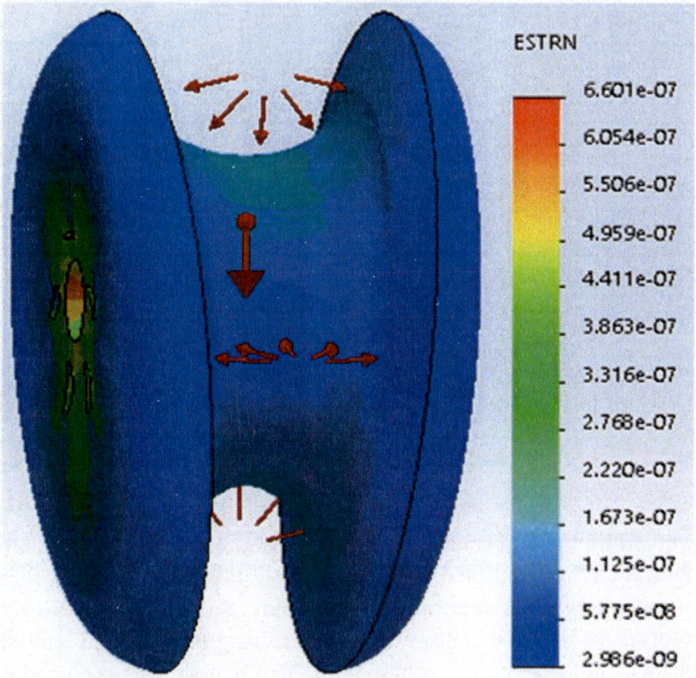

Fig. 40.3 Stress test results

severe loads. This arose due to the fact that the main part of the mass of the robot and side loads fell on one shoulder, the length of which did not exceed 40% of the width of the robot. With increasing shoulder length, the ability to quickly land the robot on the wire is lost.

The second version of the robot took the form of "π". This form simplifies the process of preparing for the operation of the robot, calculating the center of mass and piloting as an unmanned aerial vehicle. However, this led to an increase in the mass of the robot, which will complicate its modernization to an aircraft. Flow Simulation study also showed a deterioration in the aerodynamic properties of the robot.

Figure 40.4 shows color plots of the displacement distribution in millimeters. The materials examined showed insignificant deformations of the robot casing, which made it possible to reduce its walls and partitions by 2 times. Material savings allowed to reduce the mass of the robot from 4.5 to 2.5 kg with the same indicators of resistance to loads.

The assembled model does not provide protection against electromagnetic influences, since the test sample was tested on 220/380 V. lines. When using the robot on high and ultra-high-voltage lines, it is necessary to protect the microcontroller and the motor driver from electromagnetic radiation generated by power lines.

An Arduino Nano board with an ATmega328P processor is used as the controller. Control signals are received from the operator from the mobile device via Bluetooth

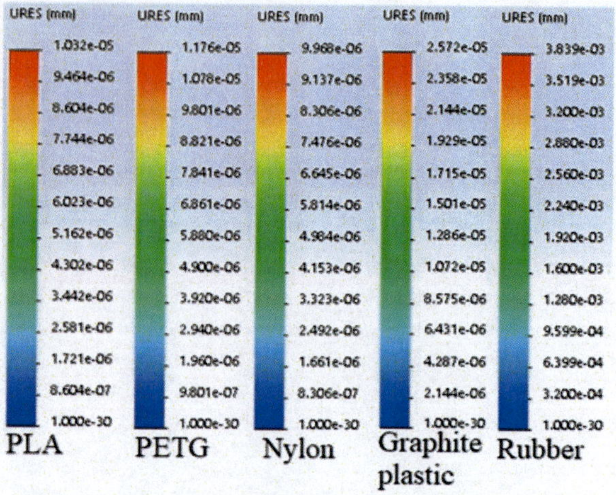

Fig. 40.4 Comparison of deformation materials

and are received by the HC-06 communication module. In addition, control signals are sent to the BTS7960 engine driver via digital outputs. Engines can rotate in both directions (depending on the operator's command) and at different speeds.

The prototype of the robot showed stability during lateral wind and vibration of the wire but was not convenient to maintain. The battery was replaced only when the case was completely disassembled, and the electronic part was not sufficiently protected from moisture and dust.

The assembled model does not provide protection against electromagnetic influences, since the test sample was tested on 220/380 V. lines. When using the robot on high and ultra-high-voltage lines, it is necessary to protect the microcontroller and the motor driver from electromagnetic radiation generated by power lines.

An Arduino Nano board with an ATmega328P processor is used as the controller. Control signals are received from the operator from the mobile device via Bluetooth and are received by the HC-06 communication module. In addition, control signals are sent to the BTS7960 engine driver via digital outputs. Engines can rotate in both directions (depending on the operator's command) and at different speeds.

The prototype of the robot showed stability during lateral wind and vibration of the wire but was not convenient to maintain. The battery was replaced only when the case was completely disassembled, and the electronic part was not sufficiently protected from moisture and dust.

The straight walls of the robot allow you to install cameras inside the housing with the output of the lens to a section of wire. Further modernization involves the installation of several cameras: One camera transmits an image to the operator, who performs a visual analysis of the state of electrical equipment. If a malfunction is detected in the form of cracks, corrosion and chips, he can connect a second chamber. This is an ultraviolet scanner that allows you to see the width and depth of even those cracks that are not visible to the eye. Based on the information received, the operator can decide on the replacement or repair of electrical equipment.

If icing is detected, the operator can turn on the cleaning of the cable with hard brushes. The power cable is cleaned with hard plastic brushes to avoid shorting the cleaning system to the motor. Brushes rotate in a circle and are pressed to the cable from several sides.

The next upgrade step is to replace the Bluetooth module with Wi-Fi. This will increase the radius of the robot and accelerate the transfer of information [17].

- To ensure increased automation of the processes of inspection and maintenance of power lines, it is planned to install engines with blades for installing the robot on overhead power lines without human intervention: unmanned takeoff and landing on a cable. This will allow one person to install the robot at the beginning of the power line and diagnose it without interrupting the manual movement of the robot between supports and insulators. This update will add designs weighing up to 2 kg but will significantly reduce the time of putting the robot into operation. The battery capacity does not need to be significantly increased, since the takeoff time and bypass time are from 15 to 40 s, provided that the power transmission towers are installed at a distance of 300–400 m. In the study of 5 km of power lines, the flight time will be 5 min [18].
- The creation of wireless charging of robotic batteries from the electromagnetic field of power lines is considered. However, calculations showed the inefficiency of such charging due to the large weight of the secondary winding.

Arduino board will be replaced by Node MCU ESP-8266 with built-in Wi-Fi module. For the robot to work efficiently, it will be necessary to install additional Wi-Fi routers on the transmission towers in order not to lose communication with the operator even at long distances.

To move the center of mass below the plane of passage of the cable, separate insulated battery compartments are provided at the very bottom of the robot. One battery is used to power the cables of moving motors, servos and brush motors and power the controller and sensors. The second battery is connected to the flight engine control board. If the battery level approaches a critical value, a warning will be sent to the operator. He will be able to remotely connect batteries in parallel to increase capacity and return to base. Figure 40.5 shows a block diagram of the conditions for the transition from the mode of movement of the robot through the wire to flight mode. Figures 40.6 and 40.7 show the interaction algorithm of the aircraft controller with actuators and communication channels with the operator.

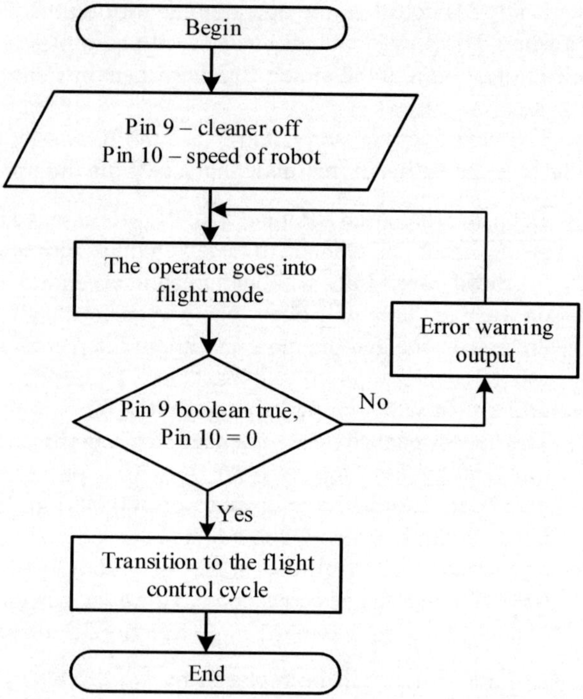

Fig. 40.5 Transition condition flowchart

40.4 Conclusion

In the process of working on robotic systems, statistics on accidents on power lines in Russia were collected. The main reasons are weather conditions and conditions that are not dependent on human activities. However, about 20% of accidents can be prevented by commissioning robots for inspection and maintenance.

In the process of research, the shape of the wheels of the robot and the shape of the body of the robot were selected, allowing further modernization of the prototype to an unmanned aerial vehicle. Materials were selected for the case and wheels, satisfying the conditions of strength, having insulating properties and resistance to deformation. For the assembled prototype PLA and PETG plastics, rubber and nylon were tested at a voltage of 220\380 V; however, with further research and testing on high-voltage lines, they can be replaced with modern polymer and composite materials.

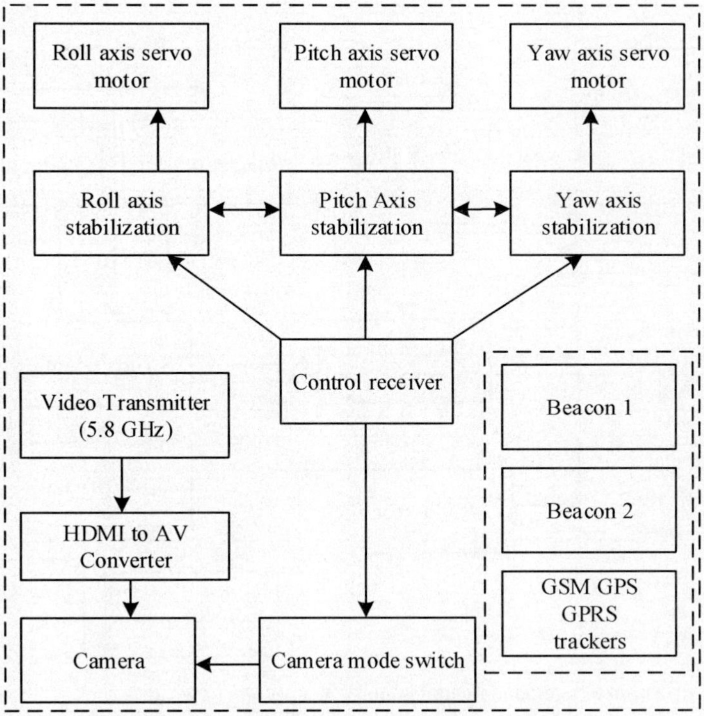

Fig. 40.6 Algorithm of interaction with executive devices and operator

When using the robot, it will be necessary to create a database for storing information and further analysis. In the future, in order to exclude subsequent accidents due to the human factor, it is planned to introduce an artificial intelligence system that will have a large sample with images of insulators of varying degrees of wear, various degrees of cable corrosion, areas most prone to icy power lines. This will make it possible to shift the decision on timely repair or replacement of equipment from the operator to the computer. The total number of accidents will decrease, and as a result, the reliability of energy supply to consumers of category II and I will increase.

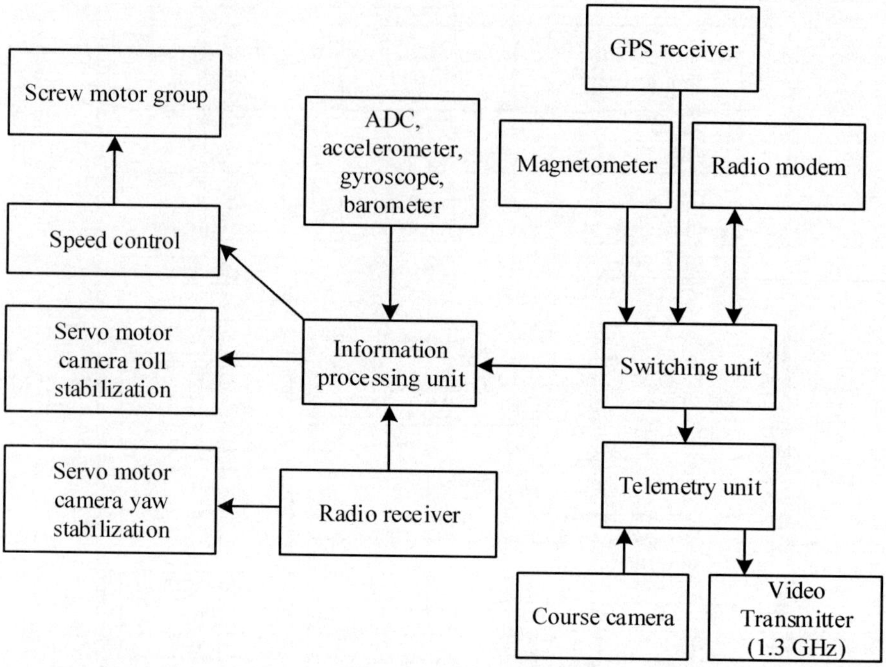

Fig. 40.7 Algorithm of interaction with executive devices and operator

Acknowledgements This research is supported by the RFBR Project No. 20-08-1056A.

References

1. Ovcharenko, K.L., Eremeev, I.Yu., Sazonov, K.V., Semenjuk, S.S., Abakumov, A.N.: Differential ranging method of locating satellite communication system earth stations using repeater on an unmanned aerial vehicle. SPIIRAS Proc. **18**(1), 176–201 (2019)
2. Samad, A.M., Kamarulzaman, N., Hamdani, M.A., Mastor, T.A., Hashim, K.A.: The potential of Unmanned Aerial Vehicle (UAV) for civilian and mapping application. In: 2013 IEEE 3rd International Conference on System Engineering and Technology, pp. 313–318 (2013)
3. Muchiri, N., Kimathi, S.: A review of applications and potential applications of UAV. In: Proceedings of Sustainable Research and Innovation Conference, pp. 280–283 (2016)
4. Hameed, I.A., la Cour-Harbo, A., Osen, O.L.: Side-to-side 3D coverage path planning approach for agricultural robots to minimize skip/overlap areas between swaths. Robot. Autonom. Syst. **76**, 36–45 (2016)
5. Aguilar, W.G., Morales, S.G.: 3D environment mapping using the Kinect V2 and path planning based on RRT algorithms. Electronics **5** (2016)
6. Ganganath, N., Cheng, C.T., Chi, K.T.: A constraint-aware heuristic path planner for finding energy-efficient paths on uneven terrains. IEEE Trans. Ind. Inform. 601–611 (2015)
7. Dijkstra, E.W.: A note on two problems in connexion with graphs. Numerische Mathematic **1**, 269–271 (1959)

8. Kim, H., Kim, B.K.: Online minimum-energy trajectory planning and control on a straight-line path for three-wheeled omnidirectional mobile robots. IEEE Trans. Industr. Electron. **61**(9), 4771–4779 (2013)
9. Ganganath, N., Cheng, C. T., Chi, K. T.: Finding energy-efficient paths on uneven terrains. In: 2014 10th France-Japan/8th Europe-Asia Congress on Mecatronics, pp. 383–388 (2014)
10. Matraji, I., Al-Durra, A., Haryono, A., Al-Wahedi, K., Abou-Khousa, M.: Trajectory tracking control of skid-steered mobile robot based on adaptive second order sliding mode control. Control Eng. Practice **72**, 167–176 (2018)
11. Yu, W., Chuy Jr., O.Y., Collins Jr., E.G., Hollis, P.: Analysis and experimental verification for dynamic modeling of a skid-steered wheeled vehicle. IEEE Trans. Rob. **26**(2), 340–353 (2010)
12. Salgado, I., Cruz-Ortiz, D., Camacho, O., Chairez, I.: Output feedback control of a skid-steered mobile robot based on the super-twisting algorithm. Control Eng. Practice **58**, 193–203 (2017)
13. Johnson, J.W.: Adapting mask-rcnn for automatic nucleus segmentation. arXiv preprint arXiv: 1805.00500 (2018)
14. Remondino, F., Barazzetti, L., Nex, F., Scaioni, M., Sarazzi, D.: UAV photogrammetry for mapping and 3d modeling–current status and future perspectives. Int. Arch. Photogram. Remote Sens. Spatial Inform. Sci. **38**(1), C22 (2011)
15. Hart, P.E., Nilsson, N.J., Raphael, B.: A formal basis for the heuristic determination of minimum cost paths. IEEE Trans. Syst. Sci. Cybern. **4**(2), 100–107 (1968)
16. Rubin, F.: The Lee path connection algorithm. IEEE Trans. Comput. **100**(9), 907–914 (1974)
17. Miroshnichenko, S.Y., Titov, V.S., Dremov, E.N., Mosin, S.A.: Hough transform application to digitize rectangular spatial objects on aerospace imagery. SPIIRAS Proc. **6**(61), 172–196 (2018)
18. Goodwin, G.K., Grebe, S.F., Salgado, M.E.: Design of control systems; Binomial, p. 912. Laboratory of Knowledge, Moscow (2016)

Chapter 41
Construction of Land Base Station for UAV Maintenance Automation

Igor Lebedev, Anton Ianin, Elizaveta Usina, and Viktor Shulyak

Abstract This paper proposes a construction of a land-based base station for automated unmanned aerial vehicle (UAV) maintenance. The station is intended for UAV storage, protection from poor weather conditions, battery replacement and information transfer. UAV data transfer at the station allows to reduce UAV energy consumption in flight. The base station has two levels: the top one and the bottom one. The construction of base station allows to store UAVs with large dimensions, weighting up to 12 kg. The top level of the station consists of a retractable roof and meteo-sensors. The roof design provides for solar panel installation on it, what allows reducing base station energy consumption from batteries. Materials for the roof were chosen with intention to minimize roof weight and reduce load on drives. The bottom level of the base station contains the landing site, the battery replacement mechanism, the landing site lifting mechanism, the base station control system, as well the energy distribution system. The landing site features an ArUco-marker with backlight and includes mechanisms for UAV centering and fixation. The marker with backlight includes a plexiglass sheet, light emitting diodes (LEDs), a reflective substrate and an opaque scatterer. The outline lighting ensures marker detection in low light conditions. In cases, when the marker cannot be detected by active lighting, dependency of the required photosensitivity of the camera from the UAV flight altitude is presented.

I. Lebedev (✉) · E. Usina
St. Petersburg Institute for Informatics and Automation of the Russian Academy of Sciences, 39, 14th Line, 199178 St. Petersburg, Russia
e-mail: igorlevedev@gmail.com

E. Usina
e-mail: lizzzi96@mail.ru

A. Ianin · V. Shulyak
St. Petersburg State University of Aerospace Instrumentation, 67, st. Bolshaya Morskaya, 190000 St. Petersburg, Russia
e-mail: anton.ianin8@gmail.com

V. Shulyak
e-mail: Vitx-on@yandex.ru

A. Ronzhin and V. Shishlakov (eds.), *Proceedings of 15th International Conference on Electromechanics and Robotics "Zavalishin's Readings"*, Smart Innovation, Systems and Technologies 187, https://doi.org/10.1007/978-981-15-5580-0_41

41.1 Introduction

Unmanned aerial vehicles (UAV) became ingrained not only in military service, but also in everyday life. Many large companies such as Amazon are starting to use UAVs to perform tasks in their daily workflows, such as cargo delivery. UAV maintenance tasks are usually human-controlled. Such maintenance can be time-consuming and reduce the operating range of the system.

To perform automated UAV motion control and flight mission completion, special base stations are used [1–3]. For example, the base control station [4] has a CPU, a telemetry module, a user control module, a graphical user interface and a wireless data subsystem. The wireless data subsystem is configured for remote communication with UAVs. The telemetry module is connected to the base control station and configured to download on-board data from the UAV to the base station, as well as to download commands from the base station to the UAV.

One possible approach to cost reduction of UAV service maintenance consists in automation of such tasks, as battery recharging or replacement. Such mechanism can be integrated into a base station. This way the problem of manual battery replacement is solved, what further facilitates UAV performance automation. In [5], an automated recharging base station is described. Several station designs have been proposed. Test results showed that the development of low-tech and inexpensive platforms increases the requirements for UAV piloting. In turn, when using base stations for automated maintenance, the navigation requirements become less crucial. A similar solution to the problem of automated battery replacement is given in [6]. Authors designed a station for UAV recharging and a conceptual model of a platform for battery replacement. This station design allows to compensate for orientation and positioning errors, as well accounts for navigation errors, weather conditions or UAV damage. The proposed solution can be used to establish a network of base stations, designed to maintain an UAV swarm in action [7].

In contrast to previous work, the base station in [8] is equipped with a camera and an infrared sensor for precision landing of UAVs. The station, proposed here, includes a housing, a roof, a landing pad, a battery replacement mechanism and battery slots. The mechanism for battery replacement and landing pad lifting is driven by stepper motors.

Integration of sensors into the landing system allows to improve UAV landing accuracy onto the platform. In [9], authors proposed to use an IR-camera within the landing site detection system. Infrared emission sources are embedded into the landing site for orientation purposes. Algorithm for detection of such infrared markers includes Gaussian-based filtration, image binarization, morphological closure operation, segment outline detection and marker discovery. Though, this research contains no data that would prove the efficiency of such approach in real-world setting. In [10], an analogous system is presented. The system includes an IR-camera, an ultra-wideband radar that measure distance between UAV and ground-based system, a

PAN-Tilt Unit (PTU). The landing marker detection method is based on the temperature difference between UAV in motion (i.e., temperature of oil drops, exhaust pipe) and the environment. UAV altitude detection error was 0.1 m.

A wider range of tasks is performed by commercial base stations. For example, the company Nightingale Security [11] offers a base station that has two sliding sash-covers that move along the guides in opposite directions. This platform is designed to store a small UAV. When the UAV approaches, the roof flaps are moved apart, after landing, if necessary, weather protection, the flaps are closed. The disadvantage of this platform is the need for a large free area around the landing platform to open the roof flaps. In addition, this station is not intended to replace the battery. The station has dimensions: 150 × 150 × 90 cm and accommodates UAVs with dimensions 80 × 80 × 40 cm.

The Percepto base station [12] has a two-leaf roof. Opening the roof is carried out using a system of levers. Battery replacement is not provided for in this concept, but there is a wireless power transmission system. This platform has a system for maintaining temperature and humidity inside. While the UAV is at the station, it sends all the footage to cloud storage. The station has the following dimensions: 100 × 150 × 150 cm and accommodates UAVs with dimensions of 80 × 80 × 40 cm.

Most of the presented stations have a limited functionality. Often, they are intended only to replace the battery, but not to store information and protect the UAV from adverse environmental conditions. Storage of UAVs at the base station will increase the working range of the vehicle and reduce the duration of missions, because there will be no to return to the base station for battery recharge. Commercial solutions of base stations have some disadvantages, such as massive roofs, lack of battery replacement mechanisms and data transfer between the station and the UAV. Thus, the base station should have the following minimal functionality: UAV storage, battery charging, battery replacement, two-way data transfer between the UAV and the base station. The objective of this research is to develop a base station for automated UAV maintenance.

41.2 Interaction of UAV with Base Station

For design of base station-related engineering solutions, an interaction diagram was proposed, describing UAV and base station interoperability (see Fig. 41.1). The diagram illustrates possible scenarios of autonomous and collaborative actions of UAV and base station.

Base station-specific diagram blocks describe obtaining and processing of weather data from add-on sensors or publicly available data from the Internet to ensure reliable take-off and landing of UAV. The proximity sensor [13] can be used to determine the distance from UAV to the station. The UAV obtains the data via two-way transmission. LED backlight of the landing marker is activated after the device gets notified of an approaching UAV. For eventual replacement of UAV battery, the diagram includes

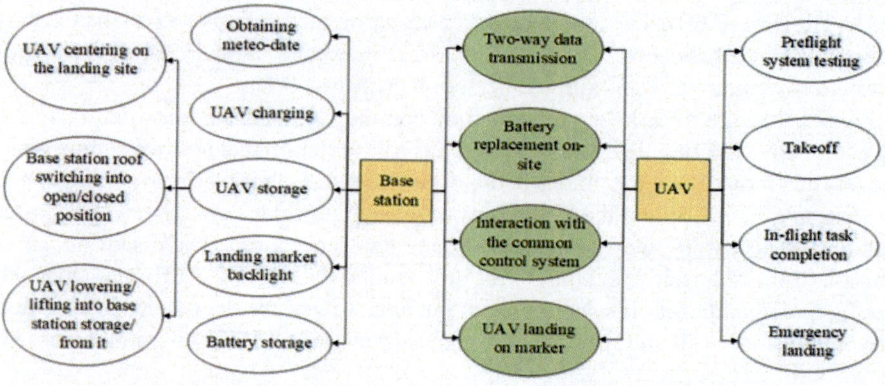

Fig. 41.1 Interaction diagram, depicting UAV and base station interoperability

dedicated blocks for battery recharge and storage. Storage block includes «UAV centering on landing site», «roof opening/closing» «UAV take-off/landing».

The above diagram of UAV and base station interoperability also contains shared blocks. Two-way data transmission is intended to transfer data, obtained in flight, from UAV to the station, as well transfer of proximity data and landing/take-off the other way. In-flight data transfer is intended for information storage on base station and its subsequent redirect onto server. This allows to reduce UAV in-flight power consumption, because no data is sent to server during the flight.

Battery replacement block allows to shorten UAV mission duration, because it is no urgent need for the vehicle to return to base station for recharge. The block «UAV landing on marker» opens the base station roof and activates the landing marker backlight if necessary. After the UAV landing on base station has been successfully completed, instructions from «UAV storage» come into play.

Dedicated functional UAV-related blocks describe basic actions of the vehicle, intended to complete in-flight tasks: preliminary system testing, take-off, flight along the route, in-flight task completion and landing. Preliminary testing includes battery charge-level check, power plant test run, inertial navigation sensor system health check and, finally, ensuring the presence of the specified number of satellites to flight along the route.

The developed diagram enables composition of scenarios of interactions between UAV and base station. These scenarios can be applied in algorithm design for UAV and base station control.

41.3 Base Station for Automated UAV Maintenance

Based on the presented list of functional features, a 3D model of base station was designed. This model provides for visualization of possible scenarios andd interactions between the UAV and the base station.

This model is depicted in Fig. 41.2 and consists of two levels. The base station housing is fabricated of aluminum profiles and PVC panels. These materials were chosen because of the requirement to minimize the overall structure weight.

The top-level elements of the station are mounted above the housing; there are essentially the roof and meteo-sensors, which can read data on environmental conditions. Thermometers, barometers, anemometers and hygrometers can be used as such sensors.

The bottom level of the station is established inside the housing and includes a landing site, landing site elevation unit, battery replacement unit, station control unit, station power supply unit, as well power distribution system. The landing site consists of an ArUco-marker with backlight, as well UAV fixation and centering system. Elevation mechanism is intended to raise the UAV before take-off and accept it after landing. The landing site mechanism consists of aluminum strips, fastener parts, sleeve bearings, lead screw and sliding carriage. The elevation mechanism is driven by step motor. Fastener parts were fabricated of stainless steel because of its durability and corrosive resistance.

Battery replacement mechanism consists of two aluminum rails, along which aluminum carriage runs. Battery extraction and insertion is effected via an aluminum

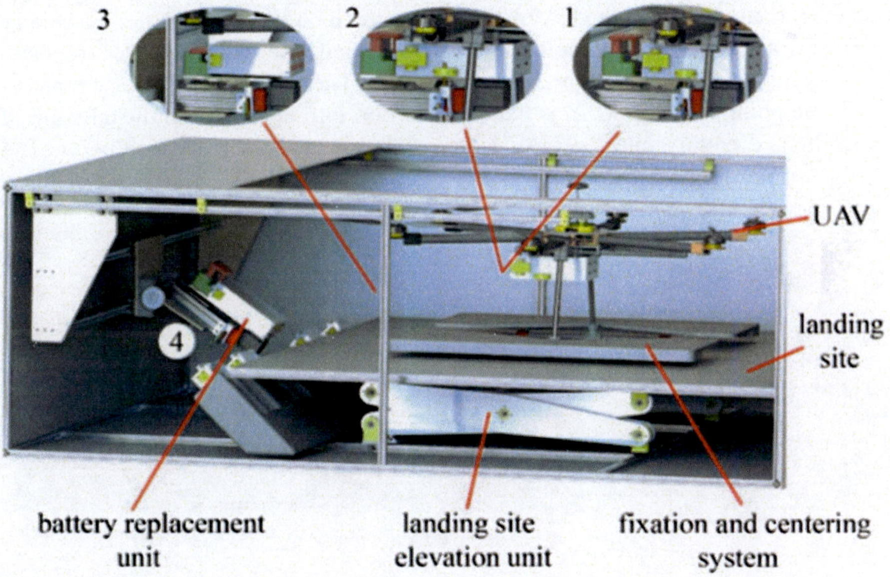

Fig. 41.2 Base station design

drawer carriage. The battery replacement mechanism is driven by a servo with three step motors, which enable mechanism motion in three planes. Battery storage is 3D-printed from PETG plastic and has three storage cells.

Step motor control is effected via a control system, which includes a single-board computer and a driver array for step motors. For power supply of motors and servo, an accumulator battery and power distribution system are used.

Technical features of the developed base station:

- outer dimensions 170 × 200 × 100 cm;
- battery replacement duration: from 5 to 10 min;
- number of reserve UAV batteries 5 items;
- autonomous function duration: up to a week;
- load capacity (UAV weight) up to 12 kg;
- inner dimensions for UAV storage: 140 × 140 × 60 cm.

The developed base station conforms to functional requirements, claimed for it. The proposed design provides for UAV battery replacement, two-way data transmission, UAV storage. After UAV landing on the platform, it is fixed and centered (see Fig. 41.3). Further, the landing site is lowered to the bottom level, where battery replacement is made when necessary, as well large data batches are transmitted, accumulated during flight task completion.

Figure 41.3 illustrates how the centering mechanism shifts along the rails laterally. When the mechanism touches the UAV base, the UAV shifts laterally (see Fig. 41.3b). Figure 41.3c shows the final step of mechanism and UAV motion toward the center of the site.

Battery replacement is fulfilled without aid of technical vision or other UAV discovery methods. After the UAV has been centered on the landing site, the battery will be extracted, and battery replacement is performed, afterward the battery replacement mechanism shifts to point 3 (see Fig. 41.2). The final step of battery replacement (see point 4 in Fig. 41.2) is battery insertion into dock station. Installation of the recharged battery into the UAV is performed in reverse order (see points 1–4

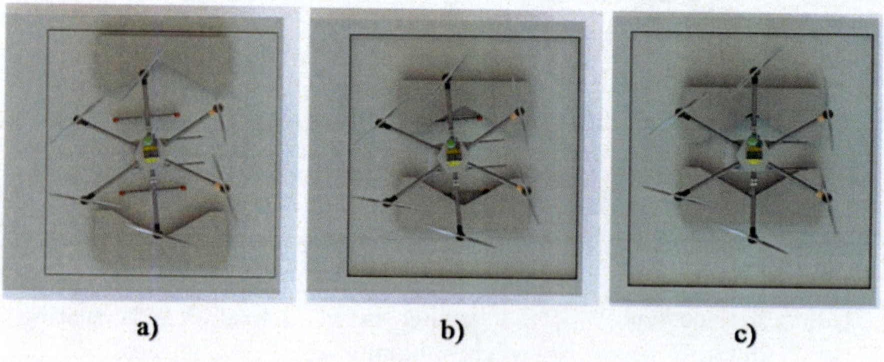

Fig. 41.3 Stages of UAV positioning on landing site: **a** initial, **b** middle, **c** final

in Fig. 41.2). After completion of functional block «UAV battery replacement», the station switches to mode «Battery storage» until the next replacement mission.

Two-way data transfer module enables data exchange at distance up to 50 km with operational frequency grids of 2.3/2.4/2.5 and 5.2–5.8 GHz. Maximum data transfer speed is 12 Mbit/s. Video stream can be transferred with resolution of 1920 × 1080p, frequency 30 frames/sec. and delay 125 ms. To obtain publicly available meteodata, a 4G modem is used.

41.4 Storage and Positioning Modules of Base Station

41.4.1 Retractable Roof

To implement functional blocks «UAV landing», «UAV storage» and «UAV take-off», a retractable roof was developed (see Fig. 41.4). The roof protects the vehicle from poor weather conditions during storage of the UAV, as well mechanical and electronic components of the station.

The retractable roof consists of five polycarbonate rectangular sections, where subsequently solar panels should be placed to reduce base station energy consumption from the battery. Roof retraction is performed along the stainless steel rails with two step motors and cord drive.

The main advantages of the developed retractable roof are the following: compactness in retracted state, protection of UAV and bottom level mechanisms from poor conditions, possibility to mount solar panels on the roof, low energy consumption. Energy consumption is reduced due to lightweight structure, what, in turn, reduces motor load.

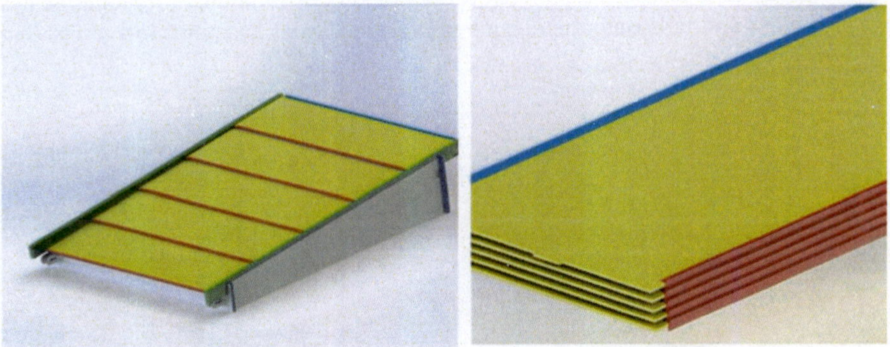

a) b)

Fig. 41.4 Retractable roof in expanded (**a**) and retracted (**b**) state

41.4.2 ArUco-Marker with Backlight

To implement the functional block «UAV landing» in low light conditions, a marker backlight was developed. Autonomous landing of multirotor is performed bearing on ArUco-marker. Marker detection rate and reliability heavily depend on the lighting conditions. There are several approaches to ensure sufficient lighting: mounting of a light source on the UAV, mounting an external light source on the base station, full array lighting and outline lighting. Light source mounting on UAV is not feasible without auxiliary optics, because of wide-angle LED light scattering. If the scattering angle is wider than 120°, lighting level is insufficient to detect UAV in flight at high altitude. However, light ray direction from auxiliary optics depends on UAV position. This may complicate marker detection by high wind velocity.

Mounting of an external light source on the platform can ensure necessary lighting level, though, this approach has a number of downsides: excessive image lighting, light obscuration by the landing UAV. LED array backlight assumes that the lighting diodes should be placed under the whole marker area. Exactly, such LED layout enables uniform lighting. The downside of such approach is a relatively thick marker. Conversely, with outline lighting approach, LEDs are positioned along the marker perimeter. In this case, light is distributed across the whole marker using scattering devices. The marker with such type of backlight is significantly thinner and consumes less energy compared to array-like backlight. Additionally, a dielectric substrate allows installing a wireless energy transfer device, such as [14, 15] onto the marker. Hence, when developing the landing site, naturally, outline backlight solution was chosen.

The developed backlight-aided marker includes a rectangular plexiglass sheet, LED, light-reflective substrate of polished aluminum, covered by white paint and an opaque polycarbonate scatterer (see Fig. 41.5).

The LED light falls onto plexiglass and, due to microlenses, partially reflects from the two medium interface and partially refracts. A certain amount of light permeates the plexiglass and falls onto the opaque scatterer to avoid excessive image lighting

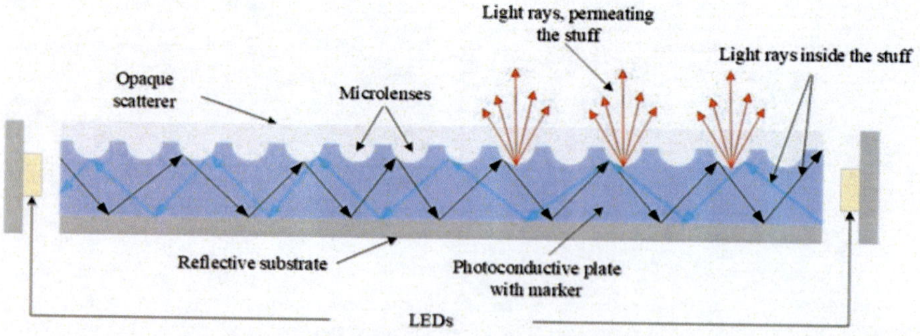

Fig. 41.5 Backlight-aided marker design

during marker detection. The opaque scatterer enables more uniform lighting. Marker image is made on the opaque scatterer using laser engraving.

41.5 Experiments and Results

To assess the efficiency of the developed backlit marker, a series of experiments was performed in the Gazebo virtual environment [16]. Onto a base station model a landing site was mounted, equipped with a backlit marker. A multirotor with a Logitech C920 camera on it was given real-world parameters. Further, it flew to the target and landed on the base station, bearing on the marker. Marker detection process includes obtaining a video stream from the cameras and consecutive processing of each video data frame in the loop. If no marker is present on the image and detection workflow is completed, the algorithm finishes. Otherwise marker outline is detected, and the marker is identified.

Marker detection was performed at altitudes from 4 to 0.25 m in natural light, without lighting (as natural, as backlight) and without natural light, but with backlit marker. No experiments in natural light with backlit marker were performed, as it seems impractical due to arising excessive marker lighting. Outcomes of virtual experiments with altitude value of 3 m are shown in Fig. 41.6.

Marker detection under natural light was successfully performed at any landing stages. In flight at altitudes from 4 to 0.5 m without lighting, the marker was not detected, whereas the backlight provided for successful marker detection.

Having reached the flight altitude of 0.5 m without lighting, the system could detect the marker (Fig. 41.7a), what was found to be impossible with active backlight (Fig. 41.7b) without additional adjustment of the camera.

Based on the designed computer model, a landing module prototype was assembled. This module was tested in laboratory conditions. Marker detection with active

a) b) c)

Fig. 41.6 Marker detection results by UAV flight altitude 3 m: **a** under natural lighting, **b** without lighting, **c** with backlight

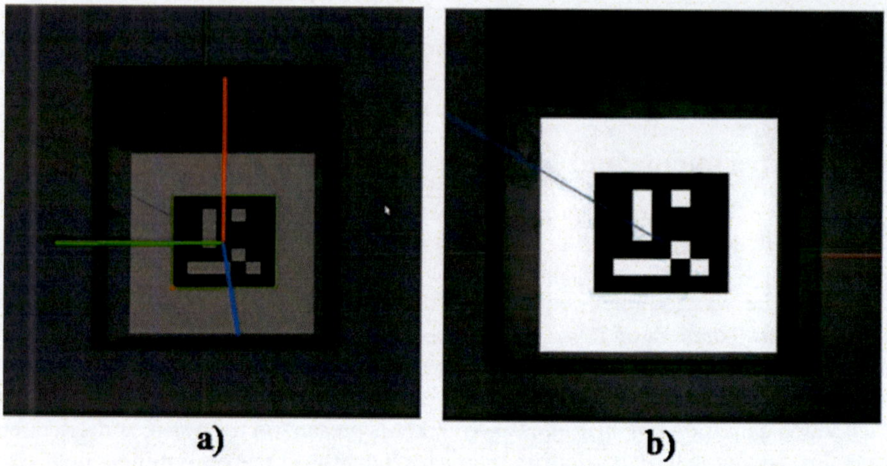

Fig. 41.7 Marker detection results by UAV flight altitude 0.5 m: **a** under natural lighting, **b** without lighting, **c** with backlight

backlight without any other light was made using Logitech C920 camera, the parameters of which were previously used in computer simulation. Virtual experiments showed that default settings of the camera fail to ensure successful marker detection on all stages of UAV descent. Consequently, it was decided to adjust the camera photosensitivity (in ISO units) in ISO range from 3 to 128. The experiments were performed at altitudes from 4 to 0.25 m.

Figure 41.8 shows the experimental results, obtained in real-world conditions at altitude of 3 m and ISO values of 128, 69, 3.

As evident from Fig. 41.8a, by camera photosensitivity value of ISO 128, the marker is excessively flooded by light, what precludes its detection. When the ISO

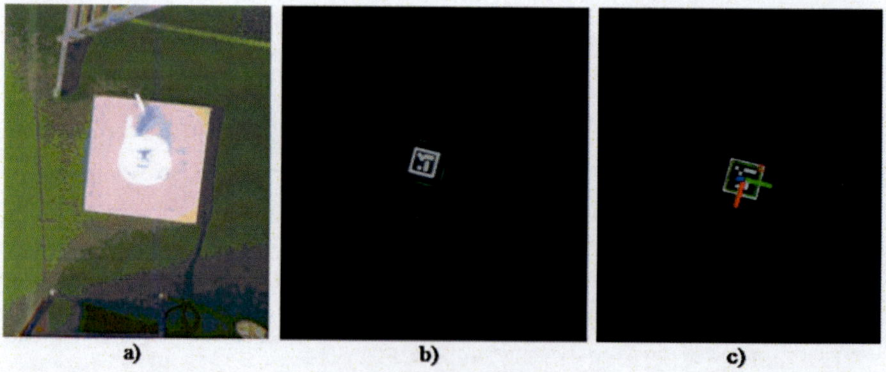

Fig. 41.8 Marker detection results by UAV flight at the altitude of 3 m with backlight: **a** default camera parameters, **b** with ISO 69, **c** with ISO 3

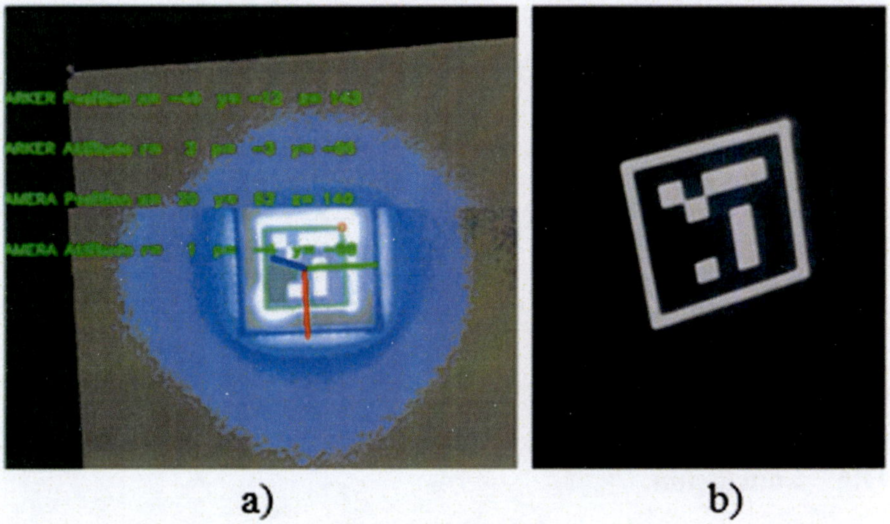

Fig. 41.9 Marker detection results by UAV flight altitude of 0.3 m with aid of backlight by the following photosensitivity values: **a** ISO 128, **b** ISO 3

value is decreased to 69, the light becomes not so excessive (Fig. 41.8b), but the marker still cannot be detected. Successful marker detection was achieved by minimal ISO value of 3.

However, when decreasing the flight altitude to 0.3 m, the optimum photosensitivity value ISO 3 (Fig. 41.9b) does not ensure successful marker detection, because it is too dark. In turn, excessive lighting, that previously arose by ISO 128, became less pronounced (Fig. 41.9a), and the marker was successfully detected. This data leads to conclusion that photosensitivity value should be increased for successful marker detection. It was empirically revealed that for successful marker detection at altitude of 0.3 m, ISO value should be increased to 59.

Therefore, for successful marker detection at any given altitude, automated adjustment of camera photosensitivity parameter has to be implemented programmatically. To obtain data, additional simulated and field experiments were performed, where altitude values were changed in the range from 4 to 0.25 m. The photosensitivity value was adjusted experimentally with step of 0.1 m. Figure 41.10 illustrates the dependencies of camera photosensitivity from UAV flight altitude, required for successful marker detection in real-world and simulation environment.

For marker detection in simulation environment at altitude of 4 meters, the camera photosensitivity value is set to ISO 128, and, upon UAV descent, at altitude of about 1 m the photosensitivity has to be increased to ISO 180 for correct marker detection, and then decreased to ISO 100 at altitude of about 0.3 m. Similar correlation is also found in real-world conditions. Photosensitivity curves differ in real-world and simulation environments because of Gazebo assumptions: natural light without overcast, perfect air transparency, lack of reflective surfaces.

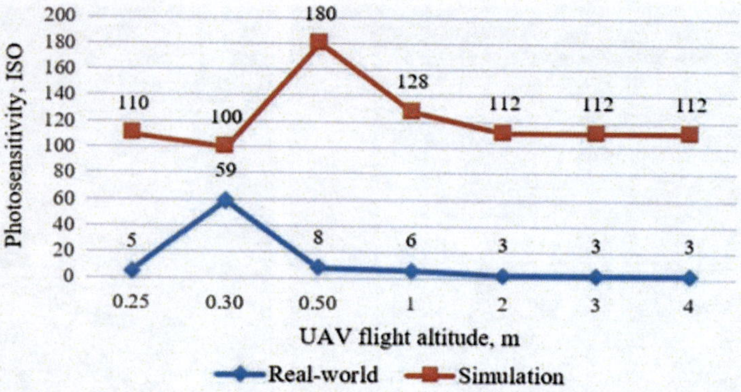

Fig. 41.10 Required dependencies of camera photosensitivity from UAV flight altitude

41.6 Conclusion

This paper presents the design of base station for automated UAV maintenance. Unlike analogs, the proposed station allows to perform a number of functions: UAV storage, UAV battery replacement, data transfer during flight tasks. Bidirectional data transfer allows to transmit data from the UAV to station during flight tasks, and from the UAV station about the approach and the UAV station about t of take-off or landing clearance. The developed station design allows to store large UAVs. To ensure an accurate UAV landing on the landing site, ArUco-marker with backlight was developed. The backlight includes an organic glass plate, LEDs, a reflective backing and a matte diffuser. The proposed marker with backlight was tested in real conditions and simulation. When flying a UAV at a height of 3 m, the backlight allows to detect a marker, unlike the case without lighting. With UAV flight altitude decreases, the marker with backlight may not be detected. In this case, it is necessary to adjust the photosensitivity of the camera. To determine the required value of photosensitivity, ISO the dependence of photosensitivity on UAV flight altitude was presented.

Acknowledgements This research is supported by the RFBR Project No. 20-08-01056 A.

References

1. Cicek, C.T., Gultekin, H., Tavli, B., Yanikomeroglu, H.: UAV base station location optimization for next generation wireless networks: Overview and future research directions. In: 2019 1st International Conference on Unmanned Vehicle Systems-Oman (UVS), pp. 1–6. IEEE (2019)
2. Scherer, J. et al.: An autonomous multi-UAV system for search and rescue. In: Proceedings of the First Workshop on Micro Aerial Vehicle Networks, Systems, and Applications for Civilian Use, pp. 33–38 (2015)

3. Popov, S.G., Zaborovsky, V.S., Kurochkin, L.M., Sharagin, M.P., Zhang, L.: Method of dynamic selection of satellite navigation system in the autonomous mode of positioning. SPIIRAS Proc. **18**(2), 302–325 (2019)
4. Rath, M., Shekarappa, N., Ramachandra, V.: U.S. Patent Application No. 11/403,472 (2007)
5. Hong, Y., Fang, J., Tao, Y.: Ground control station development for autonomous UAV. In: International Conference on Intelligent Robotics and Applications, pp. 36–44 (2008)
6. Kemper, F.P., Suzuki, K.A., Morrison, J.R.: UAV consumable replenishment: design concepts for automated service stations. J. Intell. Rob. Syst. **61**(1–4), 369–397 (2011)
7. Suzuki, K.A., Kemper Filho, P., Morrison, J.R.: Automatic battery replacement system for UAVs. Analysis and design. J. Intell. Robot. Syst. **65**(1–4), 563–586 (2012)
8. Liu, Z.N., Liu, X.Q., Yang, L.J., Leo, D., Zhao, H.W.: An autonomous dock and battery swapping system for multirotor UAV. Unpublished, https://www.researchgate.net/publication/325077351. Last accessed 2020/01/10
9. Ivannikov, K.V., Gavrilov, A.V., Boev, A.S., Shoshin, I.S.: A method of landing an unmanned aerial vehicle of a helicopter type using an infrared camera. Herald East Kazakhstan Concern Almaz-Antey **3** (2016)
10. Zhou, D. et al.: Autonomous landing of a helicopter UAV with a ground-based multisensory fusion system. In: Seventh International Conference on Machine Vision (ICMV 2014), International Society for Optics and Photonics 9445, pp. 94451R (2014)
11. Nightingale Security: https://www.nightingalesecurity.com. Last accessed 2020/01/10
12. Percepto: https://percepto.co/solutions/. Last accessed 2020/01/10
13. Pavliuk, N., Cherskikh, E., Pshchelko, N., Shabanova, A.: Circuit schematics of a capacitive proximity sensor. In: 2019 1st International Conference on Control Systems, Mathematical Modelling, Automation and Energy Efficiency (SUMMA), pp. 486–490. IEEE (2019)
14. Krestovnikov, K., Cherskikh, E., Pavliuk, N.: Concept of a synchronous rectifier for wireless power transfer system. In: IEEE EUROCON 2019-18th International Conference on Smart Technologies, pp. 1–5. IEEE (2019)
15. Krestovnikov, K., Saveliev, A., Shabanova, A., Vatamaniuk, I.: Comparative study of synchronous and non-synchronous rectifiers for use in the receiving part of a wireless charging system. In: Proceedings of 14th International Conference on Electromechanics and Robotics "Zavalishin's Readings", pp. 675–685 (2020)
16. Lavrenov, R.O., Magid, E.A., Matsuno, F., Svinin, M.M., Suthakorn, J.: Development and implementation of spline-based path planning algorithm in Ros/Gazebo environment. SPIIRAS Proc. **18**(1), 57–84 (2019)

Chapter 42
Combined Capacitive Pressure and Proximity Sensor for Using in Robotic Systems

Konstantin Krestovnikov, Ekaterina Cherskikh, and Eldar Zimuldinov

Abstract Circuit solution and primary transducer of combined capacitive pressure and proximity sensor are described. Principle of operation of the developed interface circuit is shown; its principal electrical circuitry is explained. Design and structure of a primary capacitive transducer is considered for further use within the combined sensor. The dependence of the output signal from the force, applied to the prototype sensor, is empirically obtained and shows linear pattern in the working range of the sensor from 0 to 2.6 kg, whereas the sensitivity to the applied pressure was 2252 value/kg in output values of ADC. Graphical dependences of output signal are obtained from the distance between the sensor and the things being observed and consisting of different types of stuff. Mean value of interface circuit sensitivity to approaching object is 6.7 value/mm on the decreasing span of the characteristic curve in the range between 0 and 10 mm. The presented sensor can be employed in manipulators to control object grip, as well in feet of humanoid robot for zero moment point calculation and detection of approach to bearing surface.

42.1 Introduction

Installation of pressure and proximity sensors into different joints and parts of robotic devices allow to provide control units with necessary data about working environment and facilitate interaction with it. The relevant task consists in embedding of pressure sensors into humanoid robot feet to compute zero moment point (ZMP) and implement resilient walk algorithms [1, 2]. Proximity sensor, being used in

K. Krestovnikov · E. Cherskikh (✉) · E. Zimuldinov
St. Petersburg Institute for Informatics and Automation of the Russian Academy of Sciences, 39, 14th Line, 199178 St. Petersburg, Russia
e-mail: katy0419@mail.ru

K. Krestovnikov
e-mail: open56it@gmail.com

E. Zimuldinov
e-mail: eld.zimuldinov@gmail.com

A. Ronzhin and V. Shishlakov (eds.), *Proceedings of 15th International Conference on Electromechanics and Robotics "Zavalishin's Readings"*, Smart Innovation, Systems and Technologies 187, https://doi.org/10.1007/978-981-15-5580-0_42

robot foot, allows the control system to track, when the foot touches the bearing surface and further stabilize robot walk on complex surfaces. Simultaneous usage of pressure and proximity sensors in humanoid robot pedipulators provides for more comprehensive control of its walk and equilibrium control. The practical possibilities of combined pressure and proximity sensors are not limited to embedding of such devices in limbs of walking robots. Manipulators grip things more reliably, when the grip force is controlled, and, having provided a proximity sensor in the manipulator gripper, it is possible optimally grip the object, track the touch instant, what is especially important for fragile things, and prevent such things to slip out. Combined structure of pressure and proximity sensors could simplify object location detection in isolated areas of cyber-physical space [3], poorly observed by other means and establish a virtual three-dimensional object representation.

Within the research, outlined above, a proximity sensor is developed, able to measure distance [4], this paper presents circuit solution of the sensor and established characteristic curves, showing dependence of output voltage from distance to observed objects consisting of various stuffs.

This paper is the follow-up of the said research toward composite pressure and proximity sensor for further installation in robotic systems. Further, we will review already existing solutions and describe circuit design of the developed composite sensor and its specifics.

42.2 Related Works

Composite capacitive array sensor is presented in [5]. This sensor is designed as a multi-layered structure, where a polyethylene-terephthalate sheet with parallel electrode strips attached to it is used as a substrate. The sensor consists of cells, arranged at overlapping of electrode strips (AgNWs); the area of each overlapping piece is 1 cm^2. Theoretically derived values and empirically obtained values conform with each other in the range 0–4 cm, with further increase of distance to object the deviation of measured distance values from calculated ones increases with distance. The sensor can switch between working modes, thereby ensuring accurate measurement of pressure and distance to object.

Hybrid approach in measurement of pressure and proximity values, employed in resistive-capacitive pressure sensor zPatch, is shown in [6]. This device is a metal overlay with a 4×4 sensor array; it is made of piezoresistive sheets, inserted between tent fabric layers, further sewn in dielectric fabric. The sensor is suitable for user gesture recognition (touch, tap) and could be used for music player control, for text input or as a game joystick. The approach employed here allows to decrease the number of false positives and improves sensor's capability in touch gesture recognition.

Currently, soft robotics becomes ever more popular, where soft materials are used for fabrication of robots or their parts. In [7], authors present development of two arrays, mounted on aluminum foil substrate, suitable for measurement of object-related pressure and proximity parameters. In these arrays, organic thin-film transistors are used. Cells from the sensitive surface of the pressure sensor are arranged as a 6×10 array, where the size of each cell is 27.5×27.5 mm. Maximum allowable force, that may be applied to the sensitive area of the sensor, is 100 kN.

Infrared sensor array, measuring distance to object, consists of cell and has overall dimension 10×10, where the size of each cell is 1×1 cm. Microlenses, attached over the infrared sensors, allow to gather light and increase sensitivity. This allows to decrease the sensor surface size.

In experiments, sensors with simplified array design were used, which do not include microlens array. The maximum distance, at which an object was successfully detected with such device, was 115 mm, where the sensitive area surface was 3.1 cm^2. Due to substrate flexibility, the prototype device can be mounted on curved surfaces.

Composite flexible proximity and pressure sensor find practical use on robot grips. Such sensor, attached to gripper tips, defines distance to object and the pressure, exerted by the object on manipulator gripper, what prevents slip-out of the object and provides for more accurate object gripping. Authors of [8] presented a proximity and pressure sensor with the following design: the proximity sensor is mounted in the 6-mm through hole in the pressure sensor, whereas the overall dimensions of the pressure sensor are 60×30 mm.

The pressure sensor is essentially a conductive rubber layer, inserted between two layers of conductive film. The processed output signal of the sensor allows to determine the center point, at which external pressure is applied to it and to quantify it.

Proximity sensor is essentially a phototransistor array, where each transistor has a corresponding IR-emitter. When an object approaches to the array, IR radiation is reflected from the object and falls on a phototransistor, changing its dynamic resistance and thereby altering sensor resistance.

During experiments, pressure was applied to sensor with 5 mm interval. It was revealed that sensor readings at pressure application point are displaced from the through hole, whereas pressure application point detection error does not exceed 2 mm.

Using on humanoid robot body an array structure, which consists of pressure and proximity sensors, it is possible to control complex manipulations with objects, robot walk stability and ensure processing of distributed touch data.

In the paper [7], considered above, square-shaped arrays of proximity and pressure sensors were used. Authors of [9] demonstrate a system, featuring triangle-shaped sensors. As humanoid robot body can include spherical and round surfaces, the shape, proposed by authors of this paper, provides the most complete coverage of humanoid robot body, opposed to square-shaped sensor array. Each sensor has 12 capacitive elements, whose signal is processed by programmable controller (AD7147), transforming capacitance changes into digital signal. Along each side of the triangular sensor, one switching port is situated, to enable communication with neighbor sensors. Therefore, a structure of several sensors can be assembled. As every sensor has a flexible substrate, such sensor assembly could be mounted on a curved surface.

The system is connected to a microcontroller, managing the rate of changes and sensor mode switching.

Pressure and proximity sensors are conventionally based on resistive and capacitive principle of operation, but inductive principle of operation also can be used to perform this task. In [10], a flexible double-mode sensor is shown, discovering proximity of an object and pressure, applied to it, thereby working in inductive and capacitive modes and based on carbon microcoils with substrate of soft dielectric elastomer. Sensing cell of such device consists of two layers: electrode and dielectric ones. Dielectric layer includes silicone elastomer and carbon coils. This layer increases sensor sensitivity, when determining distance to object, and protects electrodes from direct physical contacts. Electrode layer includes a flexible printed board with electrodes in one layer. The sensor can swap working modes depending on that, if pressure is applied to it, or an object approaches to sensor, but does not touch it.

The sensor is fabricated as a 10 × 10 array, and it can measure pressure up to 330 kPa, as well determine distance to metal object in the range from 150 mm to direct contact with the object.

Proximity, pressure and object surface deformation liquid-metal sensor, based on eutectic gallium-indium alloy (EGaIn), is presented in [11]. This solution has been implemented with a custom architecture, consisting of nine capacitive sensors, layered in two tiers, and a helical pressure sensor. Due to capacitive elements, the sensor discerns approaching things and contact pressure, comparing changes of capacitance and resistance. The capacity is measured using CapSense controller with one receiving electrode and one transmitting electrode. When a conductive object approaches the electrodes, a capacitance arises between the electrodes and the conductive objects. When the conductive object touches the sensor panel, mutual line-column capacity decreases, what indicates presence of a conductive object.

The sensor enables distance measurement in the range from 0 to 8.7 cm, detect moving object in the horizontal plane. Maximum allowable contact pressure, exerted on the sensor—110 kPa.

One of the domains where combinations of proximity and pressure sensors are relevant is development of stuffs for electronic skin. Development of sensor devices, akin to human skin enables robot interaction with environment. Issues, related to design complexities, poor extensibility and sensitivity, exclude usage of conventional technologies in fabrication of artificial skin, sensitive to pressure and proximity. In [12], authors presented Tribo-skins stuff. Such stuff can react on approaching object, contact with the object and on external pressure. The operational principle of Tribo-skins is based on triboelectrical effect, consisting in buildup of electrical charge in the stuff after friction contact with other stuff. The sensor is fabricated of natural rubber, a triboelectrically negative stuff (it accumulates negative charge after frictional contact with other stuff). When some object comes into contact with Tribo-skins, electrons drift from object surface to silicone surface due to electronegativity difference. When the object moves away, two oppositely charged surfaces come apart and create potential difference. Unshielded negative charges on silicone surface drive free electrons and flow from silver layer into ground, thereby inducing buildup of positive charge and, consequently, generating output current and EMF. When the object is at greater distance, electrostatic equilibrium is achieved, and electrons stop moving. When the object approaches again, distance between Tribo-skins and the object and their potential difference alike decrease. This process generates reverse flux of electrons from ground to the embedded electrode and generates oppositely directed voltage. When the stuff is under pressure, output voltage of the sensor increases together with contact pressure and rebuilds up to the baseline, when the pressure ends. The device can react as to dielectric, as to conductive objects.

Contrary to the example, detailed above. Electronic skin, presented in [13], consists of fibers based on silver nanothreads with helical microstructure, used as main electrodes to confer the electronic skin decent extensibility and deformation sensing capability. The demonstrated prototype can switch between two modes: capacitive one for pressure measurement and resistive one for measurement of surface deformation of the object. The maximum measurable contact pressure amounts to 50 kPa, whereas the minimum measurable limit is 1.5 Pa. Distance detection range is from 0 to 20 cm. Characteristics of previously detached fiber electrodes enabled electronic skin to detect lateral pressure and reveal deformation direction.

The sensors, examined here, are prototypes of composite sensing devices, intended for measurement of pressure force and distance to object. The most solutions, considered here, are based on tensoresistive and capacitive principles. One of the presented sensors employ inductive working principle and can measure pressure up to 330 kPa, as well define distance to metal object in the range from 0 to 150 mm. Flexibility of some considered prototypes enables to mount them on uneven surface. The downside of the considered prototypes is their structural complexity and usage of exotic stuffs in fabrication of such devices.

42.3 Electrical Circuit and Principle of Operation

To implement readout from the composite primary capacitor transducer (PCT) to measure pressure and distance to object, an interface circuit has been developed, shown in Fig. 42.1. High-frequency (HF) pulse generator obtains power from the stabilized power supply with output voltage of 5 V. Rectangular generator of HF pulses is built on a TLC555IDR chip, frequency-setting circuits R1C1 and R2C2 set the working frequency $F = 280$ kHz. PCT $C(D)$ is connected to the generator chip output.

When altering distance from PCT to object and exerting external pressure to PCT, its capacity and, hence, reactance, change. Voltage, supplied to non-inverting input of the operational amplifier, depends on the reactance level of the PCT and of $R3$ resistor value. Resistance of $R3$ influences capacity discharge rate $C(D)$ and is calculated according to formula (42.1). The discharge rate time constant τ should be taken as a quarter of oscillation period of HF pulse generator:

$$R3 = \frac{\tau}{4C_{\max}}. \tag{42.1}$$

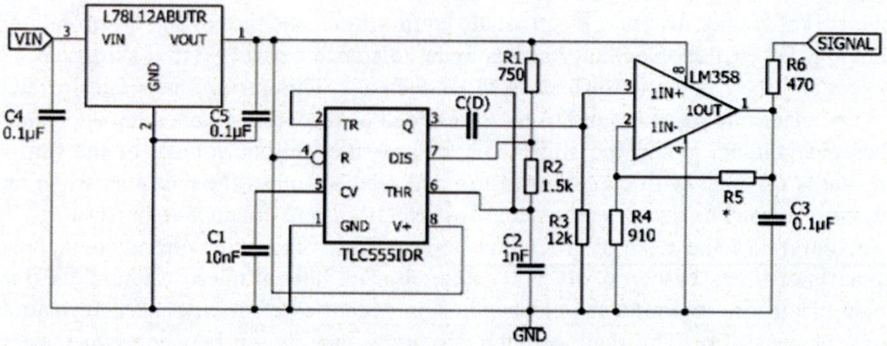

Fig. 42.1 Layout diagram of the composite pressure and proximity sensor

Maximum PCT capacitance value corresponds to the minimal distance between its electrodes and is calculated according to formula (42.2):

$$C_{\max} = \frac{\varepsilon\varepsilon_0 S}{d_{\min}}, \tag{42.2}$$

where ε is dielectric permeability of PCT gap; S is area of electrodes; $d_{\min}$ is minimum distance between the electrodes, corresponding to the maximal pressure on PCT.

Voltage U_{in} from the divider, established by the reactance $C(D)$ and $R3$, is supplied to non-inverting input of operational amplifier (OA), acting as non-inverting amplifier. Amplification factor OA depends on resistance values of voltage divider $R4$, $R5$ and is calculated according to formula (42.3):

$$K = 1 + \frac{R_5}{R_4}. \tag{42.3}$$

OA output signal value is calculated according to formula (42.4):

$$U_{\text{out}} = K U_{\text{in}}. \tag{42.4}$$

Output signal of operational amplifier is effectively the sensor output signal and is the value of constant voltage, whose oscillations are flattened using $C3$ capacitor.

Operational amplifier usage in the developed interface circuit simplifies output signal range tuning. The tuning consists in fitting of $R5$ resistance value such a way, that, by maximum pressure on the PCT, OA saturation would not occur. This feature allows execute tuning on a single cell, when transiting to a sensor array of this type, where each cell has minor deviations in capacitance parameter and then use the $R5$ nominal value as the reference for all other cells. This circuit solution fits for PCT usage in an array better, compared to other solutions, previously developed by us, where transistors were used for signal amplification.

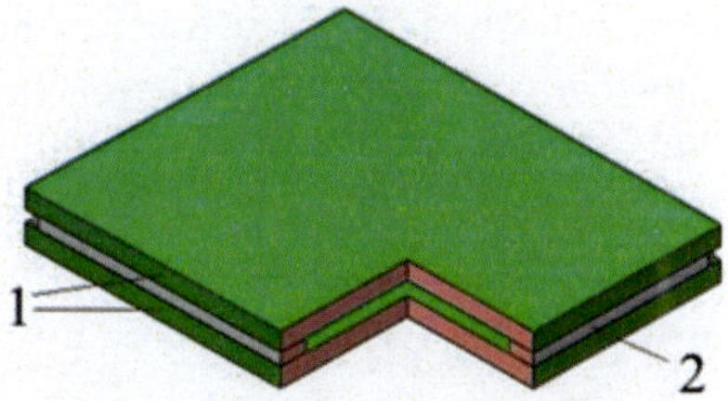

Fig. 42.2 PCT structural diagram

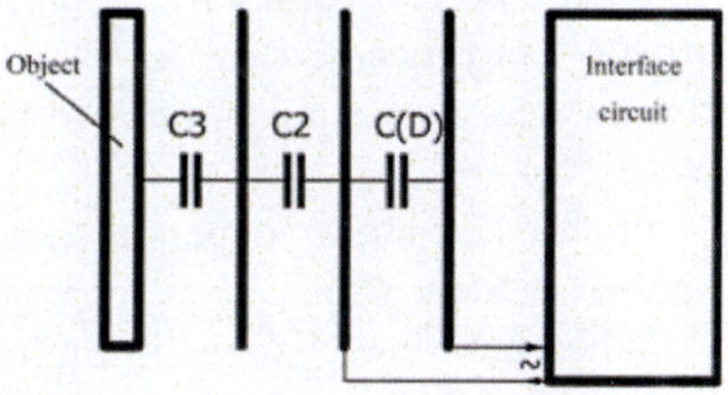

Fig. 42.3 Equivalent PCT circuit

Linear change of distance between PCT electrodes is caused by elastic deformation of the stuff, laid down the perimeter of the electrodes; here, this is assumed density silicone, whose solidity alters depending on requirements to working range of the sensor. The PCT structural diagram is depicted in Fig. 42.2.

PCT is fabricated as a three-layer structure with outer dimensions 30 × 25 × 3.8 mm. The top and bottom layers are made of metal-clad glass textolite (42.1), where the top textolite layer is covered by copper foil on both sides for better PCT sensitivity to distance changes. Internal layers of copper foil act as electrodes $C(D)$ of a capacitive PCT. The middle layer (42.2) is a 2 mm wide and 0.6 mm thick silicone strip, running down the perimeter of the electrodes. Equivalent PCT circuit is presented in Fig. 42.3.

$C(D)$ is a variable capacity between the metal-clad glass textolite plates, $C2$—capacity between top and bottom copper foil layers of the top plate. $C2$ can be calculated according to formula (42.2), where dielectric permeability of metal-clad glass textolite is taken as $\varepsilon = 5.5$ (GOST 12652-74). By this value ε and metal-clad glass textolite thickness $d = 1.6$ mm, $C2 \approx 22.83$ pF. Capacitive reactance is calculated according to formula (42.5):

$$\text{Xc} = \frac{1}{2\pi FC}. \tag{42.5}$$

By generator working frequency $F = 280$ kHz, capacitive reactance $Xc \approx 24.9$ kΩ. When an external object approaches to sensor, capacitive coupling arises, denoted as $C3$ on the diagram. This value shows hyperbolical dependence from the distance between object and PCT. The charges, accumulated on $C(D)$ electrodes, start flowing onto the external object, thereby changing the output signal of the sensor. At the moment, when the object touches the PCT, $C3$ capacity approaches to infinity; consequently, capacitance X_{c3} approaches to null. If pressure is further exerted on PCT, the output signal of the sensor will depend on $C(D)$ change.

42.4 Experiments

To ensure high accuracy in the performed experiments, analog output signal of the sensor was processed by ADC with high effective bit number, and experimental results were presented in relative units, without converting the values into voltage level ones.

Figure 42.4 shows dependence of the output signal of the system from pressure, applied to the sensor.

When pressure changes in the range from 0 to 2.6 kg, the output signal shows linear pattern. This is because, that to obtain pressure data from PCT, the changing parameter under consideration (capacity) is measured not directly, but indirectly; the changing capacity in question arises between electrodes of PCT. Having plugged (42.2) into (42.5), we get distance between electrodes in the numerator (42.6):

$$\mathrm{Xc} = \frac{d}{2\pi F \varepsilon \varepsilon_0 S}. \tag{42.6}$$

Further decrease of output signal growth by pressure increase is caused by operational amplifier saturation and reaching of operational limits of the sensor by pressure parameter.

Figure 42.5 shows experimentally obtained dependence curves of system output signal from approaching objects, fabricated from different stuffs.

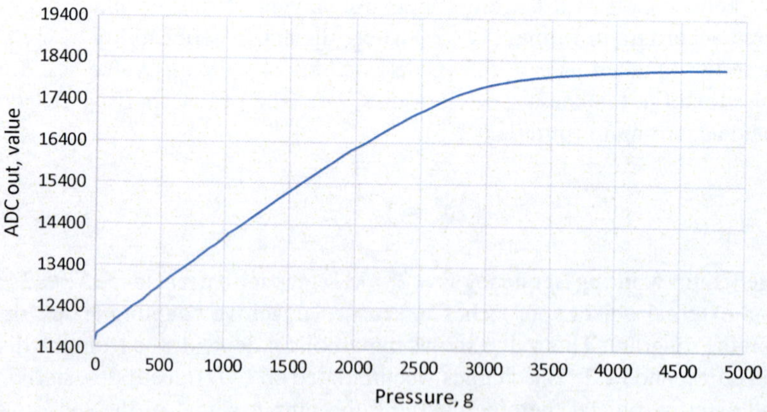

Fig. 42.4 Dependence of the output signal of the system from pressure

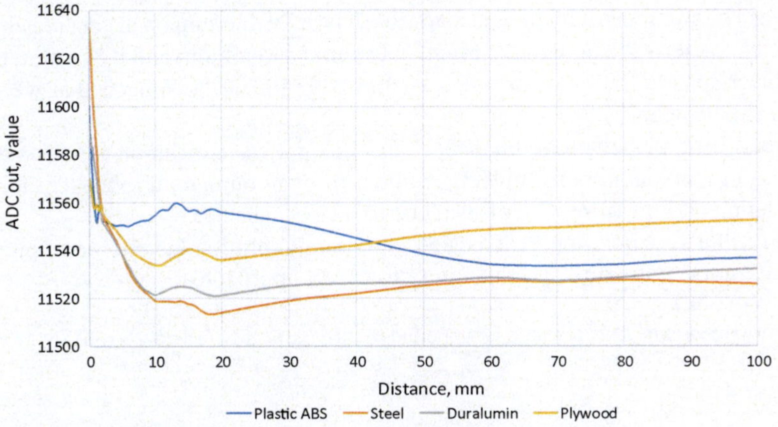

Fig. 42.5 Dependence of the sensor signal from the distances to objects

Using ADC with high effective number of bits allowed to improve accuracy of distance measurement between PCT and object, but also caused increased sensor susceptibility to environmental noise. To flatten out the noise, recursive moving average filter was applied to ADC output data (42.7):

$$y_n = \frac{1}{N}\sum_{K=0}^{N-1} x_{n-k}, \tag{42.7}$$

where y_n is arithmetic mean between respective ADC output value—x_n and $N-1$ previous output values of.

The filter has been applied to $N = 4$. The resulting curves have similar shapes; consequently, the maximum sensitivity of the sensor to object approach lies in the range between 0 and 20 mm and further smooth decrease of sensitivity with distance increase.

42.5 Conclusion

A distinctive feature of the interface circuit solution and PCT design presented here is the possibility of simultaneous measurement of pressure and distance to object. Operational amplifier usage in the interface circuit allows to employ it with PCT arrays, without need for individual tuning for every PCT cell, what represents an advantage in contrast to similar transistor-based interface circuits.

The prototype of the developed interface circuit solution with PCT design described above has sensitivity to applied pressure of 2252 value/kg on the linear span of the characteristic curve in the range between 0 and 2.6 kg. Mean value of interface circuit sensitivity to object approach is 6.7 value/mm on the decreasing span of the characteristic curve in the range between 0 and 10 mm and 0.37 value/mm on the linear span in the range between 10 and 100 mm for the objects that were used in the experiments.

The sensor, presented in this paper, may be applied in different robotic systems, such as in manipulators for object grip control or in humanoid robot feet for ZMP calculation and measurement of distance to bearing surface.

In further research, the PCT design and circuit solution have to be improved to increase noise immunity and sensitivity to object approach.

Acknowledgements This research is supported by RSF project No 16-19-00044P.

References

1. Kovalev, A., Pavliuk, N., Krestovnikov, K., Saveliev, A.: Generation of walking patterns for biped robots based on dynamics of 3D linear inverted pendulum. In: International Conference on Interactive Collaborative Robotics, pp. 170–181 (2019)
2. Gorobtsov, A.S., et al.: Features of solving the inverse dynamic method equations for the synthesis of stable walking robots controlled motion. SPIIRAS Proc. **18**, 85–122 (2019)
3. Soldatenko, S.A., Yusupov, R.M., Colman, R.: Cybernetic approach to problem of interaction between nature and human sosiety in context of unprecedented climate change. SPIIRAS Proc. **19**, 5–42 (2020)
4. Pavliuk, N., Cherskikh, E., Pshchelko, N., Shabanova, A.: Circuit schematics of a capacitive proximity sensor. In: 2019 1st International Conference on Control Systems, Mathematical Modelling, Automation and Energy Efficiency (SUMMA), pp. 486–490 (2019)
5. Zhang, B., et al.: Dual functional transparent film for proximity and pressure sensing. Nano Res. **7**(10), 1488–1496 (2014)
6. Strohmeier, P., Knibbe, J., Boring, S., Hornbæk, K.: zPatch: Hybrid resistive/capacitive etextile input. In: Proceedings of the Twelfth International Conference on Tangible, Embedded, and Embodied Interaction, pp. 188–198 (2018)
7. Fattori, M. et al.: Flexible pressure and proximity sensor surfaces manufactured with organic materials. In: 2017 7th IEEE International Workshop on Advances in Sensors and Interfaces (IWASI), pp. 53–58 (2017)
8. Hasegawa, H. et al.: Development of intelligent robot hand using proximity, contact and slip sensing. In: 2010 IEEE International Conference on Robotics and Automation, pp. 777–784 (2010)
9. Cannata, G., Maggiali, M., Metta, G., Sandini, G.: An embedded artificial skin for humanoid robots. In: 2008 IEEE International Conference on Multisensor Fusion and Integration for Intelligent Systems, pp. 434–438 (2008)
10. Nguyen, T.D., et al.: Highly sensitive flexible proximity tactile array sensor by using carbon micro coils. Sens. Actuators A **266**, 166–177 (2017)
11. Rocha, R. et al.: Soft-matter sensor for proximity, tactile and pressure detection. In: 2017 IEEE/RSJ International Conference on Intelligent Robots and Systems (IROS), pp. 3734–3738 (2017)

12. Lai, Y.C., et al.: Actively perceiving and responsive soft robots enabled by self-powered, highly extensible, and highly sensitive triboelectric proximity-and pressure-sensing skins. Adv. Mater. **30**(28), 1801114 (2018)
13. Cheng, Y., Wang, R., Zhai, H., Sun, J.: Stretchable electronic skin based on silver nanowire composite fiber electrodes for sensing pressure, proximity, and multidirectional strain. Nanoscale **9**(11), 3834–3842 (2017)

Chapter 43
Piezoelectric Micropumps for Microrobotics: Operating Modes Simulating and Analysis of the Main Parameters of the Fluid Flow Generation

Ildar Nasibullayev, Oleg Darintsev, Elvira Nasibullaeva, and Dinar Bogdanov

Abstract The operation of a piezoelectric micropump, consisting of a resilient elastic microtube of circular cross-section and a system of ring piezoelectric actuators placed on it, which forms a flow of the working fluid by radial compression, is simulated. The flow created by the oscillations of the piezoelectric actuators according to certain schemes is numerically studied: with a symmetric scheme; the average time flow rate of the liquid is zero; with an asymmetric scheme, it is possible to generate a nonzero time-average flow in the positive direction. It was found that the operation mode of the device according to an asymmetric scheme will allow using it as a piezoelectric micropump having a small size and allowing to pump small volumes of liquid. It is also shown that in an asymmetric mode of operation, with an increase in the frequency of oscillations of the piezoelectric actuators and with an increase in their quantity, the average pumped volume of liquid increases, and with a decrease in the radius of the channel, it decreases. An approximate analytical formula is obtained that relates the main parameters of the operating mode (time to reach the operating mode and average fluid flow rate) with the system parameters (the number and frequency of oscillations of the piezoelectric actuators, the radius of the microchannel), which makes it possible to construct an element of the computational module for evaluating the operating modes of the device in question, as well as manage the device itself in real time.

I. Nasibullayev (✉) · O. Darintsev · E. Nasibullaeva · D. Bogdanov
Mavlyutov Institute of Mechanics, Ufa Centre of the Russian Academy of Sciences, 71, ave. Oktyabrya, 450054 Ufa, Russia
e-mail: sp.ishn@gmail.com

O. Darintsev
e-mail: ovd@uimech.org

E. Nasibullaeva
e-mail: elvira@anrb.ru

D. Bogdanov
e-mail: dr_bog@mail.ru

A. Ronzhin and V. Shishlakov (eds.), *Proceedings of 15th International Conference on Electromechanics and Robotics "Zavalishin's Readings"*, Smart Innovation, Systems and Technologies 187, https://doi.org/10.1007/978-981-15-5580-0_43

43.1 Introduction

The development of microfluidics has been actively continuing for the past three decades (see, for example, review [1]). Interest in this area is associated with the features of fluid flow at the micro- and nanoscale, which allows the development of new devices that have new properties or even have no analogues in traditional macrosystems. With a decrease in the scale of the system under consideration, the Reynolds and Peclet numbers, which characterize the dynamic properties of the system, become small, and the influence of surface tension becomes significant. A small Reynolds number implies the laminar character of the flow, which corresponds to both most technical microdevices and biological systems [2]. And a small Peclet number means that during heat transfer, the contribution of molecular thermal conductivity prevails over convective heat transfer; therefore, in microdevices (micro heat exchanger, cooling systems, micro-gripper) both heat transfer mechanisms must be taken into account [3]. As a result of strong surface tension, capillary forces become comparable to gravitational forces, which allows the creation of unique microdevices, such as ink printing and 3d printing systems [4], as well as drug dosing systems [5].

The last decade is characterized by intensive growth in the production of microelectromechanical systems, the widespread introduction of microrobots and microtechnological complexes, which is a significant incentive for the development of promising new devices that provide gentle handling of microcomponents. The authors of this work develop capillary micro-grips [6] for working with flat micro-objects, membranes and other components with a sufficient area of the flat side, the capture of which by finger grips is difficult. Creating a prototype of this micro-grip, built on the basis of a miniature version of the thermoelectric module (Peltier element), involves the use of liquid cooling to remove heat from the hot surface of the Peltier element. The removal of thermal power from the Peltier element requires developers to design miniature pumps, and the provision of micro-grip operating modes requires the development of an appropriate mathematical model that takes into account the specifics of microfluidics.

Existing micropumps cover a wide range of applications [7]: biological systems and medicine (drug discovery systems), cooling systems (pumping coolant through micro heat exchanger), space exploration (high vacuum pumps), high-speed liquid transfer (high Reynolds number centrifugal pumps), pumping special fluids (electrohydrodynamic pumps, electroosmotic pumps and magnetohydrodynamic pumps). The disadvantage of most existing micropumps is the limited range of applications. Existing devices are difficult to use for a micro-grip cooling system, since a micropump with low dimension, low flow rate and flow rate control system is required.

Theoretical models developed for existing micropumps are based on empirical data, which does not allow for a complete geometric optimization of the design and determine optimal operating modes. On the other hand, the generation of fluid flow in an elastic microchannel belongs to the field of multiphysics, namely fluid-structure interaction (FSI) [8]. All this requires when developing new technical devices for determining and optimizing their operating modes the construction of new mathematical models linking hydrodynamics, elasticity theory, heat transfer process, parametric optimization and device control theory.

Computer simulation of fluid flow in microchannels, induced by external influences (such as pressure drop, translational movement of the boundary or deformation of the channel), has been actively developing recently. For example, the authors of this work have developed computer models that can be used as an element of a computing module for two separate elements of a technical device (piston and hydrodynamic resistance). The piston model for the oscillating pressure drop was constructed in the work [9] by the methods of a second-order factorial computational experiment taking into account five parameters (amplitude and frequency of the pressure drop, piston friction, spring stiffness, ratio of rest friction force and sliding friction).

With the development of mathematical and computer models, numerical simulation software using the finite element method (FEM) has appeared, which allows modeling devices with almost arbitrary geometry [10]. In the work [11], the FSI axisymmetric computer model was constructed for the interaction of Newtonian fluid with hyperelastic incompressible solid and a numerical study was performed for the stability of this model in the framework of FreeFem++ software (FEM differential equations solver) [12].

In the work [13], a technique for constructing an element of a computational module in the form of an analytical formula based on three-dimensional FEM modeling of fluid flow through an elastic microchannel deformed by a piezoelectric actuator according to a harmonic law is presented, and also options are proposed for using the results in practical applications. A three-dimensional computer model for controlling the fluid flow induced by the pressure drop applied along the channel using the pressure drop and the compression frequency of the tube with a piezoelectric actuator was constructed. This model is the first part of a liquid cooling system of the micro-grip.

In the present work, the numerical study of the effect of the oscillations of piezoelectric actuators on the fluid flow in the microchannel as applied to the development of a new technical device, the piezoelectric micropump, is continued and also its operation modes are analyzed. The advantages of the proposed micropump model are as follows:

– A simple design with the ability to expand its functionality. The micropump consists of a tube and several circular piezoelectric actuators, which allows one to use the required number of piezoelectric actuators, to install a tube with microvalves, etc.

- Low dimension. Length of 10 mm and a radius of 5 mm with five piezoelectric actuators.
- Low flow rate. The volume is enough to supply coolant to the micro-grip chamber.
- Average flow control by number of piezoelectric actuators and frequency.

43.2 Problem Statement and Basic Equations

The axisymmetric fluid flow through an elastic microchannel with internal R_1 and external R_2 radii and length L, the central part of which is located inside the piezoelectric system, is considered. Figure 43.1 shows the geometry of the problem with this notation. In axisymmetric geometry (see Fig. 43.1), the Oz axis is the axis of symmetry of the channel, and the origin is located in the center of the Oz axis. The Or axis defines the radial coordinate.

The fluid flow is described by the Navier–Stokes equations and the continuity equation in axisymmetric geometry [14]:

$$\rho\left(\frac{\partial u_r}{\partial t}+u_r\frac{\partial u_r}{\partial r}+u_z\frac{\partial u_r}{\partial z}\right)=-\frac{\partial p}{\partial r}+\mu\left[\frac{1}{r}\frac{\partial}{\partial r}\left(r\frac{\partial u_r}{\partial r}\right)+\frac{\partial^2 u_r}{\partial z^2}-\frac{u_r}{r^2}\right], \quad (43.1)$$

$$\rho\left(\frac{\partial u_z}{\partial t}+u_r\frac{\partial u_z}{\partial r}+u_z\frac{\partial u_z}{\partial z}\right)=-\frac{\partial p}{\partial z}+\mu\left[\frac{1}{r}\frac{\partial}{\partial r}\left(r\frac{\partial u_z}{\partial r}\right)+\frac{\partial^2 u_z}{\partial z^2}\right], \quad (43.2)$$

$$\frac{1}{r}\frac{\partial}{\partial r}(ru_r)+\frac{\partial u_z}{\partial z}=0, \quad (43.3)$$

where ρ is the density of the medium; $\mathbf{u} = (u_r, u_z)$ is the fluid velocity vector consisting of the radial and axial components; t is the time; p is the pressure; μ is the dynamic viscosity of the fluid.

The deformation of an elastic axisymmetric microchannel is described by the Lame equations [15]:

$$2(1-\nu)\left(\frac{\partial^2 s_r}{\partial r^2}+\frac{1}{r}\frac{\partial s_r}{\partial r}-\frac{s_r}{r^2}\right)+\frac{\partial^2 s_z}{\partial r\partial z}+(1-2\nu)\frac{\partial^2 s_r}{\partial z^2}=0, \quad (43.4)$$

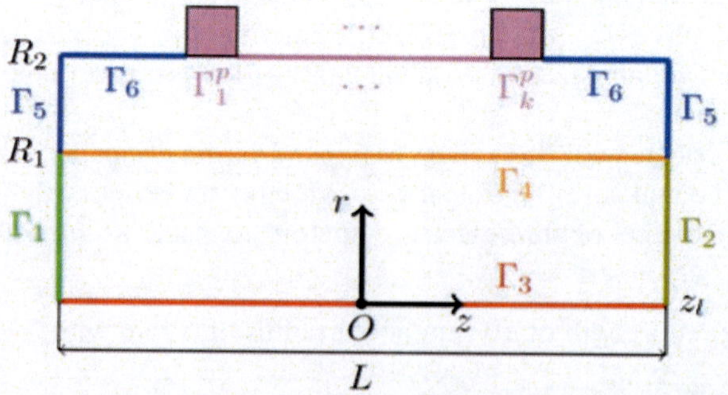

Fig. 43.1 The geometry of the microchannel with a system of piezoelectric actuators: Γ_1 and Γ_2 are the inlet and outlet of the channel; Γ_3 is the axis of symmetry; Γ_4 and Γ_5 are the inner and outer walls of the channel; Γ_k^P is the contact area of the k-th piezoelectric actuator with the outer wall

$$2(1-\nu)\frac{\partial^2 s_z}{\partial z^2}+\left(\frac{\partial^2 s_r}{\partial r \partial z}+\frac{1}{r}\frac{\partial s_r}{\partial z}\right)+(1-2\nu)\left(\frac{\partial^2 s_z}{\partial r^2}+\frac{1}{r}\frac{\partial s_z}{\partial r}\right)=0, \quad (43.5)$$

where $\mathbf{s} = (s_r, s_z)$ is the deformation vector consisting of the radial and axial components; ν is the Poisson's ratio.

Equations (43.1)–(43.5) were solved numerically by the finite element method in the FreeFem++ numerical simulation software [12]. Time discretization was performed by the Euler method with linearization of the nonlinear convective term according to the Picard scheme $(\mathbf{u}^{n+1} \cdot \nabla)\mathbf{u}^{n+1} \rightarrow (u^n \cdot \nabla)\mathbf{u}^{n+1}$.

The initial conditions correspond to the liquid at rest and the absence of deformations on the outer wall $\mathbf{u} = 0$ and $\mathbf{s} = 0$.

The boundary conditions for the fluid velocity $\mathbf{u}$ and deformations $\mathbf{s}$ have the following form:

$$\Gamma_1 : u_r = 0, p = \Delta p/2; \Gamma_2 : u_r = 0, p = \Delta p/2; \Gamma_3 : u_r = 0; \Gamma_4 : u_r = 0, u_z = 0;$$
$$\Gamma_5 : s_r = 0, s_z = 0; \Gamma_k^p : 2(1-\nu)\frac{\partial s_r}{\partial r}+\nu\left(\frac{s_r}{r}+\frac{\partial s_z}{\partial z}\right)=-p_k^p/E,$$

where p_k^p is the function that determines the radial pressure on the external wall of the microchannel by the k-th piezoelectric actuator; E is Young's modulus. The dependence of the pressure of the piezoelectric actuator on the outer wall of the microchannel is determined in the form $p^p(t) = \max(p^p) f_p(t)$, where the waveform is a function normalized in the region [0, 1]. The harmonic function is chosen as the main function (f is the compression frequency of the piezoelectric actuator):

$$f_p = \frac{1}{2}(1-\cos(2\pi f t)). \quad (43.6)$$

43.3 Results

In the model under study, an elastic tube with a length of $L = 10$ mm and a circular cross-section with an internal $R_1 = 1$ mm and an external $R = 1.75$ mm radius (see Fig. 43.1) is considered. Symmetrically with respect to the ends of the tube, N_p ring piezoelectric actuators ($N_\mathrm{p} = 1, 3, 5$) with a width of $L_\mathrm{p} = 1$ mm are worn on it. The tube material is silicone rubber with Young's modulus $E = 5$ MPa and Poisson's ratio $\nu = 0.49$. Water is used as a working fluid at a temperature $T_0 = 20$ °C with a bulk density $\rho = 998$ kg/m^2 and a dynamic viscosity $\mu = 1.002$ mPas. Several methods of creating a flow in a channel are investigated: using a pressure drop applied along the layer $G = \Delta p/L$ and/or using periodic compressions N_p of piezoelectric actuators.

Simulation takes place on two axisymmetric matched computational grids in the Ozr plane. First, the elastic equations are solved on the grid defining the tube wall. Based on the obtained deformation field, a computational grid for the channel is constructed (the deformed inner side of the tube determines the local radius of the channel). On the computational grid for the channel, the equations of hydrodynamics were solved.

For one piezoelectric actuator, the boundary conditions in the area of contact between the piezoelectric actuator and the tube can be specified in the form of Dirichlet (maximum compression of the tube) or Neumann (maximum pressure on the tube). If more than one piezoelectric actuator is used, then it is difficult to set the boundary conditions in the form of Dirichlet. It was shown in [16] that the Dirichlet boundary conditions can be replaced by the Neumann boundary conditions using the following relation $p_p = A \cdot \Delta d$, where Δd is the maximum compression of the tube with a piezoelectric actuator; A is the dimensional coefficient determined during the simulation (for the selected tube material, geometry, and $\Delta d = 2.8\ \mu\text{m}$, we have $A = 71.4$ GPa/m). The local deformation of the internal ΔR_1 and external ΔR_2 radii is related by the linear relation $\Delta R_1 = 0.8 \cdot \Delta R_2$. Therefore, the maximum pressure exerted by the piezoelectric actuator on the tube is $p_\text{p} = 0.2$ MPa. From the side of the piezoelectric actuator, the force $F_\text{p} = p_\text{p} \cdot S_\text{p}$ acts on the tube, where the contact area is $S_\text{p} = 2\pi R_2 L_\text{p}$. Thus, during operation of the piezoelectric actuator, the force F_p does not exceed 2.2 N, which lies within the operational limits of the annular piezoelectric actuator (~10 N).

Figure 43.2 shows the dependences of the fluid flow rate Q and the total fluid volume V at the outlet of the microchannel versus time t for one piezoelectric actuator with a constant pressure gradient G. If the piezoelectric actuator does not compress, then a Poiseuille fluid flow in a cylindrical pipe with a constant flow rate $Q(G)$ is obtained. When the piezoelectric actuator is compressed with a frequency f, the fluid flow begins to oscillate with the same frequency around the value of $Q(G)$. With increasing frequency f, the amplitude of the fluid flow increases, which leads to the

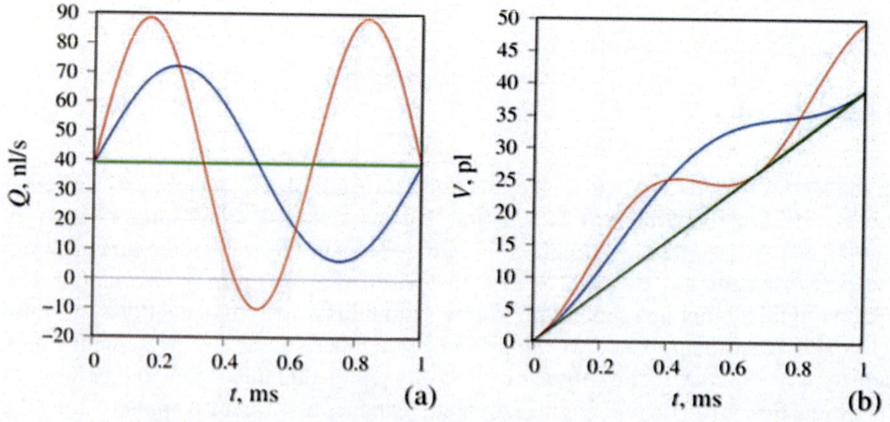

Fig. 43.2 The fluid flow rate (**a**) and the total fluid volume (**b**) at the outlet of the channel versus time. For $N_\text{p} = 1$: $f = 1$ kHz (red line) and $f = 1.5$ kHz (blue line); $N_\text{p} = 0$ (green line for numerical results, black line for analytical Poiseuille formulae [14]); $G = 100$ Pa/m

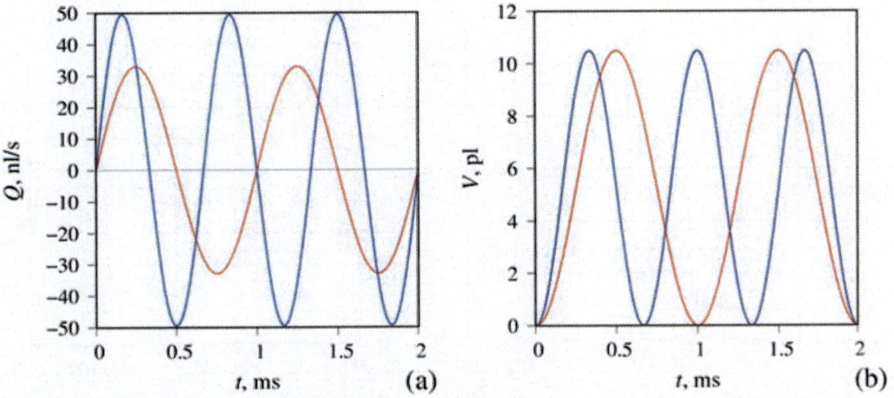

Fig. 43.3 The fluid flow rate (**a**) and the total fluid volume (**b**) at the channel outlet versus time for $f = 1$ kHz (red line) and $f = 1.5$ kHz (blue line). $N_p = 1$; $G = 0$ Pa/m

fact that part of the period; the flow rate can be negative (fluid flows in the opposite direction). This case was considered in more detail in [13].

Consider the options for creating the flow of micropumps due to the oscillations of the piezoelectric actuators in the absence of a pressure gradient.

Figure 43.3 shows the dependences of the fluid flow Q and the total fluid volume V at the inlet and outlet of the microchannel versus time t for a single piezoelectric actuator, compressed according to the law (43.6) with different frequencies f. It is seen that with increasing frequency, the amplitude of the fluid flow increases, but the amplitude of the volume change does not change since the entire displaced volume is equal to the change in the volume of the microchannel, which depends on the value of axial deformation and does not depend on the compression frequency of the piezoelectric actuator. The fluid flow rate at the inlet and outlet is the same; the total volume of fluid over the period is zero. Obviously, in the absence of a pressure drop, it is impossible to create a nonzero flow for a period using one piezoelectric actuator.

Figure 43.4 shows the dependences of fluid flow Q and the total volume of fluid V at the outlet of the microchannel versus time t for three and five piezoelectric actuators, compressed according to the schemes that are displayed in Tables 43.1 and 43.2, respectively, with different frequencies f in the absence of a pressure drop. The half-period number j corresponds to the time interval $(j{\cdot}T/2, \ldots, (j+1){\cdot}T/2]$, that is, the complete cycle of the operation of the piezoelectric actuators is $T_s = 2T$ for three and $T_s = 3T$ for five piezoelectric actuators. Then the cycle repeats. Over time the fluid volume V flowing through the microchannel increases, i.e., there is an average positive fluid flow V_a. Although the change in the average value of V over the T_s period is small, over several periods this change accumulates. V_a changes relative to max(V) by 10% for $N_p = 3$ and $f = 1$ kHz; 40% for $N_p = 3$ and $f = 2$ kHz; 20% for $N_p = 5$ and $f = 1$ kHz; 80% for $N_p = 5$ and $f = 2$ kHz over time period 0.1 s. Thus, with an increase in the frequency f, a nonlinear increase in V_a occurs, and with an increase in N_p, an approximately linear increase in V_a occurs.

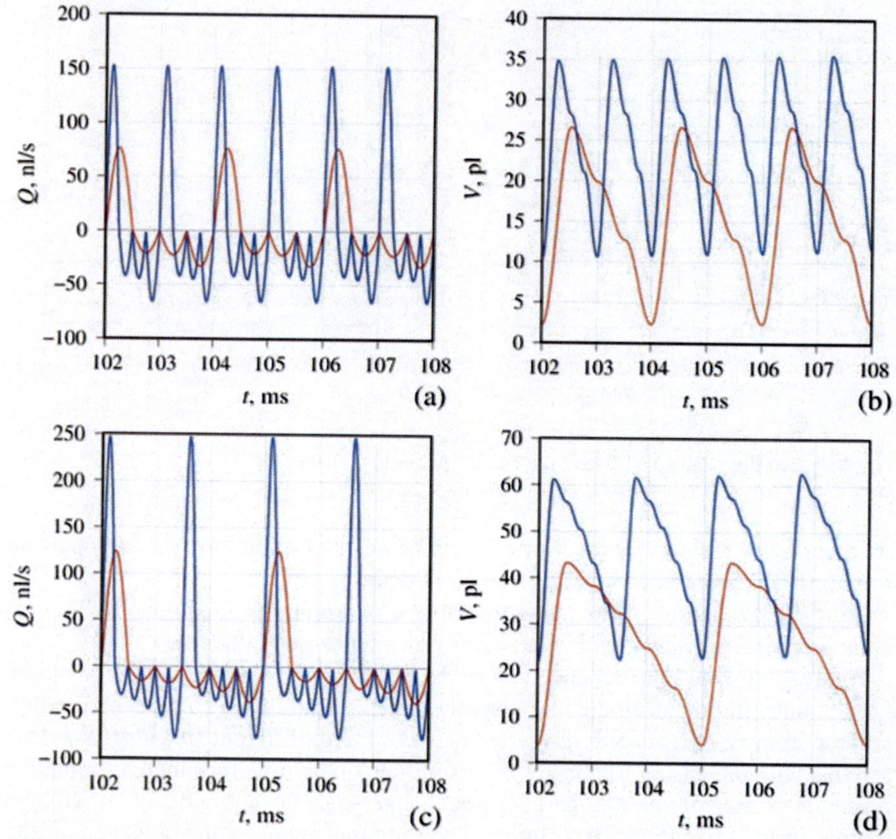

Fig. 43.4 The fluid flow Q (**a**, **c**) and the total fluid volume V (**b**, **d**) at the outlet of the microchannel versus time for $N_p = 3$ (**a**, **b**) and $N_p = 5$ (**c**, **d**): $f = 1$ kHz (red line) and $f = 2$ kHz (blue line); $G = 0$ Pa/m

Table 43.1 The order of operation of the piezoelectric system with $N_p = 3$

j	p_1^p	p_2^p	p_3^p
0	+	+	+
1	–	1	1
2	–	–	1
3	0	0	–

j is the number of half-periods; p_i^p $(i = 1, 2, 3)$ is the pressure on the i-th piezoelectric actuator
Legend "0" means no compression; "1" means maximum compression; "+" and "–" mean compression and tension according to the formula (43.6)

Table 43.2 The order of operation of the piezoelectric system with $N_p = 5$

j	p_1^p	p_2^p	p_3^p	p_4^p	p_5^p
0	+	+	+	+	+
1	–	1	1	1	1
2	–	–	1	1	1
3	0	0	–	1	1
4	0	0	0	–	1
5	0	0	0	0	–

j is the number of half-periods; $p_i^p (i = 1, \ldots, 5)$ is the pressure on the i-th piezoelectric actuator

The growth of V_a slows down and, for a characteristic relaxation time of the velocity $\tau_u \sim \rho R_1^2/\mu$ (in the geometry under consideration, τ_u ~ 1 s), leads to a linear function. It is convenient to introduce the average fluid flow rate $Q_a = dV_a/dt$, which is a quantitative characteristic of the micropump and allows one to clearly demonstrate the effectiveness of various operating modes of the device. Figure 43.5 shows the dependences of V_a and Q_a on time for various frequencies f and number of piezoelectric actuators N_p. During the time τ_u, the average volume of the leaked liquid reaches a straight line, and the average liquid flow rate reaches a constant. When the micropump reaches the operating mode, the steady-state flow $Q_a(\tau_u)$ (see Fig. 43.6a) depends on the frequency approximately quadratically, and a change in the number of piezoelectric actuators from 3 to 5 doubles Q_a.

Figure 43.6b shows the dependence of $Q_a(\tau_u)$ on the radius of the channel R_1. It is seen that with a decrease in the radius of the channel, the fluid flow rate decreases quadratically.

When constructing an element of the computing module for the device under consideration, second-order polynomial products were chosen as a simple analytical

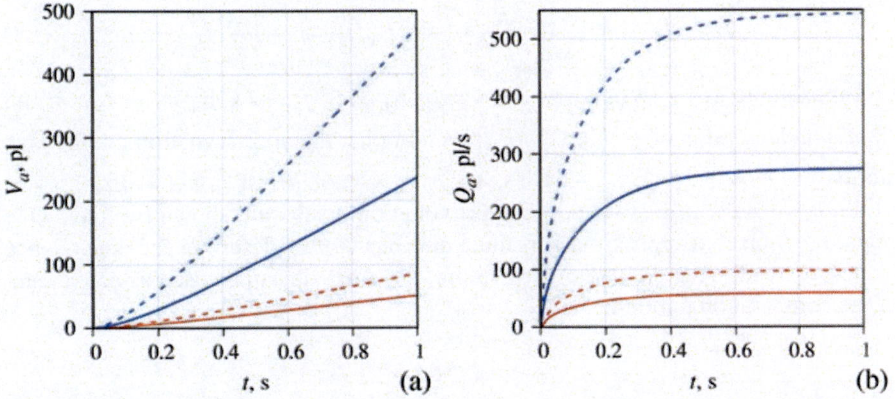

Fig. 43.5 Time dependence of V_a (**a**) and Q_a (**b**) for the frequency $f = 1$ kHz (red line) and $f = 2$ kHz (blue line) with three (solid line) or five (dashed line) piezoelectric actuators for $G = 0$ Pa/m

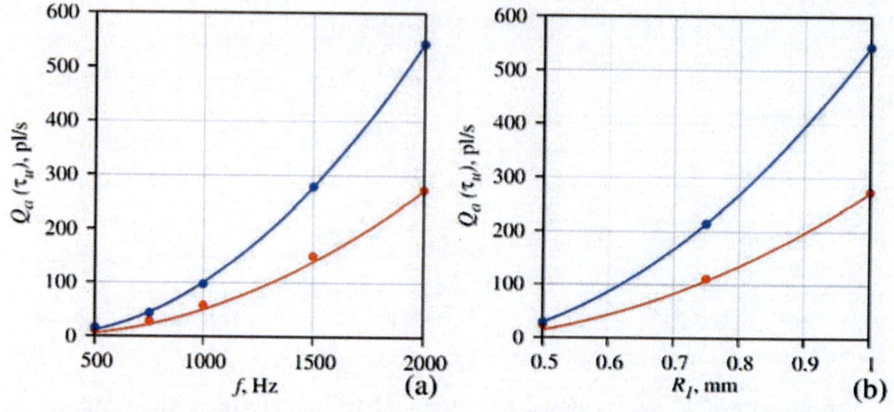

Fig. 43.6 The steady flow $Q_a(\tau_u)$ vs. the frequency f for $R_1 = 1$ mm (**a**); dependence of $Q_a(\tau_u)$ vs. the radius of the channel R_1 for $f = 2$ kHz (**b**). $N_p = 3$ (red symbols) and $N_p = 5$ (blue symbols); $G = 0$ Pa/m. Symbols show numerical calculations; lines show the approximation by the formula (43.7)

approximation. Polynomial coefficients were determined by the least-squares method according to numerical calculations (indicated by symbols in Fig. 43.6). The values for which the calculations were made are denoted by the notation "~" above the variable. According to the steady-state flow rate of fluid Q_a, corresponding to the parameters $\tilde{N}_p$, $\tilde{f}$ and $\tilde{R}_1$, one can approximately estimate the steady-state flow rate of the fluid for new values of N_p, f and R_1 using the following formula:

$$Q_a(N_p, f, R_1; \tau_u) \approx \frac{N_p - 1}{\tilde{N}_p - 1}\left(a_0 + a_1 f_1 + a_2 f_1^2\right)\left(b_0 + b_1 R + b_2 R^2\right) Q_a\left(\tilde{N}_p, \tilde{f}, \tilde{R}_1; \tilde{\tau}_u\right),$$

$$f_1 = f/\tilde{f},\ R = R_1/\tilde{R}_1,\ \tau_u = \rho R_1^2/\mu,\ \tilde{\tau}_u = \rho \tilde{R}_1^2/\mu. \tag{43.7}$$

For example, for a set of parameter values $\tilde{N}_p = 5$, $\tilde{f} = 2$ kHz, $\tilde{R}_1 = 1$ mm and corresponding value of $Q_a\left(\tilde{N}_p, \tilde{f}, \tilde{R}_1\right) = 544$ pl/s, the approximation coefficients are $a_0 = 3.8 \times 10^{-2}$, $a_1 = -3.16 \times 10^{-4}$, $a_2 = 6.48 \times 10^{-7}$, $b_0 = 0.223$, $b_1 = -1.43$, $b_2 = 2.21$. In Fig. 43.6, the approximation results are shown by lines. One can see the selected approximate formula quite accurately describes the dependence of $Q_a(\tau_u)$ on all three parameters; therefore, this formula can be used as an element of the computational module.

43.4 Conclusion

In present work, various modes of the working fluid flow in a piezoelectric micropump, which is an elastic circular microtube with a system of ring piezoelectric actuators located on it, are numerically studied. The micropump operation scheme is considered in which all the piezoelectric actuators are simultaneously compressed, and then the piezoelectric actuators are expanded one by one (from left to right). After each complete cycle T_s, more volume of fluid flows in the positive direction than in the opposite direction, i.e., the average fluid flow rate is positive. Over time, the average liquid flow rate increases until the system reaches saturation (operating mode). With an increase in the frequency of oscillations of the piezoelectric actuators and with an increase in their quantity, the average pumped volume of the fluid increases, and with a decrease in the radius of the channel, it decreases. A micropump with a length of 10 mm and a radius of 5 mm with five piezoelectric actuators during a time of reaching the operating mode of about 1 s creates a flow with an average fluid flow rate of several nl/s within the operating parameters of the device (the oscillation frequency is several kHz).

Based on the results of numerical calculations, an approximate analytical formula that relates the main parameters of the operating mode (time to reach the operating mode and average liquid flow rate) with the system parameters (number and frequency of oscillations of the piezoelectric actuators, radius of the microchannel) is obtained. This formula allows one to build an element of the computing module to define the operating modes of the considered device, as well as to control the device itself in real time.

The obtained results are planned to be further verified and clarified on the base of experimental data. A further area of research is the study of the behavior of hydraulic systems when changing the sizes of elements, operating frequencies, the range of deformation of piezoelectric actuators, etc. to determine the optimal values for increasing work efficiency, reducing energy consumption and reaction speed to control actions.

Acknowledgements This work was supported within the framework of state assignment No. 0246-2018-007.

References

1. Convery, N., Gadegaard, N.: 30 years of microfluidics. Micro Nano Eng. **2**, 76–91 (2019). https://doi.org/10.1016/j.mne.2019.01.003
2. Kakaç, S., Kosoy, B., Li, D., Pramuanjaroenkij, A.: NATO Science for Peace and Security Series A: Chemistry and Biology. Springer, Dordrecht (2010)
3. Cotta, R.M., Knupp, D.C., Naveira-Cotta, C.P.: Analytical Heat and Fluid Flow in Microchannels and Microsystems. Springer, Cham (2016). https://doi.org/10.1007/978-3-319-23312-3_3
4. Zapka, W.: Handbook of Industrial Inkjet Printing: A Full System Approach, 2nd edn. Wiley-VCH Verlag GmbH & Co, KGaA (2018)

5. Dietzel, A.: Microsystems for Pharmatechnology. Manipulation of Fluids, Particles, Droplets, and Cells. Springer, Cham (2016). https://doi.org/10.1007/978-3-319-26920-7
6. Darintsev, O.V., Migranov, A.B.: Capillary Micro-Grip with Feedback. Patent RF No. 2261795 RU (2005)
7. Laser, D.J., Santiago, J.G.: A review of micropumps. J. Micromech. Microeng. **14**, R35–R64 (2004). https://doi.org/10.1088/0960-1317/14/6/R01
8. Bungartz, H.-J., Schäfer, M.: Fluid-structure Interaction: Modelling, Simulation, Optimization. Springer, Heidelberg (2006). https://doi.org/10.1007/3-540-34596-5
9. Nasibullayev, I.Sh.: The development of a computer model for the main element of the fuel metering unit. Computational Technologies. **21**(2), 26–41 (2016)
10. Bruus, H.: Theoretical microfluidics. Lecture notes third edition. MIC Department of Micro and Nanotechnology Technical University of Denmark, Denmark (2006)
11. Chiang, C.-Y, Pironneau, O., Sheu, T., Thiriet, M.: Numerical study of a 3D Eulerian monolithic formulation for incompressible fluid-structures systems. Fluids **2**(2), 34–53 (2017). https://doi.org/10.3390/fluids2020034
12. Hecht, F.: New development in FreeFem++. J. Numer. Math. **20**(3–4), 251–265 (2012). https://doi.org/10.1515/jnum-2012-0013
13. Nasibullayev, I.Sh., Nasibullaeva, E.Sh., Darintsev, O.V.: Simulation of fluid flow through a elastic microchannel deformed by a piezoelement in microgrip cooling systems. Mekhatronika, Avtomatizatsiya, Upravlenie **20**(12), 740–750 (2019). https://doi.org/10.17587/mau.20.740-750
14. Landau, L., Lifshits, E.M.: Theoretical Physics. Nauka, vol. 6 (1988)
15. Durand, B., Delvare, F., Bailly, P.: Numerical solution of Cauchy problems in linear elasticity in axisymmetric situations. Int. J. Solids Struct. **48**(21), 3041–3053 (2011). https://doi.org/10.1016/j.ijsolstr.2011.06.017
16. Nasibullayev, I.Sh., Nasibullaeva, E.Sh., Darintsev, O.V.: Study of fluid flow through a channel deformed by piezoelement. Multiphase Systems. **13**(3), 1–10 (2018). https://doi.org/10.21662/mfs2018.3.001

Chapter 44
Vibration Amplitude and Frequency Parameters of Technological Equipment Drives

Dmitry Ershov and Irina Lukyanenko

Abstract The paper studies the effective non-uniform operation mode of technological equipment by the example of numerically controlled milling machine which is described as a multi-mass technological system. The mathematical models presented can be used in studies of the vibration amplitude and frequency parameters of the torque and the angular velocity of the drive conditioned on the drive dynamics. The dynamic ratios of the technological system appear conditioned on the ratio of electromagnetic to mechanical time constants of the motor powering the technological equipment drive. The vibration amplitude and frequency parameters of the drive obtained are analyzed and conditioned on the external harmonic disturbances. The resonance-inducing conditions are outlined. These properties are required for the drive dynamic parameters to be analyzed in order to correctly evaluate the drive vibration amplitude and frequency parameters.

44.1 Introduction

Technological equipment drive often operates in the effective non-uniform operation mode which results from the condition of loads periodically applied to the drive, i.e., with forced vibrations in the drive links occurring near the uniform-motion operation mode with a certain median velocity. The moment of resistance forces in a wide class of technological equipment can be approximated by the time periodic function [1–3].

D. Ershov (✉)
Department of Mechanical Engineering, St. Petersburg Mining University, 21 Line, 2, 199106 St. Petersburg, Russia
e-mail: fetcat@mail.ru

D. Ershov · I. Lukyanenko
Department of Advanced Mathematics and Mechanics, St. Petersburg State University of Aerospace Instrumentation (SUAI), Bolshaya Morskaya str., 67, 190000 St. Petersburg, Russia
e-mail: irina.n.lukyanenko@gmail.com

A. Ronzhin and V. Shishlakov (eds.), *Proceedings of 15th International Conference on Electromechanics and Robotics "Zavalishin's Readings"*, Smart Innovation, Systems and Technologies 187, https://doi.org/10.1007/978-981-15-5580-0_44

For instance, when machining the surface of the workpiece or boring the hole with evenly distributed along the circumference slots, the load moment can be approximated as the time periodic function [4–7]. The below simplified load moment equation allows to determine and analyze basic properties of the operation mode under discussion:

$$M_L(t) = M_L^0 + \Delta M_L^0 \cos \Omega t, \quad (44.1)$$

where M_L^0, ΔM_L^0 stand for the constant component and the amplitude of the moment harmonics, and Ω stands for frequency of external disturbance.

44.2 Dynamic Model of the Drive

The studies [8–10] describe mathematical models conditioned on the external dynamic disturbances under various modes of operation. Following the above studies and the drive dynamic model as given in Fig. 44.1, we will investigate the parameters of the mode of operation under discussion.

The rotor with the moment of inertia J_E is used to power the actuator with the moment of inertia J_L by rigidly linked shaft. The motor torque M_E applies to the rotor, while the load moment $M_L(t)$ applies to the actuator. The torque on the shaft applied to the rotor is shown as M_{12}, and the torque applied to the actuator is shown as M_{21}. The equality $M_{12} = - M_{21}$ will hold.

The study of the effective motion mode of the technological equipment drive will require calculation of the motor torque, angular velocity, and the torque variance on the shaft [11, 12]. Calculation of the motor torque variance will guarantee dynamically strong links in the drive [13].

The differential equations for motion calculation will be:

$$\begin{gathered} J_E \dot{\omega}_E = M_E + M_{12},\ J_L \dot{\omega}_L = M_{21} - M_L, \\ T_e \dot{M}_E + M_E = \beta_E \omega_0 - \beta_E \omega_E \end{gathered} \quad (44.2)$$

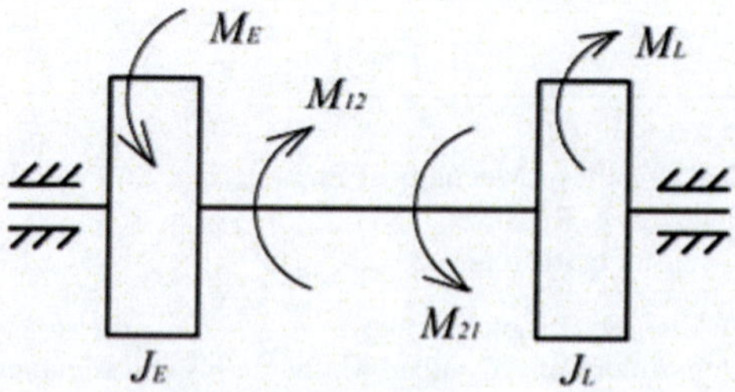

Fig. 44.1 Dynamic model of the drive

where $\dot{\omega}_E = \dot{\omega}_L$ stands for acceleration of the rotor and actuator, respectively; $\beta_E = \frac{\mathrm{d}M_E}{\mathrm{d}\omega_E}$ stands for the static rigidity module. The system of Eq. (44.2) allows conditioning on the motor dynamic properties. Value T_e determines the electromagnetic time constant of the motor.

44.3 Motor Torque Variance

We further examine the motor torque M_E variance. To convert the system (44.2), we combine the first and the second equations as follows:

$$(J_E + J_L)\dot{\omega}_E = M_E - M_L; \tag{44.3}$$

$$\dot{\omega}_E = \frac{M_E - M_L}{J_E + J_L} \tag{44.4}$$

Then we calculate the time derivative for both left-hand and right-hand parts of the third equation in the system (44.2) as:

$$T_e \ddot{M}_E + \dot{M}_E = -\beta_E \dot{\omega}_E \tag{44.5}$$

In calculating the variable, we assume $\omega_0 = \text{const}$, by substituting (44.4) into (44.5):

$$T_e \ddot{M}_E + \dot{M}_E = -\beta_E \frac{M_E - M_L}{J_E + J_L}$$

and multiplying both left-hand and right-hand parts of the equation by $(J_E + J_L)\beta_E^{-1}$ we arrive at:

$$T_e T_m \ddot{M}_E + T_m \dot{M}_E + M_E = M_L = M_L^0 + \Delta M_L^0 \cos \Omega t, \tag{44.6}$$

where $T_m = (J_E + J_L)\beta_E^{-1} = J_E \beta_E^{-1}(1 + k) = T_{m0}(1 + k)$ stands for the drive mechanical time constant; $k = \frac{J_L}{J_E}$ stands for the dimensionless value, which will determine the ratio of the actuator moment of inertia and the rotor.

Thus, Eq. (44.6) will be solved as:

$$M_E(t) = M_E^0 + \Delta M_E(t). \tag{44.7}$$

By substituting function (44.7) and its derivatives $\dot{M}_E(t)$ and $\ddot{M}_E(t)$ into differential Eq. (44.6), we arrive at the following equations:

$$T_e T_m \Delta \ddot{M}_E + T_m \Delta \dot{M}_E + \Delta M_E = \Delta M_L^0 \cos \Omega t,$$

$$M_E^0 = M_L^0 \tag{44.8}$$

It follows from the second equation in the system (44.8) that the constant of the motor torque is determined by the constant component of the load moment.

Since we discuss the effective modes of motion, the particular solution of the first equation in the system (44.8) can be given as:

$$\Delta M_E(t) = \Delta M_E^0 \cos(\Omega t + \alpha_M), \tag{44.9}$$

where ΔM_E^0 and α_M stand for the amplitude and phase variance to be calculated.

By substituting function (44.9) and its derivatives $\Delta \dot{M}_E(t)$ and $\Delta \ddot{M}_E(t)$ into differential Eq. (44.8),

we obtain two equations for unknown ΔM_E^0 and α_M values:

$$\left(1 - T_e T_m \Omega^2\right) \Delta M_E^0 = \Delta M_L^0 \cos \alpha_M, \; T_m \Omega \Delta M_E^0 = \Delta M_L^0 \sin \alpha_M.$$

The solution will be:

$$\Delta M_E^0 = \frac{\Delta M_L^0}{\sqrt{\left(1 - T_e T_m \Omega^2\right)^2 + T_m^2 \Omega^2}}, \; \alpha = \mathrm{arctg} \frac{T_m \Omega}{1 - T_e T_m \Omega^2} \tag{44.10}$$

Finally, the function which determines variance of motor torque values in the effective uniform-motion operation mode will be as:

$$M_E(t) = M_L^0 + \Delta M_E^0 \cos(\Omega t - \alpha_M) \tag{44.11}$$

44.4 Vibration Amplitude and Frequency Parameters of Motor Torque

Using the first Eq. (44.10), we calculate the vibration amplitude and frequency parameters of moment $M_E(t)$.

$$W_M(\Omega) = \frac{\Delta M_E^0}{\Delta M_L^0} = \frac{1}{\sqrt{\left(1 - T_e T_m \Omega^2\right)^2 + T_m^2 \Omega^2}} \tag{44.12}$$

Then we calculate such value of Ω, at which the function (44.12) will be at its maximum. Function $W_M(\Omega)$ will reach its maximum if the radicand is minimum:

$$f(\Omega) = \left(1 - T_e T_m \Omega^2\right)^2 + T_m^2 \Omega^2,$$
$$f'_\Omega = 2\left(1 - T_e T_m \Omega^2\right)(-2T_e T_m \Omega) + 2T_m^2 \Omega = 0.$$

The equation solution will produce two roots:

$$\Omega_1 = 0, \Omega_2 = \frac{1}{\sqrt{T_e T_m}} \sqrt{1 - \frac{T_m}{2T_e}} \tag{44.13}$$

The second root appears real, if condition $1 - \frac{T_m}{2T_e} > 0$ holds, which gives:

$$\frac{T_e}{T_m} > 0.5. \tag{44.14}$$

Then we calculate the second function $f(\Omega)$ and find its sign at $\Omega = \Omega_1$ and $\Omega = \Omega_2$.

Then the second derivative will be:

$$f''_\Omega = 2\left[-2T_e T_m \left(1 - \frac{T_m}{2T_e}\right) + 6T_e^2 T_m^2 \Omega^2\right].$$

Value f''_Ω at $\Omega = \Omega_1 = 0$ is calculated by:

$$f''_\Omega(\Omega_1) = -4T_e T_m \left(1 - \frac{T_m}{2T_e}\right).$$

If condition (44.14) holds, the second derivative is $f''_\Omega(\Omega_1) < 0$, while the first derivative $f'_\Omega(\Omega_1) = 0$. Hence, at $\Omega = \Omega_1$ function $f(\Omega)$ will be the maximum and function $W_M(\Omega)$ will be the minimum.

The second derivative at $\Omega = \Omega_2$ will be defined as:

$$f''_\Omega(\Omega_2) = 8T_e T_m \left(1 - \frac{T_m}{2T_e}\right).$$

If condition (44.14) holds, the second derivative is positive $f''_\Omega(\Omega_2) > 0$, while the first derivative $f'_\Omega(\Omega_2) = 0$. Thus, the inequality (44.14) appears condition for the maximum of vibration amplitude and frequency parameters at frequency $\Omega = \Omega_2$, i.e., the condition under which rigidly linked drive will possess resonance properties.

It follows from (44.13) that frequency Ω at which function $W_M(\Omega)$ has an extremum is determined by time constants T_e and T_m. Condition (44.14) allows determining these to numbers for k, under which the drive has resonance properties.

$$k < 2\frac{T_e}{T_m} - 1 = k^* \tag{44.15}$$

Value $k^*$is the threshold value for k, at which frequency $\Omega = 0$. If the ratio of the moment of load inertia to the moment of rotor inertia (parameter $k = \frac{J_L}{J_E}$) satisfies inequality (44.15), the drive will have torque resonance properties. Plotting

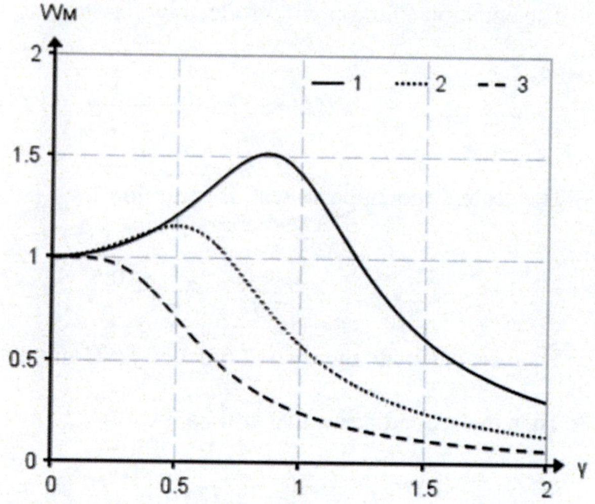

Fig. 44.2 Vibration amplitude and frequency parameters conditioned on motor torque

the graph for vibration amplitude and frequency parameters of motor torque $W_M(\Omega)$, we convert function (44.12) with dimensionless frequency of external disturbance $\gamma = \Omega\sqrt{T_e - T_m}$:

$$W_M(\gamma) = \frac{1}{\sqrt{\left[1 - (1+k)\gamma^2\right]^2 + \frac{T_m}{T_e}(1+k)^2\gamma^2}} \qquad (44.16)$$

Figure 44.2 shows the graphs for function (44.16) at $\frac{T_e}{T_m} = 2$ and three various values of parameter k: curve 1 ($k = 0$), curve 2 ($k = 1$), and curve 3 ($k = 3$). If the threshold value $k = k^* = 3$, vibration amplitude and frequency parameters $W_M(\Omega)$ at $\gamma = \Omega\sqrt{T_e - T_m}$ will be the maxima. Higher values for parameter k will lower the resonance maximum. If $k \geq k^*$, function $W_M(\Omega)$ satisfies condition $W_M(\Omega) \leq 1$ within the whole range of external disturbance frequency variance.

44.5 Drive Angular Velocity Variance

Further, we analyze the variance of angular velocity ω_E. In order to achieve this, we solve Eq. (44.3) for motor torque:

$$M_E = (J_E + J_L)\dot{\omega}_E + M_L \qquad (44.17)$$

Differentiating by time Eq. (44.17):

$$\dot{M}_E = (J_E + J_L)\ddot{\omega}_E + \dot{M}_L \qquad (44.18)$$

and substituting functions (44.16) and (44.17) into the third equation of the system (44.2):

$$T_e(J_E + J_L)\ddot{\omega}_E + T_e\dot{M}_L + (J_E + J_L)\dot{\omega}_E + M_L = \beta_E\omega_0 - \beta_E\omega_E,$$

we convert the latter into:

$$T_e(J_E + J_L)\beta_E^{-1}\ddot{\omega}_E + (J_E + J_L)\beta_E^{-1}\dot{\omega}_E + \omega_E = \omega_0 - M_L\beta_E^{-1} - T_e\beta_E^{-1}\dot{M}_L.$$

Then we substitute function (44.1) and its derivative $\dot{M}_L = -\Omega\Delta M_L^0 \sin\Omega t$ into differential equation:

$$T_eT_m\ddot{\omega}_E + T_m\dot{\omega}_E + \omega_E = \omega_0 - M_L^0\beta_E^{-1} - \Delta M_L^0\beta_E^{-1}\cos\Omega t + T_e\Omega\beta_E^{-1}\Delta M_L^0\sin\Omega t.$$

Then $\omega_0 - M_L^0\beta_E^{-1} = \omega_E^0$ can be used for calculating the constant component of angular velocity. If angular velocity $\omega_E(t)$ is taken as the sum of the constant ω_E^0 and the variable $\Delta\omega(t)$ components, then:

$$\omega_E(t) = \omega_E^0 + \Delta\omega_E(t) \tag{44.19}$$

Holding it that $\dot{\omega}_E = \Delta\dot{\omega}_E$ and $\ddot{\omega}_E = \Delta\ddot{\omega}_E$, we then convert differential equation as:

$$T_eT_m\Delta\ddot{\omega}_E + T_m\Delta\dot{\omega}_E + \Delta\omega_E = \Delta M_L^0\beta_E^{-1}\sqrt{1 + T_e^2\Omega^2}\sin(\Omega t + \delta_m) \tag{44.20}$$

where $\delta_m = \text{arctg}\frac{1}{T_e\Omega}$.

The stationary solution for Eq. (44.20) will be found as:

$$\omega_E(t) = \Delta\omega_E^0\sin(\Omega t - \delta_m - \alpha_\omega) \tag{44.21}$$

In expression (44.21), amplitude $\Delta\omega_E^0$ and phase variance α_ω are unknown values and are to be calculated.

Then we substitute solution obtained and its derivatives $\Delta\dot{\omega}_E = \Delta\omega_E^0\Omega\cos(\Omega t - \delta_m - \alpha_\omega)$, $\Delta\ddot{\omega}_E = -\Delta\omega_E^0\Omega^2\sin(\Omega t - \delta_m - \alpha_\omega)$ into differential Eq. (44.20) converting the right-hand part of equation:

$$\Delta\omega_E^0 = \Delta M_L^0\frac{\sqrt{1 + T_e^2\Omega^2}}{\beta_E\sqrt{\left(1 - T_eT_m\Omega^2\right)^2 + T_m^2\Omega^2}},$$
$$\alpha_\omega = \text{arctg}\frac{T_m\Omega}{1 - T_eT_m\Omega^2} \tag{44.22}$$

Employing (44.19), (44.21), and (44.22), we can obtain the function which will determine the variance of angular velocity under the effective modes of motion.

$$\omega_E(t) = \omega_E^0 + \Delta M_L^0 \frac{\sqrt{1 + T_e^2 \Omega^2}}{\beta_E \sqrt{\left(1 - T_e T_m \Omega^2\right)^2 + T_m^2 \Omega^2}} \sin(\Omega t - \delta_m - \alpha_\omega) \quad (44.23)$$

The variance of the drive angular velocity $\omega_E(t)$ is conditioned on the sum of two motions: the drive revolution at constant angular velocity ω_E^0 and the drive vibration under the uniform revolution at the external disturbance frequency Ω. The velocity of uniform revolution ω_E^0 will be determined by the constant component M_L^0 of the moment of load and the static rigidity β_E of mechanical parameters.

44.6 Vibration Amplitude and Frequency Parameters of Drive angular Velocity

This section studies the harmonics of the drive angular velocity. The drives of technological equipment should meet strict requirements as to the variance ratio δ, which is proportional to doubled angular velocity vibration frequency $\Delta\omega_E^0$. Thus, the analysis of the effective non-uniform mode of motion makes its essential to reduce the drive angular velocity vibration amplitude. Here, we introduce into equation the drive angular velocity vibration amplitude and frequency parameters. From (44.22), we arrive at:

$$W_\omega(\Omega) = \frac{\Delta\omega_E^0}{\Delta M_L^0} = \frac{\sqrt{1 + T_e^2 \Omega^2}}{\beta_E \sqrt{\left(1 - T_e T_m \Omega^2\right)^2 + T_m^2 \Omega^2}} \quad (44.24)$$

Function $W_\omega(\Omega)$ will determine the drive dynamic flexibility under varying angular velocities.

Now the angular velocity vibration amplitude can be given as:

$$\Delta\omega_E^0 = W_\omega(\Omega) \Delta M_L^0 \quad (44.25)$$

Further, we analyze the vibration amplitude and frequency parameters by determining the frequency of external disturbance, at which function $W_\omega(\Omega)$ will be the maximum and calculating the derivative of expression (44.24):

$$\frac{dW_\omega}{d\Omega} = \frac{T_e^2 \Omega\left[\left(1 - T_e T_m \Omega^2\right)^2 + T_m^2 \Omega^2\right] + \left(1 + T_e^2 \Omega^2\right)\left[\left(1 - T_e T_m \Omega^2\right)(2 T_e T_m \Omega) - T_m^2 \Omega\right]}{\beta_E \sqrt{1 + T_e^2 \Omega^2}\left[(1 - T_e T_m \Omega)^2 + T_m^2 \Omega^2\right]^{\frac{3}{2}}}.$$

By equaling it to zero, we arrive at the below equation:

$$\Omega_\omega \left(\Omega_\omega^4 + \frac{2}{T_e^2} \Omega_\omega^2 - \frac{1}{T_e^2 T_m^2} - \frac{2}{T_e^2 T_m} + \frac{1}{T_e^4} \right) = 0,$$

where Ω_ω determines the value of frequency Ω, corresponding the function $W_\omega(\Omega)$ extremum. The roots in the equation are:

$$\Omega_{\omega 1} = 0, \Omega_{\omega 2} = \frac{1}{\sqrt{T_e T_m}} \left(\sqrt{1 + 2\frac{T_m}{T_e}} - \frac{T_m}{T_e} \right)^{\frac{1}{2}} \tag{44.26}$$

In further plotting graphs for function (44.24), we determine the conditions at which the roots calculated will correspond to the maximum or minimum vibration amplitude and frequency parameters. Root $\Omega_{\omega 2}$ is real, and function $W_\omega(\Omega)$ will be the maximum if the following condition holds:

$$\sqrt{1 + 2\frac{T_m}{T_e}} - \frac{T_m}{T_e} > 0$$

which is converted as:

$$\frac{T_e}{T_m} > 0.41 \tag{44.27}$$

We determine the range of k values, within which function $W_\omega(\Omega)$ has its extremum. From (44.27), it goes that:

$$k\langle \left(1 + \sqrt{2}\right) \frac{T_e}{T_m} - 1 = k^{**} \tag{44.28}$$

Plotting the graph for vibration amplitude and frequency parameters, we convert functions (44.24) and (44.26) with dimensionless frequency γ.

$$W_\omega(\gamma) = \frac{\sqrt{1 + \frac{T_e}{T_m}\gamma^2}}{\beta_E \sqrt{\left[1 - (1+k)\gamma^2\right]^2 + \frac{T_m}{T_e}(1+k)^2\gamma^2}} \tag{44.29}$$

The dimensionless value of the stationary value $\Omega_{\omega 2}$ of frequency Ω equals:

$$\gamma = \frac{1}{\sqrt{1+k}} \left[\sqrt{1 + 2\frac{T_m}{T_e}(1+k)} - \frac{T_m}{T_e}(1+k) \right]^{\frac{1}{2}} \tag{44.30}$$

Employing functions (44.29) and (44.30) in Fig. 44.3 allows to plot the graph of the drive vibration amplitude and frequency parameters in coordinates γ and $W_\omega(\gamma)$ β_E.

Curve *1* is plotted for values of $k = 0$, $\frac{T_e}{T_m} = 1$ and $k^{**} = 1.41$ and at $\gamma = 0$ has the minimum, while at $\gamma = 0.85$ it has the maximum. Conditions (44.27) and (44.28) hold and the drive will be characterized by the parameter of angular velocity.

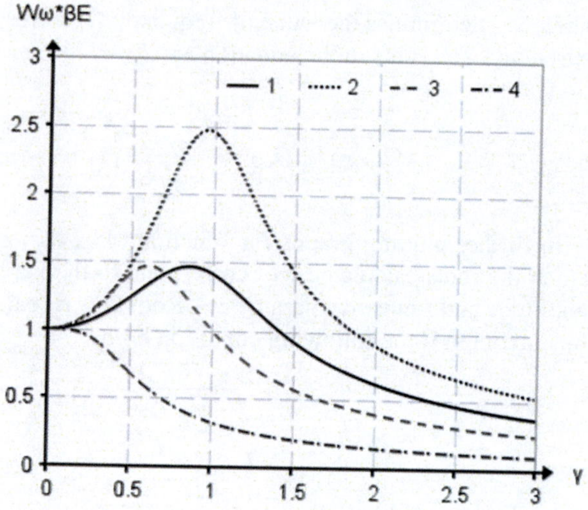

Fig. 44.3 Drive vibration amplitude and frequency parameters conditioned on angular velocity

Curve 2 is plotted at $k = 0$, $\frac{T_e}{T_m} = 2$. Vibration amplitude and frequency parameters at frequency $\gamma = 0.96$ will be the maximum. Higher ratio $\frac{T_e}{T_m}$ of the drive will result in more substantial resonance properties. It follows from comparison of curve 1 and curve 2 that the resonance maxima can be lowered if to choose a motor with a lower ratio $\frac{T_e}{T_m}$. Curve 3 in Fig. 44.3 is plotted at $k = 1$, $\frac{T_e}{T_m} = 2$, and $k^{**} = 3.82$. It follows from comparison of curve 2 and curve 3 that higher moments of inertia of the drive load will result in lower resonance maxima of the drive vibration amplitude and frequency parameters. Curve 4 is plotted for values of $k = 4$, $\frac{T_e}{T_m} = 2$, and $k^{**} = 3.84$. At any value of external disturbance frequency γ vibration amplitude and frequency parameters of angular velocity will satisfy the condition of $W_\omega(\gamma)\beta_E \le 1$. In the case studied, conditions (44.27) and (44.28) do not hold and the drive possesses no resonance properties.

Ultimately, the function $\omega_E(t)$ can be given as follows:

$$\omega_E(t) = \omega_E^0 + W_\omega(\Omega)\Delta M_L^0 \sin(\Omega t - \delta_m - \alpha_m).$$

Low values of angular velocity ω_E^0 under the effective non-uniform motion and high values of vibration angular velocity amplitude $\Delta\omega_E^0$ may induce periodical reversing of the function $\omega_E(t)$ sign. The inequality:

$$W_\omega(\Omega)\Delta M_L^0 \le \omega_E^0$$

makes it impossible for the drive to come to halt and then to reverse its motion.

44.7 Conclusion

The analysis of motor torque and the drive angular velocity under harmonic disturbances allows for the following conclusions to be made:

1. Variance of function $W_{\omega}(\Omega)$ conditioned on T_m is not of monotonous character—at certain value of $T_m = T_m^*$, it will have its maximum. This means that prior to adding to the system flexibility by the flywheel, the inequality $T_m < T_m^*$ holds and after the flywheel the value T_m will approximate the value T_m^* resulting in increased values of effective dynamic error $\Delta\omega_E^0$ in velocity calculation. Hence, the flywheel will not compensate for the variance in revolution, it will rather add to it.
2. The vibration amplitude and frequency parameters of motor torque $W_M(\Omega)$ will be at the resonance maximum if condition (44.14) holds for the frequency calculated by the function (44.13). The vibration amplitude and frequency parameters of angular velocity $W_{\omega}(\Omega)$ will be at the resonance maximum if condition (44.27) holds for the frequency calculated by the function (44.26). These mean that if electromagnetic and mechanical time constants T_e and T_m satisfy the inequality (44.27) as the stronger of the two conditions (44.14) and (44.27), then evaluation of the technological equipment drive will require conditioning on the motor dynamic properties such as the ratio to angular velocity moment of inertia under the effective non-uniform operation mode of the drive.

References

1. Ershov, D.: Mathematical modeling of multi-mass mechanical systems. In: Proceedings of the 4th International Research and Practice Conference "Innovations in Transportation and Machine-Building", vol. 3, pp. 31–35 (2016)
2. Evgrafov, A., Karazin, V., Kozlikin, D., Khlebosolov, I.: Some characteristics of linear acceleration reproduction with flexible harmonics: In: 2018 Lectures in Mechanical Engineering, vol. 5, pp. 71–81 (2018)
3. Evgrafov, A., Petrov, G.: Drive selection of multidirectional mechanism with input excess. In: 2016 Lectures in Mechanical Engineering, vol. 1, pp. 31–37 (2016)
4. Maksarov, V., Khalimonenko, A., Timofeev, D.: Machining quality when lathing blanks with ceramic cutting tools. Agron. Res. **12**(1), 269–278 (2014)
5. Maksarov V., Krasnyy V., Viushin R.: Simulation of dynamic processes when machining transition surfaces of stepped shafts. Proc. Sci. Eng. IOP Conf. Ser. **327**(022047), 1–6 (2018)
6. Efimov, A., Maksarov, V.: Simulation modelling of dynamic characteristics of machining in LabView software environment to improve processing technique of a rod component. Int. Conf. Innov. Prospects Dev. Mining Machin. Electr. Eng. **194**(22021), 1–6 (2018)
7. Alekseeva, L., Maksarov, V.: Evaluation of effect of oil film of rotor bearing. IOP Conf. Series-Mater. Sci. Eng. **327**(022003), 1–6 (2018)
8. Ershov, D., Lukyanenko, I.: Three-mass dynamic models in studies of vibrations in electromechanical systems. Mod. Issues Machin. Theor. **4**, 71–74 (2016)
9. Ershov, D., Lukyanenko, I.: Causes of dynamic processes in machining technological systems. J. Adv. Res. Tech. Sci. **4**, 39–43 (2017)

10. Ershov, D., Zlotnikov, E., Koboyankwe, L.: Dynamic processes in technological systems of machining and the nature of their origin. IOP Conf. Ser. Earth Environ. Sci. **87**(82016), 1–6 (2017)
11. Ershov, D., Zlotnikov, E., Timofeev, D.: Analysis of proper fluctuations of technological systems. In: IOP Conf. Ser. Mater. Sci. Eng. **560**(012015), 1–6 (2019)
12. Ershov, D., Lukyanenko, I.: Dynamic model of technological equipment drive with s DOFs. In: 2019 Proceedings of 14th International Conference on Electromechanics and Robotics "Zavalishin's Readings". Smart Innovation, Systems and Technologies, vol. 154, pp. 171–182. Springer, Singapore (2020)
13. Maksarov, V., Olt, J.: Improving the precision of manufacturing power hydraulic cylinders of powered roof supports based on a vibration-damping tooling system. J. Mining Inst. **214**, 71–84 (2015)

Author Index

A. Ronzhin and V. Shishlakov (eds.), *Proceedings of 15th International Conference on Electromechanics and Robotics "Zavalishin's Readings"*, Smart Innovation, Systems and Technologies 187, https://doi.org/10.1007/978-981-15-5580-0